D0026158

OAKTON COMMUNITY COLLEGE

DES PLAINES, ILLINOIS   60016

PATTEN'S FOUNDATIONS OF
EMBRYOLOGY

McGraw-Hill Series in Organismic Biology

CONSULTING EDITOR

Professor Melvin S. Fuller, *Department of Botany, University of Georgia, Athens*

Barker and Breland: A Laboratory Manual of Comparative Anatomy
Carlson: Patten's Foundations of Embryology
Gunderson: Mammalogy
Jones and Luchsinger: Plant Systematics
Kramer: Plant and Soil Water Relationships: A Modern Synthesis
Leopold and Kriedemann: Plant Growth and Development
Ralph: Introductory Animal Physiology
Ross: Biology of the Fungi
Weichert and Presch: Elements of Chordate Anatomy

PATTEN'S FOUNDATIONS OF
EMBRYOLOGY

Fourth Edition

Bruce M. Carlson, M.D., Ph.D.

*Department of Anatomy
and Biological Sciences
University of Michigan*

McGraw-Hill Book Company

New York St. Louis San Francisco Auckland Bogotá Hamburg
Johannesburg London Madrid Mexico Montreal New Delhi
Panama Paris São Paulo Singapore Sydney Tokyo Toronto

This book was set in Times Roman by Black Dot, Inc.
The editors were James E. Vastyan and Stephen Wagley;
the production supervisor was Charles Hess.
New drawings were done by Margaret Croup.
The cover was designed by Scott Chelius.

PATTEN'S FOUNDATIONS OF EMBRYOLOGY

34567890 HDHD 89876543

Library of Congress Cataloging in Publication Data

Patten, Bradley Merrill, dates
    Patten's Foundations of embryology.

    (McGraw-Hill series in organismic biology)
    Bibliography: p.
    Includes index.
    1. Embryology. I. Carlson, Bruce M. II. Title.
QL955.P23 1981   599.03'3   80-23943
ISBN 0-07-009875-1

To Jean and the kids

Contents

Preface

This edition of the *Foundations of Embryology* differs considerably from the previous edition with respect to both organization and content. Yet the principal objectives and manner of approach have not substantially changed. The fundamental goal is still to provide a coherent description of normal vertebrate embryonic development, so that the student will acquire an organismal frame of reference for understanding more advanced concepts of mechanisms of both normal and abnormal development. To facilitate the transition from pure morphology to the conceptual or mechanism-oriented approach toward development, this edition emphasizes more than previous ones the systems that are widely used as experimental models by contemporary developmental biologists.

In previous editions of the *Foundations* the text was divided into three sections—the first dealing with early development up to germ layer formation, the second using the embryonic chick as a basis for describing how the basic plan of the body is laid out, and the third covering organogenesis in mammals. The second section, on chick development, has been used in a number of courses as the basis for laboratory study of the chick embryo. This subdivision by species is adequate for a purely descriptive text, but it causes difficulties when one wishes to integrate descriptive and experimental material, for often, particularly in organogenesis, the best morphological descriptions are based upon mammalian material, whereas the experimental analysis has been carried out largely upon amphibian and avian embryos.

In order to accomplish integration of the text and to reduce redundancy without eliminating material that is important for laboratory study, I have reorganized the book. The main body of the text follows the course of vertebrate embryonic development from gametogenesis through organogenesis. Because of the pronounced differences among the classes of vertebrates, I have covered amphibian, avian, and mammalian embryos in the chapter dealing with early development. In the chapters on organogenesis, the descriptive embryology is largely mammalian, but more experimental material from other forms has been added. The morphological descriptions of chick development found in Part II of the previous edition have been brought together as an appendix designed to serve as a basis for laboratory study.

Within the reorganized text considerable changes have been made. Over half of the book has been completely rewritten, with the intent of preserving the best of the descriptive portions written by Dr. Patten but updating many sections and adding more experimental material. In the early chapters, the sections on endocrinology and gametogenesis have been greatly changed and the coverage of cleavage and germ layer formation has been expanded, with the addition of more information on amphibians and early mammalian embryos. A new Chapter 6 has been added to give greater emphasis to neurulation and somite formation, particularly in amphibians and birds. The previously separate chapters on extraembryonic membranes in birds and mammals have been combined into a single chapter (7).

A largely descriptive chapter (8) on the basic body plan of mammalian embryos sets the stage for the remainder of the book, in which organogenesis is emphasized. Because of the considerable interest in the developing limb as a model system and because of my own research interest in this area, a new chapter (10) on limb formation has been added. Although the remaining chapters have been extensively revised, their order has not been changed from the previous edition.

In order to standardize terminology and to lessen the problems of students who use laboratory manuals to accompany this text, several changes have been made. The spelling of "entoderm" has been changed to "endoderm" to conform with common usage and the official nomenclature. In designation of the extraembryonic membranes, the term "serosa" has been changed to "chorion." Also, "omphalomesenteric," as applied to the major arteries and veins connecting the vasculature of the yolk sac to that of the body, has been changed to "vitelline." In making the latter two changes, I have been persuaded that for an introductory text conformance with common contemporary usage is more important than comparative implications of the terms which the new words replace.

I would welcome any comments on or criticisms of either the organization of the text or specific items contained therein. The experience of those who have used the book is invaluable in making future editions more effective in meeting the needs of the courses for which this text is used.

Thanks are in order to a number of people whose help and advice have greatly facilitated the preparation of this revision. A number of colleagues in the field have given me information and advice about new developments and interpretations in specific areas. The comments and suggestions of the reviewers of this revision were particularly helpful. Thanks are due to Sandie Baldauf for the typing of the manuscript and verification of the bibliographic references. Much of the effectiveness of a book like this is due to the illustrative material. I feel very fortunate to have been able to work with Margaret Croup, who prepared all the new artwork for this edition. Both of us thank William Brudon, who did the illustrations for the third edition, for his continuing advice and support.

Bruce M. Carlson

Chapter 1

Embryology—Its Scope, History, and Special Fields

Historical Background   How we develop before we are born has always been a matter of intriguing interest. "Where did I come from?" is one of the first thoughtful questions a child asks. Primitive peoples—peoples in their cultural childhood—have always had this urgent and intense interest; but for them, groping curiosity has turned only toward speculation and mysticism. Aristotle's work on embryos is now significant for us, but not because of the information he secured, surprisingly accurate as some of it was. Rather, his work is for us a symbol of the beginning of the turning of the human mind away from superstition and conjecture, toward observation. Unfortunately, such an approach did not take firm root. Through much of the Middle Ages the spark that the better of the Greek and Roman scholars had been attempting to fan was smothered by bigotry and authoritarianism.

But the manner of approach was not the only reason for the lag in the growth of our knowledge of embryology. The early phases of development involve exceedingly minute structures, and curiosity and the willingness to learn by observation were not enough. Galen, it is true, had learned much about the structure of relatively advanced fetuses, but it was not until the close of the seventeenth century, when the microscope was being developed into an efficient instrument, that the early stages of embryology could be studied effectively.

1

The human sperm was first seen by Hamm and Leeuwenhoek in 1677, shortly after ovarian follicles were described by de Graaf (1672). Even then the significance of the gametes in development was not understood. Two camps grew up, one contending that the sperm contained the new individual in miniature (Fig. 1-1), which was merely nourished in the ovum, the other arguing that the ovum contained a minute body, which was in some way stimulated to growth by the seminal fluid. For a time, the ovists seemed to gain the ascendency when Bonnet (1745) discovered that the eggs of some insects could develop parthenogenetically. But preconceived ideas are persistent, and the war between the homunculists and the ovists continued to be bitter and vituperative. Ardor for each cause was not dampened even by the absurdity of the inevitable implication of the encasement concept—the implication that each miniature must in turn enclose the miniature of the next generation, and so on for as many generations as the race was to survive.

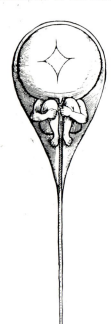

Fig. 1-1 Reproduction of Hartsoeker's drawing of a spermatozoon showing a pre-formed individual (homunculus) in the sperm head. (From *Essay de Dioptrique,* Paris, 1694.)

This bootless controversy continued into the next century, until it was finally laid to rest by the studies of Spallanzani (1729–1799) and Wolff (1733–1794). The work of Lazzaro Spallanzani is of special interest to us, for it was an initial step in bringing the experimental method to bear on embryological problems. By an ingeniously planned series of experiments, he demonstrated that both the female and the male sex products were necessary for the initiation of development.

Spallanzani's contemporary, Kaspar Friedrich Wolff, at the age of 26, wrote a brilliant thesis setting forth his conception of *epigenesis*. This idea of development by progressive growth and differentiation rapidly replaced the old encasement theories. Although this was an important step forward, it rested too largely on theoretical grounds to give a lasting impetus to the subject. For more than half a century there was little published to advance our knowledge of the early stages of development, even though accurate observation and recording were becoming more common.

The important work of Karl Ernst von Baer (1828) first emphasized the fact that the more general basic features of any large group of animals appear earlier in development than the special features that are peculiar to different members of the group. This concept is sometimes referred to as *von Baer's law*. It was von Baer, also, who gave us the foundations of our knowledge of the germ layers in embryos. But the real significance of these layers, and of the sex elements from which they arose, could not be grasped until the cellular basis of animal structure became known. With the formulation of the cell theory by Matthias Schleiden and Theodor Schwann (1839), the foundations of modern embryology and histology were simultaneously laid. The knowledge that the adult body was composed entirely of cells and cell products paved the way for a realization of the basic fact of embryology: the body of the new individual is developed from a single cell, the cell formed by the union, in fertilization, of a germ cell contributed by the male parent with a germ cell contributed by the female parent. Thus, although there had been curiosity since before the dawn of written history, and critical observation had begun, with Aristotle, to replace conjecture, it was not until the development of the microscope, the advent of the experimental method, and the discovery of the cellular structure of the body that embryology began to become a science.

Embryology   Essentially all higher animals start their lives from a single cell, the fertilized ovum (*zygote*). As its name implies, the zygote has a dual origin from two gametes—a spermatozoon from the male parent and an ovum from the female parent. The time of fertilization, when the spermatozoon meets the egg, represents the starting point in the life history, or *ontogeny*, of the individual. In its broadest sense, ontogeny refers to one's entire life-span.

A century ago the great German biologist August Weismann (1834–1914) made the important distinction between the *soma* (body) and the germ cell line (*gametes*). From the standpoint of perpetuation of the species, he considered the germ cell line to be all-important, whereas the soma was a vehicle for protecting and perpetuating the germ plasm. In more contemporary biological thought, particularly with the ever-increasing emphasis on sociobiology, this viewpoint may seem somewhat restrictive, but it does provide a convenient framework for looking at the perpetuation of life (Fig. 1-2). Once an individual has passed the reproductive years, the remainder of its ontogeny does not provide direct physical input into the generative process. Strictly speaking, embryology is usually regarded as the period starting with fertilization and ending with metamorphosis in amphibia, hatching in birds, and birth in mammals. Books on vertebrate embryology, however, also deal with the development and maturation of gametes (*gametogenesis*). Thus, this text covers most of the phases of ontogeny shown below the dashed line in Fig. 1-2. It is important, however, to recognize that hatching, birth, and metamorphosis merely represent convenient landmarks in a continuing process and that development is in reality an uninterrupted series of correlated events.

The study of embryology now encompasses a bewildering array of approaches, facts, techniques, and concepts. It is impossible to cover them all in a single course or in a single book. Over the years various special fields have developed within the general subject of embryology. This development has been

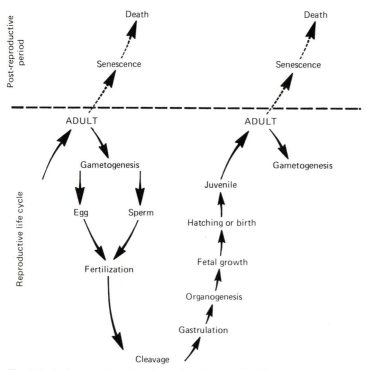

Fig. 1-2 A diagram illustrating the major phases of the life cycle of a typical vertebrate. Continuity of the germ plasm is indicated by the solid arrows.

the logical outcome of progress, both conceptual and technical, in the natural sciences as a whole.

Earlier studies were chiefly concerned with learning the basic structural pattern of the embryonic body. Interest gradually shifted, however, from general body configuration to more detailed studies of the structure and arrangement of the minute internal organs of the embryo. Work of this type received a great impetus and gained much in accuracy from new techniques that were developed between 1880 and 1890—the making of serial sections and the waxplate method of making accurate scale reconstructions from such sections (His and Born). Toward the turn of the century, attention began to shift to the finer, cellular structure of embryos. But there was still relatively little physiological or experimental work, and publications were concerned primarily with drawings and explanations of the structural features of embryos of various ages. Work of this type is generally characterized as *descriptive embryology*.

Having its roots in the same type of descriptive work, the field of *comparative embryology* arose late in the nineteenth century. A driving force behind the development of this field was the great interest in evolution, which was the dominating factor in biology during this time. Early comparative studies were carried out on the most readily available forms and on those that were easiest to deal with by simple techniques. The embryos of marine invertebrates were, and continue to be, very popular objects of study. With improving methodology, a wealth of detailed information has been gathered from the study of many different types of embryos. This information was initially very important in the study of early stages of human embryonic development. With the availability of more human material, the relationships between early developmental stages of humans and many other animals have been strikingly well established.

The gradual acquisition of detailed and accurate information on the structural stages embryos pass through in the course of their development paved the way for the growth of *experimental embryology*. This branch of the subject aims at ascertaining the factors which activate or regulate developmental processes. Descriptive embryology tells us when and how a process is carried out; experimental embryology seeks to find out why a process occurs at that specific time and in just that particular manner. Put in another way, experimental embryology seeks to ascertain the controlling and regulating mechanisms in development. One of the pioneers in this field was Wilhelm Roux (1850–1924), who coined the German word *Entwicklungsmechanik* for such studies. Its literal translation into English is *developmental mechanics*. Waddington (1956) felt that this term carries the unfortunate implication that the phenomena involved are largely physical and machinelike. He preferred the term *epigenetics*, which expresses the concept that "development is brought about through a series of causal interactions between the various parts; and also reminds one that genetic factors are among the most important determinants of development" (Waddington, 1956, p. 10).

Recent spectacular advances in molecular biology have given new impetus to the field of *chemical embryology*. Earlier in this century chemical studies involving embryos were largely descriptive (Needham, 1931). Biochemical investigations are currently playing a basic role in broadening our knowledge of the physiology of the embryo. Most exciting of all is the way these investigations are helping us to understand how, through the activities of deoxyribonucleic acid (DNA) and ribonucleic acid (RNA), the information contained in the genetic material of the fertilized ovum presides over the fabrication of the specific chemical and structural components of the embryo.

Teratology is the branch of embryology concerned with the study of malformations. Drawings and images of abnormal individuals are among the oldest known biological records. In ancient times the birth of a "monster" was supposed to portend events to come. As a matter of fact, the word *monster* is derived from the Latin verb meaning *to show*. This usage was based on the belief that the birth of a malformed infant was a supernatural way of showing what future happenings to expect. In the Middle Ages the writings on teratology seemed to degenerate into contests to discover who could assemble the most bizarre cases of malformations. When an author was falling behind his competitors he apparently had no compunctions about drawing on his imagination to fabricate weird monstrosities. Those interested in this phase of teratology can find fascinating illustrations in Gould and Pyle's *Anomalies and Curiosities of Medicine* (Fig. 1-3). Recently, work in teratology has taken on an entirely new aspect. With birth defects having moved well up among the top 10 causes of death in countries with advanced levels of medicine, great effort and much money are being spent both to identify and to eliminate factors causing congenital defects. Investigations into the mechanisms by which *teratogens* (substances that cause birth defects) interfere with normal development are becoming increasingly biochemical and pharmacological in orientation.

The extremely rapid growth in recent years of research related to problems of conception and contraception has led to the establishment of a discipline that is commonly called *reproductive biology*. In addition to more practically oriented problems involving techniques of fertilization and contraception, this field places heavy emphasis on normal gametogenesis, the endocrinology of reproduction, the transport of gametes and fertilization, early embryonic development, and the implantation of the mammalian embryo.

A currently popular way of looking at embryonic development is through the approach known as *developmental biology*. Broad in scope, it includes not only embryonic development but also postnatal processes such as normal and neoplastic growth, metamorphosis, regeneration, and tissue repair at levels of complexity ranging from the molecular to the organismal. The focus of developmental biology is upon processes and concepts rather than upon specific morphological structures, and both plant and animal systems are studied. Ideally, developmental biology and the more classically oriented methods of embryology should not be looked upon as competing methods of studying the embryo, but rather as complementary approaches, each offering exciting insight into the way that development is accomplished.

Fig. 1-3 Early drawings purporting to illustrate cases of human malformation. (A) The bird-boy of Paré (about 1520). (B) Single monsters, part human and part animal (Schwalbe, 1906-1909).

FUNDAMENTAL PROCESSES AND CONCEPTS IN DEVELOPMENT

Although a large part of this text is devoted to the morphology of embryonic development, it is important to recognize that the organs and structures which we see grossly or in microscopic preparations are the end products of dynamic processes that are as much an integral part of development as the structures themselves. In this section we shall describe briefly some of the processes and concepts that are of vital importance in explaining why an embryo develops the way it does. Specific examples of many of these fundamental processes will be described later in the text as they relate to characteristic morphological events in embryogenesis.

Intracellular Synthesis and Its Regulation From the moment of fertilization, embryonic development at all levels is a direct or indirect result of synthetic activities within cells. The DNA within the nucleus is the repository of much of the genetic information within the cell, and the transcription of this information from DNA to RNA and its subsequent translation into proteins are familiar subjects to all students of biology. One of the most important aspects of embryonic development is the nature of the regulatory mechanisms that restrict or permit the synthesis of specific proteins and other macromolecules. The intricacies of the mechanisms controlling nucleic acid and protein synthesis are beyond the scope of this text, but a review of some intracellular synthetic and regulatory pathways relating to development is in order. Figure 1-4 represents a

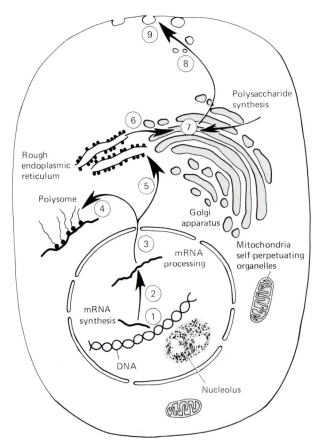

Fig. 1-4 Generalized model of a cell, emphasizing the pathway of protein synthesis. In protein synthesis, messenger RNA is first synthesized on the DNA template (1). After processing within the nucleus (2), the mRNA leaves the nucleus through nuclear pores (3). The synthesis of intracellular proteins (4) is accomplished on polysomes, which consist of molecules of mRNA associated with ribosomes. Synthesis of proteins for export from the cell is accomplished on the rough endoplasmic reticulum (5). From there it is transported to the Golgi apparatus (6), where it may be complexed with newly synthesized polysaccharides (7). Small membrane-bound vesicles containing the protein leave the Golgi apparatus (8) and, when they reach the cell membrane (9), fuse with it and release the protein molecules by a process called exocytosis.

generalized model of a cell; it stresses only intracellular structures and pathways that will be referred to later in the text.

In an *interphase cell* (one between mitotic divisions) certain portions of the nuclear DNA molecule are free of restricting proteins that bind to it and can direct the synthesis of messenger RNA (mRNA). This step is known as *transcription* (Fig. 1-4, one). After further processing in the nucleus, the newly formed mRNA molecules migrate from the nucleus into the cytoplasm of the cell via pores in the nuclear membrane. Once in the cytoplasm, the mRNA molecules may follow either of two chief pathways, depending upon the type of molecule and the type of cell. For the formation of protein molecules that are destined to function within the cell (structural proteins and most enzymes), the mRNA molecules link up with ribosomes to form polyribosomes, the length of which varies according to the size of the protein that is being made. If, on the other hand, the mRNAs are coding for proteins that will be secreted from the cell (e.g., collagen, immunoglobulins), the mRNA forms complexes with the rough endoplasmic reticulum. The polypeptide chains that are formed in the rough endoplasmic reticulum are then transported to the Golgi apparatus, where they are commonly linked with polysaccharide molecules. From the Golgi complex, the finished proteins are then brought to the cell membrane within vesicles and emptied into the medium surrounding the cell.

Regulatory mechanisms operate at almost every level of the protein synthetic pathway. Some are strictly intracellular, whereas others are extracellular influences that are effected through intracellular pathways. It is becoming increasingly apparent that many of the extracellular influences are mediated by receptor molecules located at the cell surface. According to the contemporary viewpoint, the cell membrane is a semifluid bilayer of lipid molecules with various membrane protein molecules embedded in it (Fig. 1-5). Some of the proteins are found in either the inner or the outer layer of lipid, and other, larger proteins traverse both layers. Many of these proteins, particularly in the outer layer, are complexed with polysaccharide molecules and act as receptor molecules for substances such as hormones or growth factors.

An important enzyme found at the cell membrane is adenyl cyclase, which catalyzes the reaction of adenosine triphosphate into cyclic adenosine monophosphate (ATP→cAMP). Cyclic AMP acts as a general intracellular messenger and final common pathway for the effects of a number of compounds that act on the cell surface. The enzyme adenyl cyclase controls the rate of cAMP synthesis. According to one hypothesis, when a substance reacts with a specific cell-surface receptor, the complex migrates along the cell membrane until it binds to the adenyl cyclase molecule. This activates the adenyl cyclase, resulting in an increase in the synthesis of cAMP. The cAMP, acting as a "second messenger," can influence, usually by activation, a number of intracellular processes. Prominent among them is the activation of *protein kinases* (enzymes which phosphorylate proteins), which in turn may stimulate metabolic pathways, such as the breakdown of glycogen. Other activities stimulated by increasing cAMP

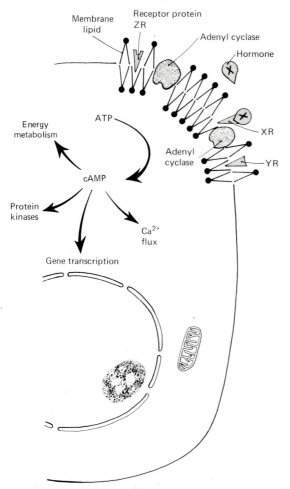

Fig. 1-5   Diagram of the cell membrane and the cyclic AMP system within a generalized cell. The cell membrane is a lipid bilayer with protein molecules interspersed in different areas and at different levels. When a hormone or some other substance joins up with a specific receptor (X-XR in this case), the receptor-hormone complex may then move to an adenyl cyclase molecule, which becomes activated and catalyzes the formation of cAMP from ATP. Several intracellular activities influenced by cAMP are shown here.

levels are gene transcription, energy metabolism, and calcium flux (Fig. 1-5). Other molecules that interact with receptors on the cell surface apparently bypass the cAMP pathway. Their influence is directed to the nucleus by other intracellular mechanisms.

Cell Division   Cell division is one of the fundamental properties of living systems, and it is of vital importance in many developmental processes. An increase in cell number is an obvious consequence of cell division. This is one of the basic mechanisms underlying growth, whether in embryonic or in postembryonic systems. Less obvious, however, is the fact that a certain minimum number of cells is sometimes required for the development of certain structures in the embryo. In the embryonic limb bud, for instance, a less than normal number of cells will typically result in the formation of a hand with less than the normal number of digits rather than a hand with the normal number of smaller digits.

For other developmental events the process of cell division itself seems to be crucial. It now seems that most, if not all, embryonic tissue interactions involving a qualitative change in the structural or functional state of groups of cells take place in populations that are characterized by a high rate of cell division. Mitosis may act as a destabilizing agent that permits the activation of certain groups of genes that previously had been tightly repressed.

Cell division is one component of the *cell cycle*. The life history of a cell can be conveniently divided into four periods. Immediately after mitosis and the separation of the dividing cell into daughter cells, the G_1 (gap 1) period, often called the *interphase*, commences. Its length is extremely variable. In rapidly cleaving embryos just after fertilization, the G_1 phase is very short and sometimes may not even exist. At the other extreme, the G_1 phase of mature neurons persists throughout the remainder of the life of the cell because further cell division does not occur. Cells of this type are called *postmitotic cells*. During the G_1 period the cell carries out its normal set of activities, such as specific synthesis, secretion, conduction, and contraction.

If a cell is in a dividing population, the synthesis of DNA, preliminary to mitosis, will occur. The period of DNA synthesis is called the S phase of the cycle. This is followed by a G_2 (gap 2) phase, which constitutes the period between the end of DNA synthesis and the beginning of mitosis itself (M phase). The process of mitosis is illustrated in Fig. 3-4.

Gene Activation   Genes are not active in the zygote, where they are tightly complexed with basic proteins called *histones*. The chromosomal DNA plus its enveloping histones is called *chromatin*, and the densely staining chromatin (*heterochromatin*) that can be seen within the nucleus at both the light and electron microscopic level represents inactivated, or *repressed*, genetic material. As development begins, certain groups of genes become activated, or *derepressed*, by becoming freed from their associated histones. Derepressed DNA represents functional genes. The first genes to become derepressed are those involved with the proliferative and general metabolic activity of the cell. As cleavage progresses and the embryo enters the stage of gastrulation, the first tissue-specific genes become activated. Later, during the period of organogenesis and histogenesis, other genes controlling more specific activity of differentiated cells come into play (Fig. 1-6).

It is estimated that at any given stage of development not more than 5 to 10 percent of the genes are active; the rest remain repressed. Studies on giant *polytene* chromosomes in insects have shown that at a given stage of develop-

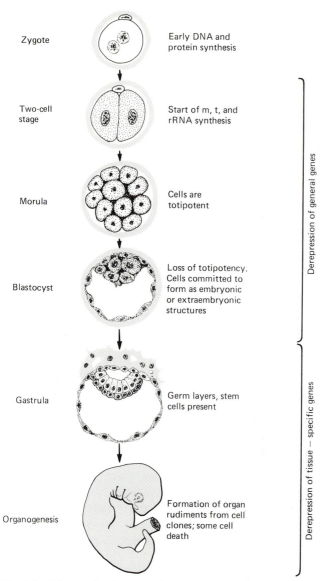

Zygote — Early DNA and protein synthesis

Two-cell stage — Start of m, t, and rRNA synthesis

Morula — Cells are totipotent

Blastocyst — Loss of totipotency. Cells committed to form as embryonic or extraembryonic structures

Derepression of general genes

Gastrula — Germ layers, stem cells present

Organogenesis — Formation of organ rudiments from cell clones; some cell death

Derepression of tissue — specific genes

Fig. 1-6   Scheme of early mammalian development, stressing important properties of the embryos and the varieties of genetic regulation. [Adapted from B. Konyukhov, 1976, *Genetic Control of the Development of Organisms* (Russian), Znanie, Moscow.]

ment certain genes are activated, whereas at another stage those same genes are repressed and other genes are activated (Fig. 1-7).

Restriction and Determination   Within the fertilized ovum lies the capability to form an entire organism. In many vertebrates the individual cells resulting

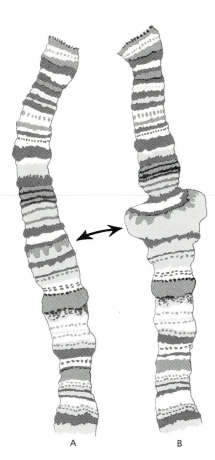

Fig. 1-7 Drawing of a segment of a giant polytene chromosome in the fly, *Sarcophaga*, showing the banding pattern. (A) One of the bands *(arrow)* has just begun to puff. (B) Two days later the puffing is much larger, indicating activation of the genetic material in that part of the chromosome.

A   B

from the first few divisions after fertilization retain this capability. According to the jargon of embryology, such cells are described as *totipotent*. As development continues, the cells gradually lose their ability to form all types of cells that are found in the adult body. It is as if they become funneled into progressively narrower channels. The reduction of the developmental options permitted to a cell is called *restriction*. Very little is known about the mechanisms that bring about restriction, and the sequence and time course of restriction vary considerably from one species to another. Nevertheless, an example representing a general pattern of restriction during development may serve to clarify the concept (Fig. 1-8).

Shortly after fertilization the zygote undergoes a series of cell divisions, called *cleavage*, during the early phase of which the cells commonly remain totipotent. The period of cleavage comes to an end when certain of the cells in the embryo undertake extensive migrations and rearrange themselves into three *primary germ layers* during a process known as *gastrulation*. Named on the basis of their relative position, the outermost layer is the *ectoderm*, the innermost is

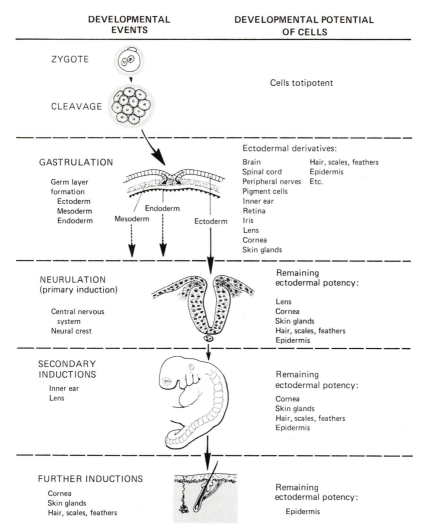

| | DEVELOPMENTAL POTENTIAL |
| DEVELOPMENTAL EVENTS | OF CELLS |

ZYGOTE

CLEAVAGE

Cells totipotent

GASTRULATION

Germ layer formation
Ectoderm
Mesoderm
Endoderm

Endoderm

Mesoderm

Ectoderm

Ectodermal derivatives:

Brain   Hair, scales, feathers
Spinal cord   Epidermis
Peripheral nerves   Etc.
Pigment cells
Inner ear
Retina
Iris
Lens
Cornea
Skin glands

NEURULATION
(primary induction)

Central nervous
system
Neural crest

Remaining
ectodermal potency:

Lens
Cornea
Skin glands
Hair, scales, feathers
Epidermis

SECONDARY
INDUCTIONS

Inner ear
Lens

Remaining
ectodermal potency:

Cornea
Skin glands
Hair, scales, feathers
Epidermis

FURTHER INDUCTIONS

Cornea
Skin glands
Hair, scales, feathers

Remaining
ectodermal potency:

Epidermis

Fig. 1-8 Diagram illustrating restriction during embryonic development. The column to the right of the figures demonstrates the progressive restriction of the developmental capacity of cells along one track, ultimately leading to the formation of epidermis. The column to the left of the figures describes major developmental events that remove groups of cells from the epidermal track.

the *endoderm*,[1] and between the two is the *mesoderm*. By this time at least one stage of restriction has usually occurred, so that cells of the three germ layers are now locked into separate developmental channels that are no longer freely interchangeable with one another. The potential options open to the cells of the ectodermal channel are shown in Fig. 1-8. In the next major developmental

[1]Over the years the name of the innermost germ layer has been spelled in two different ways, *entoderm* and *endoderm*. The latter spelling has been approved by the International Anatomical Nomenclature Committee (*Nomina Embryologica*, 1974) and will be used throughout this book.

event, part of the ectoderm becomes thickened and is henceforth committed to forming the brain, the spinal cord, and other associated structures. This stage of development is commonly called *neurulation*. The remainder of the ectodermal cells can no longer form these structures and have thus undergone another phase of restriction. Soon, as the result of tissue interactions with the newly forming brain, groups of ectodermal cells become committed to forming lens and inner ear, whereas the remainder of the ectoderm ultimately loses this capacity.

Subsequent developmental events see the ectoderm further subdivided into groups of cells destined to form cornea; hair, scales, or feathers; cutaneous glands; or simply epidermis. When restriction has proceeded to the point at which a group of cells becomes committed to a single developmental fate (for example, the formation of cornea), we say that *determination* of these cells has taken place. Thus, determination represents the final step in the process of restriction. The mechanisms that bring about determination of various groups of cells are receiving intensive study, but, as is the case with restriction, much remains to be learned. Usually, however, tissue interactions called *inductions* (see page 17) shortly precede the process of determination (and some phases of restriction) and are almost certainly involved in some manner.

Differentiation   Whereas restriction and determination signify the progressive limitation of the developmental capacities of cells in the embryo, *differentiation* refers to the actual morphological or functional expression of the portion of the genome that remains available to a particular cell or group of cells. Differentiation is really the process by which a cell becomes specialized, and the final product is called a *differentiated cell*. Although in many respects differentiation is a cellular event, rarely does a cell undergo differentiation in isolation. It is becoming increasingly apparent that the differentiation of many tissues in the embryo will not occur unless a minimum critical number of cells is present. Typically, differentiation in vivo is a communal process that occurs within groups of similar cells. Nevertheless, much of the most incisive analysis of differentiation has been performed in vitro, albeit even under these conditions usually with groups of cells rather than individual cells.

There are many ways of looking at differentiation. From the biochemical standpoint, differentiation may be viewed as the process by which a cell chooses one or a few specialized synthetic pathways, for example, the synthesis of hemoglobin by erythrocytes or of specific crystallin proteins by the lens. Functional differentiation can be looked upon as the development of contractility by muscle fibers or as the development of conductivity along a nerve. From the morphological standpoint, final differentiation is represented by a myriad of specific cell shapes and structures. A comparison between the histological properties of differentiated and undifferentiated cells is given in Table 1. Although exceptions can be given for every category, this table should prove useful as a general guide.

Definitions of differentiation vary greatly, and it is beyond the scope of this section to treat them in any detail. The most restrictive definition would limit

Table 1   Characteristics of Undifferentiated vs. Differentiated Cells

Characteristic	Undifferentiated cells	Differentiated cells
Nuclear size	Larger	Smaller
Nucleocytoplasmic ratio	High	Low
Nuclear chromatin	Dispersed	Condensed
Nucleolus	Prominent	Less prominent
Cytoplasmic staining	Basophilic	Acidophilic
Ribosomes	Numerous	Less numerous
RNA synthesis	Greater	Lesser
Mitotic activity	Great	Reduced
Metabolism	Generalized	Specialized

differentiation to the maturation of a cell during a single cell cycle—often the terminal cycle. Other, broader definitions would include the maturation of a cell and its descendants over the span of several cell cycles. Irrespective of the working definition, differentiation can follow several general pathways. One type of differentiation pathway, which is without question a terminal one, results in a population of highly specialized cells which have lost their nuclei. Examples of this are the platelets and erythrocytes in the bloodstream of higher vertebrates and the cells of the outer layer of the epidermis. For other cells that retain their nuclei, differentiation may be expressed by the synthesis of highly specialized intracellular molecules, such as contractile proteins in muscle, or by the secretion of extracellular substances, such as hormones or collagen fibrils.

At the tissue level, differentiation can often be recognized as characteristic morphological changes occurring in groups of cells in certain locations and at certain times. The process by which individual tissues take on a characteristic appearance through the differentiation of their component cells is called *histogenesis*. At this level it is often difficult to separate histogenesis from morphogenesis, the subject of the next section.

Morphogenesis   The entire group of processes which mold the external and internal configuration of an embryo is included under the general term *morphogenesis*. Morphogenesis remains one of the major mysteries of biology, and our present knowledge of this field is so slight that it can be compared with the state of genetics before the rediscovery of Mendel's laws.

A bewildering array of phenomena can be included under the overall umbrella of morphogenetic events in the vertebrate embryo. Consider, for example, the branching of the lungs, the form of the limbs, the shape of the eyeball, the pattern of blood vessels in the thorax, the intricate structure of a feather, or the complex loops and whorls on the fingertips. All are the result of morphogenetic processes.

How, then, does one approach the study of morphogenesis? In the absence of a general theory, it is first necessary to have some facts in hand. This has been most effectively done by identifying morphogenetic processes that are relatively easy to characterize and then subjecting these processes to genetic or experimental analysis. Some classical models for morphogenetic analysis have been slime molds, bacteriophages, and the regenerating amphibian limb. Within the sphere of vertebrate embryology, the formation of the neural tube, limb development, and the branching of internal glands have been subjected to intensive analysis. With increasing information about morphogenesis in certain systems, attempts are now being made to categorize morphogenetic phenomena and to construct hypotheses that will explain them.

Many types of processes can contribute to the morphogenesis of a structure. One of the most fundamental, but least understood, is *pattern formation*. There is evidence that very early in the development of many structures, prior to the onset of cell differentiation, an invisible pattern or blueprint is laid down and that further development uses this pattern as a guide. The pattern is not always fixed and firm because at certain periods experimental manipulations or natural events can result in parts of the plan not being followed by the developing structure. Whether or not developing internal organs follow the same sorts of patterns as external structures or the body as a whole remains to be seen.

Assuming that some sort of pattern has been set, the next phase in morphogenesis consists of realization of the pattern. This is accomplished by employing familiar processes in special ways. Some of these processes are illustrated in Fig. 1-9. They may include cell proliferation, migration, aggregation, secretion of extracellular substances, change in cell shape, and even localized cell death. How the cells within the primordium of a structure communicate with one another to carry out the instructions inherent in a pattern is still poorly understood. A currently popular way of looking at cell communication during morphogenesis is the concept of *positional information*. A simplified explanation of this concept is that a given cell is able to (1) recognize its position on a coordinate system that is set up within the primordium of an organ and (2) differentiate according to its position. For additional details, as well as applications to specific developing systems, the reader is referred to two reviews by Wolpert (1969, 1971).

Induction One of the most remarkable features of embryology is the precision with which developmental signals are generated and transmitted to the appropriate receptor. These signals may be of many types, and their effects may be made manifest in a variety of ways. One of the most important systems of embryonic signal calling is the process of *induction*. By induction we mean an effect of one embryonic tissue (the *inductor*) upon another, so that the developmental course of the responding tissue is qualitatively changed from what it would have been in the absence of the inductor. One of the classic examples of embryonic induction is the formation of the lens of the eye as a result of the inductive action of the optic cup upon the overlying ectoderm (Spemann, 1901, 1912). Details of this inductive system will be presented later in the text (Chap. 12). Another very important system of induction was found by Spemann and Mangold (1924), who transplanted a piece of the dorsal lip of the blastopore (Fig. 5-9) from one early amphibian embryo to another. Under the organizing influence of the transplanted tissue, a secondary embryo developed

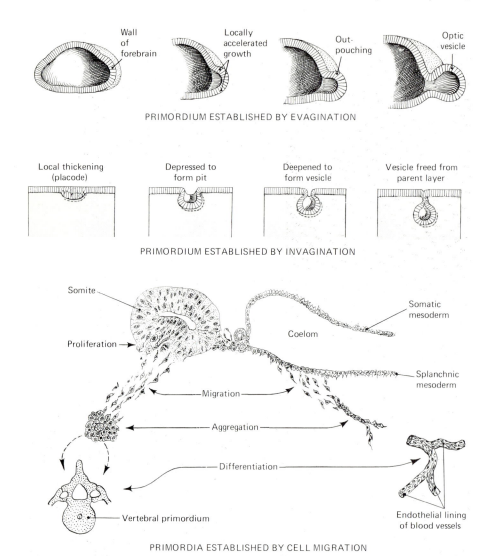

Fig. 1-9 Diagrams illustrating some of the different ways in which primordial cell groups may arise from parent cell layers.

along the flank of the host. The earliest inductive event in the embryo takes place as the germ layers are becoming established during gastrulation. Part of the mesodermal layer (specifically the chordamesoderm, see page 166) acts upon the overlying ectoderm, resulting in the appearance of a thickened plate of cells that will ultimately form the central nervous system. This specific interaction is called *primary embryonic induction*. (For an excellent review, see Saxén and Toivonen, 1962.) Subsequent inductive events in the embryo are commonly called *secondary inductions*.

The nature of the inductive stimulus and its mode of transmission have been the subject of intensive research. Several varieties of inductive mechanisms have been suggested. One is the extracellular diffusion of inductive substances secreted by the inducing tissue. There is considerable evidence in favor of this mechanism in primary embryonic induction, resulting in the formation of the nervous tissues (Saxén and Toivonen, 1962; Toivonen et al., 1975). Other forms of induction appear more likely to be contact-mediated, either by direct cell-to-cell contact or through the extracellular matrix secreted by cells involved in the interaction (Lehtonen, 1976; Hay, 1977). In 1956, Grobstein reported that direct contact between inducing and responding tissues in the kidney was not required; placing a porous filter between the two tissues did not halt induction. Inductive reactions do not occur through a nonporous membrane. More recent research, however, has shown that in many transfilter induction experiments close cell contact does occur by means of small cellular processes growing into the pores of the filter from both sides of the membrane (Lehtonen and Saxén, 1975).

During the 1930s several groups of investigators attempted to define chemically the nature of the inductive effect evoked by the dorsal lip of the amphibian blastopore. It was soon found that a wide variety of killed tissues could duplicate the inductive effect of some of the natural inductors. Several classes of chemicals, ranging from proteins and nucleoproteins to steroids, produced inductive effects similar to those produced by the cells of the dorsal lip of the blastopore. As more agents—including inorganic ions and even slight damage to the cells of the responding tissue—were found to produce inductive effects, embryologists turned their attention to the responding tissues.

There is some evidence that certain inductors may be specific, to a greater or lesser extent, in directing the fate of the responding tissues. This sort of induction is often called *instructive interaction*. It is also apparent that many inducing agents merely act as nonspecific triggers or *evocators* to release a response already encoded in the cells of the responding tissue. This type of induction is called a *permissive interaction*. Despite considerable research, little is known about how an inductive stimulus is received and processed by the responding tissues.

Intercellular Communication   One of the fundamental properties of living things, whether a flock of birds or a collection of organelles within a cell, is the ability of components of a biological community to generate signals and to respond, in turn, to signals from other members of that community. The developing embryo can be looked upon as a community of cells whose integrity and activities depend upon a well-developed system of intercellular communication. We have already seen how one type of communication, namely embryonic induction, can bring about profound qualitative changes in subsequent development. It has been recognized for many years that cellular communication must exist in induction and in the reaggregation of dissociated embryonic cells. Not until recently, however, have embryologists been able to approach this phenomenon with any degree of real understanding.

Steps are now being made toward recognizing some of the means by which individual cells communicate with one another. In certain cases, for instance, it has been shown that very small electric currents, inorganic ions, or even relatively large molecules can pass from one cell to its neighbors (Lowenstein, 1970). Such intercellular communication takes place in localized regions called *gap junctions*, where the membrane of one cell is in intimate contact with that of another (Fig. 1-10). Recent research involving the formation of *cell hybrids* by artificially joining together cells of different types is adding another dimension to our knowledge of intercellular communication.

The behavior of cells acting as a group is responsible for many of the segregative phenomena characterizing early development. In the establishment of the primary germ layers, for instance, large numbers of individual cells are somehow guided through relatively extensive migrations or rearrangements in a highly predictable fashion.

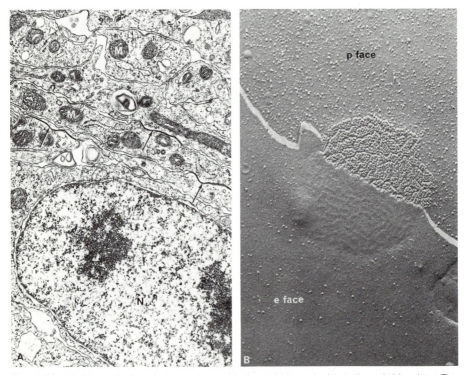

Fig. 1-10 (A) Transmission electron micrograph through the apical ectodermal ridge (see Fig. 10-7) of the limb bud in a chick embryo. Gap junctions between adjacent cells are indicated by arrows. ×19,000. Abbreviations: *N*, nucleus; *M*, mitochondria. (B) Freeze-fracture replica of a gap junction within the apical ectodermal ridge of a quail embryo at a stage comparable to the chick in B. In making a freeze-fracture preparation the tissue is frozen at very low temperatures and then cleaved with a special apparatus. This procedure splits membranes into their inner and outer components along the interface between the hydrophobic ends of the lipid molecules that constitute the two sheets of the membrane. The fractured membranes are then examined with the electron microscope. The *P* face shows the surface of the inner portion of the fractured cell membrane whereas the *E* face refers to the outer portion of the membrane. The gap junction itself is the large aggregate of particles in the center of the photograph. (From J. F. Fallon, and R. O. Kelly, 1977, *J. Embryol. Exp. Morph.* **41**:223. Courtesy of the authors and publisher.)

Cell Movements   At numerous periods during embryonic life cells or groups of cells move from one part of the embryo to another. Some movements are short migrations of individual cells, whereas others involve the massive dislocation of groups or sheets of cells over relatively great distances. Individual cells in embryos commonly migrate by amoeboid movements. Although these cells are mesenchymal in appearance, they may originate from any of the three germ layers. In amoeboid movement the cell is continually testing its surroundings, and its activity is characterized by the presence of a ruffled membrane along the leading surface (Fig. 1-11). A unique form of individual cell movement occurs in early bird embryos, in which the primordial germ cells move from the wall of the yolk sac into the bloodstream and are carried via the blood to the gonads (see Chap. 3). Examples of individual cells moving by amoeboid movement are the migration of cells away from the neural crest (ectoderm), the spreading out of mesodermal cells during germ-layer formation, and the migration of primary germ cells (endoderm) from the yolk sac to the gonads in mammalian embryos. These processes will be dealt with in greater detail later in the text.

Movement as a sheet is principally a property of epithelial cells, particularly those of the ectodermal germ layer. The migration of sheets of cells during gastrulation in amphibians and the spreading of cells over the yolk in bird embryos are good examples of this phenomenon. Little is known about what causes sheets of cells in an embryo to move (for reviews, see Gustafson and Wolpert, 1963; and Trinkhaus, 1969). The movement of cells as sheets is not confined to embryos. A simple cut in the skin of an adult vertebrate mobilizes the epidermis on either side of the defect, and within hours the wound is covered by a new layer of epidermal cells.

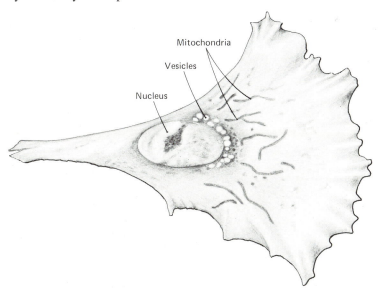

Fig. 1-11 Drawing of a mesenchymal cell moving in culture. The advancing edge (right) is ruffled, whereas the trailing edge (left) is tapered.

Cell Death It may seem paradoxical that destructive processes, even the actual death of cells, should play a vital role in the development of embryos. Nevertheless, cell death is an altogether necessary component of many phases of development (Glücksmann, 1951). Although perhaps most spectacularly represented in some postembryonic events, such as resorption of the tail, intestine, and opercular membrane of metamorphosing tadpoles, or the liquefaction of most internal organs of a metamorphosing insect larva, cell death also occurs in many regions of the avian or mammalian embryo. For example, separation of digits in the embryonic hand or foot is preceded by well-defined areas of cell death. Details of the way this process is involved in sculpturing the chick wing are given later in the text.

Although the exact mechanism responsible for cell death is still poorly understood, the process appears to be genetically determined. In the chick (Saunders et al., 1962) the death of certain groups of cells becomes irreversibly fixed; if they are transplanted to another location, they still die according to a predetermined schedule.

Hormones sometimes play an important role in stimulating the death of cells. The primitive female (Müllerian) genital ducts in the embryo regress in the presence of the male gonad and its secretions, whereas the male ducts, which lie alongside, are stimulated to further growth. In the case of the central nervous system, death is the fate awaiting those motor nerve cells that fail to make functional contact with a muscle fiber.

The Clonal Mode of Development Within recent years it has become increasingly clear that many structures in the embryo arise from the descendants of small numbers of cells (Mintz, 1971). A group of cells arising from a single precursor is called a *clone*. This concept arose from immunological studies, in which it has been shown that following the introduction of a foreign antigen into the body, a single immunologically competent cell undergoes a massive proliferative response and subsequently produces antibody against the antigen. This represents the basis for the "clonal selection" theory of Burnet (1969). It is now becoming apparent that many tumors also arise as clones descended from a single malignant cell. Some examples of clonal development in the embryo are the apparent formation of the body of the mammalian embryo from only three cells of the 64-cell embryo (page 127) and the origin of large portions of the central nervous system from well-defined cells of the early embryo (Fig. 11-3).

An important consequence of clonal selection in the embryo is that many cells in the early embryo are destined not to participate in subsequent development. Why these cells are not selected for further proliferation instead of the precursors of the clones is not known; presumably they ultimately die, but their fate remains obscure. Likewise unknown are the specific times when the clonal precursor cells of embryonic structures are selected and the mechanisms of selection.

Regulation and Regeneration During early development of the entire organism or of specific organ systems, most vertebrate embryos have an uncanny ability to recognize whether or not the structure is intact. If part of a structure is lost, either by accident or experimental manipulation, the loss is recognized and reparative processes are set in motion. If this occurs before differentiation of the structure has set in, the restoration of the missing material is called *regulation*.

Regulation is the basis for the development of identical twins. In mammals, including human beings, twinning usually results from the subdivision of embryos during early stages of cleavage (Fig. 1-12). Each half of the embryo is able to compensate for the lost tissues and develop into a perfectly normal individual. Occasionally separation of the two portions of the embryo is incomplete. This results in the formation of *conjoined twins*, one variety of

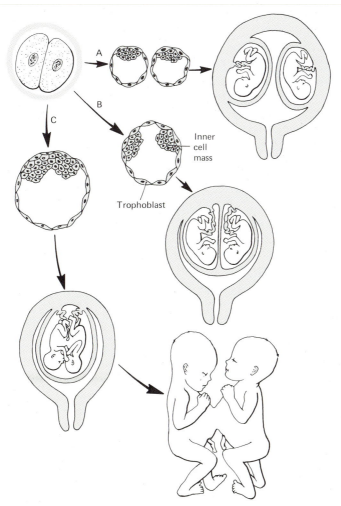

Fig. 1-12 Modes of monozygotic twinning. (A) A cleaving embryo may split at an early stage of cleavage, allowing the two portions to develop as completely separate embryos. (B) At a later stage of development the inner cell mass may split into two separate masses, both enclosed within the same shell of trophoblast. This is the most common mode of development of human twins. (C) If the inner cell mass does not become completely subdivided, conjoined twins may result.

which is the Siamese twins (Fig. 1-12). Commonly, when entire individuals or parts of organs are incompletely separated, one structure is a mirror image of the other. The reason for this reversal of symmetry is not known. In the normal development of the armadillo, the embryo breaks up at the four-cell stage, producing identical quadruplets.

Portions of the body that are able to reconstitute lost portions are sometimes called *morphogenetic fields* (Gurwitsch, 1944; Weiss, 1939). A morphogenetic field is a region of the body, such as that surrounding an appendage bud, in which the cells as a group are somehow cognizant of the overall nature of the structure to be formed. Thus, if cells are removed from the field or extra cells are added to the field, the primordium, as a whole, adjusts to the change and the cells establish a harmonious relationship with one another, resulting in the formation of a normal structure. Morphogenetic fields have boundaries, which can be defined experimentally but not anatomically, and if all the cells within a field are removed, the structure does not form. Chapter 10 describes regulatory properties within the limb field.

Sometimes in the late embryo or in postnatal life a missing structure can be replaced. If differentiation of recognizable structures has already occurred, the process of replacement is called *regeneration*. One of the main features of a regenerating system is the formation of a mass of primitive-appearing cells that demonstrate many of the properties of the embryonic primordium of the structure. One of the most difficult problems in either regulation or regeneration is how the cells remaining in the field are able to recognize that something is missing. Regulative activities within morphogenetic fields are now commonly interpreted on the basis of positional information of the component cells.

Growth   When one compares the bulk of the human ovum, a spheroidal cell about 0.15 mm in diameter, with that of an adult human being, it is obvious that the amount of growth involved is quantitatively astronomical. Even more striking is the example of the whale, in which an ovum about the same size as the human ovum produces an adult body weighing several tons. *Growth* can be defined in many ways, but perhaps the simplest definition is *an increase in mass*. This implies a concomitant increase in the number of cells, and in embryonic systems this is, indeed, the case. Normally accompanying the increase in mass is an increase in linear dimensions, but in some circumstances involving changes in form as well, length of a structure may increase in the absence of an increase in mass. Moreover, mass may increase in the absence of cell divisions if the cells undergo *hypertrophy*. Hypertrophy resulting from functional demand or from pathological processes is more common in postnatal life, but the growth in length and mass of the cartilaginous models of long bones at one phase is due principally to the hypertrophy of existing chondrocytes and their secretion of an extracellular matrix.

One would expect from the varied characteristics of the tissues themselves and from the different ways in which they grow that they would not all increase at a constant rate. What is meant by *differential growth* goes far beyond this obvious situation. In embryology this term is employed to cover different rates of growth of the same kinds of tissue in different locations and at different times. One of the striking features of young embryos is the rapid growth of the cephalic region. This results in the formation of a disproportionately large head in the embryo and fetus. Later, when growth in the cephalic region becomes relatively less rapid, the rest of the body catches up and adult proportions are established (Fig. 1-13). This is a manifestation of the effect of differential growth on the proportions of the body as a whole. A specific term for the disproportionate growth of body parts during postembryonic life is *allometric growth*.

How is growth controlled? This is another of the major mysteries of biology. With respect to overall growth of the body, there are two major patterns of growth. In *determinate growth* the body grows up to a certain point, characteristic of the species and sex, and then growth ceases. This is the characteristic pattern of growth in birds and mammals, but how to account for the enormous difference in growth potential between a pigmy shrew and a blue whale remains obscure. The typical pattern of growth in the lower vertebrates is *indeterminate growth*, in which growth continues throughout the lifetime of the individual, although at a reduced rate later in life. It is because of this characteristic that it is possible to determine the age of fish by examining the annual growth rings on their scales or in cross sections of certain skeletal elements. Despite the two patterns of growth, the problem of different degrees of growth potential remains. The difference in size between a guppy and a whale shark is as striking an example of the range of growth potential as that between the shrew and the whale. Some aspects of growth are obviously due to *growth hormone*, but all the growth hormone in the world would still fail to produce a shrew the size of a whale.

Some progress has been made in understanding certain components of growth at the tissue level. Several specific growth-stimulating factors have been

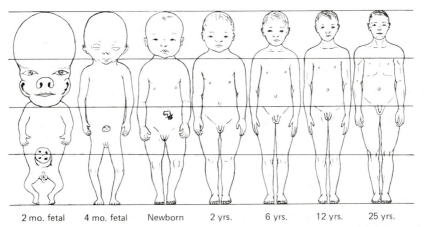

2 mo. fetal   4 mo. fetal   Newborn   2 yrs.   6 yrs.   12 yrs.   25 yrs.

Fig. 1-13 Two fetal and five postnatal stages drawn to the same total height to show the characteristic age changes in the proportions of various parts of the body. (Redrawn from Scammon.)

isolated and chemically characterized (Papaconstantinou and Rutter, 1978). *Nerve growth factor* (Chap. 11) acts specifically on sensory and sympathetic nerves. Investigations on the nature of nerve growth factor brought to light the existence of an *epidermal growth factor* (Carpenter and Cohen, 1978), and later research has uncovered a *fibroblast growth factor* (Gospodarowicz et al., 1978). The existence of a substance regulating hematopoiesis (erythropoietin) has been recognized for a number of years.

There is also evidence for the existence of tissue-specific inhibitors of growth. These have been called *chalones* (Bullough et al., 1967). The first of these substances to be recognized involved the epidermis, but chalones for several other types of tissues have been isolated as well. Chalones are glycoproteins, and they act by inhibiting or slowing down the rate of mitosis in the tissues that produce them. Chalones are characterized by (1) being formed by the same tissues in which they act, (2) being cell specific (e.g., epidermal chalone affects only epidermis), and (3) lacking species or even class specificity (e.g., epidermal chalone from the codfish acts upon mammalian epidermis).

Whether each type of tissue and cell will have its own specific stimulators and inhibitors of growth remains to be seen. This area of biology is still in its infancy, but research to date has revealed the existence of some very interesting control systems that operate in both the embryo and the adult.

Integration The word *integration* is capable of conveying such a broad range of meanings that it is not surprising to find it used by embryologists to cover a variety of concepts. Perhaps the most useful and significant way to employ it is with reference to the processes by which different tissues are brought together and combined to form organs. Take, for example, the eye. The photosensitive cells of its retina are of ectodermal origin. They arrive in the optic region by being infolded from the general surface as a part of the neural tube and thence grow out as a stalked vesicle. The primary optic vesicle then invaginates to form the optic cup, and the inner layer of the cup becomes the sensory layer of the retina. The lens of the eye is formed at a later time from cells of the superficial ectoderm situated over the opening in the optic cup. Still later the tough, fibrous connective-tissue coat of the eye is formed by regional mesenchymal cells. The muscles which move the eyeball are derived from special clusters of mesodermal cells, and the vessels which carry blood to and from the eye are fabricated from mesenchymal cells called *angioblasts*. An even finer level of integration, linking structure and function, occurs when nerve processes grow from the retina toward the appropriate parts of the brain. When this extremely precise matching-up of nerve fibers from the eye and brain is completed, the individual is ready to process a visual stimulus into an image. The bringing together of all these groups of cells in the right place at the right time, the control of their changing positional relations to each other, the regulation of all their histogenetic changes and their functional activities—the control of all these processes in such a manner that the completed organ can effectively carry out its function in the body—is *integration*. The eye is but one result of the many processes of integration which take place throughout the body of the embryo. Understanding integration at its many different levels of organization remains one of the fascinating challenges of experimental embryology.

Recapitulation The story of individual development sketches for us an approximate outline of the evolutionary changes passed through by our forebears. This concept is known as the *biogenetic law of Müller and Haeckel.* The general idea of recapitulation was first propounded by Müller (1864) on the basis of his studies of the development of invertebrates. Haeckel (1868) formulated its principles much more fully and gave it the name *biogenetic law.* In essence the law tells us that *an animal in its individual development passes through a series of constructive stages like those in the evolutionary development of the race to which it belongs.* More technically and more succinctly phrased, *ontogeny is an abbreviated recapitulation of phylogeny.*

In recent years the biogenetic law has been subjected to considerable criticism. Most of the objections to it have been directed against attempts to apply it too rigidly to details. It is readily apparent that recapitulation does not consist simply of adding more recent phylogenetic traits onto old ones. In particular, one would not expect the embryo to pass through stages in which many of the specialized structural features of present-day lower chordates are emphasized, for many of these are specific adaptations that diverge from the mainstream of chordate evolution. Rather, ontogenetic recapitulation is a conservative process which retains the basic ontogenetic stages of more primitive forms. Thus, in a mammalian embryo only the most fundamental steps of early development and the establishment of major organ systems, such as the heart and large blood vessels, would resemble those of a fish embryo. There would be a greater similarity between mammalian and reptilian or avian embryos (Fig. 1-14), and they would be apparent for a greater portion of embryonic life. In many cases ontogenetic processes leading to the formation of specialized structures in lower species are discarded or greatly reduced, and new ones are superimposed.

Often, however, a phylogenetically newer structure will make use of some component of the older one during the early phases of its development. An example of this is seen in the human placenta, the major organ of exchange between the embryo and the mother. The major blood vessels supplying and draining the placenta are homologous with those supplying the allantois, which subserves a similar exchange function in embryos of birds and some lower mammals. In the human, the allantois itself remains vestigial, but the allantoic blood vessels become incorporated into the phylogenetically newer circulatory system of the placenta.

Heredity and Environment Both heredity and environment are of vital importance in development, but in quite different ways. *Heredity* establishes an individual's physical and mental potentialities. *Environment* determines how far a person can go toward a full realization of his inheritance. Take a simple

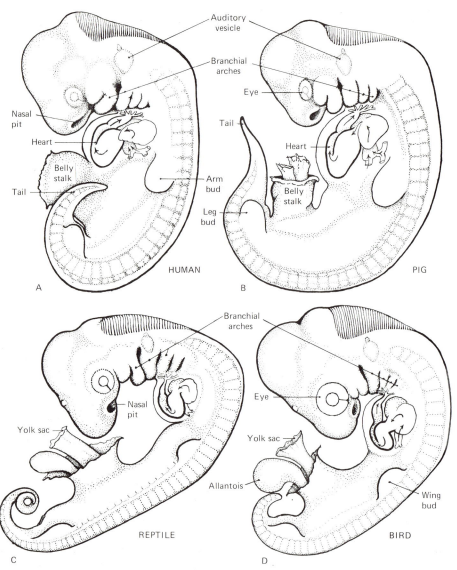

Fig. 1-14 Embryos of (A) human, (B) pig, (C) reptile, and (D) bird at corresponding developmental stages. The striking resemblance of the embryos to one another is indicative of the fundamental similarity of the processes involved in their development. (From William Patten, 1922, *Evolution*, Dartmouth College Press, Hanover, N. H.)

example involving postnatal social environment. A child endowed by heredity with a splendid physique and fine coordination may develop into an Olympic gold medalist if the child grows up in an environment which provides good nutrition, encourages habits of healthful living, and offers expert coaching. Another individual, who has an equally good heritage but lacks proper nutrition and opportunities for healthful outdoor exercise, may be prevented by environmental circumstances from realizing full potentiality.

A similar situation exists with regard to mental ability. An environment providing suitable education is necessary for the development of the intellectual capacity received from one's ancestors. A corollary, not always so well recognized, is that no amount of environmental privilege can carry a person beyond the limits of hereditary endowment. Environment can foster or handicap, but it cannot create.

From the biological standpoint, environment goes far deeper than such essentially social factors as physical and mental education. To begin with, there is a prenatal environment furnished by the mother during intrauterine life. The fertilized ovum has its hereditary potentialities already established by the qualities of the sex cells which united in its formation. But just as soon as it begins to develop, it is, perforce, subjected to whatever kind of living conditions the mother's own health provides within her gravid uterus. Within the growing body of the embryo itself are other environmental factors. Each growing organ is influenced by other organs adjacent to it. If an optic cup fails to grow close to the superficial ectoderm at the proper time, no lens is formed in that eye. If the blood vessels fail to keep pace in their own growth with the growth of the organ they are supplying, that organ will be defective in its own development.

Although environment cannot change the basic hereditary pattern in a given generation, it may, under certain circumstances, so affect the germ cells that the hereditary makeup of the offspring is altered. Fortunately, most of the ordinary things that have a deleterious effect on the body do not disturb the germ cells. However, experiments conducted on laboratory animals have shown that violent treatments, such as radiation of the gonads, may alter the chromosomes of the sex cells so that mutations from the expected parental pattern result. Like other mutations, these experimentally produced ones are inherited. Unfortunately, they tend to be of disadvantageous types involving the imperfect development of some part or parts of the body.

Then, too, there are conditions in which heredity and environment may interact. A most interesting example of this has been brought to light by the work of Frazer and his colleagues. They found that when pregnant females of a particular genetic strain of mice were fed heavy doses of cortisone, cleft palates resulted in practically 100 percent of their offspring. Under exactly the same treatment a different genetic strain showed cleft palates in only 17 percent of their offspring. When the same experimental procedures were applied to animals resulting from the crossing of these two strains, approximately 40 percent of the offspring showed the defect.

Following these dramatic results, Frazer and his colleagues studied the rate of growth of the palatal shelves in the embryos of the strain in which the incidence of the defect was high as compared with the strain in which it was low. The high-incidence strain showed a slow growth rate of the palatal shelves. They had barely enough growth energy to meet and fuse with each other if they were not disturbed in any way. In contrast, the low-incidence strain showed a high growth rate of the palatal shelves. In other words, there was enough reserve vigor in their growth so that in over 80 percent of the cases the shelves went ahead and fused in spite of the same disturbing treatment that caused 100 percent defects in the other strain. Here, then, is a situation in which neither heredity nor environment can be said to be the sole cause of the defective development. The interaction of the two determines the degree of vulnerability.

METHODS USED IN THE STUDY OF EMBRYONIC DEVELOPMENT

Over the years many methods have been devised for studying various aspects of embryonic development. These range from the examination of entire embryos with the naked eye or simple lenses to extremely sophisticated molecular probes. All techniques have their uses, and it is important to recognize that the choice of the technique is determined by the question that is being asked. This section will provide a brief survey of the major methods and techniques that are used in the study of vertebrate embryogenesis. These methods will be described to a greater or lesser extent depending upon the emphasis placed on them throughout the text. For those wishing more details on embryological techniques, a number of books dealing specifically with methods used in embryological studies are listed in the bibliography.

Direct Observation of Living Embryos   The earliest technique used in embryology was the direct observation of embryos, either with the naked eye or with simple lenses. After the introduction of the microscope, gametes and small embryos were examined in greater detail. Direct observation, particularly of a living embryo, provides one with a good overall view of the embryo and it impresses the observer with the dynamic and often sweeping changes that constitute embryonic development. A major disadvantage of direct observation is that often resolution of finer details is sacrificed in favor of the total picture. With small structures these disadvantages can sometimes be overcome by the use of special optical techniques, such as phase-contrast microscopy, or by the use of *vital dyes*, which permit the identification and tracing of specific cells or cell groups. A powerful tool for investigating the development of entire embryos or groups of cells is *microcinematography*. This technique provides a moving picture of development, usually accelerated by a considerable extent, that can later be subjected to quantitative analysis. Anyone who has seen time-lapse films of developmental processes, such as cleavage or the outgrowth of a nerve fiber, cannot fail to be amazed at the amount and precision of the changes that are taking place.

Examination of Fixed Material   It was recognized early that direct observation of living embryos is inadequate for analysis of many aspects of development. At times one wishes to arrest a process at a critical phase so that the material can be examined at leisure. This is normally accomplished by *fixation*, in which the embryo is treated with various chemicals, such as formalin or glutaraldehyde, that will preserve structures as faithfully as possible without causing undue distortion or other artifacts in the tissue. About a century ago fixed material was used to prepare serial microscopic sections of entire embryos; this was one of its earliest uses. From these sections, the three-dimensional internal structure of an entire embryo can be reconstructed. This method of observation is still commonly used in student laboratories as well as in research. With the advent of electron microscopy, attention has shifted to the finer details of embryonic structure, but the basic principles, as well as advantages and disadvantages, remain. A new dimension in the study of embryos has been added with the application of *scanning electron microscopy* to living systems. This technique produces a three-dimensional view of entire embryos or parts of embryos with a clarity and resolution that had been previously unattainable (Fig. 1-15), and it should prove immensely valuable in helping to sort out structural relationships between cells and tissues in the embryo. Aside from the obvious disadvantages of possible artifacts and the need to reconstruct an image of the structure from sections, examination of fixed tissue can be a pitfall that can lead to major errors of interpretation of developmental processes. It is common to examine a series of sections from embryos fixed over a span of time and then to try to reconstruct the dynamic nature of the process and the sequence of morphological changes on the basis of the sections. Usually this technique is valid, but occasionally misinterpretations of the nature of a developmental process, based upon the faulty interpretation of fixed material, have gotten into the scientific literature. Once so entrenched, errors of this sort are notoriously difficult to uproot.

Histochemical Methods   *Histochemistry* is a method of localizing specific chemical substances or sites of chemical activity on morphological structures that are as little disturbed as possible. Typically, the tissue or embryo is rapidly frozen in liquid nitrogen and sectioned with a special low-temperature micro-

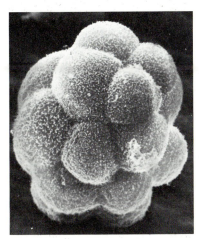

Fig. 1-15 Scanning electron micrograph of a 4-day cleavage stage (blastocyst) of a hamster embryo. The cells bulging on the surface (trophoblast) will form extraembryonic membranes rather than the embryo proper. The small projections on the surface of the cells are microvilli. ×700. (From P. Grant, B. O. Nilsson, and S. Bergström, 1977, *Fert. and Steril.* **28:**866. Courtesy of the authors and the publisher.)

tome called a *cryostat*. The tissue is then placed on a glass slide and subjected to a specific chemical reaction that leads to the deposition of a colored product at the site of enzymatic activity or at a place where some molecules are concentrated. In some cases it is now possible to obtain very fine resolution of histochemical reactions at the electron microscopic level. One disadvantage of histochemistry at the light microscopic level is that the staining methods often do not show general structural features of the embryo or organ with great clarity.

Autoradiography One of the useful by-products of the atomic age has been the widespread availability of radioactive isotopes for use in biomedical research. Precursors of macromolecules containing radioactively tagged atoms can be introduced into a biological system and then followed through a metabolic cycle by various analytical means. One such means of analysis is called *autoradiography*, which is a method that allows the localization of a radioactive isotope within cells or tissues by employing methods similar to those used in photography (Baserga and Malamud, 1969).

Typically, an embryo or a part of an embryo is placed in a solution containing a radioactively labeled amino acid or a precursor of DNA or RNA. After a given period, the embryo is removed from the radioactive solution and sectioned for microscopic examination, but in addition to the usual histological procedures the tissue sections are covered with a photographic emulsion and kept in the dark for several weeks. Radioactive emissions from the isotope, which has been incorporated into proteins or nucleic acids of the embryo, impinge upon the emulsion during the period of exposure. The emulsion is then developed in much the same manner as photographic film, and tiny grains of silver are deposited in the emulsion over the cells containing labeled atoms. With this as a guide, the labeled structures can be localized with a microscope.

Autoradiography is now routinely employed at both the light and electron microscopic level. With ordinary processing techniques, autoradiography is limited to localizing labeled macromolecules that are not dissolved out of the tissues. Newer, painstaking techniques have been developed for autoradiographic localization of soluble substances, such as steroid hormones. These require the tissues to be frozen immediately and then dried in such a way that displacement of the labeled soluble compounds does not occur.

Autoradiographic techniques have been of great help in localizing sites of nucleic acid and protein synthesis in embryos (Fig. 1-16) and in tracing the movements of cells. They do not, however, allow the direct analysis of their chemical form in the tissues. Thus, it is extremely important to employ the correct labeled precursor molecules and to recognize the chemical and technical pitfalls that can lead to misinterpretation of results.

Tracing Methods Many types of markers have been used to trace cell movements in the growing embryo. Some of the classic studies of cell movements have involved the application of nontoxic markers to small groups of cells. Certain stains, such as Nile blue sulfate or neutral red, can be applied to

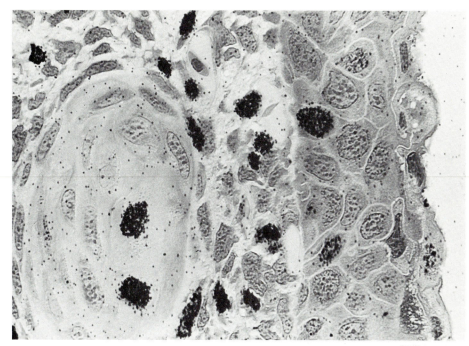

Fig. 1-16 Example of an autoradiograph of a regenerating salamander limb. The animal was injected with ³H-thymidine and the limb was fixed an hour later. Photographic emulsion is coated over the slide and developed after several weeks exposure. Dense accumulations of silver grains over a nucleus indicate that the cell was undergoing DNA synthesis when the isotope was administered. (Courtesy of Dr. T. Connelly.)

living cells without harming them. These are known as *vital dyes*. Changes of position of cells treated with these dyes can be followed through an extensive period of growth before the dye becomes so diffused that identification is no longer possible. We shall have occasion to see the application of such techniques in connection with the cell movements in amphibian gastrulation (Fig. 5-5). Marking may also be carried out by placing finely divided, physiologically inert carbon particles, such as those of blood charcoal, on a small group of cells. The results of experiments involving carbon marking will be used in discussing cell movements in the neighborhood of the primitive streak in the chick.

In the case of cells that produce highly specialized proteins, tracing can be accomplished by preparing a specific antibody to the protein. The antibody can be combined with a fluorescent dye and then applied to the embryo. The fluorescent antibody will combine only with the specialized protein, thus serving to identify those cells containing the protein.

Cells in which DNA has been heavily labeled with radioactive isotopes have been used for extremely precise localization of migrating cells. Such isotopic labeling of DNA has the disadvantage that the label is diluted by each cell division. This renders it unsuitable for long-term tracing of rapidly dividing cells.

An important category of marker, especially for the long-term tracing of cells, consists of "natural" markers. By introducing into an embryo cells which differ from those of the host by virtue of size, pigmentation, isoenzyme types, chromosomal complement, or number of nucleoli, a stable marker is effected. Sex chromatin (Fig. 3-28) has been used with some success, but the most recent major advance in tracing methods has been the use of Japanese quail cells as marker grafts into chick embryos (Fig. 10-14). Differences in nuclear size and morphology and the ease of grafting pieces of quail tissue into homologous sites in chick embryos have permitted investigators to solve a number of long-standing problems regarding the cellular origin and composition of certain tissues and organs.

Microsurgical Techniques   Much of the fundamental information about causative mechanisms in embryonic development, particularly those involving tissue interactions, has been obtained by the use of microsurgical techniques. Work with embryos, often no more than a few millimeters in length, has necessitated the development of special tools, such as glass or tungsten needles and loops of baby's hair, instead of the scalpels and forceps commonly associated with the usual types of surgery. For work with extremely small embryos or groups of cells, micromanipulators, instead of the hands, are used to hold the instruments.

Microsurgical techniques are used in many types of experiments. One of the simplest is *ablation*, or removal of part of an embryo in order to determine what effect the absence of that structure will have upon the remainder of the embryo. Such an experiment ushered in the era of experimental embryology. As a test of the concepts of preformation and epigenesis, Roux (1888) cauterized one of the blastomeres of a two-cell frog embryo with a hot needle. He wanted to learn whether the remaining cell would give rise to only half an embryo or whether that cell could, in its subsequent development, restore the deficiency. Although his experimental results proved to be somewhat misleading, other investigators, stimulated by this work, soon showed that if the cells of two-cell frog embryos are entirely separated, each cell is capable of giving rise to a complete individual. This procedure provided experimental proof of the untenability of the preformationist doctrine.

A most interesting form of ablation experiment can be performed by exposing portions of cells or tissue to tiny *laser beams* (Berns and Saleti, 1972). The laser emits a beam of coherent light that can be accurately focused on extremely small structures, which are then disintegrated by the energy of the beam (Fig. 1-18).

Transplantation and *explantation* are commonly used surgical techniques that have found wide application in embryological studies. Because of their widespread application in embryological research, transplantation techniques will be treated separately below.

Explantation consists of excising a small sample of embryonic tissue and growing it in an artificial environment. Explants may be handled in various ways.

One method is to graft the excised tissue into a host organism in such a location that it is well supplied with nutritive materials but must grow and differentiate without the influence of the other tissues of its own body which normally surround it. In working with bird embryos it is common to explant a small group of primordial cells from a young individual to the *chorioallantoic membrane* in an older host. With mammalian embryos favorable locations are the vitreous body of the eye or a vascular area of the peritoneum. Explantation experiments give us much information on how the tissue can adapt to and differentiate in a new location. Some embryonic primordia show a striking capacity for *self-differentiation*, which means that within the transplanted cells there is sufficient information to direct the development of the organ.

In embryological studies, tissues are sometimes transplanted to other sites on the same embryo (*autografting*), but often tissues or organs from a donor embryo are grafted to hosts of a different species (*heterografting*) or even to a different order (*xenografting*). Some types of transplantation experiments involve relatively minor shifts or rotations of the tissues, but in other types of transplantation an embryonic structure is moved far from its normal location. It has been found many times that transplanted embryonic tissues and the tissues of the host do not lie passively side by side but that they may exert profound influences upon the course of development of their new neighbors. As we shall see, examples of this type of influence have been particularly striking in the field of embryonic induction.

One of the most recent applications of transplantation involves not tissues or organs but components of single cells. Using the technique of nuclear transplantation developed by Briggs and King (1952), Gurdon (1962) transplanted the nucleus from the intestinal epithelium of a postmetamorphic frog (*Xenopus*) into an egg whose nucleus had been inactivated by *ultraviolet* (UV) *radiation* (Fig. 1-17). An adult frog developed from the egg. This experiment demonstrated that even the nucleus of a highly specialized intestinal epithelial cell still contained a sufficient endowment of genetic information to guide the development of an entire mature animal from the egg.

In at least one case the technique of transplantation has even been employed to economic advantage. A group of South African sheep ranchers wished to begin raising a special strain of sheep native to Scotland. To ship a sufficient number of adult sheep to South Africa by boat would have meant a long and costly trip. This problem was met by removing newly fertilized eggs from the Scottish sheep and transplanting these early embryos into the uteri of rabbits. The rabbits were then flown to South Africa where the sheep embryos were removed from the rabbits and transplanted into the uteri of local ewes. In due time normal lambs of the Scottish strain were delivered to the South African ewes and the long-distance transfer of the herd of sheep was completed (Hunter et al., 1962).

Clinical application of the technique of *embryo transfer* resulted in the first birth of a human conceived outside the uterus. A woman in England was unable to have a baby because of a blockage of her uterine tubes. This prevented

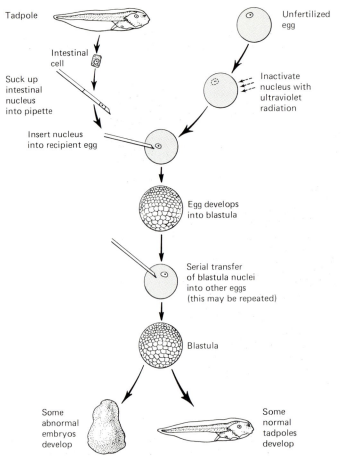

Fig. 1-17 Outline of a nuclear transplantation procedure in Amphibia (*Xenopus*). A nucleus from an intestinal epithelial cell is sucked into a micropipette and then injected into a recipient egg, the nucleus of which has been inactivated with ultraviolet light. The egg then develops into a blastula. When mature nuclei are transplanted, it is often necessary to make several serial transfers of daughter nuclei from blastulae into other eggs in order to create appropriate conditions for full development. Only a small percentage of the embryos develop normally. In the remainder, development is grossly abnormal and becomes arrested. (After Gurdon.)

ovulated eggs from reaching her uterus. Two British reproductive biologists, Robert Edwards and Patrick Steptoe, obtained an ovum from the woman's ovary and fertilized the egg in vitro with her husband's sperm. The embryo was allowed to develop to the eight-cell stage and was then transplanted to the woman's uterus. The embryo became implanted in the uterine lining, and an uneventful pregnancy and the birth of a normal baby girl was the result.

Culture Techniques One of the most interesting and instructive ways of studying embryonic development is to grow components of embryos or even whole embryos in an artificial environment. Depending upon the nature of the explanted material, the techniques are known as cell, tissue, organ, or even whole embryo culture. Each type of culture requires slightly different methods, but the principle of culture is the same. The embryonic material is placed into dishes or tubes of glass or plastic and surrounded by an artificial culture medium designed to resemble as closely as possible the environment surrounding the material in its normal site in the embryo. The ideal culture medium is completely defined chemically, but commonly it is necessary to add undefined biological factors such as serum or even extracts from entire embryos in order to provide necessary growth factors.

The culture method was developed in a revolutionary experiment by Ross G. Harrison (1907), who was looking for a method to demonstrate the growth of nerves (Fig. 11-20). In the time since his pioneer work, culture methods have contributed greatly to our understanding of developmental processes. Today, the culture of tissue or organ rudiments is often routine, and for a number of cell types (e.g., muscle; Konigsberg, 1963) it is possible to produce differentiated clones from single precursor cells. Recent years have seen an increasing interest in the culture of complex organs or even whole mammalian embryos, but progress in refining culture techniques is often slow and the work is frequently frustrating.

A major advantage of culture techniques is that the surrounding medium or the tissue itself can often be altered in defined ways that would never be possible to accomplish in vivo. A disadvantage, particularly in cell and tissue culture, is that it is sometimes difficult to distinguish processes that operate only in culture conditions (culture artifacts) from those that occur naturally in the embryo.

Cellular Dissociation and Aggregation   One of the newer methods of studying growing tissues depends on dissociating the cells of which they are composed. This can be accomplished by the use of a weak solution of certain enzymes, such as trypsin, which break down the material responsible for adhesion between cells. The dissociated cells, after thorough mixing, are grown in tissue-culture media under such conditions that they are free to move about and reassociate. Studies along such lines (Townes and Holtfreter, 1955; Moscona and Moscona, 1952; and Weiss and Taylor, 1960) indicate the amazing ability of the dissociated cells to regroup themselves with other cells of like kind and sometimes even to assume a functionally significant reorganization of tissues as complex as parts of the brain.

There are several main lines of thought regarding the mechanisms of cell aggregation, both in vitro and in vivo. According to one, cell aggregation is mediated by the action of specific intracellular compounds, known as *ligands*. Another school of thought places more emphasis on physical properties of the cell surfaces, which differ from cell to cell. These studies, plus the great surge of recent research on the properties of cell membranes, have directed increasing attention toward the role of the cell surface in developmental processes. Such investigations are being aided by the use of *lectins*, naturally derived plant compounds that specifically bind to chemically defined carbohydrate chains protruding from the cell membranes. These and other kinds of probes are demonstrating how important the cell surface is in the association of like kinds of cells or the formation of boundaries between dissimilar groups of cells.

Biochemical Techniques The biochemical analysis of embryonic tissues, particularly with the use of the newer techniques of molecular biology, is one of the most rapidly growing ways of studying development. The biochemical techniques used to study embryonic systems would include almost all of those now in use in biochemistry, so in the space allotted here only general categories of techniques will be mentioned. Amond the older techniques are those purely chemical techniques designed to determine the presence or absence of specific compounds and their amounts. Needham's (1931) treatise summarizes much of this material. Analysis of enzyme activity is frequently used in studies on metabolic properties of embryos. In these experiments a chemical reaction involving the mediation of an enzyme (as are most biological reactions) is allowed to occur and the amount of some reaction project is commonly measured—usually by spectrophotometric means—and compared with a reference curve.

Separation methods are widely used in studies on embryos. The first separation techniques were paper chromatography and electrophoresis. These techniques make use of physical properties of compounds, such as amino acids or proteins, which give them different migratory properties in solutions or in electrical fields. Among the molecules that can be separated by electrophoresis are *isoenzymes* (Markert, 1975). Isoenzymes (isozymes) are different forms of molecules with the same enzymatic activity, but with slightly different structures that enable them to be separated from one another. Because different isozyme forms of a given enzyme are often formed by separate populations of cell types at various times, they often make good developmental markers (Fig. 9-18).

Another family of separation techniques involves the centrifugation of solutions or tissue homogenates containing macromolecules. After prolonged centrifugation at high speeds, subcellular fractions or different classes of macromolecules become stratified according to their size and density.

Column chromatographic methods have been developed to separate many classes of compounds according to various physical characteristics. In typical column techniques, a glass column is loaded with beads or other special materials that have been developed to allow the differential migration of molecules, and a solution containing a family of macromolecules—commonly nucleic acids—is added to the tube. The fluid in the column is allowed to drip from the bottom into a fraction collector—a container of some sort full of tubes. At periodic intervals the tubes are moved; their content of molecular material is a reflection of the rate of passage of the molecules through the column. The fluid in the tubes may then be examined for content of the molecules in question or, if isotopic labeling is also used, as is commonly the case, the fluid in each tube is also analyzed for radioactivity.

A number of highly specialized techniques have been devised for demonstrating particular species of information-containing nucleic acids. These take advantage of the unique sequence of bases that constitute a strand of DNA or RNA that contains the information for the formation of a specific protein and involve the *hybridization* of a specific DNA or RNA molecule with a molecule of the opposite sort so that there is a matching-up of complementary sets of bases. Techniques of this sort have proved to be very valuable in demonstrating the presence or absence of very specific molecules of nucleic acids in embryonic cells.

Irradiation Techniques   Various forms of irradiation have been used in embryological studies, mainly to inflict some form of damage upon parts of the embryo. For some experiments *x-rays* are brought to focus on small areas of an embryo in order to provide a circumscribed area of tissue injury or to inactivate a group of cells. UV rays are sometimes used for the same purpose, but they lack the deep penetrating power of x-rays. Laser beams are proving to be another valuable tool in producing sharply localized lesions in embryos. Their precision is already such that small areas of individual chromosomes can be destroyed (Fig. 1-18).

Inhibitory Agents and Teratogens   In the analysis of embryonic development many types of chemical agents have been used to inhibit normal embryonic processes. In some cases the mechanism of action of a chemical inhibitor is quite

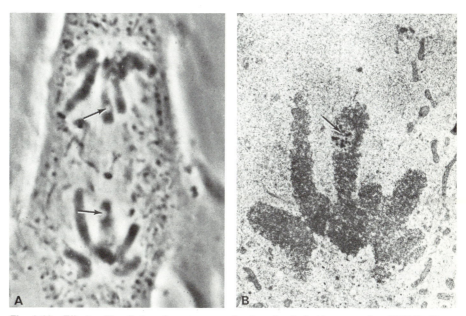

Fig. 1-18 Effects of irradiating chromosomes of cultured cells [rat kangaroo line (PTK$_2$)] with an argon laser beam. (A) In a phase-contrast micrograph, the irradiated area appears as pale spots within the chromosome (arrows). ×1300. (B) In an electron micrograph of the lower set of chromosomes, the irradiated area contains aggregates of electron-dense material (arrow). ×3100. (From J. B. Rattner, and M. W. Berns, 1974, *J. Cell Biol.* **62**:526. Courtesy of the authors and publisher.)

well defined and an alteration in development can be attributed to a specific disturbance in a metabolic pathway. For example, the antibiotic actinomycin D is known to inhibit the synthesis of RNA. If actinomycin D is administered to a very early embryo, development proceeds normally for a short time but is soon markedly inhibited (Gross and Cousineau, 1964). One can infer that the stage which was inhibited in its development required the synthesis of new RNA molecules.

In most cases the exact mechanism of the disturbing action caused by a chemical or by a radiation (such as x-ray or UV ray) is not yet known. The effects of these agents can only be interpreted at a less specific level, sometimes involving the intermediary effect of one tissue upon another rather than a direct disturbance in a chemical process.

In recent years a number of drugs have been shown to cause many types of abnormal development. Such drugs are said to have a teratogenic effect and are commonly referred to as *teratogens*. A striking example of a chemical teratogen acting upon human development occurred in West Germany and several other countries during the early 1960s. A large number of children were born with an unusual type of defect of the limbs. In severe forms of this defect the proximal segments of both arms and legs are missing and the hands and feet appear to grow directly from the body. Because such limbs resemble the flippers of a seal, the condition was called *phocomelia* (seal appendage). It was soon discovered that these deformities were caused by a supposedly safe sedative called *thalidomide*, which was commonly used by pregnant women in those countries. This drug has been taken off the market, but because of its sometimes tragic effects the screening procedures for all new drugs now include rigorous tests to determine whether the drugs exert any ill effects upon embryos.

Genetic Markers and the Use of Mutants   As our knowledge of the genetics of certain laboratory animals increases, the use of genetically defined, often lethal, *mutant* strains is assuming increasing importance in the analysis of embryonic development. By identifying where and when development first goes wrong in a mutant strain, it is often possible to pinpoint the effects of certain specific genes on developmental processes. One such mutant is the *o* gene in the axolotl. The *o* gene was first recognized when embryos homozygous for the o^-/o^- gene consistently stopped developing in the late blastula stage. Injections of small amounts of cytoplasm from mature wild-type (o^+/o^+) eggs corrected the defect. It is now known that the nucleus of the normal egg produces the o^+ substance and releases it into the cytoplasm and that during cleavage this substance, probably a protein, acts back on the nucleus and permits further cleavage (Brothers, 1976).

Not all mutants, of course, are lethal. As a general rule, one can say that the later in development that a gene begins to act, the less the likelihood is that a mutant will be lethal. In albinism, for example, the lack of color is not due to the absence of pigment cells, but rather the absence of a specific enzyme (tyrosinase) that is required in the synthetic pathway of the black pigment, melanin. As a rule, the more that an animal is studied, the more mutant genes are discovered. Among the vertebrates, the mouse is by far the best-studied species, with several hundred defined mutant genes, but mutant strains in axolotls, *Xenopus*, and the chicken have also proved quite useful in embryological studies.

Genetic markers are also useful in developmental studies, often as tracers. Particularly valuable in some work have been strains of mice that have developed isoenzymic differences in the form of certain enzymes. Identification of specific isoenzymes by electrophoretic methods has proved useful as a tracing method.

The methodological approaches listed in this section are being fruitfully utilized in investigations that are placing us on the threshold of a new era in the understanding of the controlling and regulating factors in development. Their combination is tending to break down the old boundaries between various fields of scientific research and is resulting in a more unified approach to the study of embryological problems.

Unfortunately, some of the newer methods of research involve apparatus that is too expensive to be in every laboratory. Any lack of special equipment should merely indicate the desirability of directing research efforts into channels not requiring such equipment. One should not be unduly impressed either by the availability of apparatus or by the lack of it. Rather it should be realized, as Ebert (1966, p. 50) has pointed out, that many of the most significant advances in our understanding of developmental mechanisms "were made with techniques of utter simplicity, the only requisite being glass needles and hair loops, physiological saline, and the ability to pose the right questions." It is well to remember, also, that "new methods can usually only be applied to old material; and new ideas do not suddenly emerge full-fashioned as Aphrodite was born from the chaotic sea; they are built up laboriously on the foundation of previous work" (Waddington, 1956, p. 5). A thorough and accurate knowledge of events during normal development is the foundation on which experimental embryology must rest. Without it there is no sound basis for interpreting the significance of experimental results.

Chapter 2

Reproductive Organs and Gametogenesis

REPRODUCTIVE ORGANS

Any logical account of prenatal development must start with a consideration of the phenomena which initiate that development. It is necessary to know more than the mere structure of the conjugating sex cells. We must know something of how they are produced and of the extraordinary provisions which ensure their union in such a place and at such a time as each is capable of discharging its function. Of vital importance, also, are the changes in the body of the mother that provide for the nutrition of the embryo during its intrauterine existence and for its feeding during the relatively long period after birth when it cannot subsist on such food as that eaten by its parents. Before it is possible to speak intelligibly about any of these things, it is necessary to become familiar with the main structural features of the reproductive organs.

Female Reproductive Organs The reproductive organs in the human female and their relations to other structures in the body are shown in Figs. 2-1 and 2-2. The paired gonads—the *ovaries*—are located in the pelvic cavity. Each ovary lies close to a funnellike opening (*ostium tubae*) at the end of the corresponding *uterine tube*. About this abdominal orifice of the tube are characteristic fringelike processes of highly vascular tissue, which are called

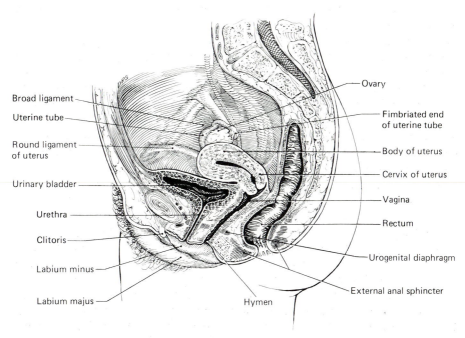

Broad ligament
Uterine tube
Round ligament of uterus
Urinary bladder
Urethra
Clitoris
Labium minus
Labium majus
Hymen
Ovary
Fimbriated end of uterine tube
Body of uterus
Cervix of uterus
Vagina
Rectum
Urogenital diaphragm
External anal sphincter

Fig. 2-1 Sagittal section, adult female pelvis. (Redrawn, with slight modifications, from Sobotta, *Atlas of Human Anatomy.* Courtesy, G. E. Stechert & Company, New York.)

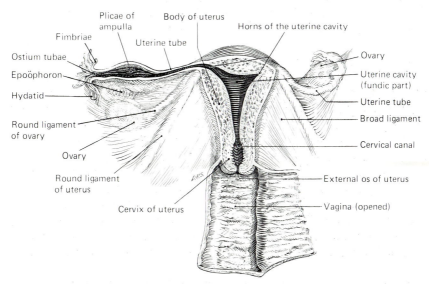

Plicae of ampulla
Fimbriae
Ostium tubae
Epoöphoron
Hydatid
Round ligament of ovary
Ovary
Round ligament of uterus
Cervix of uterus
Body of uterus
Uterine tube
Horns of the uterine cavity
Ovary
Uterine cavity (fundic part)
Uterine tube
Broad ligament
Cervical canal
External os of uterus
Vagina (opened)

Fig. 2-2 Internal reproductive organs of the female, spread out and viewed in ventral aspect. The vagina, uterus, and right uterine tube have been opened to show their internal configuration. (Redrawn, with slight modifications, from Rauber-Kopsch, *Lehrbuch und Atlas der Anatomie des Menschen.* Courtesy, Georg Thieme Verlag KG, Stuttgart.)

fimbriae. The lining of the uterine tube is thrown up into numerous complex folds (Fig. 2-3) and the epithelial surface contains many ciliated cells, whose beat causes strong fluid currents flowing toward the uterine cavity. When an ovum is liberated from the surface of the ovary, it enters the fimbriated end of the uterine tube and passes slowly along the tube to the uterus. There, if it has been fertilized, it becomes attached and nourished during prenatal development.

The human *uterus* is a pear-shaped organ, which in the nonpregnant condition has thick walls, is richly vascular, and is well supplied with smooth muscle. The body of the uterus is continuous caudally with the neck or *cervix*, a region characterized by an attenuated lumen, thick walls, and glands of a different type from those occurring in the body of the uterus. The cervix of the uterus projects into the upper part of the *vagina*, which serves the double function of an organ of copulation and a birth canal.

The external genitalia of the female are a complex of structures grouped about the vaginal orifice. Collectively they constitute the *vulva*. The outermost structures are a pair of fat-containing folds of skin known as the *labia majora* (Fig. 2-1). Within the cleft between the labia majora is a second, smaller pair of skin folds, highly vascular and devoid of fat. These are the *labia minora*. Partially enwrapped by the labia minora where they meet anteriorly is the *clitoris*, a small erectile organ which is the homolog of the penis of the male. In the vulva, about midway between the clitoris and the vaginal orifice, is the opening of the *urethra*. The vaginal orifice is located in the posterior part of the

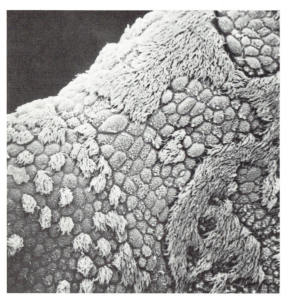

Fig. 2-3 Scanning electron micrograph of the mucosal surface of the ampullary portion of the human uterine tube during the premenstrual (late luteal) phase. Cells with long tufts of cilia are scattered among nonciliated cells. ×1000. (From H. Ludwig, and H. Metzger, 1976, *The Human Female Reproductive Tract*, Springer-Verlag, Berlin. Courtesy of the authors and publisher.)

vulva (Fig. 2-1). In the virgin the entrance into the vagina is narrowed by a thin fold of tissue known as the *hymen*.

Male Reproductive Organs The general arrangement and relationships of the male reproductive system are shown in Figs. 2-4 and 2-5. The *testes*, unlike the ovaries, do not lie in the abdominal cavity; instead, they are suspended in a pouchlike sac called the *scrotum*. Because of their location in the scrotum and the specialized arrangement of the vascular supply to the testes (a countercurrent heat-exchange system), the temperature of the testes is several degrees lower than that of the abdominal cavity. This is a requirement for the normal production of spermatozoa. The spermatozoa are produced in a large number of highly convoluted *seminiferous tubules*. The total length of the seminiferous tubules is astonishing. Bascom and Osterud (1925) estimated that the seminiferous tubules from one testis of a mature boar, if laid end to end, would extend 3200 meters. Knowledge of the size of these sperm-producing tubules makes it easier to understand how each ejaculate can contain tens or hundreds of millions of spermatozoa.

The spermatozoa must pass over a long and elaborate series of ducts before reaching the outside. From the seminiferous tubules they find their way through short, straight ducts—the *tubuli recti*—into an irregular network of slender

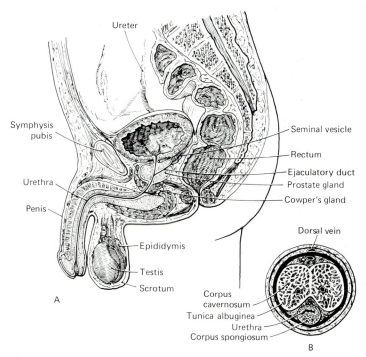

Fig. 2-4 (A) Lateral view of the male reproductive organs. (B) Cross section through the penis to show the arrangement of its masses of erectile tissue (the paired corpora cavernosa) and the unpaired corpus spongiosum.

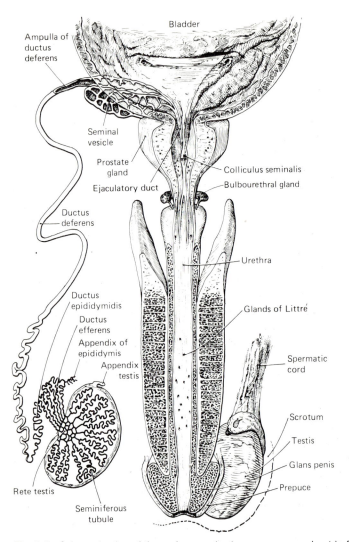

Fig. 2-5 Schematic plan of the male reproductive organs spread out in frontal aspect.

anastomosing ducts known as the *rete testis*. From the rete testis the spermatozoa are collected by the *ductuli efferentes*, which in turn pass them on by way of the much coiled duct of the *epididymis*, into the *ductus deferens*. At the distal end of the ductus ·deferens is a glandular dilation known as the *seminal vesicle*. It had been believed that, as the name implies, the seminal vesicles served as a kind of reservoir in which the spermatozoa were stored pending their ejaculation. We now know that the spermatozoa are stored in the epididymis and ductus deferens and that the seminal vesicles are glandular organs which produce a secretion that serves as a vehicle for the spermatozoa and contributes to their nutrition.

The external genitalia in the male consist of the scrotum, containing the

testes, and the penis. The *penis* consists of three rodlike masses of erectile tissue held together by dense connective tissue and covered by freely movable skin. The paired dorsal structures are the *corpora cavernosa*. The single erectile mass located mesially beneath the corpora cavernosa is the *corpus spongiosum* (Fig. 2-4B). In the shaft of the penis it is smaller than the corpora cavernosa, but distally it expands to form the *glans*. It is traversed throughout its length by the urethra. Opening into the urethra are numerous small mucus-producing glands, the *glands of Littré* (Fig. 2-5). They become active in sexual excitement and produce a lubricating fluid which facilitates intromission.

When, during the coital climax (*orgasm*), the spermatozoa are discharged, they enter the urethra by way of the *ejaculatory ducts* (Fig. 2-5). At the same time, the contents of the seminal vesicles, the *prostate gland*, and the *bulbourethral glands* (*Cowper's glands*) are forcibly evacuated into the urethra, providing a fluid medium in which the spermatozoa become actively motile. This mixture of secretions with spermatozoa suspended in it (*semen*) is swept out along the urethra by rhythmic muscular contractions culminating in *ejaculation*.

SEXUAL CYCLE IN MAMMALS

Reproduction in mammals is a closely orchestrated process, which requires the coordinated preparation of many tissues in the body of the female. Not only must an ovum be liberated from the ovary, but the tissues of the female reproductive tract must be ready to transport both eggs and sperm to a common site where fertilization can occur. In the event of fertilization, the early embryos must be carried to a portion of the uterus that is prepared both to receive the embryo and to meet its nutritional requirements throughout the duration of pregnancy. In the behavioral realm, the female must signal to the male her readiness for copulation, and the male, in turn, must be ready to respond.

Most of the preparations for reproduction are of a cyclic nature. The changes in structural and functional characteristics of both male and female reproductive tissues are mediated by hormones, often interacting in tightly controlled feedback loops. More than before, it is also recognized that there is a substantial neural influence upon reproduction and that environmental and psychic influences can exert profound effects upon reproductive patterns.

Estrous Cycle in Mammals   Sexual periodicity is, as a rule, much less strongly developed in the male than in the female. In some animals, such as those of the deer family, there is a brief period of intense sexual activity at one particular season of the year and then a long period during which there is sexual impotence and cessation of spermatogenesis. More commonly, and this is especially true among the primates, the male is sexually potent throughout adult life. A brief period of pronounced sexual activity, when it does occur in males, is known to animal breeders as the "rutting season." It always corresponds in time with the females' period of strong mating impulse, which breeders call the "period of heat" and biologists speak of as the *estrus*.

Originally the term *estrus* referred merely to the existence of a period of strong sexual desire made evident through behavior. As more information has been acquired about the concomitant changes going on within the body, it has become evident that estrus is close to the time of ovulation and that the characteristic behavior is simply an external indication that all the complicated internal mechanisms of reproduction are ready to become functional. If pregnancy does not occur at this time, regressive changes follow and another period of preparation must ensue before conditions are again favorable for reproduction. This repeated series of changes is known as the *estrous* or *sexual cycle*. Its phases in the absence of pregnancy are (1) a short time of complete preparedness for reproduction accompanied by sexual desire (estrus), (2) a period during which the fruitless preparations for pregnancy undergo regression (*post-* or *metestrum*), (3) a period of rest (*diestrum*), followed by (4) a period of active preparatory changes (*proestrum*) leading up to the next estrus, when everything is again in readiness for reproduction (Fig. 2-6).

There is wide variation among animals in the length of time occupied by this cycle. In some it occurs only once in an entire year, the estrus being so placed seasonally that when the young are born, conditions are favorable for their rearing. Species having only one breeding season in the year are said to be *monestrous*. Other animals exhibit several breeding periods in a year. They are said to be *polyestrous*. On the basis of our present knowledge it would appear

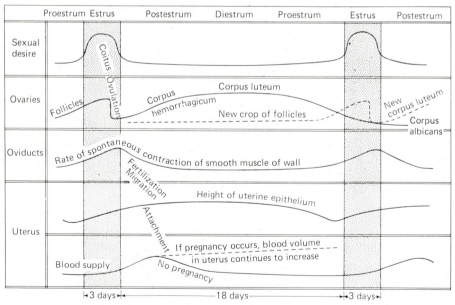

Fig. 2-6 Graph showing correlation of changes which occur during the estrous cycle in the sow. Note the correlation of the important events leading toward pregnancy (coitus, ovulation, fertilization, and the migration of the ovum through the oviduct to the uterus, and finally its attachment to the uterine mucosa) with the height of local activity as indicated by the curves. (Compiled from the work of Corner, Seckinger, and Keye.)

that a polyestrous rhythm is the underlying condition in mammals generally. Many factors mask or modify it in different cases, but in the forms which have been most fully studied it is unmistakably present. It is not unreasonable to suppose that an annual estrus such as that exhibited by the deer family has become established through the suppression of other periods, primarily because of the regular recurrence of pregnancies of long duration following what was originally merely the most favorable of several estrous periods. It is well known, furthermore, that the estrous cycle may be interrupted by many things other than pregnancy. Thus starvation, extreme exposure, or severe sickness may cause the suppression of an estrus. A contributing cause in reducing a polyestrous rhythm to a monestrous one, operative in females failing to become pregnant, might well be the severity of the conditions under which many wild animals live during the winter or during a dry season.

Many animals (sheep, for example) that have only one breeding season a year when living in their wild state develop a polyestrous rhythm when living under domestication. An underlying polyestrous condition may thus be obscured when a pregnancy of long duration follows each estrus—which occurs normally among many of the wild animals. It becomes apparent, however, when such an animal, under conditions of domestication or under experimental conditions in the laboratory, is not permitted to become pregnant. Then, after an unfruitful estrus, there appears a brief interval occupied by regression, rest, and preparation, followed shortly by another estrus. Living conditions under domestication being relatively uniform, suppression of an estrus through starvation or exposure does not occur, and the estrous periods keep recurring at fairly regular intervals until one of them is consummated by pregnancy.

Other mammals with short periods of gestation, such as the rabbit, may be polyestrous except in the winter. In these animals light is apparently a critical initiating factor. Only when the average daily amount of light gets above a certain threshold does the hypophysis become active in the production of enough *follicle-stimulating hormone* (FSH) to set the whole reproductive cycle into operation (Fig. 2-7A).

Primate Menstrual Cycle In primates the sexual cycle in females is characterized by *menstruation*, the discharge from the uterus of blood, mucus, and cellular debris at periodic intervals (approximately 4 weeks in humans). In the human female, menstruation usually commences (*menarche*) when she is 12 to 14 years old and continues until the time of *menopause*, which ordinarily occurs during the late forties. The usual duration of the menstrual discharge is from 4 to 5 days, but there is considerable individual variability in both the length of period and the interval at which it occurs.

The menstrual cycle may be divided into three major phases: (1) the *menses*, (2) the *proliferative (follicular) phase*, and (3) the *secretory (luteal) phase*. It will be easiest to follow the changes if we commence with conditions just after a menstrual period has ended. The diagram of Fig. 2-8 represents very schematically the changes occurring in the mucosal lining of the uterus during

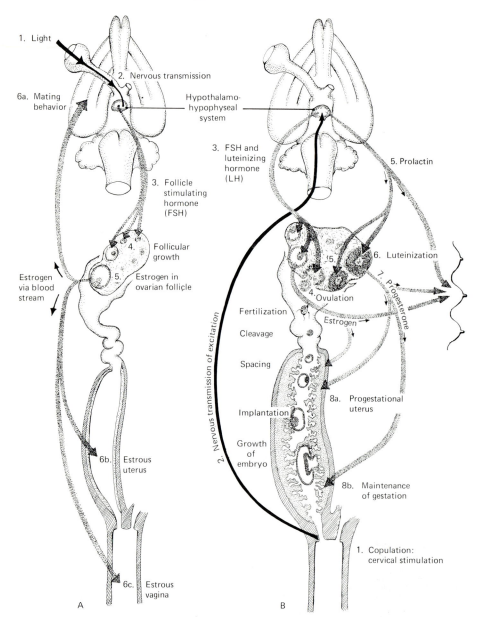

Fig. 2-7 Diagrams showing the sequence of events in the reproductive cycle of the rabbit. (A) Reactions leading to the estrous state. (B) Sequence of events following mating. (Redrawn with slight modifications from Witschi, 1956, *Development of Vertebrates*. Courtesy of the author and W. B. Saunders Company, Philadelphia.)

the cycle. The black descending bands signify the abrupt decrease in thickness which results from the menstrual sloughing. The tubular structures shown within the mucosa represent uterine glands. The more darkly shaded lower part of the

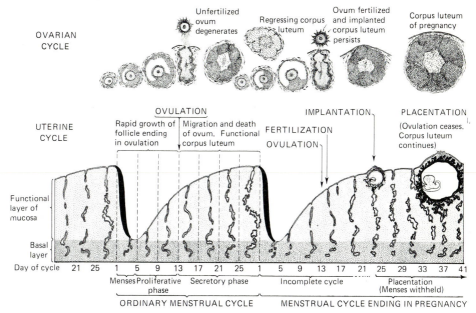

Fig. 2-8 Graphic summary of changes in the endometrium during an ordinary menstrual cycle and a subsequent cycle in which pregnancy occurs. The correlated changes in the ovary are suggested above, in their proper relation to the same time scale. (Modified from Schroder.)

mucosa is the so-called basal layer, which is below the level involved in sloughing. It is from this layer that the repair and growth processes of the proliferative phase are initiated. Restoration of the epithelial lining is accomplished with surprising rapidity by proliferation of the cells of the deep part of the glands, which remained undisturbed in the basal layer. The glands increase in length as the mucosa increases in thickness, but throughout the proliferative phase they remain slender and relatively straight, and their lumina are small and devoid of any conspicuous amount of secretion.

Following ovulation the proliferative phase gradually changes over into the secretory phase. The walls of the glands become irregular and the size of their lumen increases, and a considerable amount of secretion can be seen within the glands. There is also a striking increase in the conspicuousness of the small arteries supplying the superficial portion of the mucosa, and they extend nearer to the surface. These arteries tend to follow a spiral course, and their coiling becomes much more marked at this stage. A week after ovulation has occurred the whole histological picture reflects heightened activity. The glands are greatly distended, the small blood vessels are engorged, and the thickness of the mucosa has increased from the 1 mm or less that was there immediately after the last period to perhaps 4 or 5 mm. At this stage the uterus is fully prepared to implant and nourish a young embryo. (See cycle represented at right-hand side of Fig. 2-8.)

If the implantation of a fertilized ovum does not occur, the activities of the secretory phase end in the brief ischemic phase which immediately precedes menstruation. There is a reduced blood flow to the superficial zone of the uterine mucosa, although the blood flow in the vessels supplying the deeper layers remains uninterrupted. In the superficial zone white blood corpuscles begin to migrate into the stroma, and the tissues, deprived of an active circulation, begin to deteriorate. When this ischemic phase has lasted a few hours, spiral arteries here and there start to open up and blood pours into the superficial capillaries and soon ruptures their weakened walls, so that there is extravasation of blood into the tissues beneath the epithelial lining. In a very brief time the now necrotic superficial tissue, the extravasated blood which remains unclotted, additional blood oozing from the freshly denuded surface, and the secretion from the opened mouths of the glands all start to come away together as the menstrual discharge. Once started, this process proceeds rapidly in a given area, but by no means is the entire uterine lining simultaneously affected. During the early part of the period, area after area is involved until by the third day the uterine surface has been pretty well denuded. Repair, beginning first in the areas which were the first to be affected, is initiated promptly, and a new proliferative phase is under way almost before menstruation has ceased.

Hormonal Regulation of the Female Sexual Cycle From the earliest inquiries into the nature of the reproductive process, it was apparent that the functions and responses of the components of the female reproductive system are closely intertwined. It was soon recognized that hormones are the coordinating agents, but it has taken decades of painstaking work to understand the intricate interactions involved in the hormonal control of reproductive events.

The initial flowering of the field of reproductive endocrinology occurred between 1920 and 1940, when a group of talented investigators delineated the fundamental hormones and pathways of control between the pituitary gland and reproductive organs during various phases of the sexual cycle. Their work is well summarized in a treatise edited by Young (1961). The next major developments were the recognition of the important relationship between the brain and the pituitary gland and also the development of radioimmunoassays, by which extremely precise levels of hormones could be determined in body fluids. With the recent identification of cellular receptor molecules for hormones, it is now possible to ask significant questions about the mechanisms of action of hormones at the molecular level.

Paradoxically, the primary hormonal activator is located not in close positional association with the reproductive organs, but deep within the skull, in close association with the brain. The *pituitary gland*, or *hypophysis*, not much larger than a cherry stone in an adult human, is an insignificant-looking mass of tissue, whose function has long piqued man's curiosity. In the Middle Ages it would have been possible to become involved in a controversy about whether it was the abiding place of the soul or a special organ for eliminating mucus from the brain. Today it is known that the hypophysis produces a number of hormones which are of vital importance to the normal functioning of the body.

The hypophysis itself is subject to regulation by a part of the brain known as the *hypothalamus*. Situated immediately above the hypophysis, the hypothalamus produces a number of chemical *releasing factors* which stimulate the anterior pituitary to secrete its hormones. These releasing factors, and possibly an inhibiting factor, are transported to the anterior lobe of the pituitary gland via a specialized set of blood vessels known as the *hypothalamohypophyseal portal system* (Fig. 2-9). The hypothalamus, in turn, is connected with many other areas of the brain by tracts of nerve fibers. In this way certain stimuli impinging upon or originating in the brain can be carried to the hypothalamus and through its influence on the hypophysis can affect the hormonal activities in

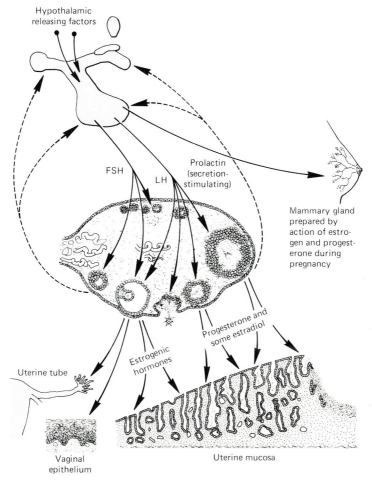

Fig. 2-9 Diagram indicating the interplay between the hormones of the anterior lobe of the hypophysis and the ovary in relation to major events of the human reproductive cycle.

various parts of the body. For example, sheep are stimulated to ovulation by a gradually decreasing photoperiod, whereas in some animals, such as the rabbit (Fig. 2-7), copulation itself is the prime stimulus for ovulation. Other sensory stimuli, such as those related to olfaction, can also affect endocrine function. It is now becoming understood that most of these exogenous stimuli are translated into changes in endocrine activity via the influence of the hypothalamus upon the hypophysis.

Several major classes of hormones are involved in control of the sexual cycle. The steroid hormones, *17β-estradiol*[1] and *progesterone*, are produced by the follicles within the ovaries. The anterior pituitary (*adenohypophysis*) secretes two *gonadotropic hormones, luteinizing hormone* (LH) and FSH. These hormones are glycoproteins with molecular weights of about 28,000 and 35,000. A third pituitary hormone is *prolactin*, a protein with a molecular weight of about 30,000. Formerly called *luteotrophic hormone* (LTH), prolactin is involved in a wide variety of regulatory functions that often differ greatly among the vertebrates. From the hypothalamus, a single decapeptide releasing factor promotes the release of both LH and FSH from the hypophysis.

In the normal sexual cycle a set of ovarian follicles (Figs. 3-18 and 3-19) begins to mature, probably because of a slight rise in pituitary FSH, just before the menstrual period begins. As the result of both FSH and LH stimulation, the follicles begin to produce estradiol. All the follicles except one soon degenerate, and the remaining preovulatory follicle secretes increasing amounts of estradiol late in the follicular phase of the cycle (Fig. 2-10). The sharp increase in estradiol secreted by the ovarian follicle acts upon the hypothalamohypophyseal axis, and one day after the peak concentration of estradiol in the blood the pituitary, probably responding to increased hypothalamic releasing factor, puts out a sharp peak of both LH and FSH (Fig. 2-10). The LH peak is the final stimulus required for follicular maturation, and ovulation then occurs within 24 hours. Even before ovulation, estrogen production by the follicle falls, possibly because of decreased sensitivity of the follicular cells to gonadotropins.

After ovulation the remains of the follicle soon become transformed into the *corpus luteum* (Fig. 3-18), mainly through the actions of LH. The corpus luteum then secretes estradiol and progesterone in gradually increasing amounts until their levels in the blood reach a broad peak, which occurs about midway through the luteal phase (Fig. 2-10). The gradual increase in steroid levels in the blood causes a feedback inhibition of LH and FSH, for the blood levels of these gonadotropins are very low. Later in the luteal phase the corpus luteum begins to regress. The cause of this regression has not yet been completely defined, but there is some evidence that there is a changing sensitivity of the luteal cells to LH. With the regression of the corpus luteum, production of estradiol and progesterone by the ovary falls, stimulating the production of releasing factor by

---

[1]Estradiol is one of a family of closely related steroids which have a similar physiological action. These substances are collectively known as *estrogens*. Estrone and estriol are other natural estrogens. The synthetic product, diethylstilbestrol, because it acts in a similar manner, is included in the same category.

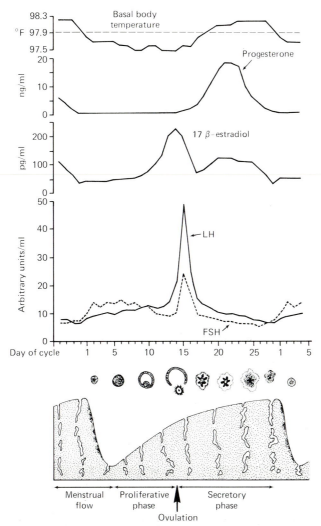

Fig. 2-10 Diagram of representative curves of basal body temperature, and daily serum concentrations of gonadotropins and sex steroids in relation to the normal 28-day human menstrual cycle. (Redrawn from S. R. Midgley, U. L. Gay, P. L. Keyes, and J. S. Hunter, 1973, in Hafez and Evans, eds., *Human Reproduction*, Harper & Row, New York.)

the hypothalamus and the gonadotropins by the pituitary. Then the cycle begins again, with the gonadotropic stimulation of a new crop of follicles shortly before the onset of the next menses.

Emphasis in the previous two paragraphs was on interactions between pituitary and ovarian hormones, but these interactions would be without purpose were it not for the effects of the ovarian steroids on other parts of the female reproductive tract. During the proliferative phase of the cycle, estrogen (estradiol) is the dominant hormone, and its actions upon the reproductive tissues seem designed to facilitate the transport of gametes and fertilization. In the uterine tubes, estrogen causes an increase in the number of ciliated cells and changes in the oviductal fluid. The rapid fall in estradiol levels just before ovulation stimulates increased motility in the smooth muscle of the uterine tube, an adaptation that allows more rapid transport of the ovulated egg.

Estradiol stimulates mitosis of the endometrial cells in the uterus and also promotes early growth of the uterine glands. Progesterone, on the other hand, prepares the lining of the uterus for implantation of the embryo, should fertilization occur, by causing the uterine lining to become thicker and more richly vascularized. The uterine endometrium is in its most receptive state about 7 days after ovulation, which is the time when implantation would occur after fertilization of the ovum.

Both the cervix and the vagina are also responsive to hormones, and around the time of ovulation, the pH of the upper vagina rises from its normally low levels to a level somewhat less inhospitable to spermatozoa. During much of the menstrual cycle the physical properties of cervical mucus act as a barrier to the passage of spermatozoa, but at the time of ovulation the viscosity of the mucus lessens and allows greater numbers of spermatozoa to pass through the cervix.

If the ovulated egg becomes fertilized, a new series of hormonal events is initiated. Of principal interest at this point is the continued maintenance of the corpus luteum by a gonadotropic hormone, *chorionic gonadotropin*, produced by the extraembryonic tissues associated with the embryo. The corpus luteum continues to grow and secrete large amounts of estrogens and progesterone. These not only help to maintain the uterine lining, but they begin to prepare the mammary gland for the eventual secretion of milk. Further details of hormonal relations during pregnancy are covered on page 226.

Hormonal Regulation of Reproduction in the Male   The principal hormone involved in sexual preparation in the male is *testosterone*, a steroid hormone produced and secreted by the *interstitial (Leydig) cells*, which are located in small clusters among the seminiferous tubules in the testis. Testosterone has a local effect in maintaining spermatogenesis, but it is also secreted into the blood, through which it acts upon a number of target organs, including the brain. In many target tissues testosterone is locally converted to a more potent form, *dihydrotestosterone*.

The interstitial cells are stimulated to produce testosterone by the pituitary gonadotropin LH, commonly called interstitial-cell-stimulating hormone (ICSH) in the male. FSH is specifically taken up by the Sertoli cells within the seminiferous tubules, but the function of this hormone in spermatogenesis remains obscure. The secreted testosterone acts upon the hypothalamohy-pophyseal axis so that there is a constant balancing between the blood level of testosterone and the production and release of the pituitary gonadotropins FSH and LH (ICSH). In general, high levels of testosterone inhibit their production and low levels stimulate it.

Chapter 3

Gametogenesis and Fertilization

GAMETOGENESIS

The reproductive cells which unite to initiate the development of a new individual are known as *gametes*—the *ova* of the female and the *spermatozoa* of the male. The gametes themselves and the cells that give rise to them constitute the individual's *germ plasm*. The other cells of the body, which take no direct part in the production of gametes, are called *somatic cells* or, collectively, the *somatoplasm*. From the phylogenetic standpoint the germ plasm is of paramount importance because it constitutes the hereditary dowry that is passed on from one generation of the species to the next (Fig. 1-2). The somatoplasm can thus be regarded as the material that protects and nourishes the germ plasm.

In species such as frogs and a number of invertebrate forms, the germ plasm can be recognized very early in the life of an individual—sometimes as regions in the vegetal pole cytoplasm of the zygote or as specific cells during the cleavage stages. Irradiation of this region of frog embryos with UV light results in the development of an embryo lacking germ cells (Smith, 1966). The early segregation of a readily recognizable germ plasm has not been demonstrated in most vertebrate embryos. Only at later stages do cells of the germ line become apparent. If these negative observations ultimately do prove to be valid, it would be necessary to reconsider the time-honored concept of the continuity of germ plasm.

Gametogenesis (*oögenesis* in the female and *spermatogenesis* in the male) is a broad term that refers to the processes by which the germ plasm becomes converted into highly specialized sex cells capable of uniting at fertilization and producing a new being. Commonly, gametogenesis is divided into four major phases. The first phase involves the origin of the germ cells and their migration to the gonads. The second phase consists of the multiplication of the germ cells in the gonads through the process of mitosis. In the third phase, meiosis reduces the number of chromosomes by one-half. During the fourth phase, the gametes undergo the final stages of maturation and differentiation into spermatozoa or ova that are capable of fertilizing or being fertilized.

The Origin of Primordial Germ Cells and Their Migration to the Gonads

Although much of the early history of the germ plasm is still unknown, the cells that are destined to give rise to the gametes are recognizable at a surprisingly early stage in development. In those species (i.e., anuran amphibians) with recognizable germ plasm it is possible to trace continuously the developmental lineage of this material from a circumscribed area in the unfertilized egg, through cleavage (in cells near the vegetal pole), and finally into certain endodermal cells during gastrulation (Fig. 3-1). It has not been possible to

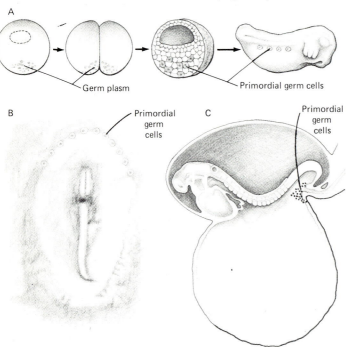

Fig. 3-1 The origin of primordial germ cells in (A) anuran amphibians, (B) the chick embryo (after Swift), and (C) the 16-somite human embryo (after Witschi).

identify germ plasm in the early embryos of amniotes (birds, reptiles, and mammals), but in all of these forms, as well as in the anurans, future gametes can be identified among the endodermal cells of the yolk or yolk sac. These cells, called *primordial germ cells*, can be recognized by their large size and clear cytoplasm (Fig. 3-2) and by certain histochemical characteristics, such as high alkaline phosphatase activity in mammals and a high glycogen content in birds.

The only major vertebrate group that deviates from an endodermal origin of the primordial germ cells is the urodele amphibians (salamanders). In the urodeles, primordial germ cells arise from mesodermal cells, which form through the inductive influence of the ventral endodermal yolk mass upon the overlying surface layer of cells. The mode of origin is so strikingly different from that of the anurans and higher vertebrates that Nieuwkoop and Sutasurya (1976) used it as the principal basis for a proposal that the urodeles had a separate phylogenetic origin from the anuran amphibians.

When the primordial germ cells are first recognizable in embryos, the gonads are either very poorly developed or are not developed at all (Fig. 3-1). That these cells ultimately colonize the gonads has been determined by three means. One of the earliest was to extirpate the endodermal area in which the primordial germ cells were found. When Willier (1937) did this in the chick, gonads without gametes formed. A second way was to trace the course of the primordial germ cells during development. This could be done because of the large size and distinctive histochemical characteristics of these cells. Studies on a number of species demonstrated the migration of these cells to the gonadal primordia. The third way made use of a mutant strain of sterile mice. Mintz and Russell (1957) used histochemical staining for alkaline phosphatase to show that

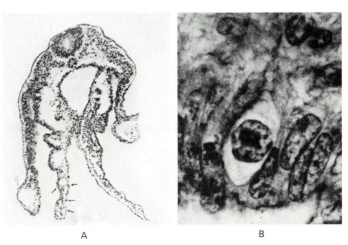

A   B

Fig. 3-2 Primordial germ cells in human embryos. (After Witschi, 1948, *Carnegie Cont. to Emb.,* vol. 32.) (A) Cross section through a 16-somite human embryo (Carnegie Collection, 8005), showing primordial germ cells (large cells indicated by arrows) in the splanchnopleure of the yolk sac. (B) A primordial germ cell (large cell with clear cytoplasm) in the splanchnopleure at the coelomic angle in an embryo of 32 somites (Carnegie Collection, 7889; photomicrograph ×1100).

only small numbers of primordial germ cells were present in the mutant embryos.

Why do the primordial germ cells originate so far from the gonads, and how do they get to the gonads? We cannot yet answer the first part of the question, but it should be noted that the germ cells are not unique in taking their origin from the area of the yolk sac. Later in development the first blood cells also come from the yolk sac, but in contrast to the primordial germ cells, they arise from the mesoderm of the yolk sac rather than the endoderm.

It is now well established that there are two principal routes by which primordial germ cells travel to the gonads (Fig. 3-3). In mammals the primordial germ cells become capable of amoeboid movements and migrate up through the dorsal mesentery and into the gonads. A different route is followed in birds. At about the time when the earliest circulation to the yolk sac is established, the primordial germ cells make their way into the blood vessels and are passively carried to the body. Initially, these cells are distributed at random throughout the body, but at later periods most are found in the gonads (Meyer, 1964). There is some experimental evidence of chemical attractants that stimulate the migration of primordial germ cells to the gonads, but further work is necessary to confirm these results. Primordial germ cells that are deposited in extragonadal sites apparently die, but it is thought that misplaced primary germ cells

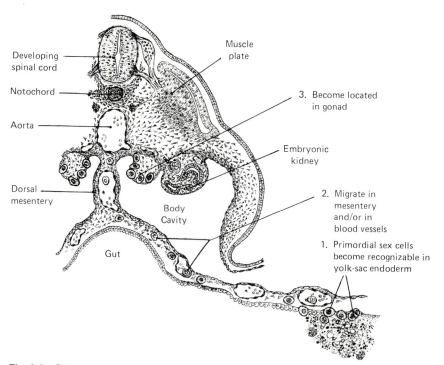

Fig. 3-3　Schematic section through the midbody region of a young vertebrate embryo, illustrating the origin of the primordial germ cells in the yolk-sac endoderm and their migration through the tissues or the blood to the developing gonads.

occasionally develop into *teratomas*. Teratomas are bizarre tumors which often contain scrambled mixtures of highly differentiated tissues, sometimes including hair and teeth!

Careful quantitative studies have shown that the number of primordial germ cells increases during their migration to the gonads. For example, in the mouse the number of primordial germ cells rises from less than 100 to about 5000 during the period of migration (Mintz and Russell, 1957). However, the major increase in the number of germ cells occurs in the gonads.

Proliferation of Germ Cells by Mitosis   The embryonic gonads are initially populated by a relatively small number of migrating primordial germ cells. Once settled in the gonads, however, the germ cells enter a proliferative phase in which their numbers increase greatly by mitosis. (For a review of mitosis, see Fig. 3-4.) Mitotically active germ cells in the female are called *oögonia*; in the male they are known as *spermatogonia*.

The pattern of mitotic activity of the germ cells in the gonads differs widely between males and females. In the human female, intense mitotic activity between the second and the fifth months of pregnancy brings the population of oögonia from a few thousand to about 7 million (Fig. 3-5). The number of

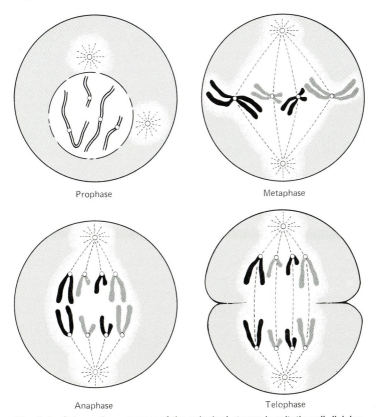

Prophase   Metaphase

Anaphase   Telophase

Fig. 3-4   Schematic summary of the principal stages in mitotic cell division.

oöcytes then falls sharply, mainly because of atresia, and by the seventh month most of the oöcytes have entered the prophase of their first meiotic division. This brings to an end the proliferative phase of gametogenesis in the female. In contrast to the human, populations of oögonia in most of the lower vertebrates are capable of dividing throughout the reproductive life cycle. When one considers that some fishes release several hundred thousand eggs at one spawning, the need for the mitotic capability of oögonia throughout life is understandable.

The germ cells of the male follow a pattern of mitotic proliferation considerably different from that of the female. Mitosis begins in the gonad of the early embryo, but it commonly persists throughout the life-span of the male. The testes always retain a germinative population of spermatogonia. Beginning at puberty, periodic waves of mitosis produce subpopulations of spermatocytes that enter meiosis as synchronous groups. This activity continues as long as the male is capable of reproduction.

Meiosis　One of the fundamental requirements in the sexual reproduction of any species is that the normal number of chromosomes must be maintained from one generation to another. This is accomplished by the reduction of the chromosomal complement of the gametes from the *diploid* (2n) to the *haploid* (1n) condition during gametogenesis. From the genetic standpoint, meiosis is essentially the same in both male and female gametes. Because of this, the general process of meiosis will be treated first. Later, the features of meiosis specific to males and females will be discussed.

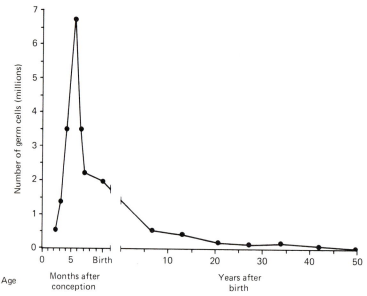

Fig. 3-5　Changes in the total population of germ cells in the human ovary with increasing age. (Modified from T. G. Baker, 1970, in Austin and Short, eds., *Reproduction in Mammals,* vol 1, p. 20.)

A major requirement of meiosis is that each haploid gamete must acquire a complete set of chromosomes. Yet meiosis is the phase during which new combinations of genetic material, some arising from maternal genes and others from paternal genes, are assembled. (Remember that each cell in the body contains one set of chromosomes contributed from the maternal side and another set from the paternal side.) Genetic recombination occurs by (1) the random distribution of maternal or paternal chromosomes to the daughter cells and (2) the exchanging of portions of homologous chromosomes by crossing-over at specific phases of meiosis.

Strictly speaking, meiosis does not involve the synthesis of DNA, because when meiotic prophase begins, DNA replication in the gamete has already taken place. The signal that tells a cell to make a change from an ordinary mitotic division to a meiotic division is poorly understood. The DNA content of the cell is double the amount in an interphase cell. Thus at the beginning of meiosis the cell can be described as 2n, 4c; in other words, the cell contains the normal number (2n) of chromosomes, but because of replication its DNA content (4c) is double the normal amount (2c). The object of meiosis is to produce haploid gametes with a 1n, 1c complement of genetic material. The reduction of genetic material involves two *maturation divisions*, in which new DNA synthesis does not occur. The first meiotic division, sometimes called the *reductional division*, results in the formation of two genetically dissimilar daughter cells (1n, 2c). In the second, or *equational*, meiotic division each of the previous two cells produces two genetically identical daughter cells (1n, 1c) that can now be properly called gametes. In contrast to mitosis, which is usually measured in terms of minutes or hours, meiosis is a leisurely process that may last from several days to as long as 45 or 50 years (in the human female).

First Meiotic Division *Prophase I* is a complex stage that is usually divided into five substages—leptotene, zygotene, pachytene, diplotene, and diakinesis. Many events of both genetic and general developmental importance occur during the first meiotic prophase.

In the *leptotene stage* (Fig. 3-6) the chromosomes are threadlike and are just beginning to coil. Each chromosomal thread actually consists of two identical sister *chromatids*, which are joined somewhere along their length by a common *centromere*. One of the chromatids is an original DNA strand, whereas the other was newly synthesized just prior to the start of meiosis. It is usually not possible to resolve the individual chromatids by standard light microscopy.

The *zygotene stage* is the period during which the homologous paired chromosomes (one set of sister chromatids from the maternal side and one set from the paternal side) come together and become closely apposed along their entire length on a point-for-point basis. This precise lining-up is called *synapsis*; it forms the basis for the crossing-over of genetic material that occurs later in the first meiotic division. The area of contact between the paired chromosomes has a specialized ultrastructure and is called the *synaptinemal complex* (Moses, 1968). In some manner the synaptinemal complex is involved in the pairing-up and possibly also in the crossing-over of chromosomes.

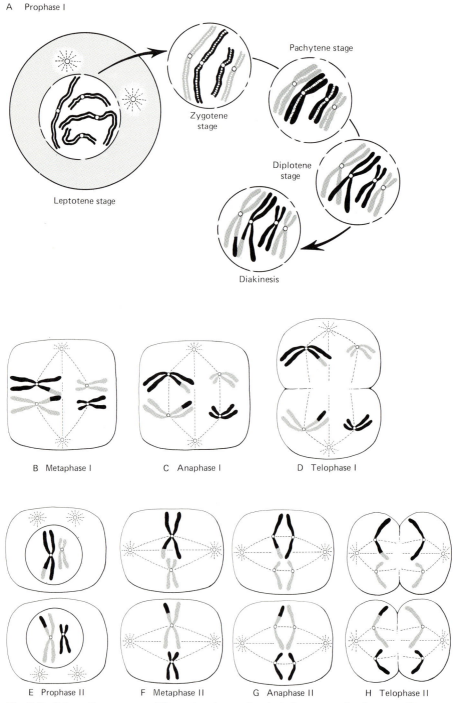

Fig. 3-6 Schematic summary of the major stages of meiosis in a generalized germ cell.

During the early *pachytene stage* synapsis is completed; the two aligned chromosomal pairs are collectively referred to as a *bivalent*. A prominent feature of this stage is the thickening of the chromosomes owing to their coiling. Within a single chromosome, the sister chromatids appear to be held together by a centromere.

Late in the pachytene stage and during the *diplotene stage* portions of the paired chromosomes overlap one another. These areas of contact are called *chiasmata*, and it is thought that they are the sites at which homologous portions of maternal and paternal chromosomal strands break and are mutually exchanged in the process of *crossing-over*. This is one of the two major ways of producing genetic differences between individuals. A major feature of the diplotene stage is the separation of the two paired chromosomes by splitting along the synaptinemal complex. The splitting occurs along most of the length of the chromosomes, but not at the chiasmata, which are now well defined. The individual chromatids are also clearly visible by this time, and it can be seen that each bivalent consists of four distinct chromatids. It can now be called a *tetrad*. During this stage the chromatids uncoil slightly, and in eggs, at least, RNA synthesis occurs in areas of uncoiling.

In the *diakinesis stage* the chromosomes shorten even more. The splitting of the chromosome pairs continues, and one component of the splitting process that is characteristic of this stage is the moving of the chiasmata toward the ends of the chromosomes. This process is known as *terminalization*. At this point the nucleolus disappears, the nuclear membrane breaks, and the spindle apparatus becomes apparent.

Metaphase I   The tetrads line up along the metaphase plate, so that for each chromosome pair, the maternal chromosome is on one side of the equatorial plate and the paternal chromosome is on the other. The alignment of chromosomes is accomplished in such a way that there is a random distribution of maternally and paternally derived chromosomes on either side of the equator. The random assortment of the chromosomes thus affected forms the basis for Mendel's second law of heredity and constitutes the most important means of ensuring that genetic differences are present among individuals.

Anaphase I   At this point the individual paired chromosomes begin to move toward opposite poles of the spindle. Both sister chromatids of each chromosome remain held together by the centromere. As the homologous chromosomes move away from one another, the chiasmata, which have moved to the ends of the chromosomes, are pulled apart and crossing-over is complete. The events in anaphase I are crucial in understanding the difference between the first meiotic division and an ordinary mitotic division. In mitosis, after the chromosomes are lined up along the metaphase plate, the centromere between the sister chromatids of each chromosome splits and one chromatid goes to each pole of the mitotic spindle, resulting in genetically equal daughter cells. In contrast, the migration of the entire maternal chromosome to one pole and the paternal chromosome to the other pole during meiosis results in genetically unequal daughter cells.

Telophase I and Interphase   In telophase I, the two daughter nuclei are separated from one another and nuclear membranes may re-form. Each nucleus now contains the haploid number of chromosomes (1n), but each chromosome still contains two sister chromatids (2c) connected by a centromere. Because the individual chromosomes of the haploid daughter cells are still in the replicated condition, there is no new replication of chromosomal DNA during the interphase between meiotic division I and meiotic division II.

Second Meiotic Division   Except for the fact that the cell is haploid (1n, 2c), the second meiotic division behaves in most respects like an ordinary mitotic division. After an atypical prophase, a mitotic spindle apparatus is set up and the chromosomes line up along the equatorial plate at metaphase II. Then, in contrast to the first meiotic division, but similar to a mitotic division, the centromere between the sister chromatid of each chromosome divides, thus allowing the sister chromatids to separate from one another during anaphase. With the completion of telophase II, meiosis is complete and the original diploid germ cell has produced four haploid daughter cells (1n, 1c). In the male, all four of the haploid cells will go on to form viable gametes, but because of asymmetric meiotic division in the female, only one viable ovum results. The remaining three daughter cells are much reduced in size and are represented only as the apparently functionless polar bodies.

Polyploidy   In rare instances animals may start their development with chromosomes in multiples of the normal number for their species. This condition is called *polyploidy*. When it occurs the individual may, for example, show three times the haploid number of chromosomes rather than the usual diploid number resulting from the union of two haploid gametes. The triploid condition does not always arise in the same manner, but it is known to be established when the reduction division fails to occur in a maturing oöcyte. The most carefully studied instances of triploidy occur in amphibia. It has been discovered that exposure of the eggs of newts or salamanders to extreme temperatures (from 0 to 3°C for 20 hours, or 37°C for as little as 10 minutes) may inhibit the carrying through of the second meiotic division. When fertilization is completed in such an ovum, the unreduced diploid chromosomal number of the ovum, joined to the haploid number of a normal sperm, results in the triploid count.

Polyploidy is not always triploid, although that is by far the most common of the types. More rarely the count may be *tetraploid*, that is, four times the haploid number for the species. It seems probable that this condition may arise during early cleavage stages as a result of chromosomal division not followed by reorganization of daughter nuclei and cytoplasmic division. However polyploidy may be initiated, cells with a greater than normal chromosome number are proportionately larger than normal in size. Surprisingly, the body size of the embryo remains essentially normal, apparently because its large cells are correspondingly fewer in number.

Spermatogenesis and Oögenesis Compared   Despite similarities in the genetic aspects, differences are prominent when one compares spermatogenesis and oögenesis. Figures 3-7 and 3-8 summarize a number of parallels between the two processes. This section will point out some of the major differences between spermatogenesis and oögenesis in the human.

In contrast to spermatogonia, each of which gives rise to four functional spermatozoa as the result of the two meiotic divisions, an oögonium produces only one viable ovum. After the first meiotic division one cell is left with the bulk of the cytoplasmic material, whereas the other, called the *first polar body*, is left with little cytoplasm. The polar body often degenerates without taking part in the second meiotic division. During the second meiotic division there is also an unequal apportionment of cytoplasm between the daughter cells, and the ovum retains the bulk of the cytoplasm, leaving the other cell to fall by the wayside as the *second polar body*.

In the human female, the first meiotic division begins in the embryo and meiosis is not completed until the onset of puberty, at the earliest, or just before menopause, at the latest. Spermatogenesis does not begin until puberty but is then continuous throughout life. There are no prolonged meiotic arrests during spermatogenesis, and the entire process is completed in somewhat more than 2 months.

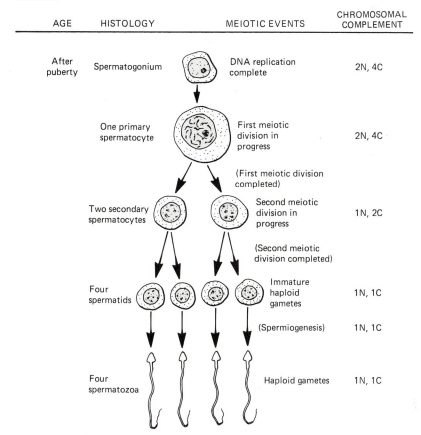

AGE	HISTOLOGY	MEIOTIC EVENTS	CHROMOSOMAL COMPLEMENT
After puberty	Spermatogonium	DNA replication complete	2N, 4C
	One primary spermatocyte	First meiotic division in progress	2N, 4C
		(First meiotic division completed)	
	Two secondary spermatocytes	Second meiotic division in progress	1N, 2C
		(Second meiotic division completed)	
	Four spermatids	Immature haploid gametes	1N, 1C
		(Spermiogenesis)	1N, 1C
	Four spermatozoa	Haploid gametes	1N, 1C

Fig. 3-7 Summary of the major events in human spermatogenesis.

AGE	FOLLICULAR HISTOLOGY		MEIOTIC EVENTS IN OVUM	CHROMOSOMAL COMPLEMENT
Fetal period	No follicle		Oögonium (Mitosis)	2N, 2C
Before or at birth	Primordial follicle		Primary oöcyte (Meiosis in progress)	2N, 4C
After birth	Primary follicle		Primary oöcyte	2N, 4C
			(Arrested in diplotene stage of first meiotic division)	
After puberty	Secondary follicle		Primary oöcyte	2N, 4C
			(First meiotic division completed, start of second meiotic division)	
	Tertiary follicle		Secondary oöcyte + polar body I	1N, 2C
			(Ovulation)	
	Ovulated ovum		Secondary oöcyte + polar body I	1N, 2C
			(Arrested at metaphase II) (Fertilization — second meiotic division completed)	
	Fertilized ovum		Fertilized ovum + polar body II	1N, 1C + Sperm

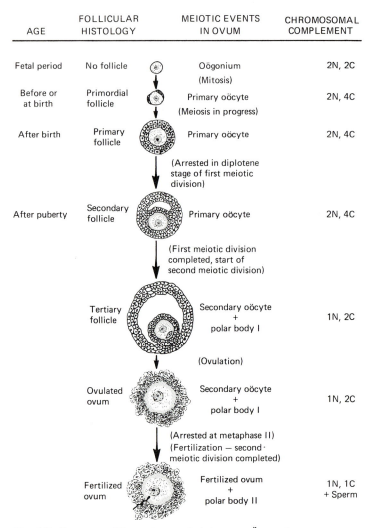

Fig. 3-8 Summary of the major events in human oögenesis.

Spermatogenesis is quite similar throughout the vertebrates and the mature spermatozoon is markedly smaller than the spermatogonium and is highly motile. Because of the wide differences both in the amount of yolk that is formed and in reproductive habits, the structural aspects of oögenesis vary considerably. Although mature ova become larger than oögonia, their relationships with the body of the mother vary with the amount of yolk that is produced. During the deposition of yolk, eggs take up large quantities of materials produced by the liver via the follicular cells. Aside from a possible contribution by Sertoli cells, developing sperm cells receive little formed material. During development the

egg stores up both energy sources and precursors of proteins and nucleic acids. Sperm cells, on the other hand, shed most of their cytoplasm and must rely upon the seminal fluid for their energy source. In anticipation of future requirements, the egg stores up much RNA, whereas there is little or no RNA synthesis during the later stages of spermatogenesis.

SPERMATOGENESIS

The transition from mitotically active primordial germ cells to mature spermatozoa is called *spermatogenesis*, and it involves a sweeping series of structural transformations. Although there is a wide variety in the morphology of mature spermatozoa, the overall process of spermatogenesis is much the same throughout the vertebrate classes. This process can be broken down into three principal phases: (1) mitotic multiplication, (2) meiosis, and (3) spermiogenesis.

Mitosis of sperm-forming cells occurs throughout life, and the mitotically active cells within the seminiferous tubules are known as *spermatogonia*. These cells are concentrated near the outer wall of the seminiferous tubules (Fig. 3-9). Spermatogonia have been subdivided into two main populations. *Type-A spermatogonia* represent the stem-cell population. Within this population is a group of dark, noncycling cells (Ad) that may be long-term reserve cells. Some of these cells become mitotically active pale cells (Ap), which ultimately give rise to *type-B spermatogonia*. These are cells which have become committed to leaving the mitotic cycle and which go on to finish the process of spermatogenesis.

After their final round of DNA duplication, these cells are called *preleptotene spermatocytes* and they are ready to pass through the meiotic phase of spermatogenesis. During the first meiotic division each *primary spermatocyte* divides into two equal daughter cells. With the onset of the second meiotic division these cells are known as *secondary spermatocytes*. The elements of meiotic division have already been described and will not be repeated here. In the human the first meiotic division lasts for several weeks, whereas the second one is completed in about 8 hours. This is why it is much easier to find primary spermatocytes than secondary spermatocytes in histological sections. Four haploid *spermatids* result from the meiotic phase of spermatogenesis.

Although they no longer divide, the spermatids undergo a profound transformation from relatively ordinary looking cells to the extremely specialized *spermatozoa*. The third phase in spermatogenesis is called *spermiogenesis*, or *spermatid metamorphosis*.

In the metamorphosis of a spermatid many radical changes occur. At the end of the second maturation division the nucleus is in typical interphase condition, having dispersed, finely granular chromatin and a reconstituted nuclear membrane (Fig. 3-10A). Almost immediately the nucleus begins to lose fluid, with a resultant decrease in its size and a concentration of its chromatin (Fig. 3-10C and D). This continues until the compacted chromatin comes to constitute the bulk of the head of the spermatozoon (Fig. 3-10D to 3-10F).

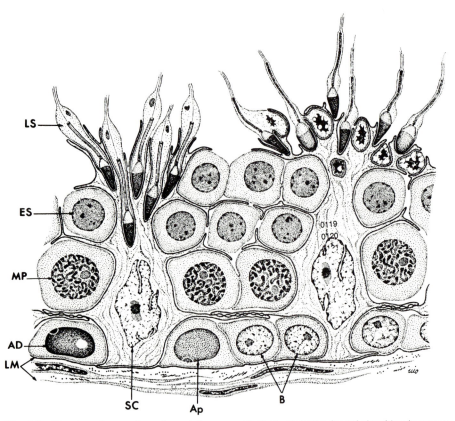

Fig. 3-9 Drawing of a section of seminiferous epithelium, showing the relationships between Sertoli cells and developing sperm cells. *Abbreviations: LM,* limiting membrane; *SC,* Sertoli cell; *AD,* dark A spermatogonium; *Ap,* pale A spermatogonium; *B,* B spermatogonia; *MP,* middle pachytene spermatocytes; *ES,* early spermatids; *LS,* late spermatids.
(Courtesy of Y. Clermont, from Dym, 1977, in Weiss and Greep, eds., *Histology,* 4th ed., McGraw-Hill Book Co., New York, p. 984.)

Concurrently there are changes in cell topography. The nucleus, which is to become the sperm head, moves to an eccentric position, that is, the cytoplasm appears to stream tailward, leaving only a thin layer covering the nucleus (Fig. 3-10D and E). Within the cytoplasm the centrioles become more conspicuous and appear to be a point of anchorage for the developing *flagellum* (Fig. 3-10B and C). The posterior centriole moves away from the anterior one and takes on the shape of a ring encircling the flagellum (Fig. 3-10D to 3-10F). *Mitochondria* begin to concentrate about the proximal part of the flagellum, which is destined to become the middle (connecting) piece of the spermatozoon (Fig. 3-10E and F). Finally, the remaining cytoplasm disintegrates, leaving the mature spermatozoon stripped of all nonessential parts and consisting only of a head containing the concentrated nuclear material bearing the genes and a tail which gives it mobility (Figs. 3-10G and H, and 3-11).

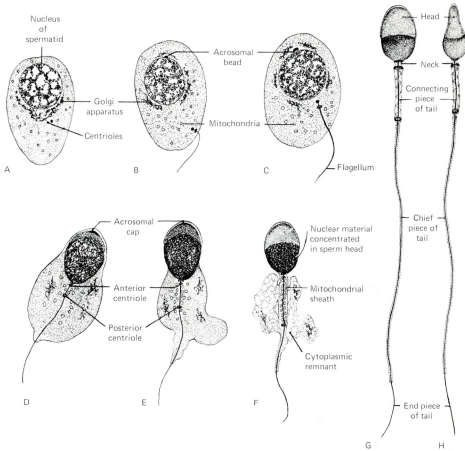

Fig 3-10 Stages in maturation of spermatids. (Modified from figures by Gatenby and Beams, 1935, *Quart. J. Micr. Sci.,* vol. 78.)

Many of the interesting details of the metamorphosis of a spermatid which are only vaguely suggested by studies with the light microscope can be brought out clearly under the electron microscope. Part of the cytoplasm containing the Golgi apparatus becomes concentrated at the apical end of the sperm head to form the acrosome within the head cap (Fig. 3-12).

With the electron microscope the mitochondria can be seen to form a spiral investment about that part of the flagellum which is destined to be in the middle piece (Fig. 3-12). This *mitochondrial helix*, as it is called, is believed to furnish the energy for the vibratile flagellum.

Distal to the middle piece, the investment of the flagellum is reduced to a thin film of cytoplasm, which disappears altogether in the end piece of the tail. The flagellum itself consists of nine axial filaments united by radial strands (Fig. 3-12).

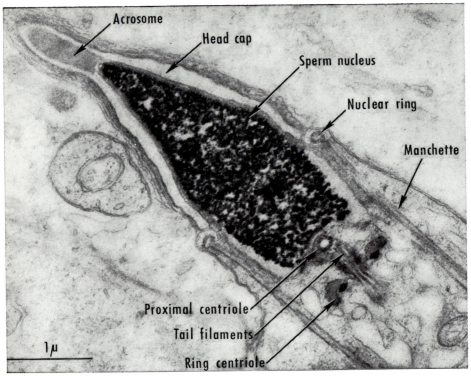

Fig. 3-11 Electron micograph of late spermatid of cat. Buffered osmic acid fixation. (From Bloom and Fawcett, *Textbook of Histology,* after M. H. Burgos and D. W. Fawcett, 1955, *J. Biophys, and Biochem. Cytol.,* vol. 1. Courtesy of the authors and the W. B. Saunders Company, Philadelphia.)

As early as the mitotic divisions of the type-A spermatogonia, the daughter cells are connected to one another by fine *intercellular bridges* of cytoplasm that are probably the result of incomplete *cytokinesis* (*cell division*) after mitosis. Clusters of interconnected cells are found throughout spermatogenesis up until the late spermatid stage. These bridges probably facilitate the synchronous differentiation and division of the sperm-producing cells. At any given stage of spermatogenesis, dozens of cells may be so interconnected.

During spermatogenesis, the cells are also closely associated with *Sertoli cells*, which lie at regular intervals along the seminiferous tubule (Fig. 3-9). Many functions have been attributed to the Sertoli cells. Among these are the nutrition of germ cells, phagocytosis of degenerated cells and parts of cells within the seminiferous tubules, secretion, processing of steroid hormones, and the release of late spermatids into the lumen of the tubule.

Spermatogenesis is not a random process; rather, within the seminiferous tubules it is highly synchronized with respect to both time and location. The earliest stages of spermatogenesis occur at the periphery of the tubule, and progressively later stages are encountered closer to the lumen. However, all

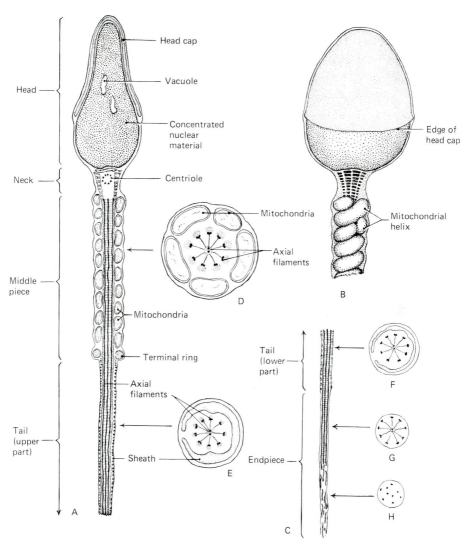

Fig. 3-12 Diagrams showing the structure of human spermatozoa as revealed by electron micrographs. (A) Longitudinal section of head, neck, middle piece, and upper part of tail. (B) Head, viewed from its flattened surface, together with neck and part of middle piece. (C) Terminal part of tail proper and endpiece. Between the parts represented in B and C, a considerable portion of the tail has been omitted. (D)–(H), More highly magnified cross sections of the middle piece and tail at the levels indicated by the arrows. (Schematized from Ånberg, 1957, and Fawcett, 1958.)

stages of spermatogenesis are not seen in the same section of the seminiferous tubule. Typically, several generations of developing sperm cells are present along any radial line drawn in a cross section of a tubule, and all cells within a given generation are in the same stage. In many mammals, such as the rat, well-defined waves of spermatogenesis can be delineated along the length of the seminiferous tubule. The distribution of cell associations in the human testis is more patchy than wavelike in nature (Clermont, 1972). In the human, the time required for a spermatogonium to develop into a spermatozoon is about 64 days.

OÖGENESIS

To a far greater extent than spermatogenesis, oögenesis varies in accordance with the reproductive habits of the animal. In species that rely upon external fertilization in the water, the number of mature eggs released at a single spawning ranges from hundreds to hundreds of thousands. Animals with internal fertilization are much more parsimonious in the production of eggs; commonly only 1 ovum, and seldom over 15 ova, mature at any one time. There is great variation not only in the number but also in the size of the eggs. Size depends principally upon whether or not the fertilized egg develops within or without the body of the mother. The ova of mammals are very small, because it is not necessary to store up in advance large amounts of yolk and other materials needed for development of the embryo. In contrast, the ova of animals developing outside the body are often quite large, since they contain within them the yolky materials needed for the embryo's development. Typically, the eggs of aquatic animals are considerably smaller than the eggs of reptiles and birds. The differences in environment also necessitate different coverings which surround the egg. In this section the development of three different types of eggs (amphibian, bird, and mammal) will be detailed.

Oögenesis in Amphibians   The reproductive patterns of amphibians, like those of most lower vertebrates, are geared towards the production of large numbers of eggs and their release at one time during the year. Typically, amphibians lay several hundred eggs at each spawning, whereas fishes, as a rule, release thousands or even hundreds of thousands of eggs. Because of both the massive number of eggs and the annual release of eggs, the pattern of oögenesis in lower vertebrates differs in some respects from that in many birds and mammals.

In amphibians the mitotic phase of oögenesis does not come to an early halt as it does in mammals. Rather, each year a new crop of eggs is generated by mitotic proliferation from a population of gametogenic stem cells. The maturation of frog eggs (*Rana pipiens*) requires 3 years (Fig. 3-13). The first batch of eggs begins to mature shortly after metamorphosis. During the first 2 years, maturation is a relatively leisurely process, but during the third summer the eggs mature rapidly, ultimately attaining a diameter of about 1500 μm in the autumn of the third year. The female then hibernates and the eggs, protected by a coating of jelly, are laid in the water early in the following spring. Because of the 3-year cycle of oögenesis, the ovary of a mature frog contains three batches of eggs at any one time.

Within the amphibian ovary, the eggs are arranged in individual follicles. They are surrounded first by a layer of follicular epithelium, next by a *theca* (a

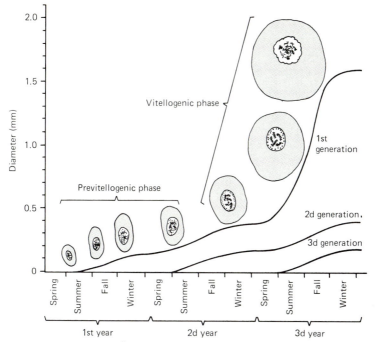

Fig. 3-13 Growth of oöcytes in the frog during the first three years of life. Ultimately, three generations of oocytes are found in the ovary. The drawings show the development of oocytes in the first generation. (Modified from P. Grant, 1953, *J. Exp. Zool.* **124**:513.)

thin layer of ovarian connective tissue containing blood vessels), and then by a layer of ovarian epithelium.

Maturation of the Amphibian Egg Maturation of the amphibian egg must anticipate the requirements of the embryo, because until the initiation of feeding, the embryo develops as an essentially closed system; that is, everything that is required for early development must be contained within the fertilized egg. Therefore, a major requirement is a sufficient supply of yolk to provide substrates for synthetic activity of the embryo. In addition to accumulating yolk, the maturing egg synthesizes a large amount of RNA—enough to accommodate most of the needs of the embryo through the period of cleavage.

Maturation of the amphibian egg can be divided into two broad phases—*previtellogenic* (before the deposition of yolk) and *vitellogenic* (the major period of yolk deposition). The previtellogenic phase usually includes the period up to the early diplotene period of meiosis. The immature (prediplotene) egg is a relatively unspecialized cell, with few cytoplasmic organelles or inclusions (Fig. 3-14). Among the earliest maturational changes are an increase in the number of mitochondria and the beginning of increased RNA synthesis, initially transfer RNA (tRNA) and the small (5S) ribosomal RNA (rRNA) molecules.

As the oöcyte enters the prolonged diplotene phase of meiosis, maturational changes begin in earnest. The nucleus becomes a site of intense synthetic

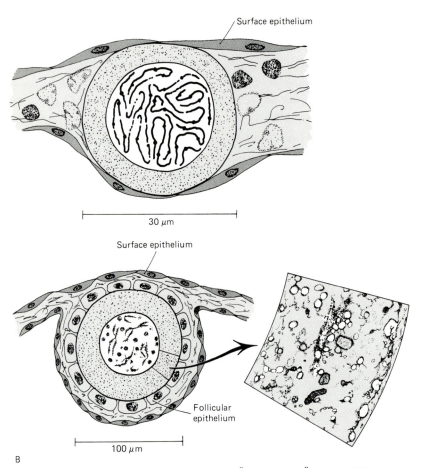

Fig. 3-14 Structural changes in the amphibian oöcyte during oögenesis. (A) Leptotene stage. (B) Very early diplotene stage. (C) Diplotene stage. (D) Maturing oöcyte in late prophase I.

activity, and with this its diameter increases by a factor of up to 7 or 8 until it reaches a diameter of about 350 μm at the end of the diplotene phase. For many years, the enlarged nucleus of the amphibian egg has been referred to as the *germinal vesicle*.

Within the nucleus the chromosomes have begun to spread out from their tightly coiled configuration in the previous (pachytene) stage of meiosis, and they soon begin to form large numbers (up to 20,000 in the newt) of symmetrical loops, which has led to their being called *lampbrush chromosomes* (Fig. 3-15). The loops of lampbrush chromosomes are areas in which the DNA has been stretched out to a length equivalent to that of many individual genes. Along the loop the polarized synthesis of RNA (presumably informational varieties) occurs (Gall and Callan, 1962), and it has been estimated that during the lampbrush

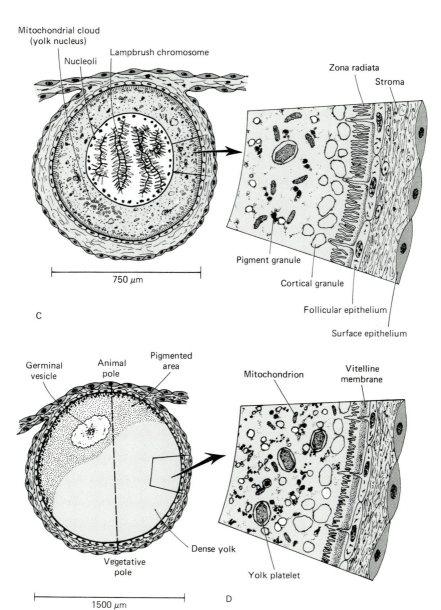

Mitochondrial cloud
(yolk nucleus)

Nucleoli

Lampbrush chromosome

Zona radiata

Stroma

750 µm

C

Pigment granule

Cortical granule

Follicular epithelium

Surface epithelium

Germinal
vesicle

Animal
pole

Pigmented
area

Mitochondrion

Vitelline
membrane

Dense yolk

Vegetative
pole

Yolk platelet

1500 µm

D

stage approximately 5 percent of the genome of the oöcyte is exposed and acting as a template for RNA synthesis.

The early diplotene nucleus is also characterized by the formation of large numbers of nucleoli (up to 1500 in *Xenopus*), which soon become distributed beneath the nuclear membrane. This large number of nucleoli is the morphologi-

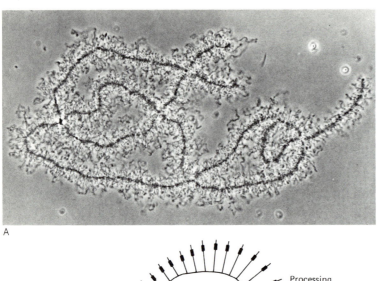

A

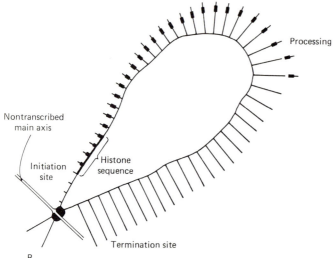

Processing

Nontranscribed
main axis

Initiation
site

Histone
sequence

Termination site

B

Fig. 3-15 (A) Photomicrograph showing lampbrush chromosomes from the oöcyte of a newt. (Courtesy of J. G. Gall.) (B) Schematized drawing of one loop of a lampbrush chromosome, showing transcription of RNA along the loop. In this loop, a histone sequence is transcribed. As synthesis proceeds around the loop (clockwise), processing occurs and the portions of the RNA molecules containing the histone transcript are cleaved off. (Modified from R. W. Olds, H. G. Callan, and K. W. Gross, 1977, *J. Cell Sci.* **27**:57–80.)

cal expression of a phenomenon known as *specific gene amplification* (Brown and Dawid, 1968). Gene amplification is an adaptive response for meeting the synthetic requirements of the egg, in this case the formation of a population of ribosomes sufficiently large to last throughout the period of cleavage in the embryo. The nucleolus is the principal structure that is involved in the synthesis of high-molecular-weight rRNAs and the assembly of ribosomes. It has been estimated that with only the number of nucleoli normally found in cells, several hundred years would be required to produce the number of ribosomes found in the mature amphibian egg. However, by selectively replicating those portions of the genome that must account for the formation of ribosomes, the time required to fulfill the production requirements of the oöcyte is reduced to only a few months.

While intense synthetic activity is occurring in the nucleus, major changes are also occurring in the cytoplasm (Fig. 3-14D). These changes are principally concerned with the formation of yolk; hence one can say that the vitellogenic phase of oögenesis has now begun. Yolk is not a specific chemical but is a collective term for several classes of chemical substances which are stored in the cytoplasm to provide nutrition for the developing embryo. In the amphibian egg, proteinaceous material is stored in the form of membrane-bound *yolk platelets*, lipid is stored as inclusions called *lipochondria*, and carbohydrate is accumulated as aggregates of glycogen granules.

Yolk platelets are the most prominent inclusions in the cytoplasm of the amphibian egg. For many years, both their origin and mode of formation remained a mystery, but now it is known that the bulk of the yolk protein is produced by cells of the liver and carried to the ovary via the blood (Flickinger and Rounds, 1956). The yolk precursor in the blood is a phospholipoprotein called *vitellogenin*. This material is cleared from the blood in the ovary and must pass through the follicular epithelium in order to reach the egg. Because vitellogenin is too large a molecule to pass through the plasma membrane of the oöcyte by diffusion, it is incorporated into the oöcyte by the process of *micropinocytosis*. In micropinocytosis, a membrane-bound vesicle, filled with the yolk precursor, forms at the surface of the egg and then pinches off into the interior of the cell. Vitellogenin cannot be detected within the mature oöcyte. Instead the protein yolk is represented by two molecules: *phosphovitin*, a protein (m.w. 35,000) with a high phosphorus content, and *lipovitellin*, a lipoprotein (m.w. ~400,000). These two proteins are packed in crystalline form within a membrane to form the yolk platelets. Although a small percentage of yolk platelets consist of the yolk proteins crystallized within mitochondrial membranes, the majority of the yolk platelets apparently arise from the coalescence of smaller pinocytotic vesicles containing the yolk proteins.

For many years, before the origin of yolk proteins in the liver was understood, yolk formation was thought to be a function of a *yolk nucleus* (Balbiani body), which is quite prominent in the developing eggs of frogs. This structure, now recognized to be a dense cloud of mitochondria possibly associated with some material of nuclear origin, is no longer considered to produce significant quantities of yolk material, but its real function remains obscure.

Cortical granules also begin to form in the cytoplasm at about the same time as the yolk platelets. Cortical granules (Fig. 3-14D) are membrane-bound inclusions comprised principally of protein and mucopolysaccharide material. These structures have an irregular distribution throughout the animal kingdom. For instance, they are found in the eggs of frogs (and humans) but not in those of salamanders. They appear to originate from the Golgi complex and ultimately become dispersed around the outer layer of cytoplasm (cortex) of the egg, just beneath the plasma membrane. As will be seen later in this chapter, the cortical granules play a role in the reaction of the egg to penetration by a spermatozoon.

By the late diplotene or the early diakinesis stage, the oöcyte is in most respects mature. The chromosomes lose their lampbrush configuration and begin to condense near the center of the nucleus. The nucleoli also move from the area adjacent to the inner nuclear membrane and become distributed around the chromosomal aggregate. These intranuclear changes presage the ultimate dissolution of the nuclear membrane (breakdown of the germinal vesicle). This results in the mixing of nuclear contents with the cytoplasm, which in many species is an essential prelude to cleavage, should the egg be fertilized.

The cytoplasm of the mature egg is packed with various organelles and inclusions. There are commonly well over a million mitochondria, in contrast to the several hundred found in most somatic cells. The large number of mitochondria is due to their autonomous division, which is based upon the replication of the circular, single-stranded mitochondrial DNA. As a result, a high proportion of the DNA of the mature oöcyte is of the mitochondrial variety. Other cytoplasmic organelles are centrioles, the Golgi apparatus (which is involved in the final packaging of carbohydrate and protein molecules), and a specialized form of endoplasmic reticulum (*annulate lamellae*), consisting of stacks of porous membranes that contain some RNA. The function of annulate lamellae is still not known.

In addition to the yolk inclusions, the mature oöcyte contains numerous pigment granules. Arising later during oögenesis than the other inclusions, the pigment granules are concentrated in one half of the egg, known as the *animal hemisphere*. The other, less heavily pigmented but more yolk-laden half is known as the *vegetal hemisphere* (Fig. 3-14D). The transitional area between these two hemispheres is commonly known as the *marginal zone*. With the appearance of an asymmetrical pigmentation of the egg, a tentative form of polarity is established. If an axis is drawn from roughly the apex of the animal hemisphere to the opposite point in the vegetal half, the *animal pole* in salamander eggs corresponds to the approximate position of the future mouth and the *vegetal pole* to the anus.

Oögenesis in Birds   The part of the hen's egg commonly known as the *yolk* is a single cell, the female sex cell or ovum. Its enormous size as compared with the other cells is due to the food material it contains. This stored food material (the true yolk, or *deutoplasm*), which is destined to be used by the embryo in its growth, is gradually accumulated within the cytoplasm of the ovum before it is liberated from the ovary. The other components of the egg (the egg white, the shell membranes, and the shell itself) are all noncellular secretions which invest the ovum in protective layers as it passes down the reproductive tract of the hen.

The early ovum is about 50 μm in diameter and grows gradually until it is about 6 mm. After this the rate of growth accelerates greatly, increasing the diameter by about 2.5 mm per day. By the time ovulation occurs, the ovum may have reached a diameter of 35 mm. This great increase in size is due largely to the accumulation of yolk materials. These are not synthesized in the ovum but are produced in the liver. From the liver they are transported via the blood to the follicular cells surrounding the ovum. The follicular cells then transfer the yolk materials to the ovum, where final morphological structuring of the transferred yolk materials occurs (Bellairs, 1964, 1965).

Under the microscope, yolk has the appearance of a viscid fluid in which granules and globules of various sizes are suspended. In the eggs of birds (Romanoff and Romanoff, 1949) the yolk is about 50 percent water, 33 percent fatty material, 16 percent protein, and only about 1 percent carbohydrate. The water carries electrolytes such as sodium chloride and the calcium salts utilized in bone formation. The yolk contains several classes of proteins which are also found in the serum of the hen. The principal ones are *vitellin* (lipovitellin) and *lipovitellenin*, proteins which bind much of the lipid found in the yolk; *phosvitin*, which binds most of the phosphorus; and a water-soluble class called *livetin*, which includes proteins identical to many of the normal serum proteins. The fatty substances are primarily in the form of neutral fats, the remainder being phosphatides, phospholipids, and cholesterol. In addition, there are vitamins A, B_1, B_2, D, and E. As these stored materials that constitute the yolk increase in amount, the nucleus and the cytoplasm are forced toward the surface; eventually the yolk comes to occupy nearly the entire cell.

Figure 3-16 shows a section of the hen's ovary, which includes several young ova and one ovum which is nearly ready for liberation. The very young ova lie deeply embedded in the substance of the ovary. As they accumulate more and more yolk, they crowd toward the surface and finally project from it, maintaining their connection only by a constricted stalk of ovarian tissue. The protuberance containing the ovum is known as an *ovarian follicle*. The bulk of the mature ovum itself is made up of the yolk. Except in the neighborhood of the nucleus, the active cytoplasm is only a thin film enveloping the yolk. The region of the ovum containing the nucleus and the bulk of the active cytoplasm is known as the *animal pole* because this subsequently becomes the site of greatest protoplasmic activity; fertilization occurs at the animal pole. The region opposite the animal pole is called the *vegetative*, or *vegetal*, *pole* because although material for growth is drawn from this region, it remains relatively less active.

Like the ova of many classes of vertebrates, the ova of birds have a highly irregular cell membrane called the *zona radiata*. This term was applied to it because, when viewed under high power with a light microscope, it seems to show delicate radial striations. After having studied the striations under the greater magnifications possible with the electron microscope, we now know that they are due to closely packed microvilli. As in amphibian oöcytes (Fig. 3-22), the functional significance of the irregular membrane is the great increase in membrane surface it affords, thereby enhancing the rate of the metabolic interchanges that can take place at the cell surface.

When a young ovum is rapidly accumulating yolk, the zona radiata is

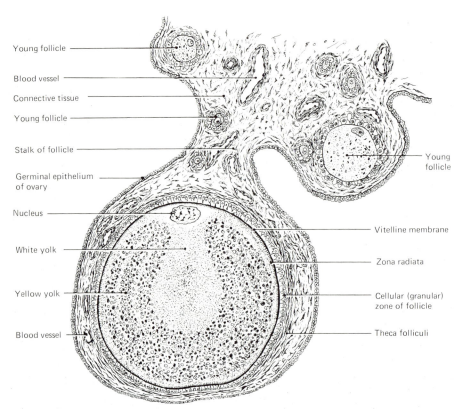

Young follicle

Blood vessel

Connective tissue

Young follicle

Stalk of follicle

Germinal epithelium
of ovary

Nucleus

White yolk

Yellow yolk

Blood vessel

Young
follicle

Vitelline membrane

Zona radiata

Cellular (granular)
zone of follicle

Theca folliculi

Fig. 3-16 Diagram showing the structure of a bird ovum still in the ovary. The section shows a follicle containing a nearly mature ovum, together with a small area of the adjacent ovarian tissue. (Modified from Lillie, after Patterson.)

relatively conspicuous. As the accumulation of yolk nears completion, the striated appearance of the cell membrane becomes less evident. This is a reflection of the reduced interchange between the follicular epithelium and the oöcyte. Between the cell membrane of the ovum and the surrounding follicular cells another membrane, the *vitelline membrane*, becomes more prominent (Fig. 3-16). The vitelline membrane of the unovulated egg of the bird, like that of the amphibian egg, is comparable to the *zona pellucida* in mammals (page 95). Immediately peripheral to the vitelline membrane is a continuous layer of follicular epithelial cells, sometimes known as the *cellular zone of the follicle*. The follicular epithelium is, in turn, surrounded by a highly vascular coat of connective tissue known as the *theca folliculi*, which can be subdivided into a *theca interna* and a *theca externa*. The abundant vascular supply is probably an adaptation required for transferring the large amounts of yolk material from the blood to the follicular epithlium.

If one compares the avian follicle with the mammalian, one cannot fail to be impressed by the similarities in the investing structures and the basic relations.

In both, the ovum is enclosed in a zone of follicle cells, which in turn is covered by a vascular connective-tissue theca. In both, the nearly ripe follicles bulge out on the surface of the ovary. The striking difference is that whereas the large, yolk-laden ovum of the bird occupies the entire follicle (Fig. 3-16), the small mammalian ovum comes nowhere near to filling it. The unoccupied territory within the mammalian follicle is the antrum filled with its liquor folliculi (Fig. 3-18).

When the full allotment of yolk has accumulated in the ovum of a bird, ovulation occurs by the rupture of an avascular band (the *stigma*) that surrounds the follicle (Fig. 3-17). Bands of smooth muscle, which run from the stalk into the follicle, contract, thereby releasing the ovum from the follicle. As is the case with the mammalian ovum, the first maturation division is almost coincident with ovulation and the second maturation does not occur unless the ovum is penetrated by a sperm cell. The significance of the maturation division in reducing the chromosomal number from diploid to haploid is the same in the two forms.

Oögenesis in Mammals   In contrast to the lower vertebrates, mammals do not replenish the stores of oöcytes present in the ovary at birth. At birth the human ovaries contain about 2 million oöcytes (many of which are already degenerating) that have been arrested in the diplotene stage of the first meiotic division. These oöcytes are already surrounded by a layer of follicular cells, or *granulosa cells*, and the complex of the ovum and its surrounding cellular investments is known as a *follicle*. Of all the germ cells present in the ovary of the embryo and the newborn human, only about 400 (one per menstrual cycle) will reach maturity and become ovulated. The remainder develop to varying degrees and then undergo *atresia* (degeneration). Why so many oöcytes are produced when so few actually leave the ovary remains, to a large extent, a mystery.

Oöcytes first become associated with follicular cells in the late fetal period, when they are going through the early stages of prophase of the first meiotic division. The *primary oöcyte* (so called because it is undergoing the first meiotic division) plus its incomplete covering of flattened follicular cells is called a *primordial follicle*. Later in the fetal period, when the follicular cells have formed a complete layer around the primary oöcyte, the complex is called a *primary follicle* (Fig. 3-18). By this time the oöcyte has entered the first of its two periods of developmental arrest, the diplotene stage of meiosis. In the human, essentially all the oöcytes, unless they degenerate, remain arrested in the diplotene stage at least until puberty. Some of these cells may not progress past the diplotene stage until the woman's last reproductive cycle (45 to 50 years). Less is known about what happens in the human oöcyte during the diplotene stage than in those of some lower vertebrates, but it is known that the chromosomes in the human oöcyte go through a lampbrush phase. Both the oöcyte and the follicular cells develop numerous microvilli at this stage. This suggests that some form of exchange is occurring between the oöcyte and the

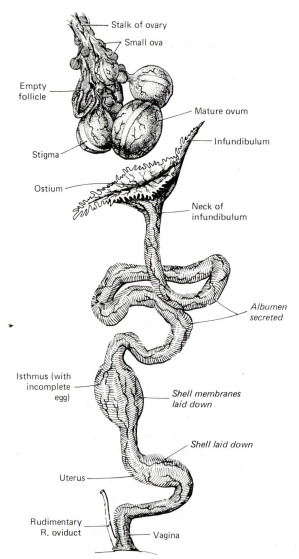

- Stalk of ovary
- Small ova
- Empty follicle
- Mature ovum
- Infundibulum
- Stigma
- Ostium
- Neck of infundibulum
- Albumen secreted
- Isthmus (with incomplete egg)
- Shell membranes laid down
- Shell laid down
- Uterus
- Rudimentary R. oviduct
- Vagina

Fig. 3-17 Drawing of the female reproductive tract of a hen. After rupture of the follicle through the stigma, the ovum passes into the left oviduct. As it passes down the genital tract, the ovum becomes successively covered with albumen, the shell membranes and finally the shell. In the bird, the right oviduct is rudimentary. (Modified from Bellairs, after Romanoff and Romanoff.)

follicular cells, but few details are yet available for mammalian follicles. The *zona pellucida*, a translucent noncellular membrane, also begins to form around the oöcyte after the first layer of follicular cells is complete.

In the prepubertal years many of the follicles enlarge as the result of (1) an increase in the size of the oöcyte, (2) formation of the zona pellucida, and (3) an increase in the size and number of the follicular cells. A thin basement membrane, called the *membrana granulosa* (Fig. 3-19F), forms around the

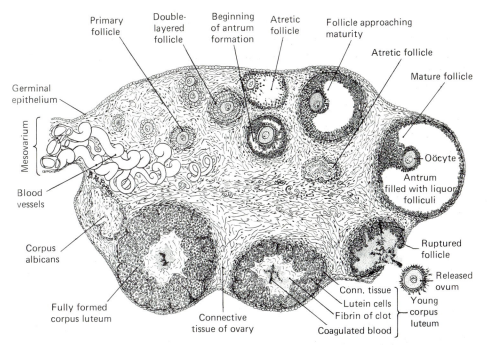

Fig. 3-18 Schematic diagram of ovary showing sequence of events in origin, growth, and rupture of ovarian (graafian) follicle and in formation and retrogression of corpus luteum. Follow clockwise around ovary, starting at mesovarium.

granulosa cells of the follicle. No blood vessels are found within the membrana granulosa, and both the oöcyte and the granulosa cells must rely upon diffusion for nourishment and oxygen.

The earliest stages of development of primary follicles (up to several layers of follicular cells) appear to occur without the mediation of sex hormones. Beyond this stage the differentiation of mammalian follicles involves the closely orchestrated interplay of several different hormones acting on the follicular cells. The next step in follicular development is the formation of a fluid-filled cavity, called the *antrum*, within the layers of granulosa cells (Figs. 3-19E and 3-20). This step depends upon the presence of pituitary gonadotropic hormones, particularly FSH (follicle-stimulating hormone), and the probable mediation of estrogens produced within the follicle. When the antrum has formed, the follicle is known as a *secondary follicle*, but the oöcyte within the follicle is still a primary oöcyte and it remains arrested in the diplotene stage. The fluid within the antrum is called the *liquor folliculi*. Initially it is formed by the secretions of granulosa cells, but later most of it arises as an exudate from the capillaries on the other side of the membrane granulosa.

The secondary follicle becomes further enveloped in a layer of modified ovarian connective tissue (*stromal*) cells. When first formed, this layer is known as the *theca folliculi* (Figs. 3-19E and 3-20), but it continues to differentiate into two layers. The inner layer, the *theca interna*, is glandular in nature and is highly

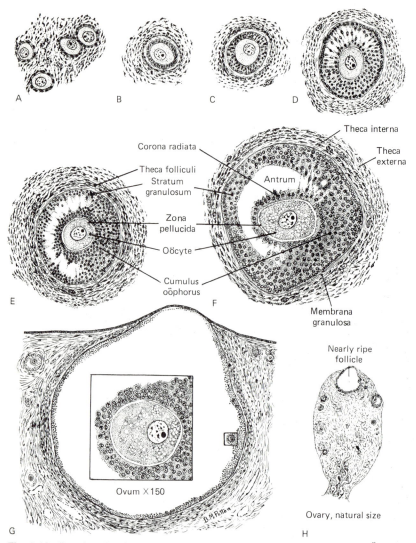

Fig. 3-19 Drawing showing a series of stages in development of the human oöcyte and ovarian follicle. (A–D) Primary follicles. (E–F) Secondary follicles, with the formation of an antrum. (G–H) Mature follicles. (Projection drawings in A–F, ×150; in G the follicle is magnified ×15, but the inset detail of the oöcyte is magnified ×150.)

vascularized, whereas the outer *theca externa* retains the characteristics of a connective-tissue covering.

Hormonal control of maturation of the secondary follicle preparatory to ovulation is complex (Richards, 1979). The thecal cells of the early secondary follicle contain LH (luteinizing hormone) receptors. When stimulated by LH, the thecal cells produce *testosterone*, which crosses the membrana granulosa into the granulosa cells. The granulosa cells contain receptor proteins for FSH and an

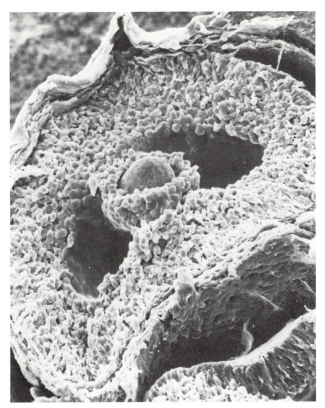

Fig. 3-20 Scanning electron micrograph of a mature follicle in the rat, showing the spherical oöcyte (*center*) surrounded by smaller cells of the corona radiata, which projects into the antrum. ×840. (Courtesy of P. Bagavandoss.)

active cAMP generating system. The latter, in particular, seems to be important in stimulating the enzymatic conversion of testosterone to *estradiol*, a potent estrogen. In addition to its systemic effects, the estradiol acts upon the nuclei of the follicular cells and also stimulates the formation of LH receptors in the granulosa cells. In this way it coordinates the receptivity of the ovarian follicle with the LH surge in the blood just prior to ovulation.

The hormonally stimulated follicle now rapidly increases in size and is known as a *tertiary* (preovulatory) *follicle*.[1] The enlarging follicle moves towards the surface of the ovary (Fig. 3-18) and the increasing liquor folliculi ultimately causes it to protrude above the general surface contour (Fig. 3-19G and H). When an ovary is exposed surgically, a nearly ripe follicle looks like a water blister. Such a follicle is nearly ready to rupture and release its contained ovum. Just before ovulation the follicle is producing large amounts of estradiol. LH receptors are now plentiful on both the thecal and granulosa cells, and the latter

[1]Formerly, tertiary follicles were designated *graafian follicles*, after the Dutch anatomist Reijnier de Graaf (1641–1673), who first described them. This term was commonly used in the older literature, and it is still seen in clinical reports.

also contain a high concentration of FSH receptors. The ovum has been released from its first meiotic block in the diplotene stage and goes on to finish its first meiotic division, releasing the first polar body as it does so. The follicle is now ready to respond to the preovulatory LH and FSH surge and complete the first stage of its cycle by releasing the ovum.

Atresia of Follicles   Only a minute percentage of the ova and follicles in the ovary reach maturity. The others undergo various degrees of maturational changes and then begin to degenerate. This process is known as *follicular atresia*, and a follicle that is involved in degeneration is said to be *atretic* (Fig. 3-18). The regulatory factors underlying atresia of follicles are not yet completely defined, but there is increasing evidence that atretic follicles are deficient in receptors for gonadotropins or estradiol.

Ovulation   The LH peak in the blood in conjunction with FSH, and perhaps with estrogen, sets in motion the final stages of follicular maturation that lead to ovulation. The follicle continues to swell as a result of both increased amounts of follicular fluid and growth of the follicle itself. The apex of the follicular protrusion is called the *stigma*, and within a day after the LH surge some characteristic changes take place in this area. The final preovulatory events begin with hemostasis of blood in the area around the stigma. Within an hour, the follicular wall in the stigma breaks down and the antral fluid along with the ovum, surrounded by the cells of the cumulus oöphorus, are expelled from the follicle (Fig. 3-21).

The precise mechanism that precipitates the rupture of the ovarian follicle is still not completely understood. In all probability several factors are involved. An early hypothesis held that increased antral fluid pressure within the follicle caused bursting of the follicular wall, but measurements have shown no increase in fluid pressure. More recent hypotheses involve weakening of the follicular wall because of ischemia or possibly the action of local lytic enzyme activity, the latter being stimulated by pituitary hormones (LH). Demonstration of smooth-muscle-like properties of the ovarian stromal cells raises the possibility that their contractile activity may play a role. No single hypothesis seems to account adequately for all the events known to occur at ovulation. This is particularly apparent when one considers ovulation in a broad range of animals, some of which (e.g., the insectivores) do not possess fluid-filled antra at the time of ovulation.

Corpus Luteum   The history of an ovarian follicle is by no means closed when it has liberated its contained ovum. Cells of both the stratum granulosum and the theca interna become involved in the formation of the *corpus luteum*. The corpus luteum, so called because of its yellow color in fresh material, grows rapidly and becomes an endocrine organ, secreting both estrogen and progesterone.

When the ovarian follicle ruptures, escape of most of the contained fluid and contraction of the stroma of the ovary reduce the size of its lumen (Fig. 3-18). Bleeding of the small vessels injured in the rupture of the follicle may partially fill the antrum with blood, which, together with the residue of the follicular contents, promptly becomes consolidated as a clot.

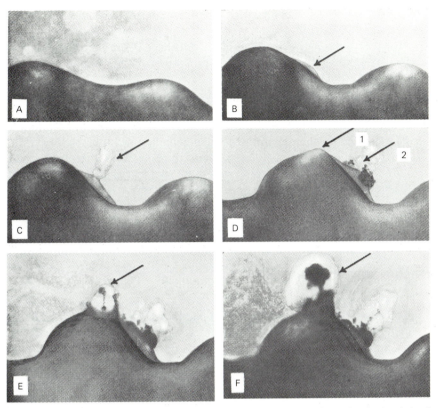

Fig. 3-21 Enlargements of single frames of a time-lapse motion picture showing ovulation in the rabbit. (A) Profile view of two follicles about 1½ hours before rupture. (B) same follicles about ½ hour before rupture. (C) Exudation of clear fluid in early phases of rupture. (D) At arrow 1, a new follicle becomes conical as the time of its rupture approaches. At arrow 2, the exudate from the follicle shown starting to rupture in (C) has become more abundant and contains some blood (dark). (E) The follicle indicated by arrow 1 in (D) is now beginning to rupture. The blood-tinged exudate from the follicle which started to rupture in (C), and showed more vigorous exudation in D *(arrow 2)* can be seen partly behind the more recently rupturing follicle. (F) The rupture of the follicle which is indicated by the arrow in E. Time elapsed between the photographs shown in (E) and (F), 8 seconds. The ovum is carried out with this final gush of fluid from the ruptured follicle. (From Hill, Allen, and Kramer, *Anat. Rec.,* vol., 63, 1935.)

Several concomitant changes occur during the transformation of the ruptured follicle to the corpus luteum. The granulosa cells swell and develop the cytological characteristics of cells that secrete large quantities of steroid hormones. The central clot is reduced by phagocytic activity while there is a massive invasion of the formerly avascular granulosa layer by blood vessels. These vessels bring in with them small cells of thecal origin, which become packed in among the more conspicuous cells originating from the stratum granulosum.

Formation and maintenance of the corpus luteum in the human require the continuous presence of LH from the pituitary. Hormonal relations vary among the mammals; for instance, in rats and sheep both LH and prolactin are required. The corpus luteum produces massive amounts of progesterone and some estrogens. One of the major functions of progesterone is to prepare the lining of the uterus to receive and implant the fertilized ovum. If pregnancy does not occur, the corpus luteum gradually loses its sensitivity to pituitary gonadotropins, probably by losing LH and FSH receptors on its cells, and it then regresses. If, however, pregnancy occurs, the corpus luteum undergoes a greatly prolonged period of growth and may attain a diameter of 2 to 3 cm in the human being. The *corpus luteum of pregnancy* is maintained by *chorionic gonadotropin*, secreted by the cells of the embryo and its surrounding membranes (Fig. 2-8).

When either type of corpus luteum begins to degenerate, the retrogressive changes are in the nature of a fibrous involution; that is, the cellular part of the organ disintegrates and fibrous connective tissue takes its place. As this connective tissue grows older and more compact, it gradually takes on the characteristic whitish appearance of scar tissue and is called a *corpus albicans* (Fig. 3-18).

ACCESSORY COVERINGS OF EGGS

After ovulation, eggs are released into a variety of environments both within and without the body. Most fishes and amphibians shed their eggs into either fresh or salt water, where the eggs must be protected from predators, disease, and environmental factors, such as extremes of osmotic pressure and pH. The animals that lay their eggs on land (reptiles, birds, primitive mammals) are faced with preventing desiccation as well as supporting and protecting the ova. In all the species mentioned above the ova are protected by layers of secretions that are applied as they pass down the female reproductive tract. The eggs of higher mammals and viviparous forms within the other classes of vertebrates remain in a physiological environment, but they too are typically covered with one or more accessory coats, some of which have less to do with protection than with other factors. Some egg coverings are secreted by the ova themselves, others by the surrounding follicular cells, and still others by the female reproductive tract after the egg has left the ovary. In this section we shall deal with the accessory coverings of three general types of eggs—those of amphibians, birds, and mammals.

The Membranes Surrounding the Amphibian Egg   Throughout its period of development in the ovary, the amphibian egg is surrounded by a follicular epithelium consisting of ovarian cells. In the very early oöcyte, both the plasma membrane of the oöcyte and the inner cell membranes of the follicular cells are quite smooth and are closely apposed. As the oöcyte grows during its first year, the plasma membranes of both the oöcyte and the follicular cells begin to form numerous minute projections, called *microvilli* and *macrovilli*, respectively. These structures increase in prominence, and the narrow space between the oöcyte and its follicular epithelium becomes filled with a homogeneous, noncellular, basement-membrane-like material that now appears to be secreted by both

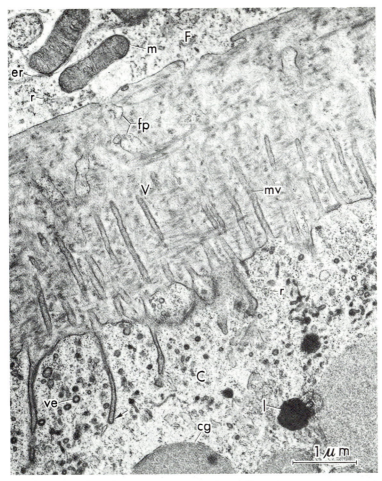

Fig. 3-22 The outer cytoplasmic zone *C* of an oöcyte of the frog *Rana pipiens* is infolded at frequent intervals (arrow points to bottom of fold). It also extends microvilli *mv* outward into the substance of the vitelline membrane *V*. In the cytoplasm are large cortical granules *cg*, occasional lipid bodies *l*, many small vesicles *ve*, and abundant free ribosomes *r*. Follicular epithelial cells *f* extend proccesses *fp* downward into the vitelline membrane. The portion of the follicle cell shown here contains mitochondria *m*, sparse endoplasmic reticulum *er*, and clusters of ribosomes *r*. (Electron micrograph by N. E. Kemp.)

the egg and the follicular cells. This noncellular membrane, the equivalent of the zona pellucida (Fig. 3-19F) in mammals, is commonly called the *vitelline membrane* in amphibians (Fig. 3-22).[2] As the oöcyte approaches the diplotene

---

[2]Over the years there has been some confusion regarding the proper usage of the term *vitelline membrane*, and definitions have often varied. The variation is the result of a lack of consensus as to when during oögenesis it is appropriate to use the word *vitelline membrane*. Some authors apply this term only after the egg is mature or actually ovulated and use zona pellucida to describe the same membrane in the unovulated egg, whereas others use this term from the membrane's first appearance.

phase of meiosis, both the size and the number of microvilli increase, and the vitelline membrane increases in thickness (Fig. 3-22). These extreme specializations of the cell membranes attest to an active exchange of materials between the follicular epithelium and the egg. After the bulk of the yolk has been laid down in the egg and as the egg approaches ovulation, the microvilli become smaller (Fig. 3-14D). At the time of ovulation a fluid-filled space, known as the *perivitelline space*, forms between the vitelline membrane and the plasma membrane of the eggs.

As the ovulated eggs pass down the oviducts, they become surrounded by a mucopolysaccharide *jelly capsule*, which is secreted by the oviduct. Upon contact with the water, the jelly swells greatly, and an inner and an outer jelly layer can be readily distinguished. The jelly coat allows the eggs to adhere both to one another and to water plants or other submerged objects.

Formation of the Accessory Coverings of Bird Eggs   At ovulation, the ovum is surrounded by a vitelline membrane, which contains an interwoven meshwork of relatively coarse, noncollagenous protein fibrils. Fertilization normally takes place just as the ovum is entering the oviduct (Fig. 3-17). The remainder of the *accessory coverings*, as the other components of the egg are called, are secreted about the ovum during its subsequent passage toward the cloaca.

First an *outer vitelline membrane*, composed of finer protein fibrils than those of the original, *inner vitelline membrane*, is laid down around the ovum when it is in the part of the oviduct adjacent to the ovary. Fibrils of this layer project along the oviduct from opposite sides of the ovum midway between the animal and vegetal poles and become enwrapped in *albumen* that is secreted in the upper oviduct. Because of the spirally arranged folds in the walls of the oviduct, the egg is rotated as it moves toward the cloaca. This rotation twists the adherent albumen into the form of spiral strands projecting at either end of the yolk; these are known as the *chalazae* (Fig. 3-24). Additional albumen, which has been secreted abundantly in advance of the ovum by the glandular lining of the oviduct, is caught in the chalazae and during the further descent of the ovum is wrapped about it in concentric layers. The albumen-secreting region of the oviduct constitutes about half its entire length.

One of the major albuminous proteins of egg white is *ovalbumin* (m.w. 43,000). The synthesis of ovalbumin and other secretions by the oviduct is a striking example of a specific response to the action of hormones. Normally the oviduct of the chicken does not become capable of secreting the components of egg white until the bird becomes sexually mature, but if a chick is treated with estrogen shortly after birth, the immature oviduct undergoes a series of rapid and profound maturational changes. The first major change is a five- to sixfold increase in mitosis that reaches a peak within 18 hours of estrogen injection. With continued injections of estrogen, tubular glands differentiate from the epithelium about 4 days after the start of hormone treatment. A couple of days later, these glands are synthesizing substantial amounts of ovalbumin and *lysozyme*, a bacteriostatic agent added to the egg white for protection of the embryo. At the same time ciliated cells become prominent in the oviductal epithelium. The next change is the differentiation of *goblet cells* from the epithelial cells. In response to a single injection of progesterone, the goblet cells suddenly begin to secrete large amounts of *avidin*, a major protein component of egg white (Fig. 3-23).

The *shell membranes*, which consist of two sheets of matted organic fibers, are added farther along in the oviduct. The *shell* is secreted as the egg is passing through the shell-gland portion of the oviduct (uterus). The entire passage of the ovum from the time of its discharge from the ovary to the time when it is ready for laying has been estimated to occupy about 25 to 26 hours. If the completely formed egg reaches the cloacal end of the oviduct during the middle of the day, it is usually laid at once, otherwise it is likely to be retained until the following day. This overnight retention of the egg is one of the factors accounting for the variability in the stage of development reached at the time of laying.

Structure of the Hen's Egg at the Time of Laying The arrangement of structures in the egg at the time of laying is shown in Fig. 3-24. Most of the gross relationships are already familiar because they appear so clearly in eggs which have been boiled. If a newly laid egg is allowed to float freely in water until it comes to rest and is then opened by cutting away the part of the shell which lies uppermost, a circular whitish area will be seen to lie atop the yolk. In eggs which have been fertilized, this area is somewhat different in appearance and noticeably larger than it is in unfertilized eggs. The differences are due to the development of an aggregation of cells known as the *blastoderm* in fertilized eggs. The blastoderm will be considered at greater length in Chap. 4.

Close examination of the yolk will show that it is not uniform throughout,

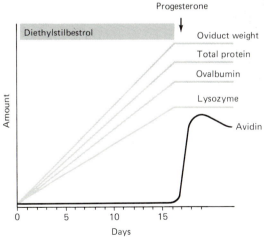

Fig. 3-23 Schematic representation of responses and patterns of molecular synthesis in the immature chick oviduct after the administration of steroid sex hormones. (Modified from B. W. O'Malley, et al., 1969, *Rec. Progr. Hormone Res.* **25**:105.)

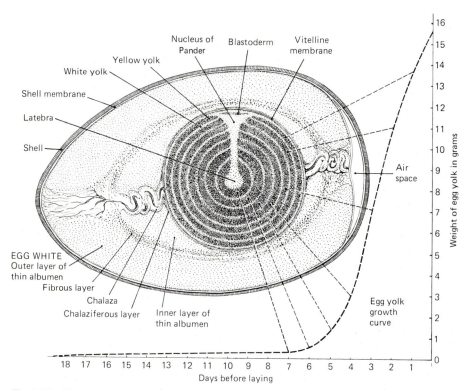

Fig. 3-24 Diagram showing the structure of the hen's egg at the time of laying. The graph indicates the rate of growth of the egg during the 18 days preceding its laying. The lines leading from the various layers of the yolk to the growth curve emphasize the time at which these layers were formed. (Redrawn, with slight modifications, after Witschi, 1956, *Development of Vertebrates.* Courtesy of the author and W. B. Saunders Company, Philadelphia.)

either in color or in texture. Two kinds of yolk can be differentiated, *white yolk* and *yellow yolk.* Aside from the difference in color visible to the unaided eye, microscopic examination will show that there are differences in the granules and globules of the two types of yolk, those in the white yolk being in general smaller and less uniform in appearance. The principal accumulation of white yolk lies in a central flask-shaped area, the *latebra*, which extends toward the blastoderm and flares out under it into a mass known as the *nucleus of Pander*. In addition to the latebra and the nucleus of Pander, there are thin concentric layers of white yolk between which lie much thicker layers of yellow yolk. The concentric layers of white and yellow yolk indicate the daily accumulation of deutoplasm during the final 7 or 8 days before ovulation. In this period the formation of yolk goes on day and night, but during the late hours of the night the yolk laid down contains only small amounts of fat but has a high protein content. The yolk deposited in the daytime has a high fat content. Its color is due to yellow carotenoids concentrated in it. Thus during the last week before ovulation there is added each day a thin layer of white yolk and a heavy layer of yellow yolk. The outermost yolk immediately under the vitelline membrane is always white.

The albumen, except for the chalazae, is nearly homogeneous in appearance, but near the yolk it is somewhat more dense than it is peripherally. The chalazae serve to suspend the yolk in the albumen.

The two layers of shell membrane lie in contact everywhere except at the large end of the egg where the inner and outer membranes are separated to form an *air chamber*. This space appears only after the egg has been laid and cooled from the body temperature of the hen (about 41°C or 106°F) to ordinary temperatures. In eggs which have been kept for any length of time the size of the air space increases because of evaporation of part of the water content of the egg. The familiar method of testing the freshness of eggs by "floating" them is done in recognition of this fact.

The egg shell is composed largely of calcareous salts, mainly in the form of calcite, a crystalline form of calcium carbonate. The calcium is ultimately derived from the food of the mother, but en route from the mother's intestinal tract it becomes involved in a curious structural adaptation found only in birds actively engaged in egg-laying. The calcium is incorporated into some specialized bony masses located within the marrow cavities of the long bones. These bony masses are called *medullary bone*. When the egg shell is being formed, the masses of medullary bone are rapidly broken down and serve as the major source of calcium for the formation of the egg shell. In the absence of medullary-bone deposits, the calcium for the egg shell would be derived from the other skeletal structures of the bird. If this were to occur, the skeleton would become seriously weakened. An interesting safeguard has been provided against this possibility: If lime-containing substances are not provided in the mother's diet, she stops producing eggs within a few days (Taylor, 1970). This is apparently due to an inhibition of the production of gonadotropic hormones by the pituitary gland. The normal egg shell is porous (about 7000 pores per egg), which enables the embryo to carry on an exchange of gases with the outside air by means of specialized vascular membranes arising in connection with the embryo but lying outside it, directly beneath the shell (Fig. A-44).

The Coverings of Mammalian Eggs   While in the ovary, mammalian eggs become surrounded by a noncellular membrane that forms between the plasmalemma of the ovum and the surrounding follicular cells. This membrane, composed principally of mucopolysaccharides and a substance that can be digested by trypsin, is commonly known as the *zona pellucida* (or *oölemma;* Boving, 1971). The zona pellucida begins to form when the ovum is surrounded by a single layer of cuboidal follicular cells. Part, and possibly all, of the zona is secreted by the follicular cells, but the oöcyte itself may contribute to the formation of this structure. At ovulation, the mammalian ovum remains surrounded by several layers of follicular cells known as the *corona radiata*. Whether or not the corona radiata plays a protective role is not known, but in some mammals (e.g., the rabbit) this layer is required for transport of the egg down the oviduct.

FERTILIZATION

There is perhaps no phenomenon in the field of biology that touches so many fundamental questions as the union of the germ cells in the act of fertilization; in this supreme event all the strands of the webs of two lives are gathered in one knot from which they diverge again and are rewoven in a new individual life-history. . . . The elements that unite are single cells, each on the point of death; but by their union a rejuvenated individual is formed, which constitutes a link in the eternal process of Life. (F.R. Lillie, in *Problems of Fertilization*)

Many biological factors must work in concert before actual union of the male and female gametes can be effected. The growth cycles of the sex cells must be such that both the eggs and sperms mature and are released from the gonads within a closely coordinated time interval. We have already seen how the act of copulation itself stimulates ovulation in some animals. In other forms changes in the length of daylight serve to coordinate both the development of gametes and the period of sexual activity. The characteristic behavior of an animal in *estrus* ("heat") is another mechanism which increases the likelihood of a functional sperm cell meeting a mature ovum.

Once the spermatozoa are released from the male, other factors play a role in ensuring the approximation of sperms and eggs. The simplest mechanism is that of marine invertebrates and most fishes. Here the female merely extrudes the eggs into the water and the male "floods" the area with millions of spermatozoa. How many of the eggs will be fertilized is a matter of chance and the water currents. Other aquatic vertebrates, such as salamanders, have adopted an elaborate breeding ritual in which the male deposits a package of sperm cells (spermatophore) upon the bottom of a pond and the female, during the courtship ritual, takes up the sperm package with the lips of her cloaca. A more efficient mechanism is the internal fertilization characteristic of mammals. In this process the male, during coitus, deposits the spermatozoa directly into the female genital tract. But even in this case many critical factors intervene before the final merging of the male and female gametes. Some of these call for special consideration because they are so important in human reproduction.

Sperm Transport in the Female Reproductive Tract of Mammals Much remains to be learned about the manner in which the spermatozoa make their way from the vagina through the uterus and the uterine tubes. In most common mammals, including man, the spermatozoa are deposited in the upper vagina at *insemination*, but in many rodents the site of insemination is the uterus. From the standpoint of the individual spermatozoon, the journey from the site of insemination to the upper uterine tube, where fertilization occurs, is an arduous one, and only a minute fraction of the spermatozoa that are deposited in the female reproductive tract ever reach the vicinity of the ovulated egg. In comparison with their size, the distance the spermatozoa must travel is great. The route may be beset with chemical hazards in the form of strong acid secretions or mechanical obstacles, such as a crooked and compressed cervical canal or uterine tubes narrowed or occluded by disease. Nevertheless, because of the enormous number of spermatozoa contained in an ejaculate of semen (200 to 300 million in man), it is probable that under normal conditions some of them will reach the uterine tube while they are still capable of penetrating and fertilizing the ovum.

The first barrier which the spermatozoa face is the natural acidity of the upper vagina. The apparent function of this acidity is to act as a bacteriostatic medium. The seminal fluid, however, acts as an effective buffer against the acidity, and within 8 seconds of insemination the vaginal pH can rise from 4.3 to 7.2 (Fox et al., 1973). In rodents, the semen coagulates shortly after insemination and forms a characteristic plug, which prevents the backflow of spermatozoa. In embryological studies on rodents, pregnancy is usually timed from the appearance of the plug.

From the upper vagina, some spermatozoa are transported extremely rapidly up the female reproductive tract, and in many mammals, including humans, they reach the uterine tube in less than 30 minutes. This form of transport is far too rapid to be accounted for by the swimming movements of the spermatozoa themselves (estimated rates are from 2 to 4 mm per minute). In fact, experimental studies have shown that nonmotile spermatozoa reach the uterine tube as quickly as motile ones during the early, rapid phase of sperm transport.

There also appears to be a slower phase of sperm transport, in which spermatozoa enter the cervix, possibly aided by their swimming movements, and lodge in the numerous irregular crypts that line the cervical canal. Normally, thick mucus fills the cervical canal, but hormonally induced changes at the time of ovulation reduce the viscosity of the mucus and allow better penetration by the spermatozoa. Once in the cervical crypts, the spermatozoa are slowly released into the uterine cavity.

The movement of spermatozoa through the uterus is less well understood, but at the height of sexual orgasm in many female mammals there are spasmatic contractions of the smooth muscle of the uterus, which may immediately draw some of the freshly deposited semen from the vagina into the uterus. Although uterine contractions may be an accelerating factor in sperm transport, it is certainly not an indispensable one, for there are innumerable well-authenticated cases, clinical and experimental, of pregnancy occurring in the absence of orgasm in the female. In such instances the traversing of the uterus must depend primarily upon the activity of the spermatozoa themselves.

The next barrier in the path of the spermatozoa is the entrance into the uterine tubes. In species that ovulate only one egg at a time, one of the barriers is purely statistical. For those spermatozoa that enter the uterine tube that does not contain an egg, mere chance has thwarted them from realizing the goal of their journey. The uterotubal junction may also act as a valve which permits or prevents the passage of spermatozoa into the uterine tube. This function is more strongly expressed in some species (e.g., mice) than in others. Once within the uterine tube, the spermatozoa continue their upward path by some combination of muscular contractions and ciliary currents of the tube and swimming of the spermatozoa themselves.

The heightened muscular activity of the uterine tubes at the time of ovulation has already been mentioned. It seems probable that the increased activity is important in sperm transportation as well as in the journey of the ova toward the uterus. Careful observation of the activity of surgically exposed tubes in living experimental animals indicates that temporary rings of contractions tend to divide the tube into a series of compartments. At any given moment in any compartment, the downward-beating cilia along the outer walls also tend to create central back eddies. In such currents and countercurrents the spermatozoa in the lumen of the tube would be scattered rapidly throughout the area between two adjacent contraction rings. When the zones of contraction relax at one level and form at another, some spermatozoa would be crowded back toward the uterus, but others would find themselves in a new compartment nearer the ovary. The formation and re-formation of such compartments by temporary rings of contraction at shifting levels would rapidly disperse the spermatozoa throughout the length of the tube.

Only at the upper end of the uterine tube does the swimming activity of the spermatozoa assume prominence. There is evidence that spermatozoa orient themselves so that they move against a gentle current, thus exhibiting what is called a *positive rheotactic response*. Presumably, the downward ciliary currents in the uterine tube serve as an effective orienting stimulus.

As the spermatozoa are transported through the female reproductive tract, they are subjected to a poorly understood influence by the maternal tissues, which enables them better to penetrate the membranes surrounding the egg. This phenomenon is called *capacitation* of the spermatozoa, and in its absence fertilization does not take place in many species of animals. The need for capacitation has been strikingly demonstrated in attempts to fertilize mammalian eggs in vitro. The ability of freshly obtained spermatozoa to fertilize eggs in vitro is often poor. If, however, the spermatozoa are first incubated for several hours close to female reproductive tissues, their success rate improves markedly. The time required for capacitation of spermatozoa varies from less than 1 hour in the mouse to 5 or 6 hours in primates and man. The change brought about by capacitation remains obscure, but there is evidence that capacitation involves the removal of proteins derived from the male genital tract, which cover the spermatozoa. There is also some evidence of changes in the plasma membrane of the spermatozoa.

The spermatozoa that are not directly involved in fertilization are ultimately removed from the female reproductive tract. Those in the uterine cavity become eventually swept through the cervix and into the vagina; those in the uterine tubes are ingested by phagocytic cells.

Egg Transport The freshly ovulated egg, surrounded by the corona radiata, lies free in the peritoneal cavity. As a means of increasing the chances that the ovum will enter the uterine tube, hormonal changes preceding ovulation result in greater muscular activity of the fimbriated ostium of the uterine tube and an increased ciliary current leading down the uterine tube. This combination causes strong fluid currents around the ovary in the direction of the ostium of the uterine tube, and in the great majority of cases the ovum is efficiently swept into the tube. This phase of ovum transport has been vividly recorded in a film by Blandau (University of Washington Audiovisuals). Within the uterine tube ciliary currents appear to be the main driving force for the ovum, for if the muscular activity of the tube is blocked by pharmacologic agents, downward transport of the egg continues at the normal rate (Halbert et al., 1976). The corona radiata (cells of the cumulus oöphorus) surrounding the ovum is very important in egg transport; without this layer the ovum makes little progress. To a large extent, this is a function of mass rather than intrinsic motility, because inert objects of the same size are also effectively moved down the uterine tube.

The structure and function of uterine tubes differ considerably among the mammals, but there is a general pattern of relatively rapid egg transport in the upper part of the tube and slower transport closer to the uterus. In the lower portions of the tube, muscular contractions assume increasing importance in transporting the fertilized egg toward the uterus.

The Viability of Ova and Spermatozoa Both egg and spermatozoa have only limited viability once they are free in the female reproductive tract. When liberated from the ovary, an ovum at once begins to undergo certain changes which can be characterized as aging or deterioration. Among other things, its cytoplasm progressively becomes more coarsely granular. This is accompanied by a general depression of metabolic activity, the course of which is reversed only if fertilization occurs. In most mammals, including man, the ovulated egg must be fertilized within 24 hours or it becomes "overripe" and nonviable.

There has been, and still is, much misinformation about the feats of travel and length of life of spermatozoa. Persistence of motility used to be equated with fertilizing capacity. We now know that motility lasts much longer than the ability to fertilize. For example, spermatozoa of the rabbit lose their ability to fertilize after about 30 hours in the female genital tract, whereas their motility may last over 2 days. Estimates now place the fertilizing power of human spermatozoa in the female genital tract at 1 or 2 days, with motility persisting for perhaps double that time. Survival of spermatozoa in the female genital tract is unusually prolonged in some species. In some bats insemination occurs in the autumn, but the spermatozoa remain dormant throughout hibernation. Not until the following spring, several months later, do ovulation and fertilization occur. In the domestic chicken spermatozoa are stored in crypts in the oviductal wall and are gradually released with the passing of eggs through the oviduct over a 3-week period.

It should be stressed that the previous statements apply to ejaculated sperm in the female genital tract. The viability of spermatozoa varies greatly under different environmental conditions. In the epididymis and ductus deferens, where they remain nonmotile, human spermatozoa retain their full capacities for many days. Their motility is aroused only when, at the moment of ejaculation, they are mixed with secretions of the seminal vesicles and the prostate and bulbourethral glands. Much of the increased metabolism of the activated spermatozoa is due to substrates supplied by the seminal fluid, such as fructose, which is produced by the seminal vesicles (Mann, 1964).

That their life after activation depends in large measure on the rate at which spermatozoa expend their limited store of potential energy is clearly indicated by experimental work in artificial insemination. The motility of the spermatozoa in freshly ejaculated semen can be checked by chilling them. Under these conditions they do not immediately dissipate their available store of energy, thus allowing the semen of pedigreed stock to be shipped thousands of miles by airplane and introduced into females by a syringe, with the successful production of offspring. In recent years the rapidly advancing techniques of cryobiology have made banks of stored human sperm a reality. Even early mammalian embryos can be frozen for extended periods of time and later resume normal development upon thawing.

Union of Gametes   In almost all vertebrates the meeting of ova and spermatozoa is a matter of chance. Only in some fishes, coelenterates, and other isolated invertebrate groups is there evidence of chemical attraction between the two. Once contact has been made, the spermatozoa must penetrate the egg coverings and finally the plasma membrane of the ovum itself before fertilization can take place.

Direct observation of fertilization in birds and mammals, where the process occurs within the body of the female, has proved to be extremely difficult. Advances in producing fertilization in vitro in a number of mammalian species, including man, have recently broadened our knowledge. Nevertheless, the vast majority of studies on fertilization are still carried out on invertebrate species or in lower vertebrates in which fertilization occurs in the water.

In mammals, fertilization occurs in the upper part of the uterine tubes. The spermatozoa must first penetrate the cells of the corona radiata and then the zona pellucida before they can make contact with the plasma membrane of the egg. To do so the spermatozoa must first undergo the *acrosome reaction*. The acrosome reaction, for which capacitation is a prerequisite, is the means by which *lytic enzymes* stored within the acrosome of the spermatozoon (Figs. 3-11 and 3-12) are released so that they can facilitate the passage of the sperm through the egg coverings. The first step in the acrosome reaction is the localized fusion of portions of the outer acrosomal membrane with the overlying plasma membrane of the spermatozoon. Those areas soon break down, allowing the release of the enzymes contained within the acrosome. A combination of the lashing of the tail of the spermatozoon and the lytic properties of the released acrosomal enzymes, particularly *hyaluronidase,* enables the spermatozoon to make its way through the cells of the corona radiata. While this is occurring, the fused plasma membrane and outer acrosomal membrane of the spermatozoon comes off entirely, leaving the inner acrosomal membrane as the leading surface of the spermatozoon (Fig. 3-25).

The next barrier that the spermatozoon must pass through is the tough, noncellular zona pellucida, which completely surrounds the egg. The exact means by which this is accomplished is not completely understood, but the spermatozoon apparently first undergoes a binding reaction to the zona pellucida (Gwatkin, 1977). The sperm cell then digests a narrow pathway through the zona with the help of an enzyme (or enzymes) called *acrosin* or *zona lysin*, which is thought to be released from the inner membrane of the acrosome.

Once through the zona pellucida, the spermatozoon finds itself in the fluid-filled *perivitelline space* between the zona and the plasma membrane of the ovum. Contact is quickly established between the egg and sperm membranes and may be facilitated by the microvilli projecting from the ovum. Where the sperm has made contact with the egg, the cytoplasm of the egg in many species bulges out in an elevated process called the *fertilization cone*. The plasma membranes of the egg and sperm fuse, and then the fertilization cone retracts, carrying the sperm head into the ovum and completing the phase of penetration.

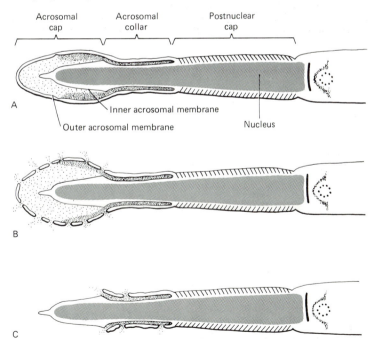

Fig. 3-25 Diagram illustrating the acrosomal reaction in hamster spermatozoa (A) before, (B) during, and (C) after the acrosomal reaction. (Modified from Yanagimachi and Noda, 1970, *Am. J. Anat.* **128**:429.)

Development and Fusion of Pronuclei   After penetration into the egg, the spermatozoon divests itself of most of the cytoplasmic structures that were involved in its approach to and fusion with the egg (Fig. 3-26). These structures include the tail, the mitochondria of the midpiece, and the remains of the

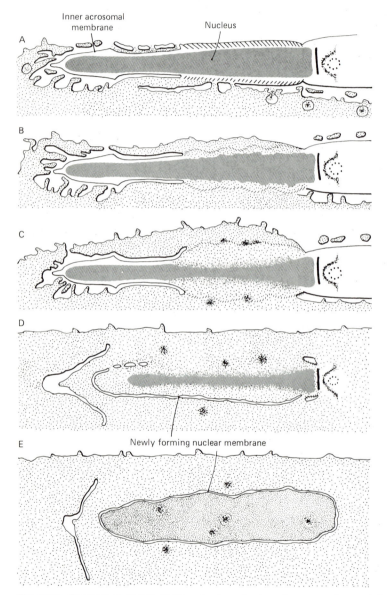

Fig. 3-26   Diagram of stages of incorporation of a hamster spermatozoon into the egg. (A, B) Fusion of the sperm head with the cytoplasm of the egg. (C) Swelling of the sperm nucleus. (D, E) Formation of a nuclear envelope around the swelling sperm nucleus. (Modified from Yanagimachi and Noda, 1970, *Am. J. Anat.* **128**:429.)

acrosome. Whether or not the centrioles are preserved remains problematic. Then the highly condensed nucleus of the spermatozoon begins to swell, as the tightly packed chromatin threads unwind and prepare themselves for the resumption of synthetic activity (Fig. 3-27).

In mammals the 12 hours after sperm penetration are occupied by the

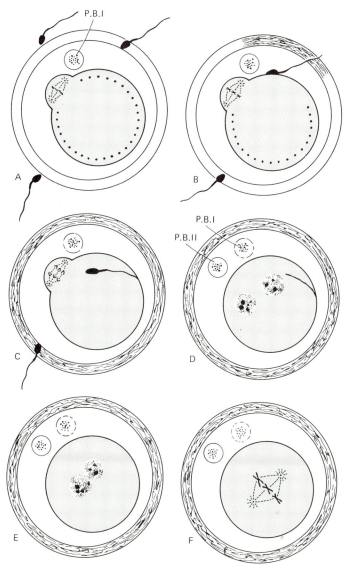

Fig. 3-27 Diagrams illustrating the process of fertilization and the formation of polar bodies. (A) Passage of spermatozoon through the zona pellucida. (B) Initiation of the cortical reaction (*disappearance of black dots*) and fertilization changes (*shading*) beginning in the zona pellucida. (C) Incorporation of spermatozoon into the egg. (D) Release of second polar body and formation of male and female pronuclei. (E) Approximation of pronuclei. (F) Metaphase of first cleavage division.

swelling of the male and female nuclei (now properly called *pronuclei*), the appearance in them of prominent nucleoli, and the migration of the pronuclei toward the center of the egg. While the pronuclei are taking shape, DNA replication occurs along the chromosomes in anticipation of the first cleavage of the fertilized egg. The nuclear membranes surrounding each of the pronuclei break up and the nucleoli lose prominence. Finally the two pronuclei lie side by side and *syngamy*, the intermingling of the maternal and paternal chromosomes, occurs. This represents the last stage of the fertilization process, which began with the fusion of the deteriorating egg and sperm cells and ends with their revitalization as they enter the prophase of the first mitotic cleavage division of the *zygote*, the newly created embryo.

The Fertilization Reaction   Although the essence of fertilization is quite straightforward, an amazing variety of mechanisms designed to deal with various aspects of the process have evolved among the animal groups. This is clearly illustrated when one examines the means by which the egg prevents the genetic material from more than one sperm cell from combining with that of the ovum. Among the urodele amphibians, reptiles, and birds the penetration of the egg by more than one spermatozoon is common, and the eggs have devised mechanisms for inactivating the extra sperm nuclei. Eggs of most of the other vertebrates prevent *polyspermy* by surface reactions which prevent the entry of more than one spermatozoon into the egg. Eggs of the latter type possess a superficial layer of cortical granules. Cortical granules are not found in the eggs of vertebrates which allow the penetration of multiple spermatozoa.

In vertebrates which allow only one sperm cell to penetrate the egg, the first reaction to the fusion of the spermatozoon with the egg is a rapid change in the electrical properties of the plasma membrane of the egg. In the frog, Cross and Elinson (1980) found that within seconds after contact with a sperm cell the charge of the membrane (membrane potential) changes from $-28\,mV$ to $+8\,mV$ and then remains positive for about 20 minutes. Further experiments showed that a positive membrane potential is inhibitory to polyspermy, whereas lowering the membrane potential in a newly fertilized egg allows polyspermy to occur.

The next block to polyspermy, occurring within a few minutes of sperm penetration, is the *cortical reaction*. Starting from the point where fusion of egg and sperm occurred, the cortical granules move up to the inner surface of the plasma membrane, fuse with it, and then empty their contents into the space surrounding the egg. There is evidence that in some species the cortical reaction may be initiated by changes in the polarization of the plasma membrane of the egg.

With release of the contents of the cortical granules, the further penetration of spermatozoa may be prevented by changes in both the zona pellucida and the plasma membrane of the egg. In many, but not all, mammals the zona pellucida becomes impenetrable to spermatozoa within minutes after the cortical reaction. The nature of these changes is still uncertain. It has been proposed that the material from the cortical granules hardens the zona pellucida, or destroys sperm receptors on the zona, or possibly directly inactivates the zona lysin of the spermatozoa (Austin, 1978). The plasma membrane of the egg also becomes resistant to sperm penetration, but in some mammals this does not occur until several hours after the reaction of the zona pellucida.

Parthenogenesis Not all eggs require the penetration by sperm for the initiation of embryonic development. In a number of invertebrate groups as well as in scattered vertebrate species (some fishes, a few lizards, and even turkeys), unfertilized eggs may become activated and develop into viable individuals as part of the normal life cycle. This process is known as *parthenogenesis*. Artificial parthenogenesis of eggs can be stimulated in the laboratory by a variety of means. Pricking with a blood-dipped needle is a classical way of producing parthenogenesis in frog eggs. A large proportion of eggs stimulated to undergo artificial parthenogenesis fail to develop normally. More than likely this is due to the unmasking of deleterious recessive genes in the haploid embryos. Parthenogenetic embryos which do survive are commonly found to be diploid, probably owing to retention of the second polar body. In mammals, parthenogenetic individuals are all genetically female because of the XX chromosomal complement of the female. On the other hand, the female in both birds and reptiles is heterogametic, allowing the development of both males and females by parthenogenesis.

Sex Determination Just before the turn of the century, Henking (1891), while studying the chromosomal pattern of certain insects, was impressed by the fact that there was one pair of chromosomes which lagged behind the others in moving toward the poles of the spindle in the maturation divisions. This was intriguing, but its significance was not at first apparent. In the manner of mathematicians, who give the unknowns in their problems the last letters of the alphabet as noncommittal designations, the members of this chromosomal pair were christened X and Y. Several years later McClung (1902) and Wilson (1905) came to the conclusion that this pair of chromosomes was involved in sex determination. Practically all the animals which have been critically studied since that time show consistent differences between the sexes in one of the pairs of chromosomes in both somatic and germ cells. In one sex, all the pairs of chromosomes are symmetrically mated (X-X). In the opposite sex, the members of the chromosomal pair are quite different from each other in size and shape (X-Y). It is now apparent that in mammals something that is carried on the Y chromosome, but not on the X, makes for maleness. In some groups of vertebrates (e.g., birds), the chromosomal pairs of the male are alike, whereas those of the female show sexual differences.

Earlier in this chapter, it was shown that during meiosis the members of the chromosomal pairs are separated. This results in the formation of equal numbers of sperm cells bearing the male-determining Y chromosome and the female-determining X chromosomes. In female mammals, the eggs always carry X chromosomes. At the time of fertilization there is a roughly equal chance of an X-bearing or a Y-bearing spermatozoon fertilizing the egg. The union of an X-bearing sperm cell with an egg produces a genetically female offspring and fertilization by a Y-bearing sperm cell gives rise to a genetically male offspring.

The union of gametes and the genetic determination of sex is just the beginning of the long process of sexual differentiation. Many aspects of this process are just beginning to be understood. The steps involved in the translation of genotypic sex into sexual phenotypes will be covered in Chap. 16. Recent experimental evidence has shown that phenotypic sex may not be as completely fixed at the time of fertilization as was first believed. There is little doubt that the chromosomal combination begins the trend of development toward one sex or another, but this trend may be inhibited or modified by massive doses of hormones or by the absence of appropriate hormone receptors. Our awareness of occasional discrepancies between genotypic and phenotypic sex stem largely from studies on the presence or absence of sex chromatin in cells.

In 1949 Barr and Bertram first presented evidence of sexual differences in fixed and stained nuclei of nondividing somatic cells. They found that in the nuclei of cells from females there was usually a conspicuous, characteristically located mass of chromatin that was absent in the nuclei of cells taken from a male (Fig. 3-28). This chromatin mass is called the *sex chromatin*. The sex chromatin is now believed to represent one of the X chromosomes, which remains highly condensed in the interphase (G_1) nucleus. Lyon (1961) has postulated that the condensed sex chromatin represents the inactivation of one of the X chromosomes. It is implied that the activity of only one X chromosome is required or permissible for normal development in either males or females. Additional X chromosomal activity is effectively eliminated by condensation of the extra X chromosome. The sex chromatin, then, represents the morphological expression of a genetic control mechanism. Sex chromatin bodies are typically not seen during early embryonic cleavage. There is evidence that both X chromosomes function during cleavage. Their role during cleavage and the genetic events associated with the inactivation of the extra X chromosome during later embryogenesis are poorly understood.

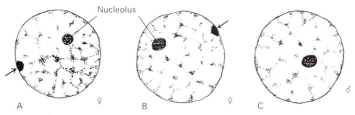

Fig. 3-28 Drawings of the nuclei of human epidermal cells to show the sex chromatin. (Schematized from the work of Moore, Grahm, and Barr, 1953, *Surg., Gyn. and Ob.,* vol. 96. (A, B) Nuclei from a female, the sex chromatin indicated by arrows. (C) Nucleus from a male, showing the absence of sex chromatin. Note that the nucleolus, although eccentric in position, does not lie against the nuclear membrane.

Establishment of Polarity in the Embryo   The vertebrate body is bilaterally symmetrical and can be viewed in the context of three polar axes—a *craniocaudal* axis, a *dorsoventral* axis, and a *mediolateral* axis. How these axes become imprinted upon the spherical egg remains one of the major mysteries of embryology. The development of polarity and axiation has been more intensely studied in amphibians than in other forms. For that reason most of this section will describe the establishment of polar axes in this class of vertebrates.

The ovum developing within the ovary is already strongly polarized morphologically into animal and vegetal halves. This is obvious to the naked eye because of the denser concentration of pigment granules in the animal half. There are gradients of other structures, as well. The nucleus (germinal vesicle) is located near the animal pole, and there is a gradient of increasing density of both ribosomes and glycogen granules toward the animal pole. Conversely both the size and the concentration of yolk platelets increase markedly toward the vegetal pole. In urodeles the craniocaudal axis of the future embryo nearly coincides with a line drawn through the animal and vegetal poles, but in anurans the two axes are not quite the same. Nevertheless, a useful generalization for amphibia is that the region of the animal pole will ultimately form the head and the vegetal pole, the tail. Although it is apparent that the craniocaudal axis is established before fertilization, the factors that lead to its being set in the ovary remain obscure. There is some evidence that early polarity resides in the egg cortex, a thin superficial layer of cytoplasm just beneath the plasma membrane of the egg.

Fertilization is the next milestone in the establishment of polar axes within the egg. Within minutes after fusion of the sperm to the egg the breakdown of cortical granules begins, starting at the site of sperm penetration. Along with expulsion of the second polar body, the cortex of the animal pole of the egg begins to contract. Later, during the first hour, DNA replication has occurred in the male and female pronuclei, and they fuse. At the same time, the cortex of the egg relaxes and the fertilization membrane, consisting of material released from the cortical granules and the innermost jelly layer, is raised up from the surface of the egg. By 2 hours after fertilization, the egg has rotated within the fluid-filled perivitelline space beneath the fertilization membrane, and in some species of amphibians a characteristic structure called the *gray crescent* appears (Fig. 3-29).

As its name implies, the gray crescent is a localized crescentic area which appears along the border between the darkly pigmented animal pole cytoplasm and the light vegetal pole cytoplasm of the fertilized egg. The location of the gray crescent is determined by the site of sperm entry, for it invariably appears on the opposite side of the egg. Its appearance is related to displacement of material in the animal pole cortex, but whether it is due to rotation or contraction of cortical material has not yet been determined with certainty. How the site of sperm entry controls the location of the gray crescent is not known, but descriptive and experimental work has shown that the middle point of the gray crescent is equivalent to the middorsal point of the body. This determines the dorsoventral axis of the future embryo. With the superimposition of the

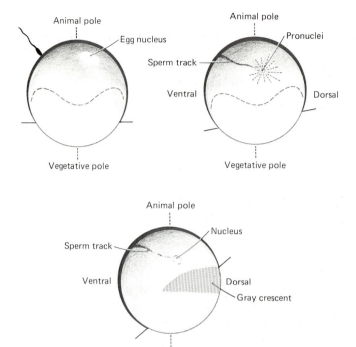

Fig. 3-29 Diagram illustrating fertilization and gray crescent formation in the amphibian egg. (A) Sperm contacting egg. (B) Approximation of egg and sperm pronuclei and early cortical shifting. (C) Formation of the gray crescent as the result of cortical shifting.

dorsoventral axis upon the previously existing cephalocaudal one, the remaining mediolateral axis is also established, simply by geometrical considerations. Thus, even before cleavage has begun, the three principal axes of the amphibian embryo are established.

The gray crescent is not only a structural landmark but is also important to the embryo in other ways. It represents the future site of formation of the dorsal lip of the blastopore (Chap. 5), a region that has been called the *embryonic organizer* because of its importance in controlling development. The gray crescent itself is of vital importance to the progression of development (Curtis, 1960, 1962). If it is removed at the one-cell stage, cleavage takes place but gastrulation does not occur (Fig. 3-30A). If, on the other hand, the gray crescent on an early embryo is split and prevented from fusing, two embryos form (Fig. 3-30B).

The above experiments suggest that the early embryo is labile with respect to the influence provided by the gray crescent. Experiments involving similar manipulations of the gray crescent in eight-cell embryos (Curtis, 1962) have shown that although the gray crescent is still able to exert an effect when transplanted to one-cell embryos (Fig. 3-30E), it can now be removed without hindering the future development of the eight-cell embryo (Fig. 3-30D). Thus

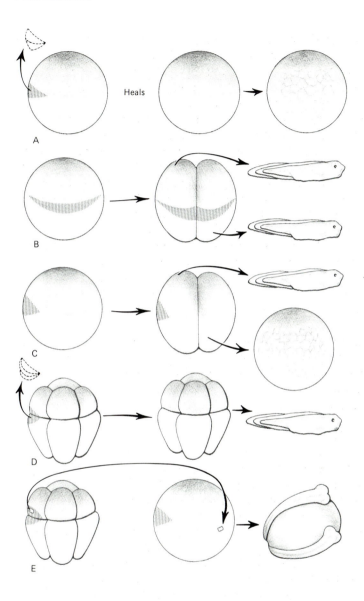

Fig. 3-30 Experiments involving manipulations of the gray crescent. (A) Excision at the one-cell stage blocks gastrulation. (B) Separation of blastomeres at the two-cell stage of a normal embryo results in the development of twins. (C) Separation of blastomeres after an abnormal first cleavage leaves the gray crescent entirely in one blastomere. The blastomere containing the gray crescent goes on to develop normally. Development of the other blastomere is arrested. (D) Removal of the gray crescent in an eight-cell embryo does not interfere with normal development. (E) Grafting of a piece of gray crescent from an eight-cell stage to a one-cell stage results in the reduction of a secondary embryonic axis. (After Curtis.)

the information provided by the gray crescent is now incorporated into the embryo in such a way that regulatory activity can compensate for loss of the gray crescent. The eight-cell embryo is sufficiently stable that grafts of gray crescent from one-cell embryos do not cause the formation of secondary embryonic axes.

Although control of dorsoventral polarity seems to reside first in the gray crescent cortex of the early embryo, it does not remain there. During the second phase of polarization, dorsoventral polarization seems to be transferred to the vegetal yolk mass. In the third phase, occurring later in development, control of dorsoventral polarity ultimately comes to rest in the presumptive mesoderm of the animal portion of the embryo. By the time neurulation occurs, the entire embryo has become polarized in the dorsoventral direction (Nieuwkoop, 1977).

The determinants of polarity are less clear in the embryos of higher vertebrates. In the chick there is evidence that some elements of polarity may be determined in part by the directions of rotation of the egg within the uterus, but the role of gravity is still being discussed. Polarity in the mammalian embryo does not seem to be fixed until considerably later in development (around the time of implantation), and polarity may be a consequence rather than a cause of differentiation.

Chapter 4

Cleavage and Formation of the Blastula

Fertilization transforms the egg from a metabolically depressed state to one of extreme vigor, characterized by a sharp increase in respiratory and synthetic activity. One of the most immediate and spectacular consequences is the initiation of cleavage. During cleavage, waves of cell division follow one another almost without pause, subdividing the unmanageably large zygote into a tightly knit mass of cells of more normal dimensions. The cells resulting from the first few cleavage divisions are, for the most part, morphologically unspecialized. They remain metabolically unspecialized, as well; their synthetic activity is geared toward the production of the DNA and the proteins required for cell division rather than for specialized molecular activity.

After a few synchronous cell divisions, during which the number of cells in the embryo doubles with each cycle, the embryo looks like a small mulberry (hence the commonly applied term, *morula*), and the cell divisions begin to lose their synchronous character. As will be seen below, many of the changes in the cleavage pattern within a given embryo, as well as differences between cleavage patterns among embryos of the various vertebrate classes, are related to the amount of yolk originally present in the egg. In time the cleaving embryo develops a central cavity (*blastocoele*) and it enters the *blastula* stage. Beneath its rather unimposing and homogeneous exterior, major changes are beginning to

111

take place. The synthesis of RNAs coding for specialized proteins gets underway in many species, and the cells of the embryo begin to communicate with one another in new and different ways. The cells themselves begin to assume different, nonequivalent properties, although their appearance does not betray these developments. In many species the cytoplasm of the egg is not homogeneous, and according to one long-standing hypothesis, different cytoplasmic areas of the early embryo act upon the individual cleavage nuclei in different ways, thus initiating their development along different lines. At the level of the embryo as whole, final polarity is becoming firmly fixed and the unseen developmental guidelines that keep the massive tissue displacements of the period of gastrulation under tight control are set up. All these changes lead toward a greater integration of the embryo as a whole and a reduction in the ability of its corresponding parts to compensate for damage or experimentally created defects.

The Cell during Cleavage In the normal embryo a cleavage division consists of nuclear division (*mitosis*) followed by cell division (*cytokinesis*). The term *cleavage* itself, however, is commonly restricted to cytokinesis, although usage of the term is not consistent. Cleavage brings to the forefront several questions of general importance to cell biology. Prominent among these are the mechanisms of cytokinesis and the formation of new membranous structures in dividing cells (Fig. 4-1). For research on both of these questions, cleaving embryos are often the material of choice, and the results of investigations into these processes are used as generalizations for other systems of cell division. Much of our knowledge in this area has arisen from investigations carried out on invertebrate forms, but the cleaving amphibian egg has also been a favored research object. Although many of the basic questions and answers in this field were set forth decades ago, modern research methods in both cytology and biophysics have allowed the identification of some concrete mechanisms by which these processes take place.

Cytokinesis occurs after mitosis, and the plane of the cleavage furrow is the same as that of the metaphase plate of the preceding mitotic figure. That there is some sort of causal relationship between the mitotic apparatus and furrow formation has been demonstrated by observations on both normal and experimentally altered eggs in which the mitotic apparatus has been displaced. The cleavage furrow forms first in the region of the cortex nearest to the mitotic spindle and then moves around the cell (Fig. 4-1). According to Rappaport (1974), the position of the cleavage furrow is established by the time of anaphase; the *asters*, or possibly some other component of the spindle apparatus, are the structures that in some way interact with the cell cortex to stimulate the formation of the cleavage furrow. The nature of the stimulus is not known, but it can be transmitted from the mitotic apparatus to the surface of the egg within a few minutes.

At the site of the cleavage furrow, the cortex of the cell contains a band of microfilaments which becomes better defined as the furrow forms. Chemically,

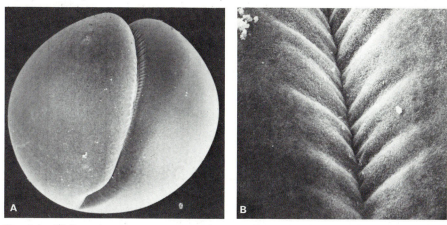

Fig. 4-1 (A) Scanning electron micrograph (×92) of a frog embryo during the first cleavage division. (B) Higher power view (×805) of the cleavage furrow, showing folds in the walls of the furrow. (Courtesy of H. W. Beams and R. G. Kessel, 1976, from *Am. Sci.* **64**:279.)

the microfilaments belong to the *actin* family of contractile proteins, and their contractile behavior appears to be the basis for the constriction of the cell during cleavage.

When a spherical cell divides into two cells, geometric necessity dictates that the total surface area of the two daughter cells will be greater. In confirmation of earlier hypotheses, recent investigations on early amphibian embryos have shown that within 7 or 8 minutes of the establishment of the cleavage furrow, areas of new membranous material form along both sides of the contractile ring. The new membrane is smooth, pale in color, and highly permeable to ions, and it seems to be deposited only in the cleavage furrow.

Effect of Yolk on Cleavage The morphology of cleavage differs considerably among the vertebrates. One of the chief factors accounting for the differences is the amount of yolk contained in the eggs. This section will outline the major forms of cleavage in vertebrate embryos as they relate to the amount and distribution of yolk.

An ovum in which the yolk is relatively meager in amount and uniformly distributed in the cytoplasm is called *oligolecithal* (*isolecithal*). An oligolecithal egg undergoes a type of division which is essentially an unmodified mitosis, with the two daughter cells equal in size but smaller than the parent cell. This type of cleavage is called *equal holoblastic cleavage* (Fig. 4-2), and it is seen both in very primitive chordates (*Amphioxus*) and in mammals.

The eggs of some species (e.g., amphibia) contain a sufficient concentration of yolk at the vegetal pole to crowd the nucleus and active cytoplasm toward the opposite (animal) pole. Such eggs, with asymmetrically distributed yolk, are called *telolecithal*. The unequal distribution of yolk in telolecithal eggs has two principal effects. First, it displaces the nucleus toward the animal pole. As has already been discussed, the cleavage furrow forms first in that part of the cell

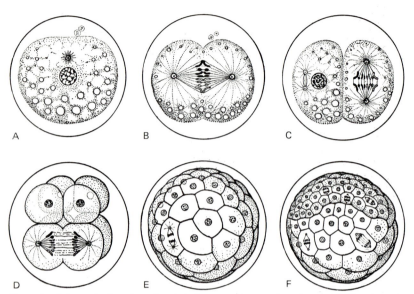

Fig. 4-2 Schematic diagrams showing the general sequence of events in the cleavage of an egg having but a small amount of yolk in its cytoplasm. (After William Patten.)

cortex closest to the mitotic spindle. Thus, in telolecithal eggs the cleavage furrow first appears at the animal pole (Fig. 4-3). The second major effect of yolk is to retard cleavage. When the major mass of yolk is encountered, extension of the cleavage furrow slows down noticeably, so much so that succeeding cell divisions begin at the animal pole before the first cleavage is completed at the vegetal pole. Because of this delay in cell divisions, which leaves some of them incomplete for a time, we call the process *meroblastic cleavage*. When the yolk thus impedes division at the vegetal pole, the cells of this hemisphere remain much larger than those at the animal pole (Fig. 4-3).

The relatively enormous mass of yolk in the eggs of birds and some reptiles produces a profound effect upon cleavage. Cleavage in bird eggs begins as it does in amphibia, but the mass of inert yolk material is so great that the yolk is not divided. The process of segmentation is limited to a small disk of protoplasm lying at the surface of the yolk at the animal pole. This extreme form of meroblastic cleavage is known as *discoidal cleavage* (Fig. 4-6). The cells formed during cleavage are known as *blastomeres*, whether they are completely separated, as in the case of holoblastic cleavage, or only partially separated, as in meroblastic cleavage.

Cleavage and Formation of the Blastula in Amphibians   The period of cleavage and blastula formation in amphibians is rapid, usually being completed within 24 hours. The first cleavage division begins at the animal pole and bisects the gray crescent (Figs. 4-1 and 4-3). In the axolotl, the cleavage furrow elongates at a rate of about 1 mm/minute in the animal hemisphere, but slows to

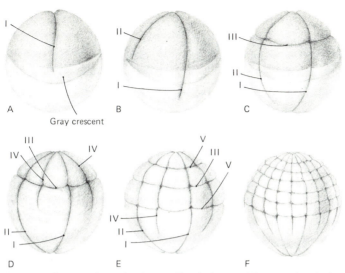

Gray crescent

Fig. 4-3 Cleavage in the frog's egg. The darkness of the upper hemisphere is due to the presence of pigment in the cytoplasmic cap, while the lighter appearance of the lower hemisphere is caused by the massing of yolk in that part of the egg. The cleavage furrows are designated by Roman numerals indicating the order of their appearance. In (A) and (B) note the retarding effect of the yolk on the extension of the cleavage furrows toward the vegetative pole. In (C) observe the displacement of the third cleavage furrow away from the yolk-laden pole, toward the center of the mass of active cytoplasm. The displacement of the center of activity from the geometrical center of the egg and the mechanical retardation of cleavage at the vegetative pole–both due to the yolk mass–result in the formation of a morula with many small blastomeres in the animal hemisphere and fewer large blastomeres in the vegetative hemisphere.

0.02 to 0.03 mm/minute as it nears the vegetal pole (Hara, 1977). The second division also begins at the animal pole, but its plane is perpendicular to that of the first cleavage plane. The third cleavage plane is horizontal and passes nearer to the animal pole, dividing the embryo into four smaller blastomeres (*micromeres*) at the animal hemisphere and four larger blastomeres (*macromeres*) at the vegetal pole. Successive cleavage divisions follow one another rapidly and synchronously. In amphibians, an embryo between the 16- and 64-cell stages is commonly called a *morula*. In subsequent cleavage cycles the waves of cleavage begin to lose their synchrony because of a lengthening of the cell cycle in the cells of the vegetative hemisphere. By the fifteenth cleavage cycle in the axolotl, the cleavage at the vegetal pole compared with cleavage at the animal pole is delayed about two cycles (Hara, 1977).

After the morula stage, a prominent cavity (*blastocoele*) appears in the animal hemisphere above the mass of yolk (Fig. 4-4B). Formation of the blastocoele in vertebrate embryos has long been of interest to embryologists, but only recently has there been any real understanding of the mechanisms involved. Kalt (1971) has traced the formation of the blastocoele in *Xenopus* back to the first cleavage division. A specialization of the cleavage furrow at the

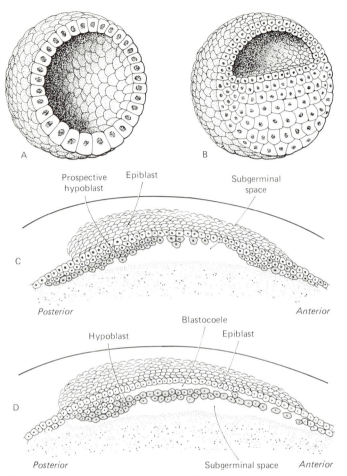

Fig. 4-4 Schematic sections comparing the blastulae of (A) *Amphioxus*, (B) amphibian, and (C, D) the chick.

animal hemisphere leaves a small intercellular cavity which becomes sealed off from the exterior by specialized close junctions between the two blastomeres. This small cavity is preserved and enlarged during subsequent cleavage cycles.

When the embryo has completed the cleavage cycle that takes it from the 64-cell morula to the 128-cell stage, it is commonly called a *blastula*. It remains in the blastula stage until successive cleavage cycles have increased the cell number to between 10,000 and 15,000 blastomeres. At this time the massive morphogenetic movements characterizing gastrulation begin. The amphibian blastula can be conveniently subdivided into three main regions: (1) A region around the animal pole, roughly including the cells forming the roof of the blastocoele. These cells correspond roughly to the future ectodermal germ layer. (2) A region around the vegetal pole, including the large cells in the interior which constitute the yolk mass. These are the future endodermal cells. (3) A marginal ring of surface cells in the subequatorial region of the embryo, including the region of the gray crescent. Cells of this zone normally form the embryonic mesoderm.

For many years pregastrulation in the amphibian embryo was considered to be a relatively unexciting period, with the main activity being the increase in number of cells. It is now known, however, that major organizational changes are taking place during the late cleavage period. Nieuwkoop (1973) has recently performed recombination experiments in which cells from the animal half of the embryo above the blastocoele were directly apposed to the yolk mass from the vegetative hemisphere. Under a presumed inductive influence by the yolk mass, the cells from the animal hemisphere formed mesodermal structures. Normally the mesoderm comes from a zone of surface cells located closer to the yolk mass (Fig. 4-5). Nieuwkoop has postulated that one of the functions of the blastocoele may be to restrict the interaction between future endodermal and ectodermal cells to the marginal ring surrounding the edges of the blastocoele.

The gray crescent, which was so prominent in the newly fertilized egg, can still be seen during early cleavage. Curtis (1962) showed that the gray crescent from the eight-cell stage of *Xenopus* embryos retains its ability to organize a secondary embryonic axis when grafted into an uncleaved egg (Fig. 3-30E). Yet, when the gray crescent is removed from an eight-cell embryo, further development continues normally, indicating that the control of dorsoventral polarity has passed from the localized gray crescent to the embryo itself (Fig. 3-30D). That

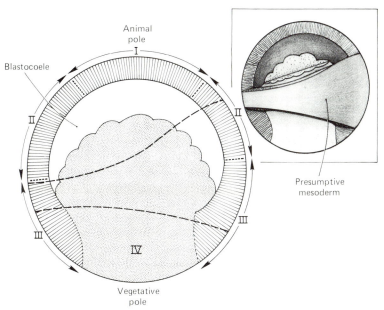

Fig. 4-5 Subdivision of the blastula of the urodelean amphibian into formative zones. Zones I and II from the animal hemisphere form ectoderm upon isolation. Zone IV, from the vegetative pole, contains yolk-laden endodermal cells. Cells from the ring-shaped subequatorial zone (III) form mesodermal cells upon isolation. (Modified from P. Nieuwkoop, 1973, *Adv. Morphogen.* **10**:1.)

this fixation of polarity supersedes the influence of the gray crescent was shown by an experiment conducted by Curtis, in which grafts of gray crescent from uncleaved eggs into eight-cell embryos failed to induce a secondary embryonic axis.

One factor which facilitates communication among the cells of cleaving embryos is the widespread distribution of low-resistance intercellular junctions which favor the passage of electrical currents and small molecules from one cell to another (Bennett, 1973). The specific relationship between cell coupling and the phenomena demonstrated by classical experimental techniques of embryology has not yet been established.

Cleavage and Formation of the Blastula in Birds   Cleavage in bird eggs has received less attention than cleavage in amphibian eggs because the entire period of cleavage occurs as the egg is passing down the oviduct. By the time the egg is laid, early gastrulation has already begun.

The newly fertilized egg, which is about to undergo cleavage, contains a whitish disk (*blastodisk* or *germinal disk*) of active protoplasm, about 3 mm in diameter, at the animal pole. Around this disk is a darker-appearing marginal area known as the *periblast*, but there is no distinct line of demarcation between the two.

The first cleavage furrow begins to appear near the center of the blastodisk during late anaphase of the first mitotic division following fertilization. As in amphibian embryos the cleavage furrow lies in the plane of the chromosomal plate during metaphase, and microfilaments are found at the base of the cleavage furrow. The sequence of avian cleavage is not always regular nor, after about the third cleavage division, is it synchronous. Nevertheless, the mitotic spindles align themselves such that the subsequent cleavage furrow forms at right angles to the preceding one (Fig. 4-6). The fourth cleavage furrow is a circumferential one which cuts off a central row from a peripheral row of blastomeres.

The blastomeres formed by the first few cleavage divisions are unusual in having their top and sides bounded by plasma membranes but their basal surfaces open to the underlying yolk. Further cleavage in the early disk of embryonic cells, now called a *blastoderm*, results in the radial extension of the embryo toward the periblast region. Nuclei are rarely seen in the open cells at the expanding periphery of the blastoderm, and it has been suggested that the presence of "accessory sperm nuclei"[1] during early cleavage may trigger off cytoplasmic division in this region (Bellairs et al., 1978).

In addition to surface cleavages, the 32-cell embryo shows cleavage planes of an entirely different character. These cleavages appear below the surface and

---

[1]Even though just a single spermatozoon takes part in fertilization in birds, other spermatozoa sometimes become lodged in the cytoplasm of the blastodisk. The nuclei of these spermatozoa migrate to the periphery, where they are recognizable for some time. These nuclei soon degenerate, and whether or not they effect any permanent changes with respect to the cleavage pattern remains unanswered.

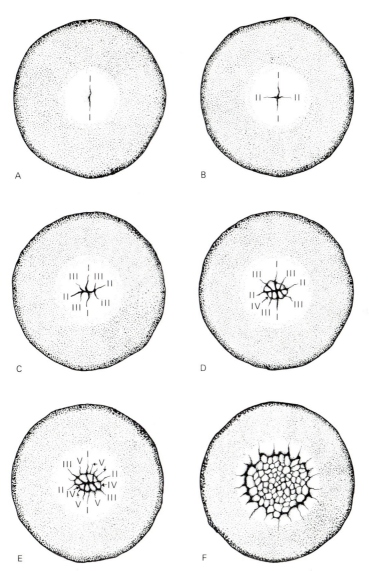

Fig. 4-6 Surface aspect of blastoderm of bird's egg at various stages of cleavage. The blastoderm and the immediately surrounding yolk are viewed directly from the animal pole, the shell and albumen having been removed. The order in which the cleavage furrows have appeared is indicated on the diagrams by Roman numerals. (A) First cleavage. (B) Second cleavage. (C) Third cleavage. (D) Fourth cleavage. (E) Fifth cleavage. (F) Early morula. (Based on Blount's photomicrographs of the pigeon's egg.)

parallel to it. They establish a superficial layer of nucleated cells which are completely delimited by plasma membranes. These superficial cells rest upon a layer of cells which on their deep faces are continuous with the yolk. Continuous divisions of the same type eventually establish several strata of superficial cells.

The division progresses centrifugally as the blastoderm increases in size but does not extend to its extreme margin. The peripheral margin, where the blastoderm abuts the periblast, remains a single cell in thickness, and the cells there lie unseparated from the yolk. By the time the embryo contains about 100 cells, the blastoderm is underlain by a *subgerminal cavity* (Fig. 4-4C).

After a number of cleavage cycles, the shedding of individual cells begins from the lower surface of the posterior region of the blastoderm and spreads toward the anterior end. The central portion of the blastoderm, thinned out by the shedding of cells and underlain by the subgerminal cavity, is called the *area pellucida*. Surrounding the area pellucida is the *area opaca*, a region where the cells of the blastoderm still abut directly onto the yolk.

At about the time of laying, individual cells, or aggregates of cells, shed from the lower surface of the blastoderm by a process of *delamination,* coalesce to form a thin disk-like layer called the *primary hypoblast*[2] (Fig. 4-4D). This process occurs first, and to a greater extent, at the posterior end of the embryo. In addition, cells moving in from the margins of the blastoderm may play a role in filling out the primary hypoblast. The primary hypoblast is separated from the outer layer of the blastoderm, called the *epiblast*, by a thin cavity, the *blastocoele*. In the primary hypoblast, which ultimately forms extraembryonic endoderm, is an inherent polarity, which it confers to the embryo proper, represented by the early epiblast. The polarity and location of the primary hypoblast determine the location and direction of the future primitive streak (Fig. 5-10) in the embryos, probably by some form of inductive interaction.

The two-layered blastoderm of the bird has been compared with a flattened amphibian blastula. The epiblast is considered the equivalent of the animal hemisphere, and the primary hypoblast shares many common properties with the vegetal hemisphere of the amphibian embryo, particularly an intrinsic polarity and an ability to induce the formation of mesoderm from the early epiblast. The epiblast, on the other hand, remains competent to react to the influence of the hypoblast by forming the mesoderm (*mesoblast*).

Cleavage and Formation of the Blastula in Mammals Despite their presumed evolutionary origin from species that laid eggs with large amounts of yolk, mammals produce extremely small eggs with almost no yolk. Freed from the encumbrance of yolk, the mammalian egg has reverted to the simple type of cleavage seen in many primitive forms. Early cleavage divisions are practically unmodified mitoses—or in more technical terms, *equal holoblastic cleavage of an isolecithal egg*. Only later in development, during gastrulation, does the pattern of morphogenetic movements of the mammalian embryo reveal a persistence of traits characteristic of large-yolked embryos.

Historically, cleavage was most thoroughly studied in pig embryos, which could be obtained from slaughterhouses. With recent improvements in laborato-

---

[2]In contemporary descriptive and experimental literature, the terms *epiblast* and *hypoblast* are almost universally used to designate the inner and outer layers of the blastoderm of birds. This convention will be followed here in the designation of the layers in early bird embryos.

ry methods, particularly in vitro techniques, early development in rodents has received much attention. The uniformity of distribution of the yolk gives no clear-cut distribution of an animal or vegetal pole in the mammalian zygote, although traditionally the site at which the polar bodies are given off has been considered to mark the animal pole. In fact, there is evidence that in some species (e.g., mouse) the definitive axis may not be determined until the time of implantation, when the embryo is in the blastula stage.

Cleavage in mammals is much slower than it is in most other vertebrates. Commonly, the first cleavage division is not completed for 24 hours, and subsequent early cleavage divisions take about 12 hours each. The point at which the polar bodies are given off at least establishes a point of reference, and the plane of the first cleavage division includes the polar bodies (Fig. 4-7A). A second mitotic spindle is formed in each of the first two blastomeres shortly after the first cleavage division is finished. In the rabbit and other mammals, the mitotic spindle of one of the blastomeres rotates 90° during the second cleavage division (Gulyas, 1975). This results in a crosswise arrangement of the

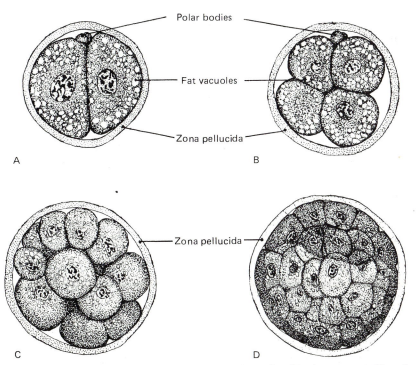

Fig. 4-7 Cleavage of the pig ovum. (A) 2-cell stage in section. Specimens secured from the oviduct of a sow killed 2 days, 3½ hours after copulation. (B) 4-cell stage in section. Probable age about 2½ days. (C) A morula of about 16 cells. Drawn from unsectioned specimen, probably age about 3½ days. (D) Blastula stage, drawn from an unsectioned specimen secured from the uterus of a sow killed 4¾ days after copulation. Note the lighter central area indicating the beginning of the formation of the segmentation cavity (blastocoele) by cell rearrangement (cf Fig. 4-10). (A, B, and D, drawn from preparations loaned by Streeter and Heuser; C, after Assheton. All figures ×400.)

blastomere at the four-cell stage (Fig. 4-8). Cleavage is more asynchronous in mammals than in amphibians, and as early as the third cleavage cycle the blastomeres do not follow an identical course. In vitro studies using early monkey embryos, flushed from the uterine tubes of the mother (Lewis and Hartman, 1933), and more recent observations on human embryos fertilized in vitro have shown that cleavage in primates follows the general mammalian pattern (Fig. 4-9).

After several cleavage divisions have taken place, the resultant group of blastomeres constitutes a compact ball of cells, still enclosed within the *zona pellucida*. The embryo is then said to be in the *morula stage* (Fig. 4-7C). Although the term *cleavage* is not ordinarily applied to cell divisions which occur after the morula stage, active cell division continues with unabated rapidity.

The transition from morula to blastula in mammalian embryos is accomplished very rapidly. In addition to the continuing increase in the number of blastomeres, the transition is marked by two main changes. One is the formation of a large fluid-filled cavity, the *blastocoele*, within the center of the embryo (Fig. 4-10). The second is the rearrangement of the cells into two structures: a thin layer surrounding the blastocoele and a smaller, more compact mass which projects into the blastocoele.

The transition from cleavage stages to the blastula stage varies with respect to time and number of cells among mammalian species, but the essential elements in this transition appear to be quite similar. The first change is the flattening of the outer layer of cells, forming a covering (*trophoblast*) of the entire embryo complex. As the cells of the trophoblast flatten, they also develop tight junctions around their intercellular borders, greatly reducing their permeability to the free passage of liquids (Ducibella and Anderson, 1975).

The other major process involved in the formation of the cavity of the

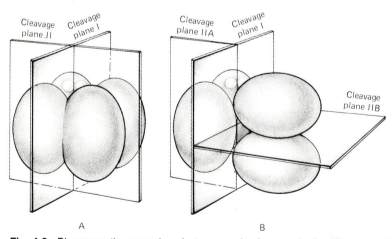

Fig. 4-8 Diagrammatic comparison between early cleavage in the (A) sea urchin and the (B) rabbit. (Modified from B. J. Gulyas, 1975, *J. exp. Zool.* **193**:235.)

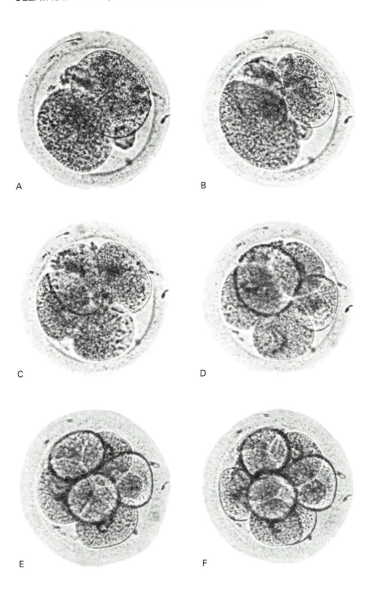

Fig. 4-9 Photomicrographs (×300) of living monkey ovum showing its cleavage divisions. The fertilized ovum was washed out of the tube, cultivated in plasma, and its growth changes recorded as a microcinematograph. The illustrations are enlargements from single frames of the film. (A) Two-cell stage, about 29½ hours after ovulation. (B) Three-cell stage, about 36½ hours after ovulation. (C) Four-cell stage, about 37½ hours after ovulation. (D) Five-cell stage, about 48½ hours after ovulation. (E) Six-cell stage, about 49 hours after ovulation. (F) Eight-cell stage, about 50 hours after ovulation. (A seven-cell stage lasting but 3 minutes intervened between the six and eight-cell stages.) (After Lewis and Hartman, 1933, *Carnegie Cont. to Emb.*, vol. 24.)

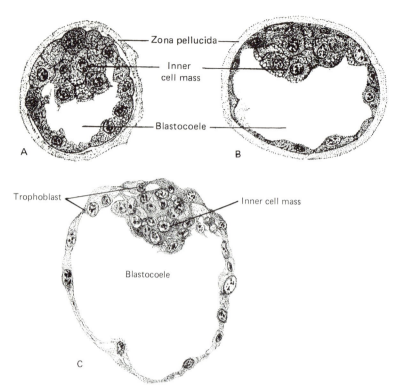

Fig. 4-10 Three stages of the blastodermic vesicle (blastocyst) of the pig, drawn from sections to show the formation of the inner cell mass. (A) Removed from uterus of sow 4¾ days after copulation (cf. Fig. 4-7D). (B) Copulation age, 6 days, 1¾ hours. (C) Copulation age, 6 days, 20 hours. (A, B, from embryos in the Carnegie Collection; C, after Corner, all, ×375.)

blastocyst, as the mammalian blastula is called, is the secretion of fluid into the blastocoele. The fluid is secreted from cytoplasmic vesicles which accumulate during late cleavage stages. Studies on the composition of the fluid of the blastocoele show concentration gradients of many inorganic ions, an indication of active transport processes (Borland et al., 1977). In addition, *uteroglobin* and other uterine proteins have been found within the blastocyst fluid of rabbit embryos (Beier and Maurer, 1975). The early mammalian blastocyst remains enclosed within the zona pellucida, but the overall size of the embryo increases to some extent because of the accumulation of fluid.

With the formation of the blastocyst, it is readily apparent that the embryo now consists of two distinct populations of cells. The outer, trophoblastic cells have assumed the configuration and many of the properties of epithelial cells. Functionally the cells of the trophoblast can both pump fluid and induce special changes in the uterine lining upon implantation. On the inner surface of the trophoblast is a small group of cells, called the *inner cell mass* (Fig. 4-10). These cells are joined to one another by communicating *gap junctions* and they retain the ability to reaggregate if separated or mixed with cells of other embryos. Cells of the inner cell mass can neither pump fluid nor evoke the decidual reaction, as can the cells of the trophoblastic layer. They are destined to form the embryo plus some of the membranes associated with it, whereas the cells of the trophoblast form a large part of the placenta (Fig. 5-22).

The results of experimental investigations have led a number of embryologists to conclude that the position of a blastomere in the morula determines whether it will become part of the trophoblast or the inner cell mass. According to the "inside-outside" hypothesis (Tarkowski and Wróblewska, 1967), cells of the morula, which have no contact with the exterior, develop in a unique microenvironment created by the external cells. These cells develop into the inner cell mass, whereas the cells located at the surface of the morula, presumably because of physiological functions required by their superficial position, are channeled into becoming trophoblastic epithelium.

Organization of the Embryo during Cleavage and Blastulation During cleavage and blastulation the embryo is simple in structure, and alterations in its cell arrangement are straightforward. Yet many properties of the embryo and its constituent cells change during this period. These changes are not, for the most part, detectable by observation alone. In order to learn about the organization of early embryos, the investigator must be able to ask incisive questions and carry out delicate manipulations.

One of the fundamental characteristics of most vertebrate embryos is their regulative properties. Viewed in its broadest sense, a survey of regulative ability can include the developmental potential (potency) of individual cells, the ability of damaged embryos or parts of embryos to produce entire individuals, and the ability of embryos to integrate into a normal structure additional tissue or cells.

In contrast to the early determination known to occur during cleavage in some invertebrates, the blastomeres of young vertebrate embryos are characterized by an astonishing amount of plasticity in their developmental fate. Some of the most incisive experimental work demonstrating this point has dealt with the properties of nuclei rather than of entire cells. In an early attempt to determine whether or not the nuclei of blastomeres retain the potentiality of guiding the entire course of development or whether their capacities become restricted, Spemann (1928) conducted the following experiment. He constricted a fertilized amphibian egg with a hair loop so that one lobe of the egg contained the nucleus and some cytoplasm and the other contained only cytoplasm. When the nucleated portion had developed to the 16-cell stage, he loosened the ligature and allowed a nucleus to move over into the lobe of nonnucleated cytoplasm (Fig. 4-11). The secondarily nucleated lobe of cytoplasm then went on to develop, thus demonstrating that the nucleus of a 16-cell embryo had still not lost its capacity to direct embryonic development from the stage of the single cell. More recent work (Briggs and King, 1952) involving the transplantation of nuclei taken from blastomeres during cleavage into enucleated eggs has

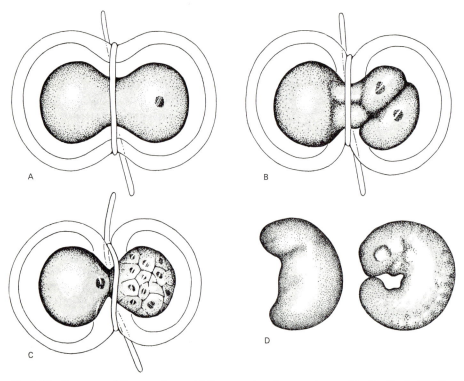

Fig. 4-11 Summary of Spemann's constriction experiment on newt eggs. Shortly after fertilization the egg is constricted with a hair loop so that one segment of cytoplasm contains a nucleus and the other does not (A). During early cleavage, the nucleated segment divides (B). If, after several cell divisions a nucleus is allowed to migrate to the nonnucleated segment of cytoplasm (c), this half of the constricted egg also begins to form an embryo which is essentially normal (D, *left*) but less mature than the embryo arising from the originally nucleated segment of the constricted egg (D, *right*). (Parts B-D after Spemann, 1938, *Embryonic Development and Induction,* Yale University Press, New Haven.)

confirmed this point. The furthest extension of this approach has been the work of Gurdon (1974), who has obtained viable frogs from nuclei of adult frogs transplanted into enucleated eggs. (Fig. 1-17).

The capacity of isolated entire blastomeres to continue development and to regulate into entire individuals has been demonstrated in several ways. Identical twins can result from the regulation and normal development of blastomeres that have become separated during early cleavage. Blastomeres isolated from mammalian embryos show a steadily decreasing ability to form complete individuals from the two-cell to the eight-cell stage.

At a different level of organization, pieces of many vertebrate embryos are able to reconstitute entire individuals by regulation. Spratt and Haas (1964) cut chick blastoderms, containing as many as 30,000 cells, almost in half. Each half underwent regulation and formed a complete chick embryo (Fig. 4-12). Once the mammalian blastocyst has become subdivided into the trophoblast and inner cell mass, these two regions cease to be equivalent, but partial or complete separation of portions of the inner cell mass results in twinning (Fig. 1-12). The nine-banded armadillo regularly produces quadruplets from a single early embryo. At the blastocyst stage, the inner cell mass forms four buds, each of which becomes organized into a separate embryo (Patterson, 1913).

A valuable experimental model for both embryological and genetic studies makes use of another regulative property of mammalian embryos. Both Tarkowski and Mintz have produced tetraparental (*allophenic*) mice by removing the zona pellucida from cleaving embryos and then combining the two embryos. The cells become reorganized into a single sphere, which is then transferred to the uterus of a foster mother. The rest of pregnancy proceeds normally, and mosaic embryos, normally formed, but containing the genetic input of four parents, are produced (Fig. 4-13). In this case the regulative properties of the embryos allowed the harmonious integration of two potentially separate individuals into one embryo. More recently, a hexaparental mouse has been produced by the fusion of three embryos derived from three different sets of parents (Fig. 4-14).

At the 64-cell stage, the inner cell mass of a mouse blastocyst contains about 15 cells. Statistical studies, based upon the frequency of appearance of mosaic characteristics in tetraparental and hexaparental mice, have provided strong indications that the body of the mouse arises from as few as three cells of the inner cell mass. These inferences are supported by other experiments in which chimeric mice have been produced by injecting single cells from one embryo into the blastocyst of another. The injected cells often fuse with the inner cell mass and become integrated into the body of the host embryo. The use of techniques such as those mentioned above is providing an immense amount of new information on the organization of early mammalian embryos (McLaren, 1976).

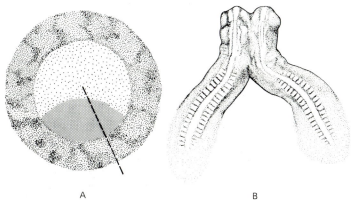

A   B

Fig. 4-12 Regulation in the early chick blastoderm. If the posterior part of the blastoderm is cut (A) and does not heal, partially conjoined twin embryos form (B). (Drawn from Spratt and Haas, 1967, *J. Exp. Zool.* **164**:31.)

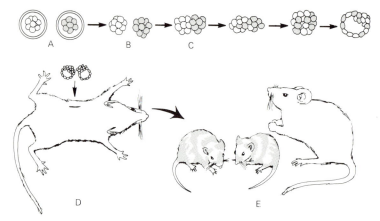

Fig. 4-13 Diagram of the experimental procedure for producing tetraparental (allophenic) mice. Two cleavage stage embryos are obtained (A). After removal of the zona pellucida (B), the embryos are allowed to fuse (C). The fused embryos are implanted into a foster mother (D) who later gives birth to the chimeric offspring (E). (Modified from B. Mintz, 1962, *Proc. Nat. Acad. Sci.* **58**:334.)

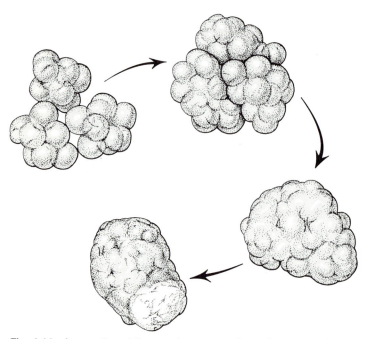

Fig. 4-14 Aggregation of three early mouse embryos from which the zona pellucida has been removed. After the embryos had fused and blastocoele had formed (*lower left*), the embryo was transferred into a foster mother. This procedure produces an embryo with genetic input from six parents. (Adapted from C. L. Markert and R. M. Peters, 1978, *Science* **202**:56.)

Another unseen, but important, aspect of the organization of embryos during cleavage and the blastula stage is their polarity. In amphibian embryos, dorsoventral polarity is first determined by the site of sperm penetration and the location of the gray crescent in the cortex of the egg. Control of polarity, however, does not reside permanently in the cortex, for it is soon shifted to the vegetative yolk mass and subsequently to the presumptive mesoderm of the animal part of the embryo before finally becoming anatomically fixed with the induction of the neural plate (Nieuwkoop, 1977). Some similarity to the control of polarity by the yolk in amphibian embryos is seen in bird embryos, with the early control of polarity residing in the primary hypoblast. Experimental reorientation of the hypoblast in relation to the overlying epiblast results in an orientation of the embryo that corresponds to the position of the hypoblast. In contrast to amphibians and birds, the mammalian embryo shows no signs of polarity (except for the position of the polar body) until well into the stage of the blastocyst. The circumstances and mechanisms of fixation of polarity in mammalian embryos have still received relatively little attention.

Molecular Activities during Cleavage and Blastulation At the time of fertilization the eggs of amphibians are endowed with a plentiful supply of yolk as well as a large amount of RNA and some proteins, which were synthesized during oögenesis in anticipation of the needs of the embryo during cleavage. In contrast, the eggs of mammals contain almost no yolk and relatively little preformed RNA and protein. These differences in the composition of the eggs exert a profound effect upon patterns of synthesis during cleavage, and early amphibian and mammalian embryos provide a clear contrast in molecular activities during cleavage. Bird embryos have been so little studied during the cleavage period that it is not profitable to include them in this discussion.

The synthesis of chromosomal DNA is the common denominator of cleavage in all species, and because of the required paternal contribution, chromosomal DNA cannot be made and prepackaged in the developing oöcyte. Although some egg proteins are carried over for usage during cleavage, others more directly involved in mitosis and cytokinesis are synthesized as needed. The importance of both protein and DNA synthesis has been amply demonstrated by experiments in which inhibitors have been applied to early embryos. Cleavage quickly comes to a halt, indicating that active synthesis of these substances is required for normal embryonic development. A good example of the carry-over of important proteins from the egg is the cytoplasmic deficiency seen in axolotl embryos which are homozygous for the o gene. Cleavage and early blastulation are normal in o^-/o^- embryos, but thereafter development becomes progressively retarded and finally stops. If mutant eggs are injected with small amounts of cytoplasm from normal eggs, the defect can be overcome (Briggs and Justus, 1968).

The patterns of RNA synthesis are much more complex than those of DNA synthesis, particularly when embryos of different species are compared. Three

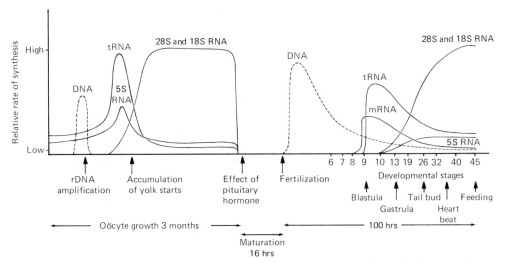

Fig. 4-15 Diagram of relative rates of nucleic acid synthesis during amphibian development. 5*s*, 18*s*, and 28*s* RNAs are forms of rRNA. The different patterns of RNA synthesis during development constitute evidence in favor of controls operating at the level of transcription. (Modified from J. B. Gurdon, 1974, *The Control of Gene Expression in Animal Development,* Harvard University Press, Cambridge.)

main classes of RNA will be dealt with here. One is *messenger RNA* (mRNA), which acts as the direct intermediary between the genetic information encoded in the DNA molecule and specific proteins, which are assembled upon mRNA molecules in concordance with this information. Associated with the mRNA are the ribosomes, which together with the mRNA form the polysomes, upon which proteins are synthesized. *Ribosomal RNA* (rRNA) is complex, consisting of three major subunits designated on the basis of their sedimentation as 5S, 18S, and 28S varieties. At the cytological level, the formation of rRNA is associated with the presence of a nucleolus. The third major class of RNA is a low-molecular-weight RNA (4S) called *transfer RNA* (tRNA). Transfer RNA attaches to and conveys amino acids to the ribosomes, where the amino acids become incorporated into protein chains.

During oögenesis, the amphibian (*Xenopus*) embryo synthesizes and stores large amounts of both tRNA and rRNA. During the pachytene stage of meiosis, the synthesis of rRNA is preceded by the amplification of the ribosomal gene sets from 2000 to roughly 2 million (Brown and Dawid, 1968). Accompanying this process is the formation of over 1000 free nucleoli within the egg.

During cleavage in amphibian embryos very little new RNA is formed. As the embryo enters the blastula stage, both mRNA and tRNA are synthesized in significant amounts (Fig. 4-15). Nucleoli are absent in the cleaving embryo, but at the period of gastrulation nucleoli reappear and the synthesis of tRNA begins again. Treatment of embryos with the antibiotic actinomycin D, an inhibitor of RNA synthesis, does not interfere with development of the embryo during cleavage or much of blastulation, but after this stage abnormalities occur and further development ceases. Similarly, *Xenopus* embryos which are genetically anucleolate develop normally through the period of gastrulation. Both the patterns of RNA synthesis and the reactions of the embryos to inhibitors of RNA synthesis or deleterious mutants allow one to conclude that in the amphibian embryo RNAs preformed in the egg are sufficient to carry the embryo through cleavage and blastulation. The major upheavals and cellular rearrangements during gastrulation, on the other hand, seem to require the combined input of both maternal and paternal genes.

In mammalian (mouse) embryos, on the other hand, RNA synthesis begins in the two-cell stage, and by the four-cell stage all major classes of RNA are being synthesized. Correlated with this is the presence of nucleoli during the cleavage period. It should be no surprise, therefore, to learn that mammalian embryos are highly susceptible to the inhibitory effects of actinomycin D during early cleavage. Although RNA synthesis begins early in mammalian embryos, the development of parthenogenetically activated and some mutant embryos up to the blastocyst stage offers some evidence that before that stage maternal genes and RNAs may suffice for many functions.

Mammalian embryos differ from amphibian embryos in their energy requirements, as well. Of necessity, the amphibian embryo must be a self-contained unit with respect to energy sources, but mammalian embryos depend upon a continuing supply of energy from their surroundings in both the uterine tubes and the uterine cavity.

What Is Accomplished by Cleavage and Blastulation?   It is easy to consider the period of cleavage as a relatively simple phase of development during which the embryo is simply increasing its number of cells. Certainly the increase in cell number is the dominant manifestation of cleavage. Without a critical number of cells, the morphogenetic movements which form the basis of gastrulation, the next major period in embryonic development, could not be accomplished in a normal fashion. Implicit in the increase in cell number is a corresponding increase in the amount of genetic material (DNA) that the embryo has at its disposal. In addition, the formation of many separate cells not only allows the segregation of different types of cytoplasm which may have been present within the egg but also allows and preserves differences in the expression of the genetic content of the embryo. This is accomplished by the formation of unique RNAs and proteins, and provides the basis for cytodifferentiation. These cellular functions are facilitated by the decrease in cell size toward a more typical nucleocytoplasmic ratio, which allows greater nuclear control over intracellular events.

At a higher level of complexity, the major embryonic axes of most embryos are definitively fixed by the late blastula stage, although the axes of mammalian embryos remain labile longer than those of most other vertebrate embryos. Lastly, during this period the cells in some embryos make some major decisions regarding their future fate. The most striking example is the mammalian embryo in which very early the cells become segregated into those that will form the embryo proper and those that will produce the extraembryonic structures. As the blastula becomes more mature, its activities are increasingly directed toward making preparations for gastrulation. These preparations are reflected not only in changing patterns of synthesis but also in surface properties and interactions of cells. In some cases they are even reflected in the movements of cells in relation to others. Thus, as with most phases of development, no sharp demarcation exists between the last blastula stages and the start of gastrulation.

Chapter 5

Gastrulation and the Establishing of the Germ Layers

Gastrulation as a Process Following the period of cleavage and the formation of a blastula, the embryo embarks upon one of the most critical periods in its development—the stage of *gastrulation* (*gastrula*, diminutive from the Greek word *gaster*, or *stomach*). Up to this time the course of development has been guided, to a considerable extent, by events which had occurred in the egg before, or at the time of, fertilization. The cells of the cleaving embryo, although increasing greatly in number, have not for the most part acquired any specific properties or geographical positions which are unique or irreplaceable. During the period of blastulation the genetic reservoirs of the cells are activated by mechanisms, still poorly known, to direct the synthesis of substantial amounts of new RNA and proteins. As was pointed out in the previous chapter, interference with these synthetic processes usually brings embryonic development to a disorganized halt before gastrulation can occur.

Gastrulation is characterized by profound but well-ordered rearrangements of the cells in the embryo. One of the major changes which occurs during early gastrulation is the acquisition by the cells of the capacity for undergoing directed *morphogenetic movements*. The details of morphogenetic movements vary considerably among the classes of vertebrates, but they result in a profound reorganization of the embryo. Whereas gastrulation itself encompasses the rearrangement of the embryo from the blastula to a stage characterized by at least two germ layers, morphogenetic movements continue for a time thereafter. One of the major consequences of this reorganization is that groups of cells, which previously may have been far removed from one another, are brought close enough together to undergo the inductive interactions which are involved in the establishment of the major organ systems.

Vertebrate embryos have adopted two principal strategies for dealing with gastrulation. The first is to carry out the gastrulation movements within the context of a sphere. This mode of gastrulation is seen in the embryos of very primitive vertebrates such as *Amphioxus*, which contain little yolk, and the amphibians, which contain moderate amounts of yolk. In *Amphioxus*, gastrulation essentially consists of an inpocketing of the blastula, forming a double-layered cup from a single-layered hollow sphere in much the same way that a hollow rubber ball can be pushed in with one's thumb (Fig. 5-1A to 5-1D). The new cavity in the double-walled cup is termed the *gastrocoele* or *archenteron*. The opening from the outside into the gastrocoele has traditionally been called the *blastopore*. Thus in gastrulation the original single layer of the blastula has been rearranged to form two layers. The outer cell layer is known as the *ectoderm*. The inner layer of the early gastrula is composed primarily of cells belonging to the *endoderm*, but its upper surface also includes cells destined to form the middle germ layer (*mesoderm*).

The second principal way in which vertebrate embryos handle gastrulation is dictated by the great amount of yolk found in the large eggs of reptiles and

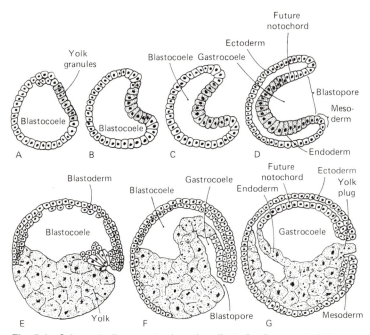

Fig. 5-1 Schematic diagrams to show the effect of yolk on gastrulation.

birds. In embryos of this type the sheer bulk of the inert yolk mass precludes the simple inpocketing mechanism used by *Amphioxus*. Instead, gastrulation occurs as the elaboration of the three germ layers as two-dimensional sheets upon one sector of an enormous sphere of completely passive yolk. Interestingly, mammalian embryos betray their origin from phylogenetic ancestors which laid highly yolky eggs by retaining the type of gastrulation movements common to birds and reptiles.

Gastrulation in Amphibian Embryos   As was just pointed out, gastrulation in the amphibian embryo is modified to a certain extent by the presence of the large yolk-laden cells in the vegetal hemisphere of the early embryos. Nevertheless, some essential features of early gastrulation in amphibians resemble quite closely those of forms having isolecithal eggs. The first external evidence of gastrulation is the appearance of a slightly curved groove (Fig. 5-2A). This groove, the blastopore, represents the site at which cells invaginate into the interior of the embryo. The appearance of the blastoporal groove is the external manifestation of a process which involves, also, changes taking place within the interior of the embryo. At the very start of gastrulation, certain cells in the region of the future blastopore exhibit a remarkable change of shape. While retaining tight connections with the external surface of the embryo, these cells undertake a pronounced inward elongation until their shape resembles an attenuated flask (Fig. 5-3). It is currently believed that this change of shape is associated with an inward-pulling movement and that by their continued

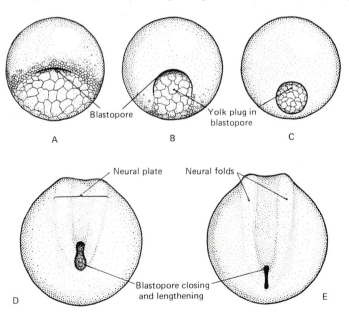

Fig. 5-2 Caudal views of amphibian embryos, showing the changing configuration of the blastopore. (Five stages selected and modified from Huettner, *Fundamentals of Comparative Embryology of the Vertebrates*. By permission of The Macmillan Company, New York.)

Bottle cells

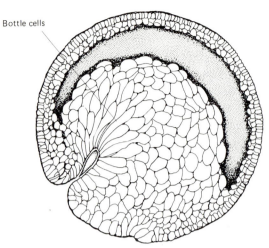

Fig. 5-3 Slightly schematized section through an advanced amphibian gastrula, showing the flask-shaped "bottle cells" moving into the interior of the embryo at the blastopore. (After Holtfreter, 1943, *J. Exp. Zool.* **94**:261–318.)

attachments to the surface layer, the concerted effort of the group of "bottle cells," as they are often called, actually pulls in cells of the surface layer to initiate the formation of the blastoporal groove.

The shape of the amphibian blastopore demonstrates clearly the effect of the yolk upon gastrulation. The inpocketing of cells during gastrulation is originally restricted to a region located dorsally to the yolk (Figs. 5-2A and 5-6A), and the upper margin of the groove is known as the *dorsal lip of the blastopore*. As the process gains momentum, the ingrowing margins or lips of the blastopore are gradually extended so they assume a circular shape as they turn inward around the yolk (Fig. 5-2B and C). The mass of yolk left presenting at the blastopore is known as the *yolk plug*. Finally, differential growth and migration crowd its lateral lips together and the closing blastopore becomes streaklike (Fig. 5-2D and E). Much of the process of gastrulation consists of surface cells moving into the interior of the embryo at the blastopore. As cells turn inward around the lips of the blastopore, they are followed by other cells which move over the surface of the embryo toward the blastopore. Most of the section on amphibian gastrulation will be devoted to emphasizing the fact that even in the early gastrula, certain groups of cells in the surface layer have begun to show evidence of commitment to their future fate, so that by the use of appropriate tracing methods they can be assigned not only to definite germ layers, but also to specific organ primordia within these layers.

In Chap. 1 we explained how movements of cells or groups of cells in embryos can be followed by the use of appropriate cell markers, such as dyes or radioisotopic tracers. Historically, one of the first uses of tracing methods was designed to determine the ultimate fate of cells in the surface layer of gastrulating amphibian embryos. Most of these methods involved the use of vital dyes (Vogt, 1929), and the findings of the early investigators have withstood the test of time remarkably well. The general principle of the tracing of cell movements during amphibian gastrulation is illustrated in Fig. 5-4.

Suppose, first, we follow the movements of a series of cells marked when they lie at the external surface of the embryo, peripheral to the blastopore, in locations such as those indicated by the black dots in Fig. 5-4A. Such marked cells can be seen to move toward the margins of the blastopore in the directions indicated by the solid arrows. When they reach the lips of the blastopore, they turn inward and begin to move away from the blastopore (dashed arrows) as constituents of the newly established inner layers. Figure 5-5 shows the extensive displacement during gastrulation of cells marked at an earlier stage of development.

Marking experiments of this type also give us the key to the elongation of the blastopore as it is closing. The inturning of cells at the upper margin of the blastopore, in other words, at its "dorsal lip," is particularly active, as indicated by the distance they travel in a given time. This is expressed graphically in Fig. 5-4B by the greater length of the dotted arrows near the dorsal midline. It is, of course, more or less figurative, and definitely oversimplifies what actually happens, but we can think of the most dorsal part of the blastopore margin as being pulled on by the very active forward growth occurring there. Concomitantly the superficial cells from either side appear to converge just a little more rapidly than the turned-in cells move away, so that the lateral margins of the blastopore tend to crowd toward the midline.

By sectioning embryos at various periods after specifically located cell groups have been marked, we can see more clearly what happens to cells from the different surface areas when they are carried to interior locations in the process of gastrulation. The sagittal section diagrams of Fig. 5-6 have been conventionally shaded to aid in following some of these movements. Gastrulation movements are not uniform among the major amphibian groups. In the following paragraphs we shall describe the morphogenetic movements which take place in the tailed amphibians (urodeles). The process is quite similar in the

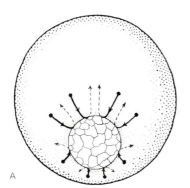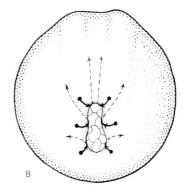

A   B

Fig. 5-4 Schematic diagrams indicating by arrows the way cells originally located in the primordial outer layer move first toward the blastopore (*solid arrows*) and then turn inward at the blastopore lips and move centrifugally (*dashed arrows*) in the newly established internal layer.

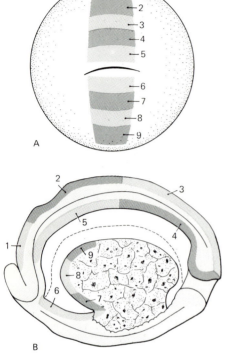

Fig. 5-5 Diagram illustrating the principle of using vital dyes as tracers in studying the displacement of cells during gastrulation in the amphibian embryo. (A) Small areas (1–9) along the midline of a blastula are marked with two different dyes. (B) Morphogenetic movements during gastrulation displace the stained areas to different regions of the embryo.

tailless amphibians (anurans), but some details, such as the endodermal participation in the early formation of the primitive gut (archenteron), differ between the two groups.

Around most of the ventral margins of the blastopore and extending down onto the ventral part of the embryo is a region of lightly stippled cells (Fig. 5-6A). This is spoken of as *prospective endoderm* because, if it is followed in the later stages, it can be seen to be rolled into the interior of the embryo and eventually comes to line its gastrocoele or primitive gut (Fig. 5-6C). Most of the cells of the dorsal lip of the blastopore are represented by small crosses. These cells as a group are often called *chordamesoderm* because they eventually give rise to the notochord and cephalic mesoderm (Fig. 5-6C). As gastrulation begins, both the chordamesoderm and the prospective endoderm move inward, and it is some of these cells which assume the characteristic bottle shape associated with early invagination (Fig. 5-3).

If we refer again to Fig. 5-6A and B, we can see that when invagination has just begun, the early gastrocoele is lined by chordamesoderm on its dorsal surface and elsewhere by endoderm. The overall dorsal displacement of the entire early gastrocoele by the yolk-laden endoderm is quite evident. As invagination continues, the gastrocoele increases in size and extends beneath the outer layer of cells toward the cephalic end of the embryo. This increase in size is made possible both by the continued inward movement of additional cells

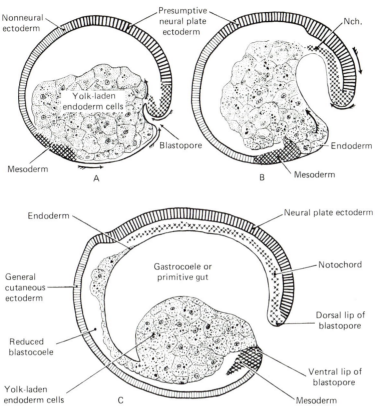

Fig. 5-6 Diagrams indicating the cell rearrangements which occur in amphibian gastrulation. (Schematic sagittal sections based primarily on the work of Vogt, 1929, *Arch. F. Entwickl-mech. d. Organ.,* vol. 120.) *Abbreviation: Nch., notochord.*

around the lips of the blastopore and by the proliferation of the cells which have already moved in. As the inward cell migrations of early gastrulation continue, the inertness of the yolk-laden endoderm finally seems to prove to be too much for the cells actively moving in around the blastopore, and some of these cells undercut ventrally the external portion of the yolk-filled endoderm (Fig. 5-6B and C). The result of this is the persistence of the externally visible yolk plug, which is surrounded on all sides by invaginating cells. In urodele embryos, part of the dorsal surface of the gastrocoele remains lined by chordamesodermal material until late in gastrulation. Finally the invaginated endodermal cells spread beneath it to form a complete endodermal lining to the primitive gut.

If we now review these early positional changes of cells destined to become internally located in the embryo, it is evident that they carry out two quite different moves in relation to the rest of the embryo. Take, for example, the most dorsally located cells in the prospective chordamesoderm (Fig. 5-6A). First they move towards the blastopore, traveling as a part of the outer cell layer. When cells move thus over the spheroidal surface of a growing embryo, the process is spoken of as *epiboly*. At the same time that some of the prospective cells are moving toward the blastopore by epiboly, other cells that have already arrived there are rolling in around the lip of the blastopore and moving away as part of a newly established internal structure. This process is called *involution*. It becomes obvious as soon as we begin to follow marked cells that gastrulation involves a certain amount of intucking of cells in its very early phases—which comes within our usual meaning when we use the term *invagination*. It is equally obvious that there is a very extensive amount of epiboly involved. This brings to the margin of the blastopore cells that are destined to roll around the blastopore lip by involution on their way to taking their places in an internal layer.

The origin of the mesoderm cannot be followed satisfactorily in sagittal sections alone, although that is the logical starting point. In sagittal sections the prospective lateral mesoderm is at first located ventrally, quite a distance from the lips of the blastopore (Fig. 5-6A). As the prospective endoderm moves along the outer surface towards the blastopore and then is carried in by involution, the prospective mesoderm follows it toward the blastopore until it, too, turns in (Fig. 5-6B and C). What is not apparent from these diagrams based on sagittal sections is that there is really a ring of prospective mesoderm girdling the embryo (Fig. 5-7A). The prospective mesodermal cells are carried back to the lips of the blastopore, where they turn in and spread anteriorly between the involuting endoderm and the outermost cell layer of the embryo (Fig. 5-8). Ultimately the leading edges of the mesoderm from either side of the embryo meet at the ventral midline and later form the heart.

Up to this point we have avoided calling the outer layer of the amphibian embryo *ectoderm* whenever possible. The reason is that until the gastrulation movements have been completed the outer layer of the embryo contains cells which later in time will be located within the mesodermal or endodermal germ layers. Sometimes, however, for practical reasons the name *ectoderm* is applied to this layer before the last of the cells which contribute to the other germ layers

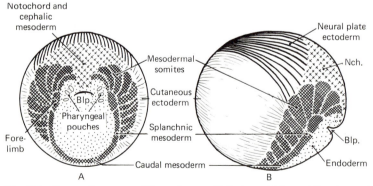

Fig. 5-7 Prospective areas of the embryos of tailed amphibians at the stage when gastrulation is just beginning. (A) Caudal aspect; (B) lateral aspect. (Modified from Vogt, *Arch. f. Entwickl.-mech. d. Organ.*, 1929, vol. 120.) *Abbreviations: Blp.*, blastopore; *Nch.*, notochord.

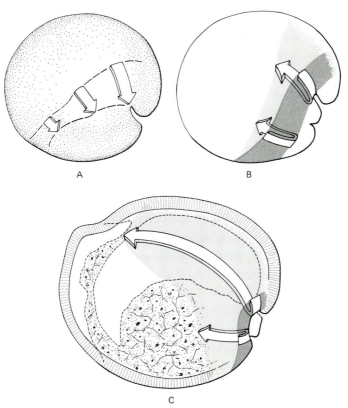

Fig. 5-8   Diagrams showing the spread of mesoderm in embryos of tailed amphibia. The arrows indicate (A) the migrations of future mesodermal cells toward the blastopore, (B) the involution of the cells around the lips of the blastopore, and (C) away from the blastopore as a discrete layer of mesodermal cells. Note that the mesoderm is extended from the entire circumference of the lips of the blastopore, but that its growth from the dorsal lip is most vigorous. (Based on Vogt, 1929, *Arch. f. Entwickl.-mech. d. Organ.*, vol. 120.)

have left it. A more exact definition would be that which remains as an external layer after the endoderm, notochord, and the rest of the mesoderm have been carried to their interior location in the process of gastrulation.

As is the case with the other germ layers, the prospective ectoderm does not remain static during gastrulation. Fate maps of early stages indicate that it can be divided into areas with different futures. One region is destined to take part in the formation of the central nervous system and is therefore designated as *neural ectoderm* (Figs. 5-6A, and 5-7B). The remainder of the ectoderm will form the epidermis of the skin, together with the specialized skin derivatives, and is, therefore, called *general cutaneous ectoderm*. By the end of gastrulation the entire outer surface of the embryo is covered by neural and general cutaneous ectoderm, whereas the future endoderm and mesoderm have come to lie entirely within (Fig. 5-6C). The limited extent of some of these prospective areas is striking; for example, note the prospective general cutaneous ectoderm as shown in Fig. 5-6A. This area must be expanded to cover the surface areas vacated by the involution of prospective endoderm and mesoderm (cf. Fig. 5-6A to 5-6C). Therefore the ectoderm must increase its area with the growing size of the embryo it clothes.

Experimental studies have shown that during gastrulation the cells of the three germ layers have developed properties characteristic of each layer and that much of the process of gastrulation as well as the stability of the arrangements of the resulting germ layers are due to these properties. Both ectodermal and endodermal cells have acquired the propensity of spreading out into sheets adjacent to the mesoderm, which is now interposed between these two germ layers. The ectoderm, which has spread out over the entire outer surface of the embryo after the involution of the endoderm and mesoderm, seems by its spreading tendency literally to hold in the endodermal lining of the primitive gut. If the ectoderm is removed, the endoderm takes advantage of its absence and tends to spread outward over the outside of the mesoderm in what is almost a reversal of its normal tendency to remain in the interior, lining the gut (Holtfreter, 1944).

The massive tissue displacements resulting from the morphogenetic movements which occur during gastrulation represent just a part of the total developmental activity during this stage. Unseen to the eye, other processes are occurring which are vital for the future normal development of the embryo. In one of the most illuminating experiments in the history of embryology, Spemann and Mangold (1924) demonstrated that the cells comprising the dorsal lip of the blastopore (Fig. 5-6C) possess the ability to "organize" much of the future development of the embryo. They removed a section of the dorsal lip from an early gastrula and implanted it into the ventral portion of the blastocoele of another amphibian embryo. As development of the host embryo progressed, a secondary embryo, quite complete in most respects, was formed in Siamese-twin fashion along its ventral surface (Fig. 5-9). The secondary embryo was composed partly of cells derived from the graft and partly from cells of the host. As a result of its ability to direct normal development in an orderly fashion, Spemann called the dorsal lip of the blastopore the *organizer*. A good share of the organizing activity of the cells originating in the dorsal lip of the blastopore consists of a series of inductions (often called *primary inductions*) in which the chordamesoderm-containing roof of the gastrocoele stimulates the overlying ectoderm to form neural tissue. This and other processes we have been considering serve as examples emphasizing the amazing migrations, regroupings, and cell proliferations which occur during gastrulation. All this intense activity results in the formation of an embryo which is fully prepared to begin the next major phase of its development, the laying down of organ systems.

Gastrulation in Birds   Gastrulation in birds is morphologically complex, and over the years it has been subject to a number of quite divergent interpretations. Compounding the inherent complexity of the process has been

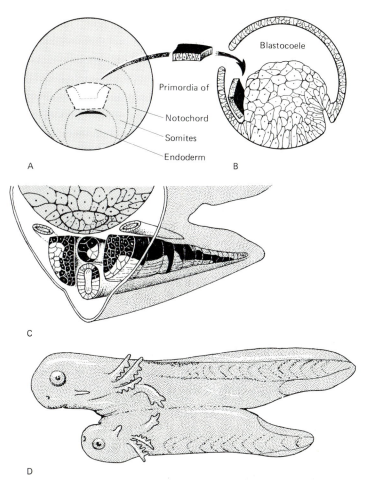

Blastocoele

Primordia of

Notochord

Somites

Endoderm

A   B

C

D

Fig. 5-9 Transplantation of the upper blastoporal lip from one amphibian gastrula into another (A, B). This procedure produces a secondary embryo (D) composed (C) of self-differentiated tissues of the graft (*black*) and induced tissues of the host (*white*). (From Holtfreter and Hamburger in Willier, Weiss and Hamburger, 1955, *Analysis of Development,* W. B. Saunders Company, Philadelphia.)

the relative inaccessibility of early developmental stages in the chick, for almost all of the pregastrulation stages occur before the egg is laid.

We have already traced the establishment of the blastula in the chick embryo as a two-layered structure, consisting of an upper layer (the epiblast) and a lower layer (the primary hypoblast), with a thin blastocoele in between (Fig. 4-4). The embryo proper occupies the transparent area pellucida and is surrounded by the area opaca, where the cells of the blastoderm lie unseparated from the yolk (Fig. A-4C).

Next, a thin sickle-shaped mass of cells (*Koller's sickle*) takes shape at the posterior end of the embryo. From Koller's sickle, a second generation of hypoblastic cells (*secondary hypoblast*) pushes anteriorly, compressing and folding the primary hypoblast ahead of it (Fig. 5-13). The exact mode of formation of the secondary hypoblast is still uncertain. Neither the primary nor the secondary hypoblast seems to form any of the embryonic germ layers. Current evidence suggests that the primordial germ cells are formed in the primary hypoblast. Their crescentic distribution along the anterior border of the blastoderm (Fig. 3-1) may be due to compression of the primary hypoblast by the expanding secondary hypoblast. The secondary hypoblast forms extra-embryonic endoderm, principally the yolk stalk.

Formation of the primary and secondary hypoblast can be considered pregastrulation phenomena. Gastrulation and formation of the definitive embryonic germ layers begin with the appearance of a condensation of cells in the posterior part of the epiblast. This condensation, seen in an embryo which has been incubated for 3 to 4 hours (Fig. 5-10A), gradually assumes a cephalocaudal elongation (Fig. 5-10B). By the seventh or eighth hour of incubation the elongation is still more definite (Fig. 5-10C), and by the end of the first half day the thickened area has assumed a shape which has led to its being called the *primitive streak* (Fig. 5-10D).

The appearance of the primitive streak is the result of an inductive interaction of the epiblast with the hypoblastic layer (Eyal-Giladi and Wolk,

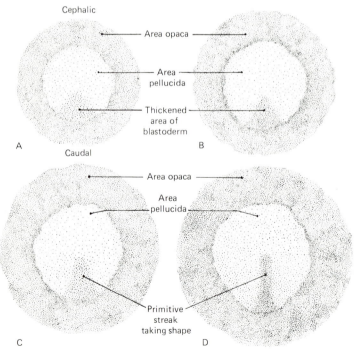

Fig. 5-10 Chick embryos showing four stages in the formation of the primitive streak. (A) 3 to 4 hours' incubation. (B) 5 to 6 hours' incubation. (C) 7 to 8 hours' incubation. (D) 10 to 12 hours' incubation. (Based in part on the photomicrographs of Spratt, 1946, *J. Exp. Zool.,* vol. 103.)

1970), and its orientation is a reflection of the intrinsic polarity of the underlying hypoblastic layer. The latter was dramatically demonstrated by Waddington (1933), who altered the orientation of the primitive streak by changing the position of the hypoblast with respect to the epiblast.

The early primitive streak initially elongates in both a cephalic and caudal direction. The carbon-marking experiments of Spratt (1946) have been especially instructive in demonstrating the processes involved in primitive-streak formation. He placed spots of carbon particles on chick blastoderms just as they were about to form the primitive streak. These marking experiments showed that throughout much of the posterior part of the blastoderm cell movements converged from the lateral areas toward the forming primitive streak (Fig. 5-11). As more cells enter the streak region, the primitive streak elongates in a cephalic direction. The cephalic extension of the primitive streak keeps pace with the

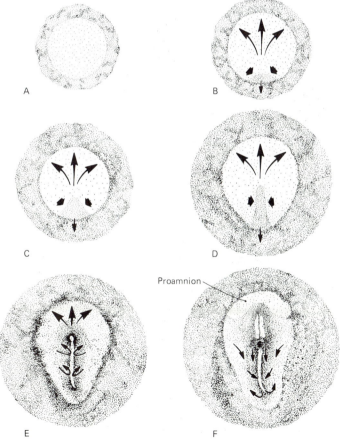

Fig. 5-11 Diagrams illustrating the patterns of cell movements involved in the formation and elongation of the primitive streak (A–D), invagination of cells through the streak (E), and regression of the streak (F) in the chick. (Based on the data of Spratt and Haas.)

expansion of the secondary hypoblast beneath it. Caudal extension moves the primitive streak into the area opaca.

After 16 hours of incubation the primitive streak has become so prominent that embryos are characterized as being in the primitive-streak stage (Fig. 5-12). A central furrow, called the *primitive groove*, now runs down the center of the primitive streak. Along either side it is flanked by thickened margins, called the *primitive ridges* (Fig. 5-12). At the cephalic end of the primitive streak closely packed cells form a local thickening known as *Hensen's node* (Fig. 5-12).[1] After the primitive streak has reached its full length at about the eighteenth hour of incubation, the cephalic end begins to regress, leaving in its wake a structure commonly referred to as the *head process*. This is a gross morphological term referring to the area where the notochord has been recently laid down (Fig. A-3).

The part of the area pellucida adjacent to the primitive streak begins to thicken and is said to constitute the embryonal area (Fig. A-3). Because of its shape, the embryonal area is frequently spoken of as the *embryonic shield*. Accompanying the formation and elongation of the primitive streak, the area pellucida undergoes a change in shape from an essentially circular disk to a pear-shaped configuration. The long axis of the future embryonic body is clearly established by the primitive streak.

With the establishment of the primitive streak and Hensen's node, the main

[1]In the experimental embryological literature the term *Hensen's node* generally implies a somewhat larger area than that defined as the node in most descriptive texts. Hensen's original description of the node in rabbit embryos described it as the slightly enlarged anterior end of the primitive streak. Recognition of varying definitions of the node is important in interpreting the experimental literature.

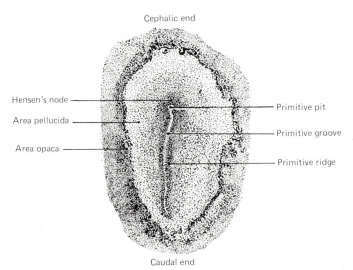

Fig. 5-12   Dorsal view ($\times$14) of entire chick embryo in the primitive-streak stage (about 16 hours of incubation).

period of gastrulation begins. The embryonic germ layers are formed by the migration of cells in the epiblast toward Hensen's node and the primitive streak, and their invagination to form the middle and lower germ layers (the embryonic mesoderm and endoderm[2]). The anterior portion of the primitive streak and the node serve as a passageway for cells even while the streak is elongating anteriorly. In birds, gastrulation is accomplished by the coordinated passage of individual cells from the exterior into the interior of the embryo rather than the immigration of integral sheets of cells.

The first cells to pass through the area of the anterior part of the primitive streak are future embryonic endodermal cells. After about 8 to 10 hours of incubation more than 80 percent of these cells will be found in the endoderm. The remainder will migrate into the middle mesodermal layer. As time goes on, a progressively greater percentage of the cells which pass through the node are destined to be incorporated into the mesoderm and a correspondingly smaller number will lodge in the endoderm. The endodermal cells which are formed in this manner enter the original hypoblastic layer and steadily displace the cells of the hypoblast outward and cephalad toward the edge of the area opaca (Fig. 5-13). Although the bulk of the endoderm has passed through the nodal region during the early, formative stages of development of the primitive streak, increasing numbers of future endodermal cells migrate through the anterior part of the primitive streak, as well. By about 22 hours of incubation, when regression of the primitive streak has commenced, essentially all of the future endodermal cells have left the epiblast.

Little formative activity of the middle germ layer (embryonic mesoderm) occurs until around the fifteenth hour of incubation, when the primitive groove becomes well established within the primitive streak (Fig. 5-12). There are two principal areas of invagination and mesoderm formation in the early chick embryo. The most extensive invagination of mesodermal cells occurs along the length of the primitive streak, where a coherent layer of mesodermal cells is formed which expands parallel to the underlying layer of the embryonic endoderm (Figs. 5-15 and 5-16). The spread of the mesoderm is shown in Fig. 5-14. The other major site of mesoderm formation is through Hensen's node, where a rod of mesodermal cells, directed cephalad, lies in the midline of the embryo in the wake of the regressing primitive streak. This mesodermal rod becomes the notochord (Fig. 5-16), which is essential for the next series of major changes that sweep over the embryo.

As the cells of the epiblast migrate toward and through the primitive streak and ultimately take their place among the other cells of the mesodermal layer, they undergo some characteristic changes in form. The epiblast is composed of cuboidal to columnar epithelial cells. As in any typical epithelium, the apical surfaces are tightly bound to one another by specialized intercellular junctions, one of which is known as a *zonula occludens*. It now appears that these

---

[2]For consistency with the designations of the germ layers in amphibians and mammals, the definitive germ layers resulting from gastrulation will be designated as *ectoderm, mesoderm,* and *endoderm.* Some authors continue to call these layers *epiblast, mesoblast,* and *hypoblast.*

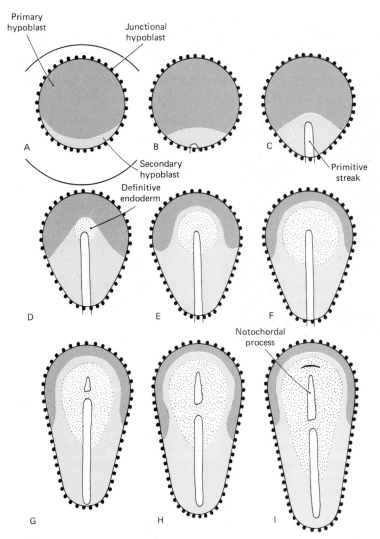

Fig. 5-13 Successive stages in the formation of the lower layer in early chick embryos. (After L. Vakaet, 1970, *Arch. Biol.* **81**:387.)

junctions, which encircle the entire cell apex, act as sealing devices to preserve differences in the environment between the inside and outside of the epithelial layer. In addition, the deeper parts of the cells are bound together by spotlike junctions, known as *gap junctions*. It is now widely assumed that gap junctions are somehow involved in cell-to-cell communication.

When the cells of the epiblast enter the primitive groove, they undergo a pronounced change of shape, becoming to some extent bottle-shaped in a manner reminiscent of the bottle cells seen in amphibian gastrulation (Fig. 5-17). The change in shape of these cells is associated with the appearance of

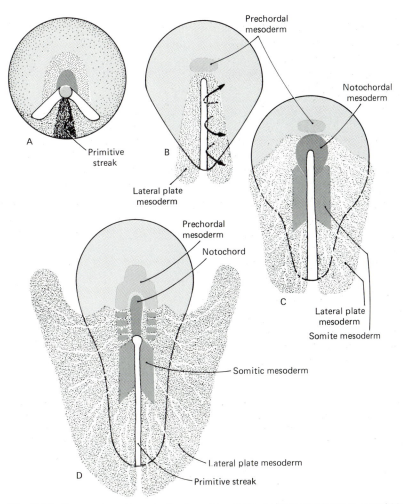

Fig. 5-14 Successive stages in the formation of the mesodermal layer in early chick embryos. (Adapted from Pasteels and Vakaet.)

orderly arrays of intracellular microtubules, which are associated with changes of shape in many varieties of cells (Fig. 5-18). During this change of shape, the zonulae occludentes begin to break up and lose their circumferential arrangement at the apex of each cell. After they have passed through the primitive streak, the cells of the mesoderm assume the stellate appearance characteristic of mesenchyme. These cells are connected to one another by small gap junctions.

Many of the cells which invaginate into the middle layer of the embryo via Hensen's node migrate cephalad as a median rodlike condensation, which will ultimately become the notochord. Soon after the first prospective notochordal cells are laid down, the primitive streak and Hensen's node undergo a regression

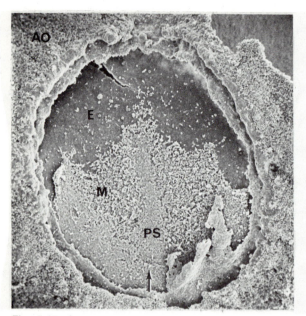

Fig. 5-15 Scanning electron micrograph of the ventral surface of an early chick embryo (Hamburger-Hamilton stage 4). The endoderm has been removed, providing a good view of the mesoderm as it extends outward from the primitive streak. ×52. The arrow points from the posterior toward the anterior end of the embryo. (From M. A. England and J. Wakely, 1977, *Anat. Embryol.* **150**:291. Courtesy of the authors and publisher.) Abbreviations: *AO,* area opaca; *E,* ectoderm; *M,* mesoderm; *PS,* primitive streak.

toward the caudal end of the embryo. Accompanying regression of the primitive streak is a corresponding elongation of the notochord (Fig. A-5). The notochord at this stage is often called the *head process*. Further morphological details of the development of chick embryos are given in the Appendix.

Prospective Areas in Chicks at the Primitive-Streak Stage   It has long been known that specific areas of early embryos will typically contribute to the formation of characteristic tissues and organs in the adult. As a result of this knowledge embryologists have been able to construct *fate maps* for embryos of some of the more intensely studied species. A fate map of the early amphibian gastrula has already been illustrated (Fig. 5-7).

Two properties of embryonic cells or groups of cells are important to embryologists who attempt to study the organization of early embryos in relation to later stages of development. One property is the *prospective significance (prospective fate)*, which can be defined as the fate of a cell or a group of cells during the course of normal development. The other property is the *prospective potency*, defined as the types of differentiation of which a cell or group of cells is capable at a given stage of development. Typically the prospective potency of a group of cells is greater than the prospective significance, particularly at early developmental stages. As developmental restriction sets in, the prospective potency decreases until at the time of final

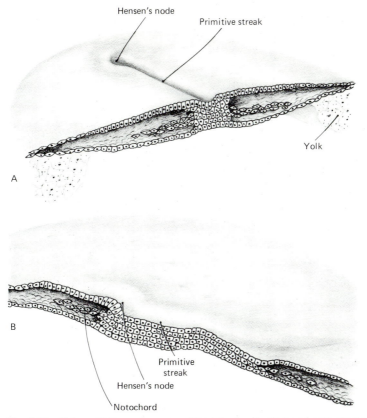

Fig. 5-16 Schematic three-dimensional models of early chick embryos, illustrating the migration of mesodermal and endodermal cells through the primitive streak. (A) transverse section, showing anterior half of the embryo. (B) Sagittal section, showing right half of the embryo. (Part A after Balinsky.)

determination the prospective potency and prospective significance are the same. Some cells always retain the potency to undergo alternate pathways of differentiation (i.e., *metaplasia*).

Several techniques have been successfully used in the construction of fate maps for embryos. One is to mark certain cells with vital dyes and to follow the stained cells as long as possible during development. This method has been very successful in the mapping of early amphibian embryos. In bird embryos the marking of cells with carbon particles was very useful in early studies (Spratt, 1946). Later, experiments involving the transplantation of radioactively labeled pieces of early embryos into equivalent locations in unlabeled hosts (Rosenquist, 1966) has allowed further refining of the early fate maps of the chick embryo. The technique of interspecific marking, using cells of the quail embryo as markers, has added still more information about the future fates of cells in early chick embryos.

Another mapping technique consists of explanting small pieces of embryos

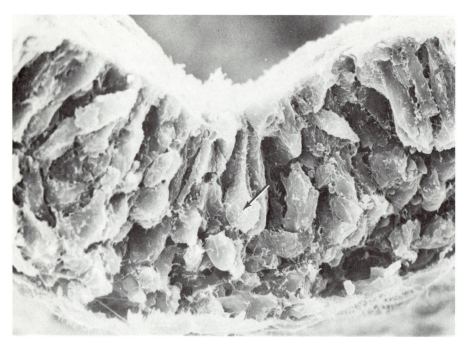

Fig. 5-17 Scanning electron micrograph through the primitive streak of a cross-fractured chick embryo (Hamburger-Hamilton stage 8). Cells from the epiblast entering the region of the primitive groove become flask-shaped (*arrow*) as they prepare to move into the interior of the embryo. ×2635. (From Solursh and Revel, 1978, *Differentiation* **11**:185. Courtesy of the authors.)

as grafts onto the chorioallantoic membrane or into the coelomic cavity. Care must be taken in the interpretation of explantation experiments, for under these conditions the cells may differentiate according to their prospective potency rather than their normal prospective fate. If an area where a particular potency (e.g., the liver) has been located is explored in more detail, it is found that there is a central part from which practically all the explants exhibit the potency in question. Explants taken farther peripherally show the potency in decreasing percentages of the grafts made. The territory where a specific potency is regularly manifested is said to be the *prospective center* for the organ in question.

Two classical maps of prospective areas for chicks of the primitive-streak stage are reproduced as Fig. 5-19 and 5-20. It should be emphasized that the sharpness of the boundaries between different prospective areas as shown in such figures is an entirely artificial device for vivid graphic presentation. In reality there are vague transition zones rather than anything like the sharp delimitations of the schematic diagram.

Comparison of Bird and Amphibian Development   There are a number of parallels between the early development of birds and that of amphibians. At the blastula stage, which in the bird can be represented by a two-layered structure

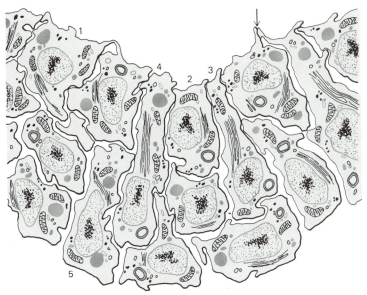

Fig. 5-18 Drawing showing the configuration of cells in the avian primitive streak. Cells labeled 1–4 show the sequence of elongation from cuboidal to bottle-shaped cells. Cell 5 has withdrawn from the free surface (*top*) of the epiblast. The apical surfaces of the epithelial cells are joined by specialized tight junctions (*arrow*). (Adapted from N. Granholm and J. R. Baker, 1970, *Devel. Biol* **23**:563.)

containing the epiblast and hypoblast, the cells that will later form the endodermal and mesodermal germ layers are located in the surface layer of the embryo. Likewise, the surface layer of the amphibian embryo at a similar stage contains cells destined to become part of the endoderm and mesoderm. There are interesting parallels, as well, between the induction of the mesoderm and control of polarity by the vegetative yolk mass in amphibians and the effects of the hypoblast on the epiblast in early chick embryos.

It also seems reasonable to regard the pre-primitive-streak thickened area of the chick blastoderm as the symbolic homologue of a blastopore that could not open because of the impeding effect of the enormous yolk mass. The movement of surface cells toward the primitive streak in the chick is suggestive of the way cells move by epiboly toward the amphibian blastopore. In both forms, cells which will comprise the future endodermal layer migrate from the exterior to the interior of the embryo first, and the inward movement of the mesodermal cells follows later. Although the early gastrula in birds is compressed, because of its disposition on top of the yolk, there is a similarity in both the size and the fate maps of amphibian and avian embryos during this period (Fig. 5-21).

If we follow the process of gastrulation in avian embryos into its later phases when the notochord and mesoderm are being established, the homology of the primitive streak with the fused lips of the blastopore becomes even more apparent. Shortly after the primitive streak has formed and the endoderm has

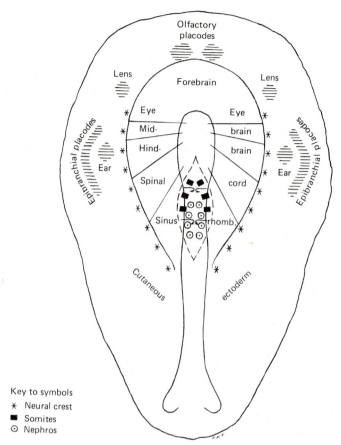

Fig. 5-19 Map of the prospective areas of the outer layer of the chick embryo in the primitive-streak stage. (Modified from Rudnick, 1944, *Quart. Rev. Biol.,* vol. 19.) *Abbreviation: rhomb., rhomboid.*

been well established, cells of the chick embryo begin to push in from the region of Hensen's node to form the rodlike notochord in the midline beneath the ectoderm (Fig. 5-16). The area where the chick notochord is formed clearly corresponds with the dorsal lip of the blastopore where the amphibian notochord arises (Fig. 5-6). Sections taken across the primitive streak caudal to Hensen's node show the chick mesoderm extending out on either side between the ectoderm and endoderm (Fig. 5-16). Those relationships are, again, very similar to those seen in amphibian embryos, with the mesoderm arising from cells turning in at the lips of the blastopore and extending between ectoderm and endoderm (Figs. 5-6C and 5-8C).

Origin of the Germ Layers in Mammals   Although the mammalian embryo contains almost no yolk, nevertheless, the morphogenetic movements and tissue displacements are remarkably similar to those which occur in birds. The inner

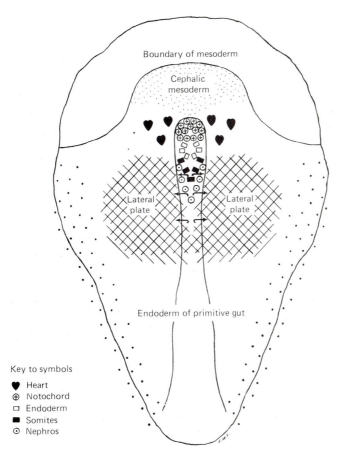

Fig. 5-20   Map of the prospective areas of the invaginated layers of the chick in the primitive-streak stage. (Modified from Rudnick, 1944, *Quart Rev. Biol.*, vol. 19.)

cell mass can be compared with the cap of blastomeres situated on the animal pole of the yolk sphere in large-yolked forms such as the chick. In view of their phylogenetic relationships, it is not surprising that the origin of the germ layers in mammals resembles the process occurring in our more immediate large-yolked ancestors rather than the simple infolding type of gastrulation seen in the small-yolked eggs of amphibians. Although the eggs of higher mammals have lost their endowment of yolk, the proliferating cells of the inner cell mass still behave as if they were crowded in on top of a larger yolk-sphere and had restricted space in which to maneuver. As a result, mammals show more of a subtle emergence of their germ layers by cell migration and regrouping than the segregation of an endodermal layer by the primitive, infolding type of gastrulation.

Recent studies of early mammalian development have indicated an origin of intra- and extraembryonic germ layers which differs in several respects from that which has been commonly presented in embryology textbooks. A scheme

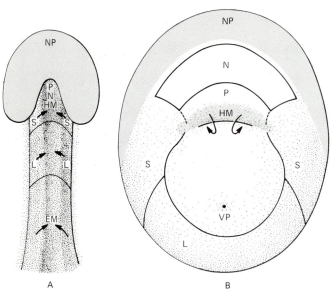

Fig. 5-21 Comparison of presumptive areas in the vicinity of the primitive streak in the chick (A) and the dorsal lip of the blastopore (*arrows*) in *Ambystoma* (B). Abbreviations: *EM*, extraembryonic mesoderm; *HM*, head mesoderm; *L*, lateral plate mesoderm; *N*, notochord; *NP*, neural plate; *P*, prechordal plate; *S*, somites; *VP*, vegetal pole. (Adapted from Nicolet, 1971, *Adv. Morphogen.* 9:231.)

summarizing the origins of intra- and extraembryonic tissues of primate embryos is illustrated in Fig. 5-22. This scheme will be useful in understanding the material presented both in this section and in Chap. 8.

Chapter 4 described how cells of the mammalian blastocyst become segregated into an embryo-forming inner cell mass, surrounded by a layer of trophoblastic cells (Fig. 5-23). The first cells to segregate out from the inner cell mass form a thin layer called the *hypoblast* (Fig. 5-24A and B). This layer forms only extraembryonic endoderm, and it is regarded as equivalent to the hypoblast of chick embryos. The hypoblast contributes the cells that will line the yolk sac. As the hypoblast forms, the remainder of the inner cell mass can be called the *epiblast*. In addition to future ectodermal cells, the epiblast contains the cells that will ultimately migrate through the primitive streak and become the definitive endodermal and mesodermal germ layers of the embryo.

Formation of the primitive streak in mammalian embryos follows a pattern quite similar to that of bird embryos. Not long after the hypoblast is established, the remaining cells of the inner cell mass become more regularly arranged and are collectively called the *embryonic disk* (Fig. 5-24C). Soon the one margin of the disk becomes thickened. The thickening occurs at the part of the disk destined to become the caudal end of the embryo.

In dorsal views of an entire embryo the thickened area is crescentic in shape when it first appears, with its convexity indicating the caudal extremity of the embryonic disk and its horns spreading out over the greater part of the caudal

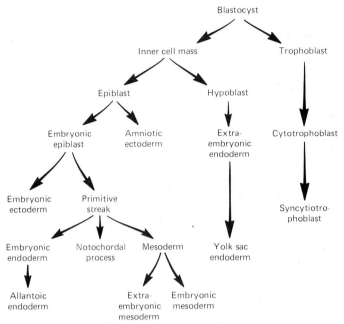

Fig. 5-22 Scheme illustrating the origin and derivatives of tissues in presomitic human and rhesus monkey embryos. (Modified from W, P. Luckett, 1978, *Am. J.. Anat.* **152**:59.)

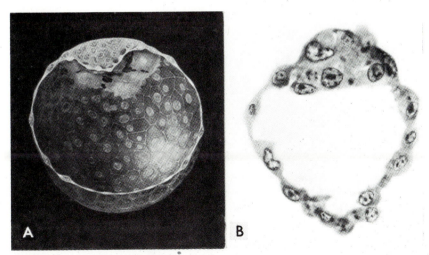

Fig. 5-23 (A) Graphic reconstruction of the blastocyst of a monkey on ninth day after fertilization. The speciman has been drawn as if opened in the midline to show the early differentiation of endoderm cells, here represented as being somewhat lighter in color than other cells of the inner cell mass. (After Streeter, 1938, *Carnegie Inst. of Washington Publication* 501.) (B) Section of the blastocyst of a human embryo estimated to be at about the fifth day after fertilization. (After Hertig, Rock, Adams, amd Mulligan, 1954, *Carnegie Cont. to Emb.,* vol. 35.)

half of the margin of the disk (Fig. 5-26A). Next the embryonic disk begins to expand asymmetrically. Its anterior margin spreads out radially in a manner indicative of a uniform rate and unspecialized direction of growth. In contrast, the thickened posterior margin of the embryonic disk expands toward a point of convergence at its caudalmost point (Fig. 5-26B). At the same time posterior elongation is occurring, this part of the embryo, including most of the future primitive streak, becomes compressed laterally. A combination of this posterior elongation of the embryonic disk itself and a concomitant anterior spreading of the cells from the thickened part of the disk (see solid arrows in Fig. 5-26B) changes the originally crescentic thickened area of the embryonic disk to an oval (Fig. 5-26C) and then pulls it out into a band lying along the long axis of the embryo. This thickened longitudinal band is called the *primitive streak* (Fig. 5-26D).

The cell movements in the region of the primitive streak of mammalian embryos have not been studied in as much detail as they have been in amphibians and birds. Little is yet known about the details of formation of the definitive endodermal layer, but indirect evidence suggests that the cells forming

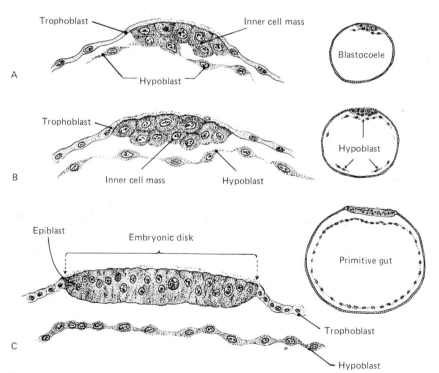

Fig. 5-24 Sections of pig blastocysts showing the first appearance and subsequent rapid extension of the endoderm. *Left,* Detailed drawings of inner cell mass (×375). *Right,* sketches of same sections, entire. The approximate ages of the embryos represented range from 7 to 8 days. (From embryos in the Carnegie Collection.)

this layer migrate through the primitive streak and become situated as the roof of the primitive gut (archenteron). It now appears that cells comprising both the embryonic and extraembryonic mesoderm pass through the posterior part of the primitive streak (Fig. 5-25B and C). Expansion of the mesoderm in relation to stages of formation of Hensen's node and the primitive streak is shown in Fig. 5-26. Much of the early mesoderm that is formed passes beyond the confines of the embryonic disk as extraembryonic mesoderm. Luckett (1978) presents considerable evidence suggesting that probably all extraembryonic mesoderm in the early mammalian embryo migrates out from the primitive streak instead of differentiating from trophoblastic cells as was once believed. Later, the embryonic mesoderm arises by the migration of cells in the epiblast towards the

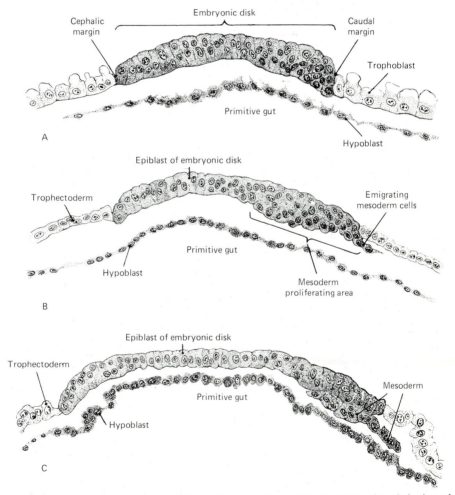

Fig. 5-25 Longitudinal sections of the embryonic disk of the pig during the ninth day of development, showing three stages in the origin of the mesoderm. (Projection drawings, ×180, from sections of embryos in the Carnegie Collection.

primitive streak. These cells then pass through the streak and spread out laterally beneath the epiblast (Figs. 5-27 and 5-28).

Except for topographical differences resulting from the absence of yolk in mammalian embryos and the very early formation of the amnion, the basic processes occurring during the primitive streak stage of mammalian and chick embryos are strikingly similar (cf. Figs. 5-14 and 5-27). It is clear that in the young mammalian embryo, as well as in the bird, the primitive streak is the homologue of the fused lips of the blastopore of lower vertebrates.

Formation of Notochord Intimately associated with the formation of the general mass of the mesoderm is the origin of an axially located cylindrical mass of cells known as the *notochord*. The notochord, both phylogenetically and ontogenetically, is of great morphological importance. In the most primitive vertebrates, it is a well-developed fibrocellular cord lying directly ventral to the central nervous system and constituting the chief axial supporting structure of the body. In sharks, ring-like cartilaginous vertebrae are formed about the notochord. Although somewhat compressed where the vertebrae encircle it, the

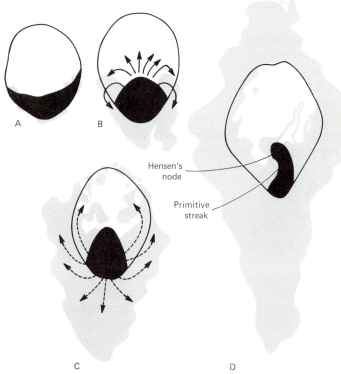

Fig. 5-26 Diagrams showing the origin and extension of the mesoderm (*gray*) in a series of pig embryos ranging from the ninth to the eleventh day of development. The solid black area in A is the thickened part of the embryonic disk from which the mesoderm first arises (cf. Fig. 5-18). Growth in this region (*see arrows in B*) gives rise to the primitive streak. The spread of mesoderm is suggested by the broken arrows in C.

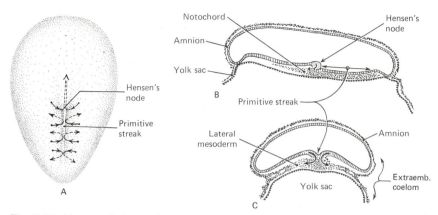

Fig. 5-27 Diagrams indicating by arrows the probable paths of cell migration in the region of the primitive streak. (A) Surface plan. (B) Longitudinal section. (C) Cross section through the primitive streak. *Abbreviation: extraemb., extraembryonic.*

notochord persists in such forms as a well-defined continuous structure extending throughout the length of the vertebral column. When, phylogenetically, cartilaginous vertebrae are replaced by more highly developed bony vertebrae, the notochord is still more compressed. But even in the higher mammals, minute canals in the centra of the vertebrae still remain to mark its existence, and the central portion of the *nucleus pulposus* of the intervertebral disk is clearly a notochordal remnant. In the early stages of development, the notochord of a mammalian embryo is a conspicuous structure, at once a record of evolutionary history and an advance indication of the location of the vertebral column.

Embryologically, the notochord in all the higher vertebrates arises in essentially the same manner. In amphibian embryos the cells which are to go into the notochord are first identifiable just cephalic to the dorsal lip of the blastopore (Fig. 5-6A). Thence they move, by epiboly, towards the blastopore lip and pass, by involution, into the interior (Fig. 5-6B and C) to form a medially located rodlike structure beneath the ectoderm. In the chick and the mammalian embryo notochord formation is modified because of the absence of an open blastopore, but there is nevertheless a basic similarity. The primitive streak, as we have seen, is the homologue of a closed blastopore, and the group of cells massed at its cephalic end as Hensen's node is homologous with the dorsal lip of the blastopore. It is through this area in birds (Fig. 5-16B) and in mammals (Fig. 5-27B) that the cells move in to take their place in the notochord. Pushing cephalad, these cells become molded into a characteristic rod-shaped mass situated medially in the growing embryo. The sheetlike masses of mesoderm spreading out from the primitive streak leave a temporarily vacant area just cephalic to Hensen's node (Fig. 5-26D). It is into this empty area that the notochord grows.

Although there has been considerable controversy concerning the germ layer to which the notochord should be assigned, most contemporary embryolo-

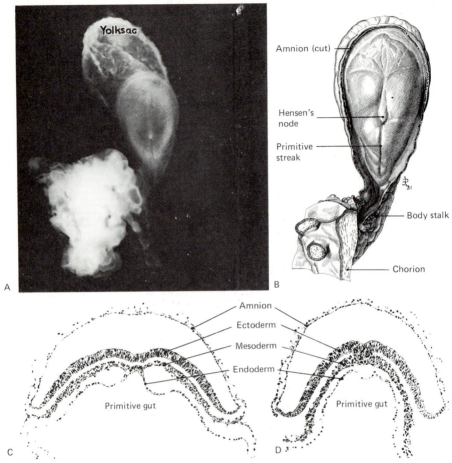

Fig. 5-28 Human embryo in primitive-streak stage—probable fertilization age of 14 to 15 days. (A) Photographed (×18) before sectioning. (B) Reconstructed from serial sections (×25). (C) Section through neural plate. (D) Section through primitive streak. (After Heuser, 1932, *Carnegie Cont. to Emb.,* vol. 23.)

gists consider it to be a component of the mesoderm. More important than its germ layer designation is the role of the notochord during normal ontogeny. The prominence of the notochord in early embryos of all vertebrates, despite its virtual disappearance in the adult body, suggests that there must be some reason for the uniform persistence of this structure in vertebrate embryos. As will be seen in the next chapter, the notochord is an initiating factor in the first major inductive event in the late gastrula. As a result of its influence the nervous system can form.

Embryological Importance of the Germ Layers The formation of three primary germ layers is a common denominator in the early development of all vertebrates. Ideas concerning the significance and importance of the germ layers have undergone gradual modifications over the years. Many early embryologists viewed the formation of germ layers as an irreversible segregation of the embryo into rigid compartments, with little interconvertibility among the germ layers. Evidence accumulated over the years, however, has shown that the differentiation into a given phenotype is not always limited to cells from a single germ layer. A good example is cartilage. Although most of the cartilage in the body is derived from cells of the mesodermal germ layer, some cartilaginous elements of the head and neck differentiate from cells of the neural crest, an ectodermal derivative.

Nevertheless the concept of germ layers is a very useful one upon which to categorize and interpret many developmental phenomena, and it provides a good framework upon which students can organize their knowledge of the formation of specific tissues and organs. A flow chart relating the differentiation of the major tissues and organs of the body to the primary germ layers is presented in Fig. 5-29. For students beginning the study of embryology, this chart will serve as a means of pointing out in a general way the direction the early processes with which we have been dealing are destined to lead. As the phenomena of development are followed further, it will be seen that each natural division of the subject centers more or less sharply upon some particular branch of this genealogical tree of the germ layers.

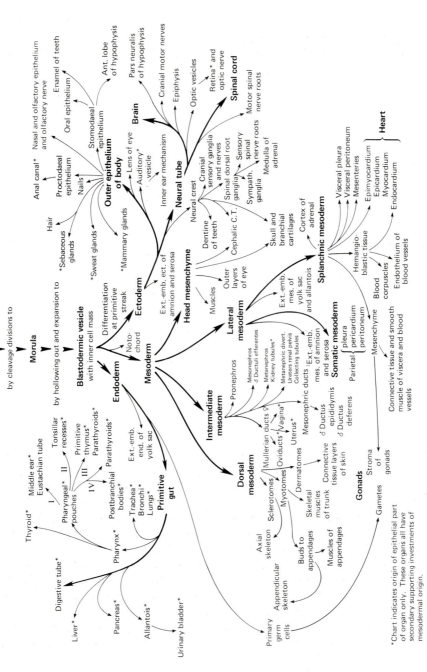

Fig. 5-29 Chart showing derivation of various parts of the body by progressive differentiation and divergent specialization. Note especially how the origin of all the organs can be traced back to the three primary germ layers.

*Chart indicates origin of epithelial part of organ only. These organs all have secondary supporting investments of mesodermal origin.

Chapter 6

Neurulation and Somite Formation

The morphogenetic movements that dominate the period of gastrulation not only result in the formation of the three primary germ layers but also cause groups of cells that were far apart in the blastula to become located close to one another. The future developmental fate of the embryo depends upon inductive interactions between some of these newly associated groups of cells. The primary inductive event is the action of the chordamesoderm or notochord upon the overlying ectoderm, resulting in the transformation of a band of unspecialized ectodermal cells into the primordium of the central nervous system. The initial response of the induced ectoderm is to form a plate of thickened cells, but soon this plate becomes transformed into a longitudinal groove and ultimately into a tube. While this is occurring, other ectodermal cells from the junction between the neural and general cutaneous ectodermal tissues form segmentally arranged aggregations, known collectively as the *neural crest*. Later, cells of the neural crest follow extensive and varied migrating and differentiation pathways throughout the body of the embryo.

Following the changes leading to the formation of the neural tube, the mesodermal layer on either side of the notochord splits into longitudinal divisions. The blocks of mesoderm on either side of the notochord soon begin to form symmetrical pairs of bricklike masses, called *somites* (Figs. 6-16 and 6-17), which are both major landmarks in the early embryo and the source of a number of important segmentally arranged mesodermal derivatives later in life. The somite pairs first take shape in the cranial part of the embryo. In successive stages additional pairs of somites are formed caudal to those already laid down. From the earliest stages of formation of the nervous system, differentiation of axial structures follows pronounced *cephalocaudal gradients*. Because of these gradients, processes which have already been completed in the cranial part of the embryo may be just beginning in the caudal part.

Commonly, neurulation is considered to be the period of development starting with the first traces of formation of the neural plate and ending with closure of the neural tube. This chapter will outline the major events involved in the development of the neural tube, the migration and differentiation of the neural crest and the development of somites.

Primary (Neural) Induction During late gastrulation, the chordamesoderm, involuting around the dorsal lip of the blastopore in amphibians, and the notochordal process, which forms from cells passing through Hensen's node in birds and mammals, push cranially, just beneath the ectoderm. While the forward movement of the notochordal tissue is taking place, the chordal cells act upon the overlying ectodermal cells, causing them to thicken and form the neural plate. This reaction, which both initiates the formation of the central nervous system and causes the central longitudinal axis of the body to be established, is commonly called *primary induction*. The inductor is the chordamesoderm (future notochordal tissue) and the responding tissue is the ectoderm. Despite decades of intensive effort, the nature of the inductive stimulus in primary induction remains poorly understood.

As in other inductive systems, it is essential that the inductor and the responding tissue be at the right place at the right time. Without the presence of the underlying notochord, the cells of the dorsal ectoderm do not form neural tissue but rather continue to differentiate as general cutaneous ectoderm. This has been demonstrated experimentally by transplanting to the ventral side of the embryo small pieces of prospective neural ectoderm before it has been acted upon by the notochord. The explants do not form neural tissue (Fig. 6-1). Yet if the same operation is performed in the late gastrula, the grafted ectoderm forms a neural plate as it would have done if it had remained in its original location. The necessity of the notochordal inductor has also been strikingly shown by the analysis of amphibian *exogastrulae*. Exogastrulae are embryos in which the normal inpocketing of the *archenteron* is exteriorized, commonly by the quantity or types of certain ions in the medium surrounding the embryos. The endodermal and chordamesodermal tissues lining the archenteron form an outpocketing from the posterior end of the embryo (Fig. 6-2). Although the tissues lining the everted archenteron undergo a certain degree of self-differentiation, the empty ectodermal hull remains non-neural.

In order for neural induction to occur, the ectoderm overlying the notochordal process must be able (*competent*) to respond to the inductive

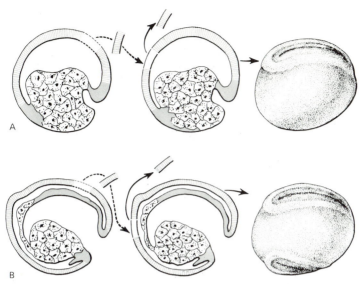

Fig. 6-1 (A) Transplantation of a piece of presumptive neural plate, before it has been acted upon by chordamesodermal induction, to the ventral part of another embryo results in integration of the graft with its surroundings. (B) After primary induction has occurred, a similar graft produces a secondary neural plate on the ventral side of the embryo. (Modified from Saxén and Toivonen, 1962, *Primary Induction*, Logos Press, London.)

stimulus. During much of the period of gastrulation in amphibian embryos not only the dorsal ectoderm but also the ventral ectoderm and that of the flank have the competence to form neural tissues when subjected to the influence of inductors. Figures 5-9 and 6-3 illustrate experiments in both the amphibian and the bird in which neural inductors grafted beneath the ectoderm induced secondary neural tubes in competent ectoderm normally not destined to form neural tissue. Later in the gastrula period the ectoderm farthest from the normal location of the nervous tissue begins to lose its capacity to respond to neural inductors, and by late in the neurula stage most non-neural ectoderm has lost its neural competence.

At least in amphibian larvae different regions of the chordamesodermal inductor impose regional specificity upon the structures differentiating as a result of this induction. This has been demonstrated by grafting different areas of chordamesoderm beneath competent ectoderm. Grafts of anterior chordamesoderm cause accessory heads to form (Fig. 6-4A), and grafts of posterior chordamesoderm induce accessory tails (Fig. 6-4B). Numerous grafting experiments have led a number of embryologists to conclude that there are two or three main inducing regions that evoke the formation of the nervous system and other axial structures of the embryo (Saxén and Toivonen, 1962). The two major regional forms of primary induction have been called *cephalic induction* and *spinocaudal induction*. The former, a neuralizing influence, results in the formation of head structures and the latter, a mesodermalizing influence, results

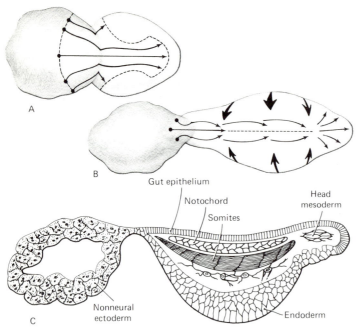

Fig. 6-2 Exogastrulation in an amphibian embryo. (A, B) Drawings showing the movements of cells (*arrows*) along the outer surface of the embryo toward the region of the blastopore (constriction), but instead of involuting, these future endodermal and mesodermal cells form a new vesicular structure (to right of constriction). (C) Sagittal section through exogastrulated embryo, showing empty ectodermal hull (*left*) and endodermal and mesodermal differentiation (*right*). (After J. Holtfreter, 1943, *J. Exp. Zool.* **94**:261.)

in trunk and tail structures. Saxén and Toivonen have postulated that the character of the induced axial tissues is a function of the interaction of two inducing gradients in the embryo—a dorsoventrally directed neuralizing gradient that extends along much of the length of the embryo and a caudocephalic gradient of decreasing mesodermalizing intensity (Fig. 6-5). Whether forebrain, hindbrain, or trunk and tail structures form in a given region depends upon the relative strengths of the two interacting gradients in that area.

Another influential viewpoint on the nature of neural induction is that of Nieuwkoop (1966) and Leussink (1970). According to this viewpoint, neural induction involves two separate phases—an activation phase and a later transformation phase. Activation, which begins during gastrulation with the invagination of cells around the dorsal lip of the blastopore, is said to evoke neural differentiation tendencies in the overlying ectoderm. Transformation, occurring somewhat later, results in the regional organization of the central nervous system. The lateral extent of the neural plate is considered to be a function of a decreasing concentration of activating factor as the result of inactivation with time and distance (Fig. 6-6).

Despite decades of intensive research, the mechanism of the inductive

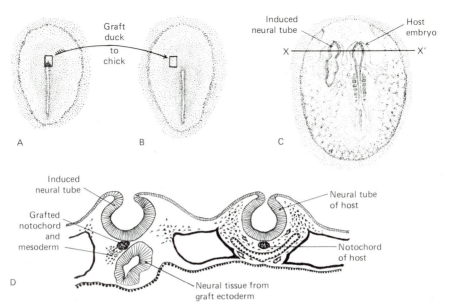

Fig. 6-3 Semischematic drawings showing the induction of an accessory neural tube as a result of grafting notochordal tissue from a duck donor into a chick host. (A) Duck embryo showing the location from which the graft was taken. (B) Chick host showing the location where the graft was implanted. (C) Embryo cultivated for 31½ hours after implanting of the graft, showing the location of the induced accessory neural tube. (D) Section at level of the line *X-X'* in (C), diagramed with the same conventional representation of the component layers as that employed in the other sectional diagrams in this text. (Based on the work of Waddington and Schmitt, *Arch. f. Entwickl.-mech. d. Organ.*, 1933, vol. 128.)

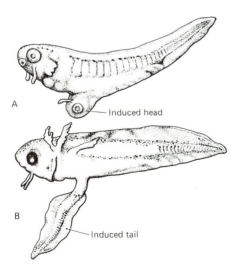

Fig. 6-4 Examples of archencephalic (A) and spinocaudal (B) induction in newt embryos. Archencephalic induction is produced by grafting anterior chordamesoderm, and spinocaudal induction results from grafts of posterior chordamesoderm. (After Balinksy, from Mangold and Tiedemann.)

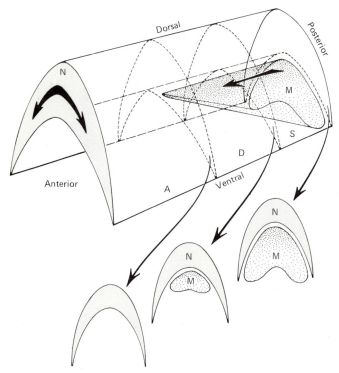

Fig. 6-5 A schematic representation of the two-gradient hypothesis of primary induction. Abbreviations: *M*, gradient of mesodermalizing action; *N*, gradient of neuralizing action; *A*, archencephalic induction; *D*, deuterencephalic induction; *S*, spinocaudal induction. (Redrawn from Saxén and Toivonen, 1962, *Primary Induction*, Logos Press, London.)

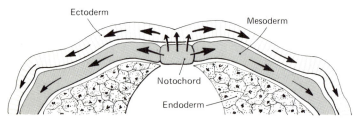

Fig. 6-6 Representation of the hypothetical course of activation events in the activation-transformation hypothesis of neural induction. The number and thickness of the arrows indicates the course and concentration of the activating factor. (Modified from J. Leussink, 1970, *Neth. J. Zool.* **20**:1.)

interaction and the nature of the inductive agent(s) remain unclear. Normally, close contact between notochordal tissues and ectoderm is the rule for neural induction, but some experiments conducted in vitro have shown that ectoderm placed in culture medium previously conditioned by the presence of mesodermal inducing agents will sometimes form neural structures in the absence of the inducing tissue itself. There is good evidence for the transfer of macromolecules from the inducing to the responding tissue (Toivonen et al., 1976), but one cannot yet rule out the influence of ions or products of cell damage (Barth and Barth, 1974).

Gallera (1971) has summarized well the major similarities and differences in primary induction between amphibians and birds. Too little is known about primary induction in mammals for meaningful comparisons to be made. In both classes of vertebrates the inducing substances are diffusible, and in both classes the time of the response to the inductive stimulus is determined by factors residing in the ectoderm. Both forms also show a similar pattern in the decline of competence of the ectoderm to the neural inductive stimulus. Initially, neural competence declines slowly, but later the rate of decline suddenly drops rapidly. Cephalic parts of the nervous system are induced first, followed later by the spinal cord. In the chick, induction of the brain is a very early event, possibly carried out by the hypoblast. The notochord of the chick loses its neural inducing capacity by the time the first pair of somites appears, whereas in amphibians its inducing capacity is preserved later into development. For neural differentiation to occur in the chick, 6 to 8 hours of close contact between inducing and responding tissues is required. Amphibians show great variability in this regard. Finally, neural induction in amphibia can be obtained by a wide variety of tissues other than chordamesoderm. These tissues, such as liver and bone marrow from guinea pigs, are called *heteroinductors*. Bird embryos are much less responsive than amphibian embryos to the effects of heteroinductors.

In response to the inductive stimulus, the ectodermal cells overlying the notochordal process proliferate, synthesize new mRNAs, and cross a restriction threshold, so that their future developmental course is channeled into the production of nervous tissues. Morphologically, the cells responding to neural induction change their shape from a cuboidal or low columnar configuration to a high columnar form. This causes the neural tissues to rise above the surrounding ectoderm as the flattened *neural plate* (Fig. 6-8).

Neurulation in Amphibians Later stages of gastrulation in amphibians are dominated by the completion of the cell movements leading to the formation of the primitive gut (Fig. 5-6) and, in urodeles, the spreading of the mesodermal mantle between the ectodermal and endodermal layers of the embryo (Fig. 5-8). Concomitant with these changes the blastopore decreases in prominence. During this period, the process of primary induction is coming to a close and the ectoderm overlying the notochordal process has begun to respond by thickening to form the neural plate.

Although anuran and urodelan embryos differ to some extent in the manner in which the endodermal and mesodermal layers become established, their basic organization is essentially similar by early neurulation. In the interior of the embryo the primitive gut is becoming completely surrounded by endodermal cells (Fig. 6-7). Surrounding the endoderm is a complete layer of mesoderm. In the dorsal midline the notochord is now a discrete rod. On either side of the notochord the dorsal part of the mesodermal sheet (*epimere*) begins to thicken and simultaneously undergo segmentation. The paired mesodermal segments will become the *somites*. The broad mass of early mesoderm (*hypomere*) filling the lateral and ventral parts of the embryo with a thin layer of cells is called the *lateral plate mesoderm*. Ultimately the lateral plate mesoderm will split into two layers—an outer layer opposed to the ectoderm (*parietal* or *somatic mesoderm*) and an inner layer surrounding the endoderm (*visceral* or *splanchnic mesoderm*). The somatic mesoderm and its overlying ectoderm are collectively called the

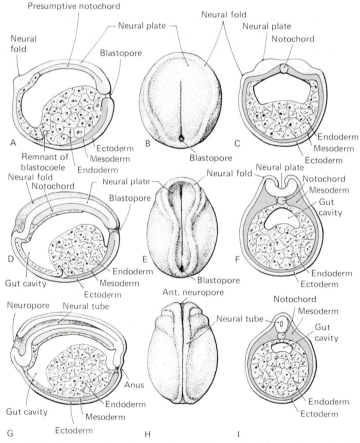

Fig. 6-7 Representative stages of neurulation in the frog embryo. (A–C) Early neurula, (D–F) middle neurula, (G–I) late neurula. *Left column*, midsagittal sections; *right column*, transverse sections cut through the embryos illustrated in the middle column. (After Balinsky.)

somatopleure, whereas the splanchnic mesoderm and the underlying layer of endoderm constitute the *splanchnopleure*. The cavity that forms between these two layers of mesodermal cells is known as the *coelom*. Joining the somites and the lateral plate mesoderm is a thin connecting link of mesodermal cells (mesomere), called the *intermediate mesoderm*. From these cells the urogenital structures ultimately arise.

In the ectodermal layer, the cells of the future nervous system have begun to follow a morphological course different from that of the rest of the ectoderm. Initially all of the ectodermal cells are arranged as a single sheet of low columnar cells (Fig. 6-8). Subsequent to neural induction, the cells in an oval area overlying the notochord and future somites undergo a pronounced elongation to form the neural plate, whereas at the same time the presumptive epidermal cells covering the rest of the embryo become flattened.

Soon the edges of the neural plate become raised, forming an elevated *neural fold* on either side of a shallow, longitudinal *neural groove*. The shape of the neural plate then undergoes a pronounced transformation, becoming attenuated posteriorly, where the spinal cord will eventually form, and remaining broader anteriorly, where the brain will form (Fig. 6-7E). The deformation of the neural plate is the result of a set of morphogenetic

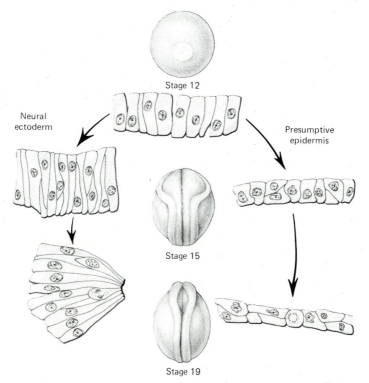

Fig. 6-8 Changes in shape of ectodermal cells during neurulation in the newt embryo. (Modified from Burnside, 1971, *Devel. Biol.* **26**:419.)

movements caused principally by changes in the shape of the cells comprising the neural plate (Burnside and Jacobson, 1968). During its deformation, the surface area of the neural plate decreases, but its volume remains roughly constant, as the shrinkage of the apical surfaces of the neuroectodermal cells is matched by a corresponding increase in their height. The shifts in positions of cells are much more pronounced in the region of the future spinal cord than in that of the brain (Fig. 6-9), and the magnitude of cellular displacement is correlated with the degree of change in shape of the individual cells. Recently, Jacobson and Gordon (1976) have been able to simulate the cell movements and changes in shape of the neural plate on a computer by taking into account both shrinkage of the apical surfaces of the cells and an anterior expansion, resulting from an intimate association between the cells of the neural plate and the notochord, which continues to extend in a cephalic direction (Fig. 6-9C).

Formation of the Neural Tube   The neural plate does not remain flat very long. Soon after it has taken shape, its lateral borders become elevated, forming the neural folds, which flank the neural groove (Figs. 6-7 and 6-10). The two lateral edges of the neural folds eventually come together in the dorsal midline to form a complete *neural tube* (Fig. 6-10). Closure of the neural tube first occurs in the upper spinal cord levels and from there proceeds both cephalad and caudad. The last remaining open portions of the nervous system are small uncovered areas called the *anterior* and *posterior neuropores* (Fig. 6-7H). Both neuropores ultimately become obliterated without leaving behind any structures of note.

The mechanism of neural-tube formation has been the subject of much speculation over the years, and even now not all aspects of the process are well understood. Modern investigations have confirmed earlier speculations that at least some of the process of neural folding can be attributed to intrinsic changes in shape of the neuroepithelial cells. Holtfreter (1974) observed that single cells isolated from the neural plate of salamander embryos complete their normal elongation in vitro. As these cells elongate, their apices constrict. Elongation of a neuroepithelial cell requires the presence of a series of intact microtubules running from the base to the apex of the cell (Fig. 6-11). The microtubules act like an internal skeleton of the cell, supporting its greatly increased height. Meanwhile, just beneath the apical surfaces of these cells are organized bundles of thin microfilaments, which can contract, resulting in a constriction of the apical end of the cell (Fig. 6-11). Integrity of the microtubules can be disrupted by exposing them to the drug *colchicine*, and the organized bundles of microfilaments can be replaced by dense granular masses through the action of *cytochalasin B*. After exposure to these inhibitors, neuroepithelial cells do not undergo their characteristic changes in shape and the neural plate remains unfolded.

It is unlikely that changes of cell shape are sufficient in themselves to account for formation of the neural tube. One must also take into account tensions caused by the growth of structures (e.g., the notochord) underlying the

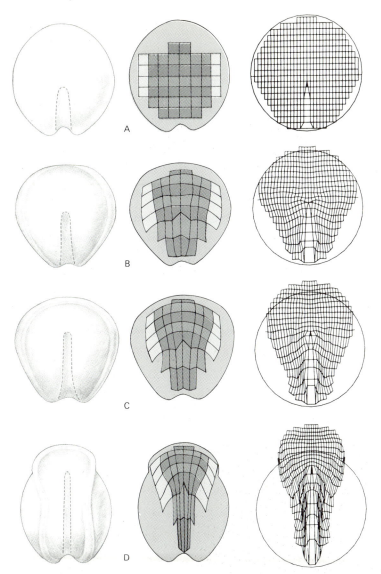

Fig. 6-9 Cell shifts during the formation of the neural plate, in the newt embryo. *Left column*, gross views of the embryo; *middle column*, changes in shape of the neural plate region, as indicated by deformation of individual components of the grid; *right column*, computer simulation of the deformation processes illustrated in the middle column. (Redrawn from papers by Jacobson and Gordon.)

neural plate, as well as changes in the aggregative and mechanical properties of the neuroepithelial cells as they relate to adjacent structures in the embryo.

The Neural Crest As the raised lateral walls of the neural folds come together to form the neural tube, a new group of ectodermal cells, emanating

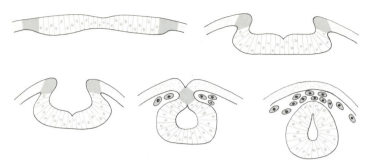

Fig. 6-10 Cross sections illustrating the formation of the neural tube and neural crest in the amphibian embryo. Black areas, neural crest. (After Balinsky.)

from the junctions between neural and non-neural ectoderm, becomes evident (Fig. 6-12). These cells, originally arranged as a longitudinal pair of loosely aggregated cells that extend along either side of the dorsal midline between the neural tube and the superficial ectoderm, constitute the *neural crest*, one of the most remarkable primordia in the embryonic body.

From the time of their first appearance, the cells of the neural crest are endowed with the ability to undertake extensive but tightly controlled migrations throughout the body. In the head the cells of the neural crest begin their

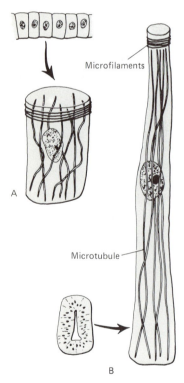

Fig. 6-11 Intracellular correlates of changes in cell shape during neural tube formation in the salamander embryo. Elongation accompanies lengthening and alignments of the microtubules; whereas narrowing of the apex is due to constriction of bands of microfilaments. (Adapted from B. Burnside, 1973, *Am. Zool.* **13**:989.)

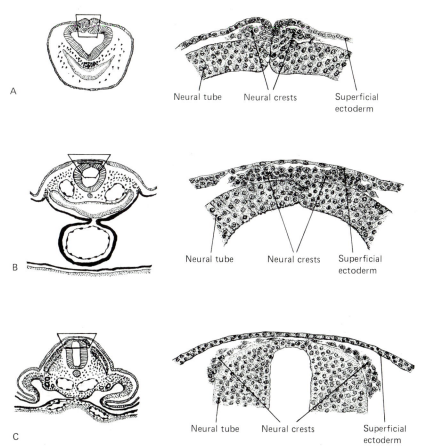

Fig. 6-12 Drawings from transverse sections to show the origin of neural crest cells. The location of the area drawn is indicated on the small sketch to the left of each drawing. (A) Anterior rhombencephalic region of 30-hour chick. (B) Posterior rhombencephalic region of 36-hour chick. (C) Middorsal region of cord in 55-hour chick.

migration as a relatively cohesive unit (Fig. 6-13), whereas in the trunk the migrations of the cells are more individualized from the start.

In the trunk, cells leave the neural crest in two major streams (Fig. 6-14). One stream is superficial and dorsally directed; the other is more ventrally directed and leads through and around the somitic mesenchyme. Cells following the dorsal migratory pathway move into the ectoderm, where they differentiate into pigment cells (melanophores, xanthophores, and iridophores), which ultimately settle down in either the epidermis or dermis according to patterns characteristic of the species.

The early cells that traverse the ventral pathway become associated with the gut as parasympathetic ganglia. Other cells among the ventrally migrating group form the sympathetic ganglia and the adrenal medulla. Still other cells remain close to the region of the original neural crest but become aggregated into

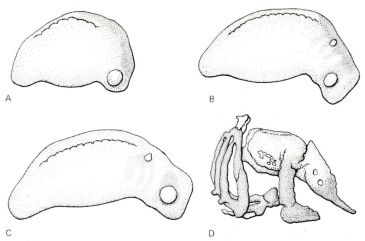

Fig. 6-13 (A–C) Migration of neural-crest cells (*gray areas*) in branchial and head region of the salamander. (D) Skull of a salamander, showing portions derived from neural crest (*gray*) and from mesoderm (*light*). (Modified from L. S. Stone, 1926, *J. Exp. Zool.* **44**:95.)

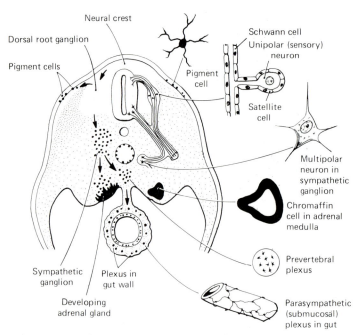

Fig. 6-14 Schematic cross section through the trunk of an embryo, showing the superficial and deep pathways of migration (*left side*) and adult derivatives (*right side*) of neural crest cells.

segmental pairs—the dorsal root ganglia of the sensory nerves. Associated with peripheral nerves are the Schwann cells, also derivatives of the neural crest. In addition to these structures, neural crest cells from the trunk appear to make some contribution to glial cells (oligodendroglia), the meninges, and, in amphibians, the mesenchyme of the dorsal fin.

Cells from the cranial division of the neural crest have a wide repertoire of differentiative capacities, including cells of the sensory roots of cranial nerves V, VII, IX, and X and their sheath cells; cells of the cranial parasympathetic ganglia; and a number of tissues traditionally considered within the province of mesodermal derivatives. The tissues include the branchial cartilages, some of the membrane bones of the head, the dental papillae and odontoblasts (which give rise to dentine), and head mesenchyme.

Although some factors intrinsic to the cells must initiate or permit the migratory behavior of neural crest cells, the pathway of migration seems, to a large extent, to be controlled by extrinsic factors. The neural tube exerts an important early influence upon the direction of migration of neural crest cells, for if the dorsoventral axis of the neural tube is reversed by 180°, so is the migratory pathway of the cells of the neural crest. Other experiments involving the translocation of pieces of neural crest from one craniocaudal level to another have shown that the migrating cells followed pathways appropriate to their new location and differentiated accordingly (LeDouarin and Teillet, 1974).

It is still not known whether cells of the premigratory neural crest have acquired some properties that would favor their migration and differentiation along certain pathways, for instance, into the adrenal medulla, or whether the appropriate cues are provided entirely by the local environment. Both intra- and interspecific grafting studies, particularly on pigment cells in birds, have shown that the donor cells differentiate in a manner appropriate to the parts of the host's body in which they find themselves but that the pattern and phenotype of the individual donor cells is appropriate to their own genotype and not their host's (Rawles, 1948).

Whatever the controlling factors, the striking diversity of morphological differentiation among neural crest derivatirves is accompanied by the differentiation of equally diverse biochemical properties. Some of the neurons derived from the neural crest produce acetylcholine as the chief neurotransmitter and others produce the catecholamines epinephrine and norepinephrine. Closely related to the latter compounds is the malanin formed by many of the pigment cells. Both melanin and the catecholamines, are synthesized along metabolic pathways originating with the amino acids phenylalanine and tyrosine. It has been suggested that an important factor in the differentiation of the cells of the neural crest may be the activation of certain key enzymes along the pathways leading from tyrosine.

The Mesoderm of the Early Embryo During the periods of germ-layer formation and early neurulation the mesoderm of the embryo consists of mesenchymatous tissue sandwiched between the epithelial layers of the ectoderm and endoderm (Fig. 5-16). *Mesenchyme* is a morphological term that refers to tissues, regardless of their germ layer of origin, which consist of aggregates of spindle-shaped or stellate cells embedded in an intercellular matrix containing varying amounts of mucopolysaccharide (glycosaminoglycan) ground substance. The mesenchyme of early embryos contains very little matrix, whereas some forms of mesenchyme in later embryos (e.g., that of the umbilical cord) are dominated by matrix. Epithelia, on the other hand, have distinct apical and basal surfaces (Fig. 6-15). The lateral walls of adjoining epithelial cells closely adhere to one another by continuous tight junctions, which seem to be very important in regulating permeability and electrical properties of the epithelia. Except for very early embryos, the basal surfaces of epithelia rest upon a basal lamina, the nature of which varies with the epithelium that secretes it. Like the term *mesenchyme*, *epithelium* is a structural classification. Cells of all three germ layers can form epithelia.

Most modern studies on mesodermal structure have been carried out on chick embryos; hence, the major emphasis in this section will be on avian mesoderm. Mesoderm originates from cells derived from the epithelial epiblast layer. These cells pass through the primitive streak and spread outward as the definitive mesoderm. Hay (1968) has described several phases in the early development of the mesoderm. When it first emerges from the primitive streak, the mesoderm consists of migrating cells arranged as a mesenchymal layer. This has been called the *primary mesenchyme*. The cells emigrating from the primitive streak appear to be polarized, with a leading edge possessing an irregular border, from which filopodia project almost like microantennae to test the local

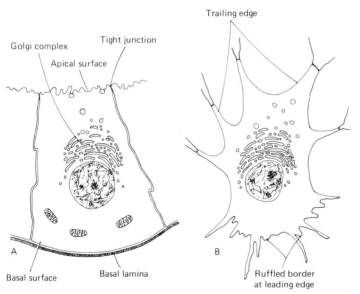

Fig. 6-15 Diagram illustrating the major structural features of epithelial (A) and mesenchymal (B) cells in the embryo. (Modified from E. Hay, 1968, in Fleischmayer and Billingham, eds., *Epithelial-Mesenchymal Interactions*, Williams & Wilkins, Baltimore.)

environment through which the cells move. The trailing edge of these cells appears to be the original apical surface when the cells were part of the epiblast layer. These mesenchymal cells are connected by numerous small gap junctions (Revel et al., 1973), which may serve as the structural basis for electrical coupling that has been demonstrated between cells in early chick embryos (Sheridan, 1966).

Once it has completed its spread outward from the primitive streak, the mesoderm of the chick becomes organized as an epithelium. Structures included in this organization are the early somites, the intermediate mesoderm, and the lateral plate. When they first form, each somite contains a central cavity, although in some species, such as the chick, it may be partially occluded with cells. The surface of the somitic cells facing the cavity of the somite is the apical surface of the somitic epithelium. Cilia are present on this surface, and the Golgi apparatus is located on this side of the cells. A basal lamina forms around the basal surfaces of these cells.

Later in development, following inductive activity by the neural tube and notochord, the somite loses its epithelial configuration, starting at its ventrome-dial border. Cells in this area (the *sclerotome*) reacquire mesenchymal proper-ties and migrate away from the main body of the somite (Fig. 6-17C and D). These cells are called *secondary mesenchyme*, and it is unlikely that they will ever again take on epithelial characteristics during their subsequent develop-mental course. A major distinguishing feature of secondary mesenchyme, in contrast to primary mesenchyme, is a greater prominence of matrix among the cells.

Secretion of Extracellular Materials in the Early Embryo   Virtually all cells in early embryos are in contact with some form of extracellular matrix, and it is not unlikely that properties of the matrix provide some of the microenvironmen-tal cues which initiate, stabilize, or change morphogenetic and differentiative processes in the embryo. Two major varieties of extracellular matrix have recently received considerable attention, particularly in the chick embryo. One is the *basal lamina*, a relatively dense sheetlike structure that is closely apposed to the basal surface of epithelia. There is increasing evidence that most epithelia secrete their own basal laminae, often as a function of a series of communica-tions between the epithelium and the underlying extracellular material. The other major variety of matrix is an amorphous form, which surrounds mesenchy-mal cells. The composition of this matrix varies in different regions of the embryo, particularly as the mesenchymal cells undergo specialization into various types of tissues.

A prominent component of the extracellular matrix is collagen. This fibrous protein is a major constituent of basal laminae, but collagen is ubiquitous throughout the body. There are many varieties of collagen in the embryonic and the adult body. Details of these varieties and their synthesis are beyond the scope of this book; for further information the reader is referred to an excellent review by Hay (1973). Another major component of the matrix is a broad class of carbohydrates, called *mucopolysaccharides* or *glycosaminoglycans*. These are polymeric structures composed of varying amounts of simpler compounds, such as hyaluronate, chondroitin and chondroitin sulfate, heparan sulfate, dermatan sulfate, and keratan sulfate. Glycosaminoglycans may be combined with proteins to form *proteoglycans*.

Collagen is first detectable in the gastrulating chick embryo as an incomplete basal lamina located beneath the epiblast (Trelstad et al., 1966). Mucopolysaccharide synthesis, as well, becomes locally prominent in the mesenchymal cells that have passed through the primitive streak and are located beneath the newly formed neural plate. It is possible that in some way the presence of basal lamina material and mucopolysaccharides is causally related to both the intracellular events and the extracellular factors that are involved in the folding of the neural plate.

At the end of neurulation, both the neural tube and notochord are arranged as epithelia, with their apical surfaces in the interior and their basal surfaces, surrounded by basal laminae and matrix material, at the periphery. There is now considerable evidence that both the neural tube and the notochord secrete collagen and glycosaminoglycans. At this stage the newly formed somites are also arranged as epithelial structures, each one surrounded by its own basal lamina. It is becoming increasingly evident that the later inductive actions of the neural tube and notochord upon the somites are mediated, at least in part, by the extracellular matrix lying between them.

The Formation and Differentiation of Somites   After the mesodermal cells of the chick have passed through the primitive streak, they migrate laterad, using the inner surfaces of the epiblast and hypoblast as substrata to assist or guide their movements. The newly formed mesoderm is a continuous sheet, several cell layers thick. The cells that will eventually form the somites are found in two thickened bands (*paraxial mesoderm*) running longitudinally along each side of the neural tube and notochord. The prospective somitic cells attach to the overlying neural plate. As Hensen's node regresses, the neural folds begin to take shape and the mesodermal cells separate from the neuroepithelium, forming pairs of blocklike somites, which are triangular in cross section. Lipton and Jacobson (1974) have noted a correlation between a shearing action of the regressing Hensen's node and the initiation of somite formation. Once they have begun to appear, new pairs of somites are formed at regular intervals in an anteroposterior direction.

The mechanism by which the segmentation of somites is accomplished remains obscure (Bellairs and Portch, 1977). Somites have been considered to be induced by Hensen's node, the notochord, or the neural plate; their segmentation has also been explained on the basis of paired somite-forming centers or regression movements of the primitive streak. More recently, Lipton and Jacobson have proposed that a somitic "prepattern" is laid down upon the unsegmented paraxial mesoderm and that it is released by the shearing effect of the regressing Hensen's node upon the mesoderm.

As the somites first take shape, they assume an epithelial configuration (Figs. 6-16 and 6-17). Each somite is surrounded by a basal lamina and around the central cavity of a somite the apical surfaces of the cells are cemented together by continuous tight junctions.

The next major morphological change in the somite is the reflection of an inductive influence emanating from the notochord and neural tube. As a result of this induction, cells of the ventromedial wall of the somites undergo a burst of mitosis, lose their epithelial characteristics, and become transformed into a secondary mesenchyme. These cells, derived from the region of the somite known as the *sclerotome* (Figs. 16-16D and 16-17D), migrate away from the somite and in time surround the notochord and the ventral portion of the neural tube. They soon begin to secrete large amounts of chondroitin sulfate and other molecules characteristic of cartilage matrix. Ultimately, the cells of the sclerotome form the vertebrae, the ribs, and the scapulae.

The inductive role of the notochord and neural tube was demonstrated by extirpation experiments (Holtzer and Detwiler, 1953). If these structures are removed from an early embryo, cartilage cells do not differentiate from the somites. Yet, there is evidence that, before the inductive stimulus, the cells of the sclerotome produce small amounts of molecules characteristic of cartilage (Lash, 1968). In this system, the role of the inductor is to increase greatly the amount of synthesis of these molecules. It now appears that various forms of collagen and proteoglycans act as the effective agents in this specific inductive process.

After the cells of the sclerotome have emigrated from the somite, the remaining dorsolateral wall gives rise to a new layer of cells, which takes shape along its inner surface (Langman and Nelson, 1968). The new inner layer is called the *myotome*, and the lateral layer, from which it arose, is called the *dermatome*. Marking experiments have shown that cells of the dermatome give rise to the dermis, whereas those of the myotomes form the musculature of the body wall and the limbs (Chevallier et al., 1977).

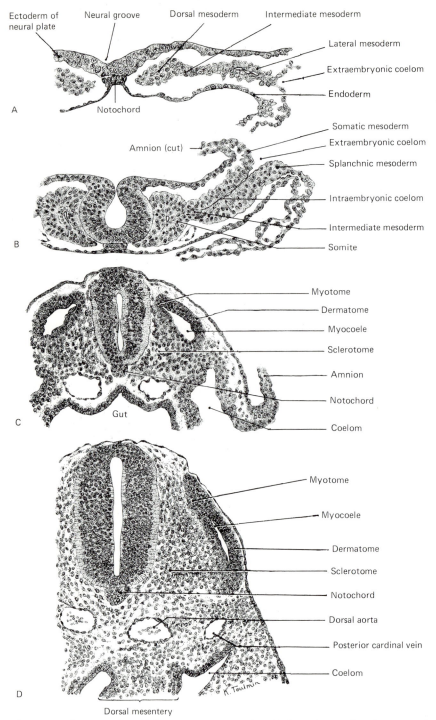

Fig. 6-16 Drawings (×150) from transverse section of pig embryos of various ages to show formation and early differentiation of somites. (A) Beginning of somite formation; (B) 7-somite embryo; (C) 16-somite embryo; (D) 30-somite embryo. (From series in the Carnegie Collection.)

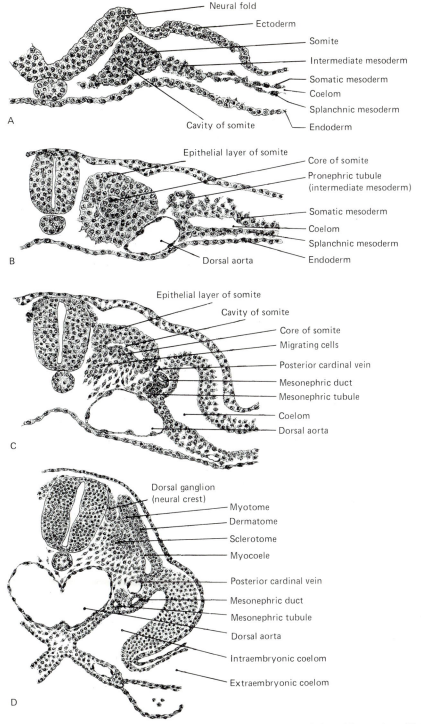

Fig. 6-17 Drawings from transverse sections to show the differentiation of the somites. (A) Second somite of a 4-somite chick (about 24 hours), (B) ninth somite of a 12-somite chick (about 33 hours), (C) twentieth somite of a 30-somite chick (about 55 hours), (D) seventeenth somite of a 33-somite chick (about 2½ days).

Chapter 7

Extraembryonic Membranes and Placenta

One of the most important evolutionary adaptations required for the independent terrestrial existence of vertebrates was the development of some means for preserving a moist, protective environment around the embryo. Some of the sharks and live-bearing fishes have evolved reproductive patterns involving the internal fertilization and development of their eggs, but in the vast majority of fishes the female lays large numbers of eggs, which are fertilized externally by sperm emitted from the males. Those eggs which are fertilized develop within the confines of simple spherical membranes, which serve to keep the embryos in a germ-free, physiologically hospitable environment. Although the amphibians solved the problem of terrestrial locomotion by evolving limbs, two major factors have forced them to remain near the water. One, of less concern to embryologists, is the incomplete adaptation of their skin and kidneys to prevent excessive water loss in a dry environment. The other is the absence of any substantial change in their mode of reproduction and embryogenesis. Each spring amphibians must return to the ponds and streams to lay eggs which, upon fertilization, develop within simple noncellular membranes in a manner not unlike the way the eggs of fishes develop.

A significant evolutionary step occurred when the first reptiles laid eggs capable of developing on land. This was made possible by the elaboration of a protective shell and a series of cellular membranes surrounding the embryonic body. These membranes, initially stemming from the body of the embryo itself, assist the embryo in vital functions, such as nutrition, gas exchange, and removal or storage of waste materials. In addition, they keep the embryo surrounded by an aquatic environment much like that of its cold-blooded forebears. Some reptiles and most mammals have dispensed with a shell in favor of intrauterine development, but the basic form and function of the extraembryonic membranes remains the same.

Four sets of extraembryonic membranes are common to the embryos of the higher vertebrates. The *amnion* is a thin ectodermally derived membrane which eventually encloses the entire embryo in a fluid-filled sac. The amniotic membrane is functionally specialized for the secretion and absorption of the amniotic fluid that bathes the embryo. So characteristic is this structure that the reptiles, birds, and mammals as a group are often called *amniotes*. Likewise, the fishes and amphibians, lacking an amnion, are collectively called *anamniotes*.

The endodermal *yolk sac* is intimately involved with the nutrition of the embryo in large-yolked forms such as the reptiles and birds. Despite the lack of stored yolk in mammalian eggs, the yolk sac has been preserved, probably because of vital secondary properties. Two of the most important are (1) the source of primordial germ cells from the yolk-sac endoderm and (2) the source of the original circulating blood cells (both red and white) from the mesodermal cells lining the yolk-sac endoderm. In addition, the endoderm of the early embryo is a participant in a number of important inductive events that lead to the establishment of some of the major organs and organ systems in the body.

The *allantois* is an endodermally lined evagination originating from the ventral surface of the early hindgut. Its principal functions are acting as a reservoir for storing or removing urinary wastes and mediating gas exchange between the embryo and its surroundings. In reptiles and birds the allantois is a large sac, and because the egg is a closed system with respect to urinary wastes, the allantois must sequester nitrogenous by-products so that they do not subject the embryo to osmotic stress or toxic effects. In mammals the role and prominence of the allantois varies with the efficiency of the interchange that can take place at the fetal-maternal interface. The allantois of the pig embryo rivals that of the bird in both size and functional importance, whereas the human allantois is reduced to a mere vestige, which contributes only a well-developed vascular network to the highly efficient placenta.

The outermost extraembryonic membrane, which abuts onto the shell or the maternal tissues and thereby represents the site of exchange between the embryo and the environment around it, is the *chorion* (*serosa*).[1] In species which lay eggs, the principal function of the chorion is the respiratory exchange of gases. The chorion in mammals serves a much more all-embracing function,

---

[1]Throughout the years there has been considerable confusion regarding the most appropriate term for the outermost membrane surrounding the embryo. In previous editions of this book it was designated the *serosa*, consistent with the usage in comparative embryology. According to this system of terminology, the composite membrane resulting from the fusion of the allantois and serosa is called the *chorion*. In contemporary usage, however, the outermost extraembryonic membrane is almost universally called the *chorion* (instead of *serosa*) and the composite membrane formed where the allantois is fused to it is called the *chorioallantoic membrane*, particularly in the literature on the experimental embryology of birds.

which includes not only respiration but also nutrition, excretion, filtration, and synthesis—hormone production being an important example of the last function.

The presence of the extraembryonic membranes introduces an added degree of complexity to the morphological study of amniote embryos. Particularly in early embryonic development it is difficult to dissociate the processes which result in membrane formation from those which are involved in the shaping of the gross form of the embryo itself. Extraembryonic membranes have different patterns of formation in different species. This chapter will describe the formation of extraembryonic membranes in three widely studied species: the chick, the pig, and the human being.

EXTRAEMBRYONIC MEMBRANES OF THE CHICK

The Folding-off of the Body of the Embryo   In early chick embryos the somatopleure[2] and the splanchnopleure extend peripherally over the yolk, beyond the region where the body of the embryo is being formed. Distal to the body of the embryo the layers are termed *extraembryonic*. At first the body of the chick has no definite boundaries; consequently embryonic and extraembryonic layers are directly continuous, having no definite point at which one ends and the other begins. As the body of the embryo takes form, a series of folds develop about it, undercut it, and finally nearly separate it from the yolk. The folds which thus definitely establish the boundaries between intraembryonic and extraembryonic regions are known as the *body folds*.

The first of the body folds to appear marks the boundary of the head. By the end of the first day of incubation, the head has grown forward and the fold originally bounding it appears to have undercut it and separated it rostrally from the blastoderm (Fig. A-1). The *subcephalic fold* at this stage is crescentic, concave caudally. As this fold continues to progress caudad, its posterior extremities become continuous with folds which develop along either side of the embryo. Because these folds bound the body of the embryo laterally, they are known as the *lateral body folds*. The lateral body folds, at first shallow (Fig. A-25D), become deeper, undercutting the body of the embryo from either side and further separating it from the yolk (Figs. 7-1 and A-31E).

A *caudal fold*, bounding the posterior region of the embryo, appears during the third day. It undercuts the tail of the embryo, forming a *subcaudal pocket* (Fig. 7-2C), just as the cephalic fold undercuts the head to form the subcephalic pocket. The combined effect of the development of the subcephalic, lateral body and the subcaudal folds is to constrict the embryo more and more from the yolk (Figs. 7-1, 7-2, and 7-4). These folds, which establish the contour of the embryo, indicate at the same time the boundary between the tissues which are built into the body of the embryo and the "extraembryonic tissues" which serve

---

[2]*Somatopleure* is the name given to a layer of ectoderm underlain by mesoderm, and *splanchnopleure* refers to a bilayer of endoderm and mesoderm (Fig. 7-1A). In keeping with this terminology, the mesoderm associated with ectoderm is called *somatic mesoderm* and that associated with endoderm is *splanchnic mesoderm*.

temporary purposes during development but are not incorporated in the structure of the adult body.

The Establishment of the Yolk Sac and the Delimitation of the Embryonic Gut   The yolk sac is the first extraembryonic membrane to make its appearance. The splanchnopleure of the chick, instead of forming a closed gut as happens in forms with little yolk, grows over the yolk surface. The primitive gut only has a dorsal cellular wall, and the yolk acts as a temporary floor (Fig. 7-2A). The extraembryonic extension of the splanchnopleure, derived originally from the primary and secondary hypoblast, eventually forms a saclike investment for the yolk (Figs. 7-1 and 7-4). Concomitantly with the spreading of the extraembryonic splanchnopleure about the yolk, the intraembryonic splanchnopleure undergoes a series of changes which result in the establishment of a completely walled gut in the body of the embryo.

The first part of the primitive gut to acquire a cellular floor is its cephalic region. Through a series of lateral folds (Fig. A-22A to A-22C) and anterior cephalic growth, the head and neck are separated from the underlying yolk by the *subcephalic pocket* (Fig. 7-2B and C). The same folding process that forms the head and neck causes the cephalic endoderm to form a tube. The part of the primitive gut which becomes tubular as the subcephalic fold progresses caudad is termed the *foregut* (Fig. 7-2B and C). During the third day of incubation the caudal fold undercuts the posterior end of the embryo. The splanchnopleure of the gut is involved in the progress of the subcaudal fold, so that a hindgut with an endodermal floor is established in a manner analogous to the formation of the foregut (Fig. 7-2C). The part of the gut which still remains open to the yolk is known as the *midgut*. As the embryo is constricted away from the yolk by the progress of the subcephalic and subcaudal folds, the foregut and hindgut are increased in extent at the expense of the midgut. The midgut is finally diminished until it opens ventrally by a small aperture which flares out, like an inverted funnel, into the yolk sac (Fig. 7-2D). This opening is the *yolk duct*, and its wall constitutes the *yolk stalk*.

The walls of the yolk sac are still continuous with the walls of the gut along the constricted yolk stalk thus formed, but the boundary between the intraembryonic splanchnopleure of the gut and the extraembryonic splanchnopleure of the yolk sac can now be established definitely at the yolk stalk.

As the neck of the yolk sac is constricted, the *vitelline (omphalomesenteric) arteries* and the *vitelline (omphalomesenteric) veins*, caught in the same series of foldings, are brought together and traverse the yolk stalk side by side (Fig. A-44). The vascular network in the splanchnopleure of the yolk sac, which in young chicks was seen spreading over the yolk, eventually nearly encompasses it (Fig. 7-5). The embryo's store of food material thus comes to be suspended from the gut of the midbody region in a sac provided with a circulatory arc of its own, the *vitelline arc*. Yolk does not pass directly through the yolk duct into the intestine; instead absorption of the yolk is effected through enzymatic activity of the endodermal cells lining the yolk sac. These digestive enzymes change the yolk into soluble material which can then be absorbed through the lining of the

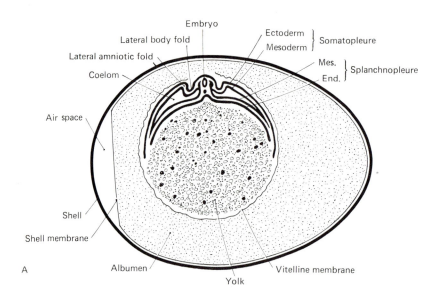

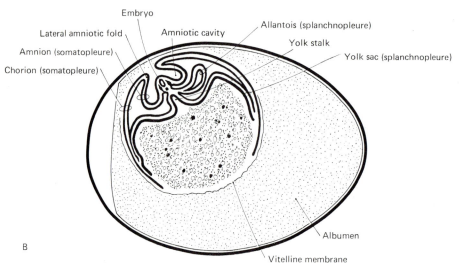

Fig. 7-1 Schematic diagrams to show the extraembryonic membranes of the chick. (After Duval.) The diagrams represent longitudinal sections through the entire egg. The body of the embryo, being oriented approximately at right angles to the long axis of the egg, is cut transversely. (A) Embryo of about 2 days' incubation. (B) Embryo of about 3 days' incubation. (C) Embryo of about 5 days' incubation. (D) Embryo of about 14 days' incubation. (See colored insert.)

yolk sac and passed on through the endothelial lining of the vitelline blood vessels to the circulating blood, by which it is carried to all parts of the growing embryo. In older embryos the splanchnopleure of the yolk sac undergoes a series of foldings (Fig. 7-1C and D) which greatly increase its surface area and thereby the amount of absorption it can accomplish.

During development the albumen loses water, becomes more viscid, and

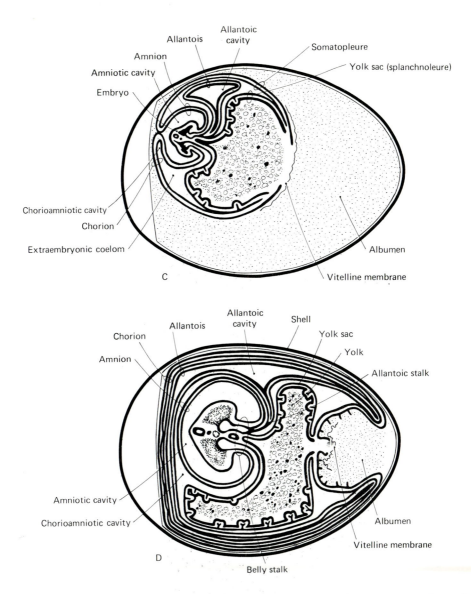

C

D

rapidly decreases in bulk. The growth of the allantois, an extraembryonic structure which we have yet to consider, forces the albumen towards the distal end of the yolk sac (Fig. 7-1D). The manner in which the albumen is encompassed between the yolk sac and folds of the allantois and serosa belongs to later stages of development than those with which we are concerned. Suffice it to say that the albumen, like the yolk, is surrounded by an extension of the yolk-sac splanchnopleure through which it is absorbed and transferred, by way of the extraembryonic circulation, to the embryo.

Toward the end of the period of incubation, usually on the nineteenth day,

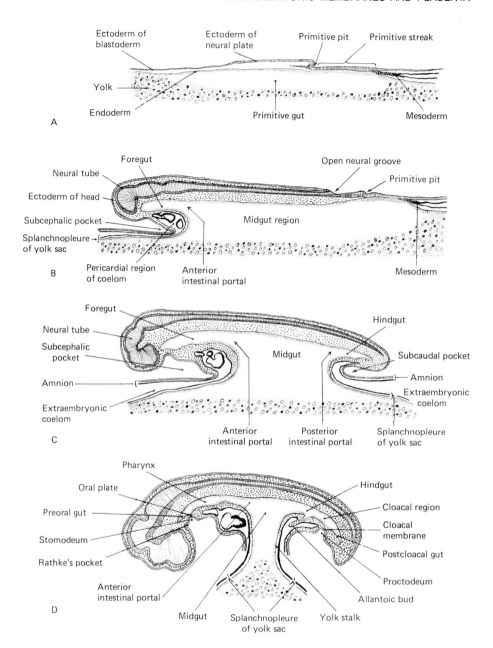

Fig. 7-2 Schematic longitudinal section diagrams of the chick, showing four stages in the formation of the gut tract. The embryos are represented as unaffected by torsion. (A) Chick toward the end of the first day of incubation; no regional differentiation of primitive gut is as yet apparent. (B) Toward the end of the second day; foregut established. (C) Chick of about 2½ days; foregut, midgut, and hindgut established. (D) Chick of about 3½ days; foregut and hindgut increased in length at expense of midgut; yolk stalk formed.

the remains of the yolk sac are enclosed within the body walls of the embryo. After the yolk sac's inclusion in the embryo, both the wall and the remaining contents of the yolk sac rapidly disappear, their resorption being practically completed in the first 6 days after hatching. The remaining yolk reserves are vital to the newly hatched chick while it is first adapting to a free-living existence and is developing its feeding behavior.

The Amnion and Chorion The amnion and chorion are so closely associated in their origin that they must be considered together. Both are derived from the extraembryonic somatopleure. The amnion encloses the embryo as a saccular investment, and the cavity thus formed between the amnion and the embryo becomes filled with a watery fluid. Muscle fibers of a primitive type develop in the amnion and by their contraction gently agitate the amniotic fluid. The slow rocking movement thus imparted to the embryo apparently aids in keeping its growing parts free from one another.

The first indication of amnion formation appears in chicks of about 30 hours' incubation. The head of the embryo sinks somewhat into the yolk, and at the same time the extraembryonic somatopleure anterior to the head is thrown into a fold, the head fold of the amnion (Figs. 7-3 and 7-4A). From a dorsal aspect the margin of this fold is crescentic in shape, with its concavity directed towards the head of the embryo. As the embryo increases in length, its head grows forward into the amniotic fold. At the same time, growth in the somatopleure itself tends to extend the amniotic fold caudad over the head of the embryo (Fig. 7-4B). By continuation of these two growth processes the head soon comes to lie in a double-walled pocket of extraembryonic somatopleure which covers it like a cap (Fig. 7-3). The free edge of the amniotic pocket retains its original crescentic shape as, in its progress caudad, it covers more and more of the embryo (Figs. A-27 and A-29).

The caudally directed limbs of the head fold of the amnion are continued posteriorly along either side of the embryo as the lateral amniotic folds. The lateral folds of the amnion grow dorsomesiad, eventually meeting in the midline dorsal to the embryo (Fig. 7-1A to 7-1C).

During the third day, the tail fold of the amnion develops about the caudal region of the embryo. Its manner of development is similar to that of the head fold of the amnion, but it grows in the opposite direction, its concavity being directed anteriorly and its progression being cephalad (Fig. 7-4B and C).

Continued growth of the head, lateral, and tail folds of the amnion results in their meeting above the embryo. At the point where the folds meet, they become fused in a scarlike thickening termed the *amniotic raphe* (Fig. 7-4C). The way in which the somatopleure has been folded about the embryo leaves the amniotic cavity completely lined by ectoderm, which is continuous with the superficial ectoderm of the embryo at the region where the yolk stalk enters the body (Fig. 7-1D).

All the amniotic folds involve doubling the somatopleure on itself. Only the

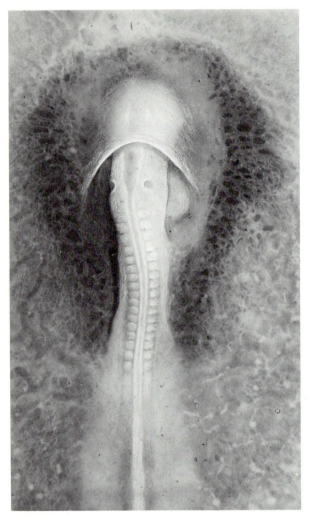

Fig. 7-3 Unstained chick of about 40 hours' incubation, photographed by reflected light to show the cephalic fold of the amnion enveloping the head of the embryo.

inner layer of the somatopleuric fold, however, is directly involved in the formation of the amniotic cavity. The outer layer of somatopleure becomes the chorion (Fig. 7-1B). The cavity between the chorion and amnion (chorioamniotic cavity) is part of the extraembryonic coelom. The continuity of the extraembryonic coelom with the intraembryonic coelom is most apparent in early stages (Fig. 7-1A and B). They remain, however, in open communication in the yolk-stalk region until relatively late in development.

The rapid peripheral growth of the somatopleure carries the chorion about the yolk sac, which it eventually envelops. The albumen sac, also, is enveloped by folds of chorion. The allantois, after its establishment, develops within the

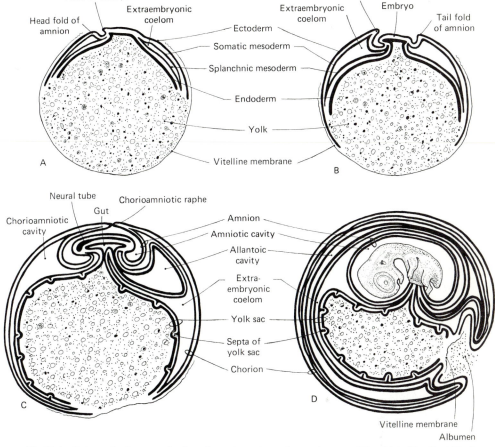

Fig. 7-4 Schematic diagrams to show the extraembryonic membranes of the chick. (D, after Lillie.) The embryo is cut longitudinally. The albumen, shell membranes, and shell are not shown; for their relations see figure 7-1. (A) Embryo early in second day of incubation. (B) Embryo early in third day of incubation. (C) Embryo of 5 days. (D) Embryo of 9 days. (See colored insert.)

chorion, between it and the amnion (Fig. 7-6). Thus the chorion eventually encompasses the embryo itself and all the other extraembryonic membranes.

Despite its simple histological organization in the normal embryo, the extraembryonic ectoderm retains considerable unexpressed developmental potential. When placed in contact with appropriate dermis from chicks, ducks, or even mice, the extraembryonic ectoderm of chick embryos can not only become cornified, like mature epidermis, but also form feathers in response to the influence of the foreign dermis (Dhouailly, 1978).

The Allantois   The allantois differs from the amnion and chorion in that it arises within the body of the embryo (Fig. 7-6A). Its proximal portion remains intraembryonic throughout development. Its distal portion, however, is carried outside the confines of the intraembryonic coelom and becomes associated with the outer extraembryonic membranes (Fig. 7-6B and C). Like the other extraembryonic membranes, the distal portion of the allantois functions only during the incubation period and is not incorporated into the structure of the adult body.

The allantois first appears late in the third day of incubation as a diverticulum from the ventral wall of the hindgut. Its walls are, therefore, splanchnopleure (Figs. 7-4 and 7-6).

During the fourth day of development the allantois pushes out of the body of the embryo into the extraembryonic coelom. Its proximal portion lies parallel to the yolk stalk and just caudal to it. When the distal portion of the allantois has grown clear of the embryo it becomes enlarged (Figs. 7-4C and 7-6C). Its narrow proximal portion is known as the *allantoic stalk*, its enlarged distal portion as the *allantoic vesicle*. Fluid accumulating in the allantois distends it, so that the appearance of its terminal portion in entire embryos is somewhat balloonlike (Fig. 7-5).

The allantoic vesicle enlarges very rapidly from the fourth to the tenth day of incubation. Extending into the chorioamniotic cavity, it becomes flattened and finally encompasses the embryo and the yolk sac (Figs. 7-1C and D, and 7-4C and D). In this process the mesodermal layer of the allantois becomes fused

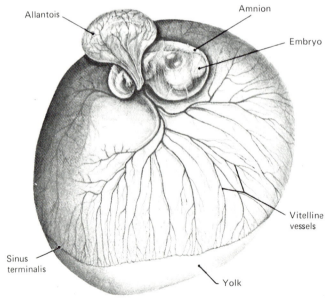

Fig. 7-5 Chick of about 5½ days' incubation taken out of the shell with yolk intact. The albumen and the serosa have been removed to expose the embryo lying within the amnion. The allantois has been displaced upward in order to show the relations of the allantoic stalk. Compare this figure with figure 7-1, C, which shows schematically the relations of the membranes in a section through an embryo of similar age. *(Modified from Kerr, 1919, Textbook of Embryology, vol. II, The Macmillan Company.)*

with the adjacent mesodermal layer of the chorion. There is thus formed a double layer of mesoderm, the chorionic component of which is somatic mesoderm and the allantoic component of which is splanchnic mesoderm (Fig. 7-6C). In this double layer of mesoderm develops an extremely rich vascular network which is connected with the embryonic circulation by the allantoic arteries and veins. It is through this circulation that the allantois carries on its primary function of oxygenating the blood of the embryo and relieving it of carbon dioxide. This is made possible by the position occupied by the allantois, close beneath the porous shell (Figs. 7-1C and A-44). It is this highly vascular fusion membrane, commonly called the *chorioallantoic membrane*, that has been used so effectively as a site for grafting small explants from younger embryos for the purpose of testing their developmental potencies.

In addition to the respiratory interchange of oxygen and carbon dioxide, the growth of the embryo, of course, involves the metabolism of proteins with the formation of urea and uric acid. Averting toxic effects from the accumulation of these waste products in an embryo growing within the confines of a shell presents some interesting problems. The allantois is again involved, for it serves as a reservoir for the secretions coming from the developing excretory organs (Fig. A-44). In the early stages of development the chick excretes mostly urea. Later the excreted material becomes chiefly uric acid. This is a significant change, for urea is a relatively soluble substance and requires large amounts of water to hold it at nontoxic levels. It can be taken care of while the embryo is young and its excretory output small, but it would present grave problems if it were produced in large quantities. In contrast, uric acid is relatively insoluble, and the large amounts of it produced by older embryos can be stored without ill effects. At the time of hatching, the slender allantoic stalk is broken and the distal portion of the allantois with its contained excretory products remains as a shriveled membrane adherent to the broken shell.

THE FORMATION OF EXTRAEMBRYONIC MEMBRANES IN THE PIG

The Yolk Sac   In the higher mammals, although virtually no yolk accumulates in the ovum, a yolk sac is formed just as if yolk were present. Such persistence of a structure, in spite of the loss of its original function, is not an uncommon phenomenon in evolution and has given rise to the biological aphorism that "morphology is more conservative than physiology." Not only does the yolk sac itself persist, but even the blood vessels characteristically associated with it in its functional condition appear on the empty yolk sac of mammalian embryos (Fig. 8-17). Therefore, when the mammalian yolk sac is spoken of as a vestigial structure, we must bear in mind that such a statement has reference to its function as a storage place for food.

The yolk sac may be defined as that part of the primitive gut which is not included within the body when the embryo is "folded off" (Fig. 8-10). Correlated with the enormous elongation of the blastocyst, the peripheral extent

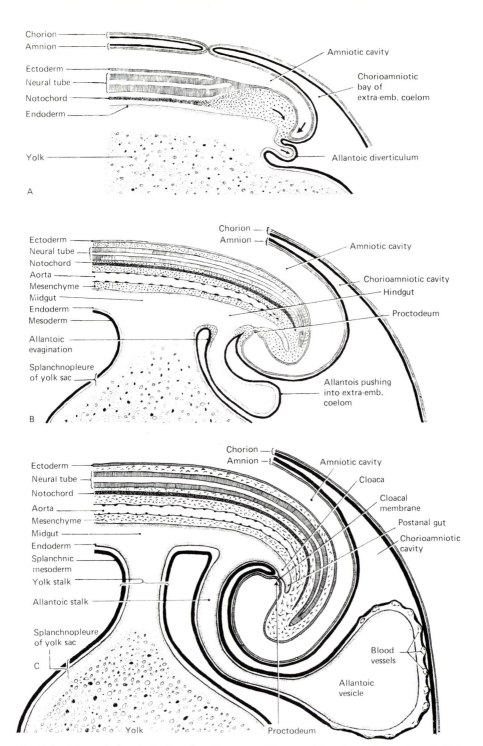

Fig. 7-6 Schematic longitudinal-section diagrams of the caudal regions of a series of chick embryos to show the formation of the allantois. (A) At about 2¾ days of incubation; (B) at about 3 days; (C) at 4 days.

of the yolk sac is unusually great in young pig embryos. From the yolk stalk, the yolk sac extends beneath the embryo, beyond its head in one direction and beyond its tail in the other, nearly to the ends of the blastocyst (Fig. 7-7A).

During this, the period of its greatest relative development, the yolk sac serves as an organ purveying nutritive material to the embryo. Separated from the uterine wall only by the thin outer layer of the blastocyst, its abundant

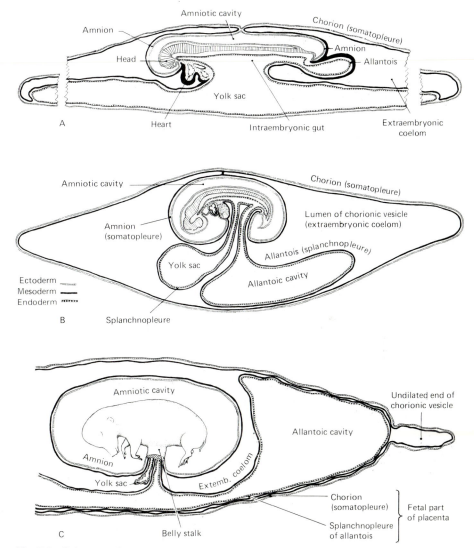

Fig. 7-7 Schematic diagrams showing the relations of the extraembryonic membranes. The great length of the chorionic vesicle of the pig precludes showing the entire vesicle in correct proportions. In A, sections of the vesicle have been omitted. In B and C, the vesicle has been shortened from its true proportions. (A) Represents conditions in embryos of 15 to 20 somites; (B) conditions in embryos of 4 to 6 mm (after flexion); (C) conditions in embryos of 30 mm or more.

vessels are in a position which makes absorption from the uterus readily possible. The food material and oxygen thus absorbed are transported by way of the vitelline circulation (Fig. 8-17) to the growing embryo. When, somewhat later, the allantois becomes highly developed, it takes over this function and the yolk sac decreases rapidly in size, finally becoming a shriveled sac buried in the belly stalk (Figs. 7-7C, and 11-11). The yolk sac, therefore, exhibits another point of special interest in addition to being an organ which persists after the cessation of its original function. The persistent yolk sac and its associated vessels are utilized by the pig embryo in a new manner. Containing no stored food materials, the yolk sac turns about and absorbs from the uterus through its external surface. The fact that the function is temporary, later being taken over by the allantois, makes this physiological opportunism in seizing on a new source of supplies none the less striking.

The Amnion   The amnion arises as a layer of somatopleure which enfolds the developing embryo in much the same manner as already described for the chick. When the amniotic sac is completed it becomes filled with a watery fluid in which the embryo is suspended. This suspension of the embryo in a fluid-filled sac, by equalizing the pressure about it, serves as a protection against mechanical injury. At the same time the soft tissues of the growing embryo, being bathed in fluid, do not tend to form adhesions with consequent malformations.

The first indication of amnion formation in pig embryos becomes evident shortly after the primitive-streak stage. The embryonic disk appears to settle into the blastocyst such that it is overhung on all sides by the extraembryonic somatopleure (Fig. 8-1). As the embryo grows in length its head and tail push deeper and deeper into these folds of somatopleure (cf. Fig. 8-10B and C). At the same time the folds themselves become more voluminous, growing over the embryo from the region of its head, its sides, and its tail. These circumferential folds of somatopleure are known as the *amniotic folds*. For convenience in description, we recognize cephalic, caudal, and lateral regions of the amniotic folds, although in reality they are all directly continuous with each other.

Progress of the amniotic folds soon brings them together above the middorsal region of the embryo to complete the amniotic sac (Figs. 8-10C and 7-7A). Where the amniotic folds close, there persists for a time a cordlike mass of tissue between the amnion and the outer layer of the blastocyst (Fig. 7-7B). In time even this trace of the fusion is obliterated and the embryo in its amnion lies free in the blastocyst (Fig. 7-7C).

The amnion is attached to the body of the embryo where the body wall opens ventrally in the region of the yolk stalk (Fig. 7-8). As development advances this ventral opening becomes progressively smaller, its margins being known finally as the *umbilical ring*. Meanwhile the yolk stalk and the allantoic stalk, in the same process of ventral closure, are brought close to each other to constitute, with their associated vessels, the *belly stalk* (Fig. 7-7C). The amnion is now reflected from its point of continuity with the skin of the body at the umbilical ring to constitute the covering of the belly stalk, before it is recurved as a free membrane enclosing a fluid-filled space about the embryo.

The Chorion   The outer layer of the mammalian blastocyst is given various names at different stages of development. In the stage of the inner cell mass it is most commonly called the *trophoblast* (Fig. 4-10C). After the formation of the hypoblast and mesoderm, the outer layer of the blastocyst is directly continuous with the ectoderm of the embryo (Fig. 5-25C). At that time it is given the more specific designation of *trophectoderm* because of the part that this layer plays in the acquisition of food materials from the uterus. Still later, after the mesoderm has split and its somatic layer becomes associated with the ectoderm, this layer, now an extraembryonic somatopleure, is called the *chorion* (*serosa*).

In the formation of the amnion in pigs, the somatopleure is thrown into folds surrounding the embryo. As we have already seen, the inner limb of the fold becomes the amnion and its outer limb is the chorion. After completion of the amnion, the chorion completely surrounds the complex of the embryo and its other extraembryonic membranes and serves as the interface between the embryonic tissues and the uterus.

The Allantois   Almost as soon as the hindgut of the embryo is established, there arises from it a diverticulum known as the *allantois* (Fig. 8-10B and C). The allantoic wall, because of its manner of origin, is necessarily composed of splanchnopleure. As the allantois increases in extent its original communication with the hindgut is narrowed to a cylindrical stalk, while its distal portion becomes enormously dilated (Figs. 8-10D, and 7-7B).

Externally the developing allantois of the pig appears at first as a crescentic enlargement directly continuous with the caudal part of the embryo. Its dilated distal portion continues to grow with great rapidity until it is a large semilunar sac, free of the caudal end of the embryo except for the stalk by which it maintains its original connection with the hindgut (Fig. 7-8).

The allantois early acquires an abundant blood supply by way of large branches from the caudal end of the aorta (Fig. 7-9). In the allantoic mesoderm these arteries break up into a maze of thin-walled vessels. As the yolk-sac circulation undergoes retrogressive changes, this plexus of allantoic vessels becomes progressively more highly developed and soon takes over entirely the function of metabolic interchange between fetus and mother.

THE RELATION OF THE PIG EMBRYO AND ITS MEMBRANES TO THE UTERUS

The Spacing of Embryos in the Uterus   Normally some of the fertilized ova are derived from one ovary and some from the other. The corpora lutea which develop from the ruptured ovarian follicles leave unmistakable evidence of the number of ova contributed by each ovary. Usually the numbers are approxi-

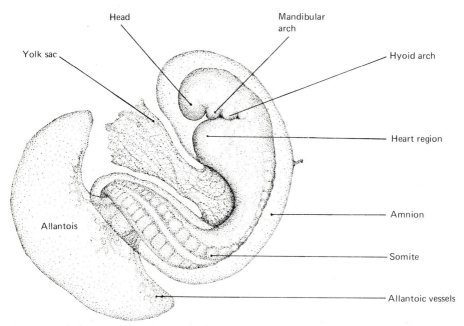

Fig. 7-8 Drawing (x12) of pig embryo having about 28 pairs of somites (crown-rump length 4 mm; age, approximately 17 days). The embryo has been removed from the chorionic vesicle with the amnion and allantois intact, but the distal portion of the yolk sac has been cut off (cf. Fig. 7-7, B).

mately equal, but occasionally, at a given estrus, one ovary is far more prolific than the other. If all the ova originating on one side remained in the corresponding horn of the uterus, there would be crowding of the embryos on one side and unutilized space on the other. Such a condition does not ordinarily occur.

The mechanism by which pig embryos are arranged within the uterus is not known, but the careful observations of Corner (1921) show that they do tend to become uniformly distributed in the two horns of the uterus regardless of the side on which the ova were liberated (Fig. 7-10). Recent work by Boving (1971) has clarified considerably the mechanism of blastocyst spacing in the rabbit. Early rabbit blastocysts are randomly arranged within the uterine horn. Shortly before implantation the blastocysts undergo a striking increase in diameter, and it is at this time that the blastocysts, whatever their number, become arranged equidistant from one another. Boving has observed that blastocyst spacing is effected by peristaltic waves of uterine contraction which are propagated not only from either end of the uterine horn but also in both directions from each area where the uterine horn is distended by a blastocyst. The peristaltic waves adjust the position of the blastocysts (as well as glass beads of similar diameter) until they become evenly spaced. As the blastocysts continue to increase in diameter, uterine contractions are no longer able to shift their positions, and implantation soon occurs.

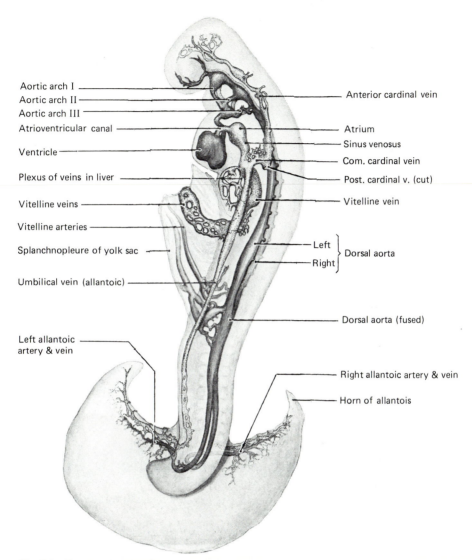

Aortic arch I
Aortic arch II
Aortic arch III
Atrioventricular canal
Ventricle
Plexus of veins in liver
Vitelline veins
Vitelline arteries
Splanchnopleure of yolk sac
Umbilical vein (allantoic)
Left allantoic
artery & vein

Anterior cardinal vein
Atrium
Sinus venosus
Com. cardinal vein
Post. cardinal v. (cut)
Vitelline vein
Left ⎫
Right ⎭ Dorsal aorta
Dorsal aorta (fused)
Right allantoic artery & vein
Horn of allantois

Fig. 7-9 Drawing showing the main vascular channels in a pig embryo of about 17 days (27 pairs of somites). (Based on cleared preparations of injected embryos loaned by Dr. C. H. Heuser and on Sabin's figures of similar material.)

The spacing of pig embryos apparently takes place before the elongation of the blastocysts occurs. Even when elongated, the blastocysts do not spread over as much space in the uterus as one might suppose. The lining of the uterus is extensively folded, and the threadlike blastocysts follow along these folds, so that a blastocyst a meter long will occupy perhaps no more than 10 to 15 cm of a uterine horn (Fig. 7-11A).

The attenuated condition of the blastocyst does not persist long. With the

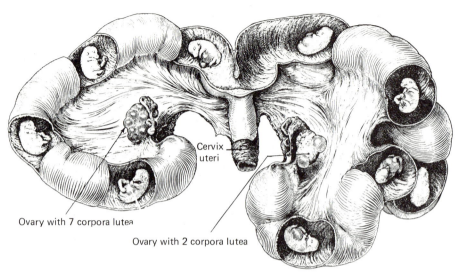

Cervix
uteri

Ovary with 7 corpora lutea

Ovary with 2 corpora lutea

Fig. 7-10 Uterus of pregnant sow opened to show distribution of embryos. Note the approximately uniform spacing of the embryos in the two horns of the uterus in spite of the fact that the corpora lutea indicate the origin of two ova from one ovary and seven from the other. (After Corner, 1921, *Johns Hopkins Hosp. Bull.,* vol. 32.)

increase in the extent of the allantois and growth of the embryo, the blastocyst becomes greatly dilated and somewhat shortened. Thus altered in shape and appearance, it is commonly called a *chorionic vesicle.* The allantois never grows to quite the entire length of the chorionic vesicle, and there remains an abrupt narrowing at either extremity of the vesicle where the allantois ends (Figs. 7-7C and 7-12).

Where these undeveloped terminal portions of neighboring chorionic vesicles lie close to each other the uterus remains undilated, sharply marking off the region where each embryo is located (Fig. 7-11B). One of the local enlargements of the uterus containing an embryo and its membranes is known as a *loculus.*

While these changes have been occurring in the general relations of the embryo and its membranes to the uterus, there have been further specializations of the fetal membranes themselves. In its peripheral growth the allantois comes into contact with the chorion (extraembryonic somatopleure) and becomes fused with it (Fig. 7-7C). Lying in intimate contact with the uterine mucosa externally and having in its double mesodermal layer the rich plexus of allantoic vessels communicating with fetal circulation, the chorioallantois is well situated to carry on the functions of metabolic interchange between the fetus and the uterine circulation of the mother.

The Placenta   In many mammals the chorion becomes fused with the uterine mucosa. In the pig, although the chorion and the uterine mucosa lie in close contact with each other, they can always be peeled apart (Fig. 7-14A).

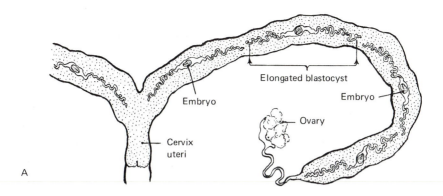

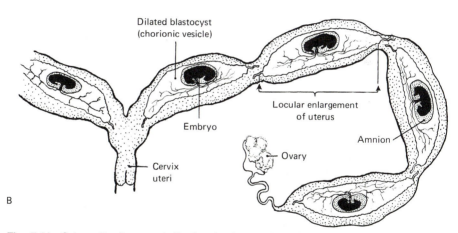

Fig. 7-11 Schematic diagrams indicating the intrauterine relations of pig embryos and their membranes: (A) in the elongated blastocyst stage; (B) when the blastocysts have been dilated to form the chorionic vesicles characteristic of the later stages of pregnancy.

When the uterine mucosa and the chorion do not actually grow together and can readily be separated from one another without the tearing of either, they are said to constitute a *contact* (nondeciduous) *placenta.* Most mammals, including humans, show a more highly specialized condition in which the fetal (chorionic) and maternal (uterine mucosal) portions of the placenta actually grow together such that they cannot be pulled apart without hemorrhage. So intimately do they become fused, in fact, that a large part of the uterine mucous membrane is pulled away with the chorion when, shortly following the birth of the fetus itself, the extraembryonic membranes are delivered as the so-called afterbirth. In contrast to the primitive contact type, such a placenta is called a *burrowing* (*deciduous*) *placenta.* An understanding of placental relationships in forms, such as the pig, where all the component parts are clearly identifiable furnishes the

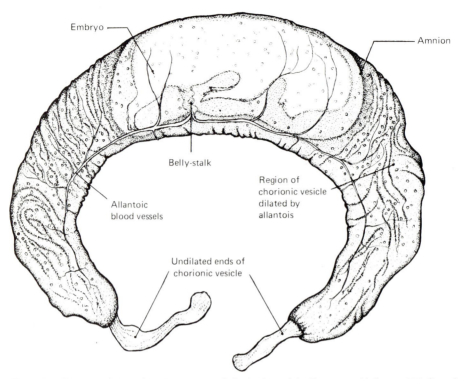

Fig. 7-12 Drawing of pig embryo in unruptured chorionic vesicle. Compare with figures 7-7, C, and 7-11, B. (After Grosser.)

best possible background for interpreting the more complex types of placentae where there are extensive fusions between fetal and maternal portions.

Even in its fully developed condition the chorion of the pig is a relatively simple structure. The endodermal lining of the allantoic cavity becomes much reduced in thickness, forming a delicate single-layered epithelial covering on the internal face of the chorion. The mesoderm of the allantoic wall fuses with that of the overlying trophoderm (Fig. 7-13), and this combined layer becomes differentiated into a primitive type of gelatinous connective tissue containing blood vessels of the allantoic plexus. With the fusion of the allantoic mesoderm to the trophoderm, these vessels invade the trophoderm and their terminal arborizations come to lie close to the ectoderm (Fig. 7-13). The ectoderm becomes a highly developed simple columnar epithelium with its surface exposure enormously increased by abundant foldings.

This plicated epithelial surface of the chorion follows the folds in the surface of the uterine mucosa, so that the chorionic and the uterine epithelia lie in contact with each other except for a thin film of uterine secretion (Fig. 7-14). Under the uterine epithelium is a loosely woven layer of connective tissue richly supplied with maternal vessels. Thus there intervenes between fetal and maternal vessels only a scanty amount of connective tissue and the epithelial

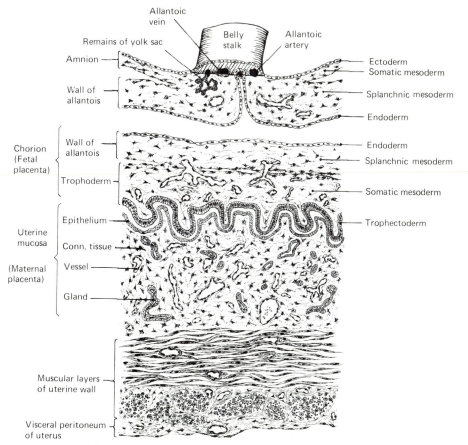

Fig. 7-13 Semischematic diagram showing the structure of the chorion and its relation to the uterine wall. Reference to figure 7-7, C, will show the general relations of the region here depicted.

layers of the uterine muscosa and of the chorion. Through these layers food materials and oxygen are passed on from material to fetal vessels, and the waste products of metabolism are absorbed from the fetal bloodstream by the maternal bloodstream.

TYPES OF PLACENTAE

Several structurally different types of placentae have appeared among the mammals. These differ in the numbers and types of cell layers intervening between the blood of the mother and that of the embryo, and they are named accordingly.

We have already seen that in the pig the chorion rests upon an intact uterine epithelium. Such a placenta is called *epitheliochorial* (Fig. 7-15A). In some hooved mammals, for example, deer, giraffes, and cattle, varying amounts of the

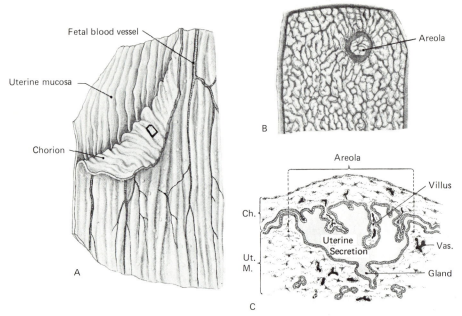

Fig. 7-14 (A) Small portion of pig placenta drawn about natural size, to show the manner in which the chorion may readily be separated from the uterine mucosa without injury to either layer. (After Grosser.) (B) Small area of chorion of 19-mm pig embryo showing its uterine surface magnified. (After Grosser.) For location of area depicted, see rectangle in A. Note the minute irregular local elevations (chorionic villi of primitive type) which are superimposed on the major folds shown in A, further increasing the surface of contact between chorion and uterine mucosa. (C) Semischematic diagram of section in region of areola. (After Zeitzschmann.) The areolae fit into depressions in the uterine mucosa near the orifice of a uterine gland. The villi of the areola are somewhat more highly developed than those of the general chorionic surface and are believed to be especially active in absorbing nutritive materials from the uterine secretions.

Abbreviations: *Ch., chorion; Ut. M.,* uterine mucosa; *Vas.,* uterine blood vessel.

uterine epithelium may be absent, thus bringing the chorion into contact with the connective tissue of the uterus. This type of placenta is called *syndesmochorial* (Fig. 7-15B) and is sometimes considered to be a variant of the epitheliochorial placenta.

When no maternal connective tissue intervenes between the endothelium of the maternal vessels and the chorionic epithelium, we say the placenta is *endotheliochorial* (Fig. 7-15C). The placenta of dogs and cats is endotheliochorial.

In primates, including humans, the endothelial continuity of the maternal blood vessels is disrupted, and the chorion is actually bathed in maternal blood. This arrangement is designated as a *hemochorial placenta* (Fig. 7-15D). Rabbits and many common rodents possess placentae in which the endothelium of the fetal vessels appears to be in direct contact with maternal blood. Called *hemoendothelial placentae*, they actually represent an extreme variation of the hemochorial type, since there is now evidence that the endothelium of the fetal

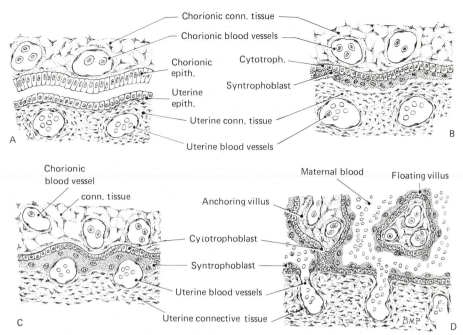

Fig. 7-15 Schematic diagrams to show four basic types of choriouterine relationships. (In part after Flexner et al., 1948, *Amer. J. Obstet. Gynec.*, vol. 55.) (A) Epitheliochorial. (B) Syndesmochorial. (C) Endotheliochorial. (D) Homochorial. For explanation, see text.

vessels is covered by a layer of trophoblastic epithelium too thin to be resolved by light microscopy (Enders, 1965).

HUMAN FETAL-MATERNAL RELATIONS

In our consideration of chick and pig embryos, we applied the term *chorioallantois* to the fetal membrane secondarily formed by the coalescence of the allantois with the chorion. In mammals which have a saccular allantois (e.g., pig; Fig. 7-16B), the chorioallantois is essentially a layer of allantoic splanchnopleure fused, mesodermal face to mesodermal face, with a layer of chorionic somatopleure. In primate embryos where the lumen of the allantois is rudimentary, the endoderm is not involved. Allantoic mesoderm and vessels, however, extend distally beyond the rudimentary allantoic lumen and spread over the inner face of the chorion, establishing the same essential relations as in less highly specialized forms (Fig. 7-16C). How far the allantoic lumen extends is of secondary importance, for the chief functional significance of this fusion between the allantois and the chorion lies in the vascular relations involved.

In the lower mammals, to which we must look for an understanding of the origin of these essential relations, the chorion is a thin membrane extending to relatively great distances from its site of origin on the ventral body walls (Fig. 7-16B). Its primary vascular supply, as might be expected from these relations, is

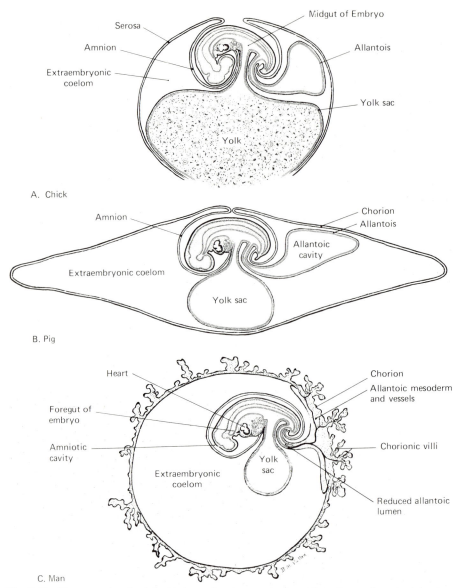

Fig. 7-16 Diagrams showing interrelations of embryo and extraembryonic membranes character-istic of higher vertebrates. Neither the absence of yolk from its yolk sac nor the reduction of its allantoic lumen radically changes the human embryo's basic architectural scheme from that of more primitive types. (See colored insert.)

meager. Moreover, the manner in which the amnion arises from the inner limbs of the same folds which form the chorion (Fig. 7-16B) is of significance, for when the amnion is pinched off as a separate sac, the original vascular connection of the serosa with the embryo is so interrupted that the maintenance of even the smallest vessels is difficult.

The allantois offers the solution to this apparent impasse. As it grows out from the hindgut of the embryo, its walls develop a rich plexus of vessels, connected directly with the main circulatory channels within the embryo by way of large arteries and veins (Fig. 8-17). The fusion of the allantoic mesoderm with the inner surface of the chorion, therefore, brings to this hitherto poorly vascularized layer an abundant circulation. In different groups of animals, there are different interrelations of the component parts of the chorioallantois, and it meets radically different environmental conditions. Nevertheless, the fundamental significance of this mechanism of vascularizing the outermost membranes of the embryo is the same. Whether it be a bird embryo dependent on this circulation for carrying on gaseous exchange with the outer air through the porous shell (Fig. A-44) or a mammalian embryo dependent on it for carrying on metabolic interchange with the uterus (Fig. 8-17) does not change the essentials of the situation. The outermost investing membrane of an embryo is inevitably the layer most favorably situated for carrying on the business of interchange with the surrounding environment, and to profit by this interchange the embryo must have adequate vascular channels communicating with the site where these exchanges take place.

Implantation   By about the sixth or seventh day after fertilization, the human embryo, which has traversed the length of the uterine tube and has been freely floating in the uterine cavity for a couple of days, is ready to become attached to the uterine lining. The embryo itself is in the blastocyst stage, consisting of an inner cell mass and a trophoblast (Fig. 7-17A). The zona pellucida, which has surrounded the embryo during the free-floating early phase of development, begins to break down just before implantation. In the monkey, embryos just prior to implantation are noticeably sticky. This characteristic is most marked in the cells directly overlying the inner cell mass, and it is this part of the embryonic surface that initially adheres to the uterine lining (Fig. 7-20).

Implantation typically takes place in the upper part of the uterine cavity, on either the anterior or posterior wall of the uterus. Few early stages of human implantation have been observed, but on the basis of studies on other primates, it is widely assumed that the pole of the blastocyst overlying the inner cell mass is the first to fuse with the endometrial epithelium (Fig. 7-20). Upon making contact with maternal tissues, the cells of the trophoblast proliferate rapidly and erode the underlying uterine mucosa (Fig. 7-21). As it expands, the original trophoblast becomes subdivided into two distinct layers. The inner layer retains a distinctly cellular nature and is therefore called the *cytotrophoblast*. The outer layer is an irregular syncytium, formed by the fusion of large numbers of cells derived from the cytotrophoblast and is appropriately called the *syntrophoblast* (*syncytiotrophoblast*). It appears likely that contact with the maternal tissues stimulates expansion of the trophoblast, because in the newly implanted embryo the portion of the trophoblast that is still uncovered in the uterine cavity remains a thin single-celled layer (Figs. 7-20A and D, and 7-21C).

While in the process of implantation, the blastocyst essentially creates a

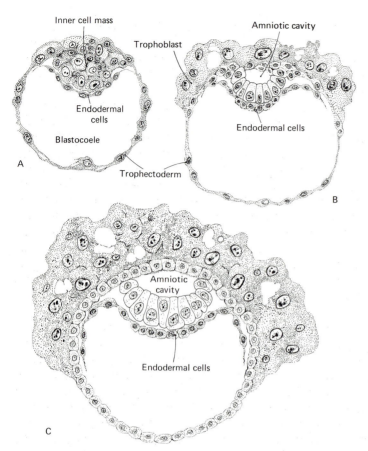

Fig. 7-17 Schematic diagrams illustrating the early steps by which young human embryos attain their basic structural plan. (A) Diagram of late primate blastocyst. (B) Diagram of primate blastocyst shortly after implantation has begun (based on the Hertig-Rock 8-day human embryo). Formation of the amniotic cavity has begun, and extraembryonic endodermal cells have begun to spread over the inner surface of the trophoblast. The expanded trophoblast represents the area that is in contact with maternal endometrium. (C) Partially implanted human blastocyst at a stage slightly later than that illustrated in B.

wound in the uterine mucosa. By 11 or 12 days the blastocyst has become almost completely embedded within the endometrium and the uterine epithelium grows back over the embryo, thus healing the wound caused by the invading blastocyst (cf. Figs. 7-18 and 7-19). During the early phase of implantation, the embryo is apparently nourished by diffusion of materials from maternal fluids and debris resulting from the destruction of endometrial cells. Soon, however, irregular lacunae appear within the expanding syntrophoblast. These lacunae become filled with maternal blood emanating from eroded uterine blood vessels (Figs. 7-18 and 7-19). This marks the beginning of the intimate relationship between maternal blood and trophoblastic tissues which forms the basis for the hemochorial placenta.

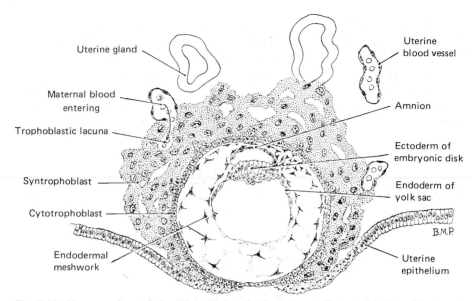

Uterine gland

Uterine blood vessel

Maternal blood entering

Amnion

Trophoblastic lacuna

Ectoderm of embryonic disk

Syntrophoblast

Endoderm of yolk sac

Cytotrophoblast

B.M.P.

Endodermal meshwork

Uterine epithelium

Fig. 7-18 Human embryo of about 9 to 10 days fertilization age. (Schematized from Hertig and Rock, 1941, *Carnegie Cont. to Emb*., vol. 29.)

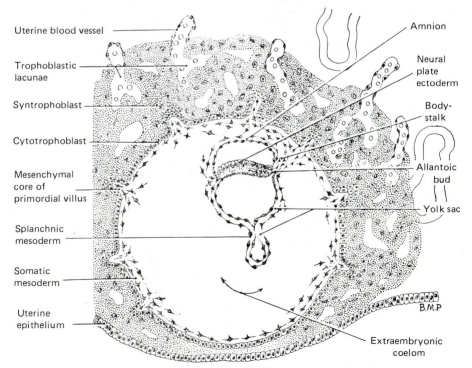

Uterine blood vessel

Amnion

Trophoblastic lacunae

Neural plate ectoderm

Syntrophoblast

Body-stalk

Cytotrophoblast

Allantoic bud

Mesenchymal core of primordial villus

Yolk sac

Splanchnic mesoderm

Somatic mesoderm

B.M.P

Uterine epithelium

Extraembryonic coelom

Fig. 7-19 Human embryo of about 13 days fertilization age. (Schematized from several sources.)

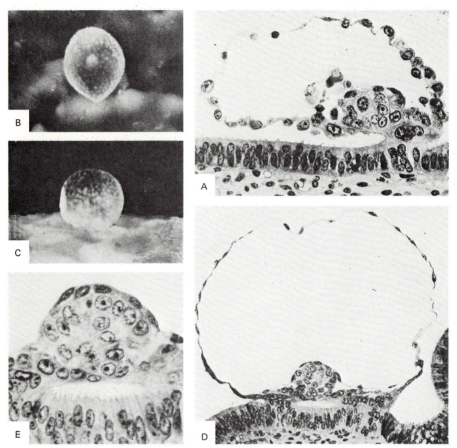

Fig. 7-20 Photomicrographs showing stages leading up to implantation of monkey embryo. (After Heuser and Streeter, 1941, *Carnegie Cont. to Emb.,* vol. 29.) (A) Section (X350) of embryo of ninth day, showing its initial adhesion to uterine epithelium. (B) A 9-day embryo attached to uterine mucosa, viewed from above (X50). (C) Same embryo photographed from the side (X50). (D) Same embryo shown in B and C after sectioning (photomicrograph, X200). (E) Inner cell mass of same embryo (photomicrograph, X500).

Whatever the importance of diffusion of maternal breakdown products into the embryo for temporary nutrition, the invasion of the endometrium rapidly paves the way for establishing the type of vascular interchange on which the embryo is to be dependent for the rest of its intrauterine life. As the trophoblast spreads out into the uterine mucosa, it inevitably comes into contact with small blood vessels and breaks down their walls. Although cells from the cytotrophoblast tend to dam the opened vessels somewhat and check excessive extravasation of blood, there must, nevertheless, be continued oozing from the invaded vessels, for the trophoblast is known to produce some substance which inhibits the coagulation of blood. In addition, there is transudation of blood serum and lymph, so that the invading trophoblast comes to lie in eroded areas of

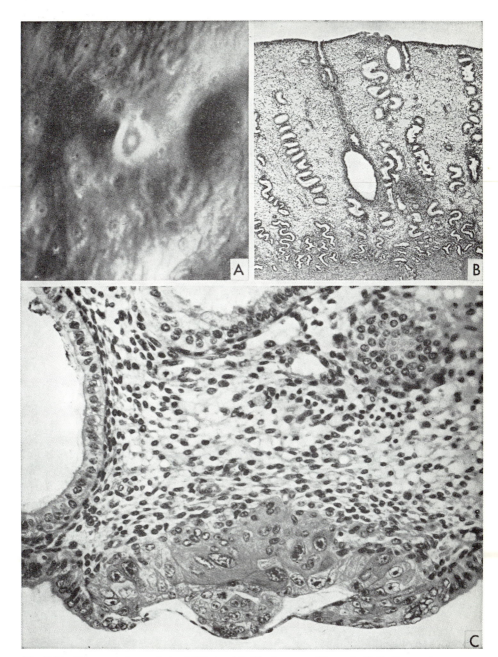

Fig. 7-21 Human embryo of about 7½ days' fertilization age. (Carnegie Collection, embryo no. 8020, after Hertig and Rock, 1945. *Carnegie Cont. to Emb.*, vol. 31.) (A) Surface view of implantation site (X27). (B) Photomicrograph (X27) of section through the uterine mucosa to show the exceedingly superficial position occupied by the recently implanted embryo. (C) Section (X300) through center of embryo and immediately surrounding endometrium. Note that the position of the embryo, as compared with that in (B), has been reversed to bring it into conventional orientation with its future dorsal side toward the top of the page.

endometrium, saturated in maternal blood and lymph. By this time, highly branched projections of the trophoblast, called *villi*, have become vascularized by terminal vessels of the allantoic circulatory arc (Fig. 8-17B). It remains only for the embryonic heart to start the blood circulating and the entire elaborately interlocking mechanism for the nutrition of the embryo is ready to go into operation early in the fourth week after fertilization, or approximately 2 weeks after implantation.

Formation of Extraembryonic Tissues in the Human Embryo   Because of the lack of well-preserved, early human material, it was difficult in the past to reconstruct the early events of human embryogenesis without recourse to inference from the findings of studies on series of primate embryos. In recent years, however, Luckett (1974, 1975, 1978) has reexamined many of the classical human preparations, as well as new embryos, and has provided an interpretation of the early steps in the formation of extraembryonic tissues that is more consonant with the findings of the comparative embryologists. Luckett's papers will serve as the basis for this section.

Formerly it was widely believed that all the extraembryonic tissues in the human arise from the trophoblast. Recent experimental studies on mammalian embryos, particularly rodents, have shown that the developmental potentialities of the trophoblast are far more restricted than the early descriptive studies suggested. According to Luckett, the original trophoblast gives rise only to cytotrophoblast and syncytiotrophoblast (Fig. 5-22). The remainder of the extraembryonic tissues come from the inner cell mass.

There are several stages in the early formation of the amniotic cavity. These stages are similar in embryos of both humans and the rhesus monkeys (Luckett, 1975). Shortly after implantation has begun, a primordial amniotic cavity forms by means of cavitation within epiblastic components of the inner cell mass (Fig. 7-17B). Then the roof of this cavity opens because of lateral spreading of the cells comprising the roof (Fig. 7-17C). During this transitory phase the primitive amniotic cavity is temporarily bounded, in part, by a region of cytotrophoblast. By means of upfolding and subsequent fusion of the lateral walls of the epiblast, the amniotic cavity again becomes completely bounded by epiblastic (ectodermal) cells (Fig. 7-18). The amniotic membrane is completed by the spreading of extraembryonic mesodermal cells, derived from the early primitive streak, around the amniotic ectoderm.

Even before the primordial amniotic cavity first appears, the inner cell mass gives to the *hypoblast* a thin layer of endodermal cells which soon spread to line the entire inner surface of the trophoblast (Fig. 7-17). These extraembryonic endodermal cells constitute the primary yolk sac. By 12 to 13 days, the primary yolk sac collapses, leaving a smaller secondary yolk sac attached to the embryo 7-19).

In human embryos, the caudal margin of the primitive streak develops precociously, giving rise to most or all of the extraembryonic mesoderm. These cells form the body stalk and continue to spread out between the extraembryonic endoderm and the overlying trophoblast. They also form the mesodermal lining layers of the amnion and yolk sac. Meanwhile, the definitive primitive streak is associated with gastrulation movements that result in the formation of the embryonic endoderm and mesoderm (see Chap. 5). The human allantois arises as an outpocketing from the hindgut much like that of the pig and chick, but the allantoic diverticulum remains a rudimentary structure embedded within the mesoderm of the body stalk.

Development of Chorionic Villi   Once implantation has occurred, expansion of the trophoblast continues. After the early establishment of lacunae, there is little in the arrangement of the trophoblast at these early stages to suggest the characteristically shaped branching villi that will be seen in later stages. Embryos that have a trophoblast in this sprawling, unorganized condition are commonly characterized as *previllous* (Figs. 7-17B and 7-18).

As embryos approach the end of the second week, the trophoblast begins to be molded into masses more suggestive of villi. These very young villi at first consist entirely of epithelium, with no connective-tissue core. In this stage they are referred to as *primary villi*. Their differentiation is very rapid, for even the previllous cell masses are already beginning to show two layers of cells. The inner layer of *cytotrophoblast* (Langhans' layer) is composed of a single, relatively regular layer of cells, each of which possesses distinct boundaries (Fig. 7-23B). Surrounding the cytotrophoblast is an outer syncytium of variable thickness and containing irregularly disposed nuclei. This layer is known as the *syntrophoblast* (*syncytiotrophoblast*) (Fig. 7-22A and B). Tracing studies, using tritiated thymidine as a marker, have demonstrated that the nuclei of the

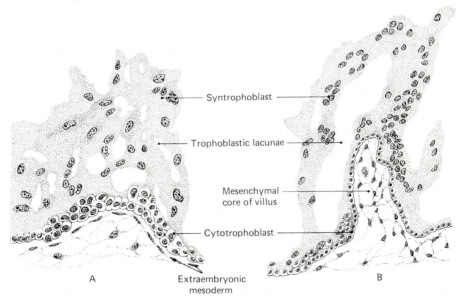

Syntrophoblast

Trophoblastic lacunae

Mesenchymal core of villus

Cytotrophoblast

A   Extraembryonic   B
mesoderm

Fig. 7-22 Early stages in development of chorionic villi. (A) Primitive trophoblastic projection without mesenchymal core. (Redrawn, X225, from Streeter's photomicrographs of the Miller Embryo.) (B) Young villus just developing a mesenchymal core. (Redrawn to about the same scale as A, from Fischel's figure of an embryo in the primitive-streak stage.)

syntrophoblast arise from the cytotrophoblastic layer. It is a general rule in development that nuclei in syncytial structures, such as the syntrophoblast or a skeletal muscle fiber, do not undergo mitotic divisions. In the case of the trophoblast, the cytotrophoblast serves as a germinative center providing both nuclei and cytoplasmic material to the syntrophoblast, which, because of its syncytial nature, is incapable of augmenting its own supply of nuclei as its volume expands (Tao and Hertig, 1965).

The phase in which the developing villi lack a mesenchymal core is very short-lived. While the primitive villi have been forming, the inner face of the blastodermic vesicle has been receiving an ingrowth of allantoic vessels and mesoderm. Early in the third week after fertilization, the mesoderm pushes into the primitive villi, so that the trophoblastic cells, instead of constituting the whole structure, become a covering epithelial layer over a framework of delicate connective tissue derived from the mesodermal ingrowth (Fig. 7-22B). These are known as *secondary villi*. Blood vessels soon appear in the connective-tissue core of the villus and push out into its newly formed branches. Such villi, with a vascular connective-tissue core, are called *tertiary villi*. This condition in which the villi are prepared for their absorptive function is reached by about the close of the third week. The villi retain this same general structural plan throughout pregnancy, although as gestation advances their connective-tissue core and blood vessels become more highly developed and there are marked regressive changes in their epithelial covering (Fig. 7-23).

Formation of Placenta   Under the influence of the presence of an embryo, striking changes take place in the endometrium. These are most marked, naturally, at the site of implantation. The uterine stromal cells around the blastocyst undergo a pronounced transformation, in which they enlarge and their cytoplasm becomes filled with glycogen and lipid droplets. This transformation is known as the *decidual reaction*. Ultimately this reaction spreads to the stromal cells throughout the endometrium. At the termination of pregnancy the endometrium containing these cells is extensively sloughed off and then rebuilt. This postpartum phenomenon of shedding and replacement has given rise to the term *decidua* (root meaning, *to shed*) for the endometrium of pregnancy.

The fact that the human embryo promptly burrows into the endometrium, instead of becoming merely adherent as is the case in certain other mammals, establishes at the outset positional relationships which shape the later course of events. As the chorionic vesicle grows, the overlying portion of the endometrium is stretched out over it, forming a layer known as the *decidua capsularis* (Fig. 7-24). The portion of the endometrium lining the walls of the uterus elsewhere than at the site of attachment of the chorionic vesicle is called the *decidua parietalis*. The area of the endometrium directly underlying the chorionic vesicle is termed the *decidua basalis*.

The ultimate absence of chorionic villi in the decidua parietalis leaves this part of the endometrium with no direct role to play in the nutrition of the embryo. The maternal blood supply is most direct and abundant in the decidua

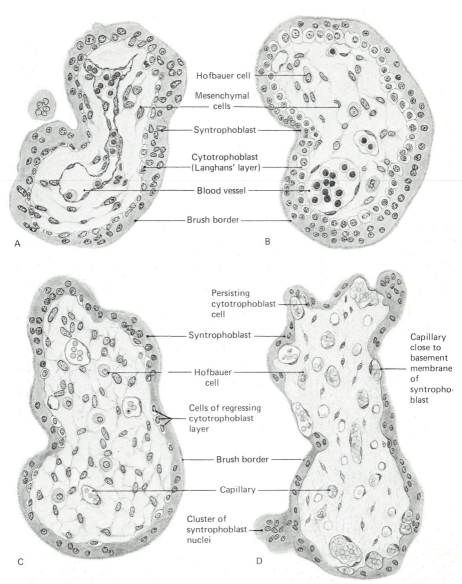

Labels in image:
Hofbauer cell
Mesenchymal cells
Syntrophoblast
Cytotrophoblast (Langhans' layer)
Blood vessel
Brush border
A
B

Persisting cytotrophoblast cell
Syntrophoblast
Hofbauer cell
Cells of regressing cytotrophoblast layer
Brush border
Capillary
Cluster of syntrophoblast nuclei
Capillary close to basement membrane of syntrophoblast
C
D

Fig. 7-23 Chorionic villi at various ages. (Camera lucida drawings, 325X.) (A) From chorion of 4-week embryo (crown-rump length, commonly abbreviated C-R, 4.5 mm). (B) Chorion from an embryo of about 6½ weeks (C-R 15.1 mm). (C) Placenta from a fetus of the fourteenth week. (D) Placenta at term. (From preparation loaned by Dr. Burton L. Baker.)

basalis. Conditions in the decidua capsularis vary considerably at different ages. At first, the chorion underlying this part of the decidua is as well supplied with villi as any other region (Fig. 7-24A), but, before long, the growth of the chorionic vesicle causes the decidua capsularis to be pushed away from the

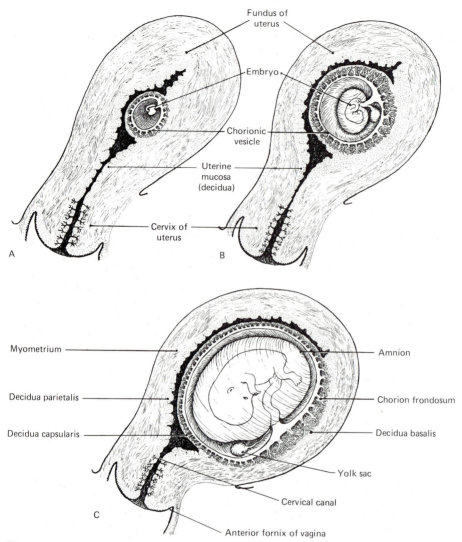

Fig. 7-24 Diagrams showing uterus in early weeks of pregnancy. Embryos and their membranes are drawn slightly smaller than actual size. (A) At fertilization age of 3 weeks; (B) 5 weeks; (C) 8 weeks.

maternal vascular supply. Moreover, the tissue of the decidua capsularis itself becomes more and more attenuated as the chorionic vesicle increases in size. These unfavorable conditions in the decidua capsularis are promptly reflected in the less exuberant growth of the chorionic villi embedded in it (Fig. 7-24C). At the end of the first trimester of pregnancy the decidua capsularis begins to undergo severe atrophy, and by midpregnancy most of the decidua capsularis has disappeared, leaving fetal chorionic tissue (chorion laeve; see next paragraph) in direct contact with the decidua parietalis of the opposite wall of the uterus (Fig. 7-31).

In contrast to the chorionic villi in the region of the decidua capsularis, the villi adjacent to the decidua basalis grow with increasing vigor. Here is, unmistakably, the part of the mechanism most effectively situated for carrying on metabolic interchange between the fetus and the mother. By the third month, when the growth of the embryo and the expansion of the amnion begin to compress the decidua capsularis and parietalis against each other, the villi begin gradually to disappear altogether from this area. Thus the chorionic vesicle, at first uniformly villated over its entire surface, has by the end of the fourth month become denuded of its villi everywhere except where they lie in the decidua basalis (Figs. 7-24 and 7-31). The part of the vesicle under the decidua capsularis which has thus lost its villi becomes known as the *chorion laeve* (smooth), and that part of the chorion next to the decidua basalis where the villi are highly developed is termed the *chorion frondosum* (tufted or bushy). The interlocked chorion frondosum of the fetus and decidua basalis of the mother constitute the *placenta*.

Later Changes in Structure of Uterus and Placenta From their first invasion of the uterine lining, the chorionic villi lie in excavated spaces in the endometrium, bathed in maternal blood and lymph. Essentially, this relationship is retained throughout pregnancy, but the extent of the blood spaces, the relations of the villi to the endometrium, and the structure of the villi themselves all vary as development progresses. During the first few weeks after implantation the invasion of the endometrium is exceedingly rapid and the area which is becoming decidua basalis is progressively extended (Fig. 7-24). In this period, the syntrophoblast is very conspicuous, forming sprawling processes extending into the endometrium far beyond the main mass of the chorionic vesicle.

Once the chorion has become well established in the uterus, the invasive process becomes relatively slow, merely keeping pace with the growth of the embryo. The slower rate of invasion is reflected in a reduction of the syntrophoblast to form a more regularly arranged covering outside the cytotrophoblastic layer of the villus. Meanwhile, the mesenchymal core of the villus has become organized into a delicate connective tissue supporting the endothelial walls of the blood vessels, so that the entire villus takes on a much more definitely organized appearance (Fig. 7-23A). Scattered in the connective tissue there appear, in variable numbers, cells which are conspicuously larger than the ordinary connective-tissue cells. These have been given the name *Hofbauer cells*, after the man who first described them. Their significance is not as yet entirely understood, but they appear to be phagocytic and are commonly believed to act as a primitive type of macrophage.

In the established parts of the placenta, the invasive function of the epithelial covering of the villi ceases to be important and the epithelial layers become relatively thinner. The cytotrophoblastic layer reaches the height of its development during the second month (Fig. 7-23B). Thereafter, it gradually loses its completeness (Fig. 7-23C). It is as if this layer spent itself in the production of the syncytial layer. During the fourth and fifth months, the cytotrophoblast undergoes still further regression. Most of the villi come to be clothed in a reduced syntrophoblastic layer with only occasional cytotrophoblastic cells persisting. During the last third of the period of gestation, attenuation of the syntrophoblast becomes more marked (Fig. 7-23D).

From the standpoint of functional significance in the development of the embryo, one would stress the exuberant development of the trophoblast during the period of invasion, followed by the gradual reduction of the epithelial layers of the villi after their invasive role has been carried out, and the thinning thereby of the amount of tissue through which the interchanges between fetal and maternal blood take place. The relations of the villi to the endometrium and to the maternal bloodstream are attained as the logical conclusion of the early invasive activity and later specifications which we have just been tracing. As pregnancy advances, the villi grow greatly in size and the complexity of their branching increases (Fig. 7-25). If we likened them to trees we should find them growing over the discoidal area of the chorion frondosum, not quite uniformly but in about 15 to 16 dense clumps. These main concentrations of villi are known as *cotyledons*. Between the cotyledons, the maternal tissue has been less deeply eroded and constitutes the "placental septa" (Fig. 7-25). Between the septa the tips of most of the villi lie free in the space which has been excavated in the uterine mucosa. The tips of other villi make contact with the uterine tissue at the bottom of the excavation. At this phase of their development the rapidly growing ends of these villi are richly cellular, consisting of a core of cytotrophoblast with syntrophoblast forming an irregular covering. This modified portion of a villus is commonly spoken of as a *cell column* (Fig. 7-26). Where such cell columns are in contact with the uterine mucosa, the trophoblastic elements spread out to clothe the eroded surfaces of the maternal tissue with trophoblast. Thus the maternal blood entering the spongy areas of the placenta from blood sinuses opened by the invading chorion comes into a maze of irregular spaces clothed on both their fetal and maternal faces by trophoblast. Some of the villi are especially intimately related to the maternal tissue and are called *anchoring villi* (Fig. 7-26). The majority of the villi, however, continue to lie more or less free in spaces in the decidua basalis. Maternal blood enters the spaces about the villi from the small vessels which were opened in the excavating process. As this blood drains back into the uterine veins, it is replaced by blood supplied by way of the uterine arteries, so that the villi are continuously steeped in fresh maternal blood.

It should be emphasized that at no time during pregnancy is there any mingling of fetal and maternal bloodstreams. The fetal circulation is, from its first establishment, a closed circuit, isolated from the maternal blood which bathes the villi by the *placental barrier*. This barrier consists of the trophoblast, its underlying basal lamina, connective tissue interposed between the trophoblast and the fetal blood vessel, the basal lamina surrounding the blood vessel, and, finally, the endothelial lining of the blood vessel itself. Through the

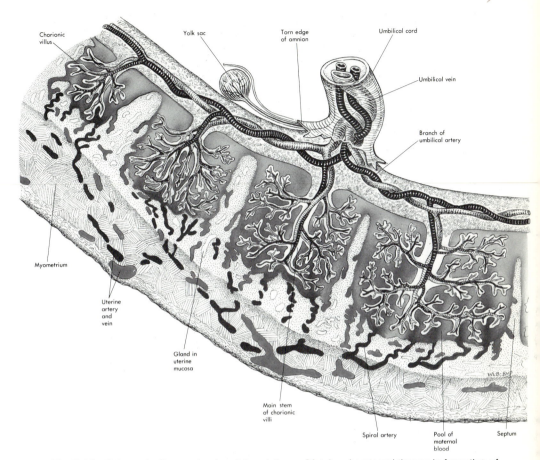

Chorionic villus

Yolk sac

Torn edge of amnion

Umbilical cord

Umbilical vein

Branch of umbilical artery

Myometrium

Uterine artery and vein

Gland in uterine mucosa

Main stem of chorionic villi

Spiral artery

Pool of maternal blood

Septum

Fig. 7-25 Schematic diagram to show interrelations of fetal and maternal tissues in formation of placenta. Chorionic villi are represented as becoming progressively further developed from left to right across the illustration. (See colored insert.)

placental barrier must pass, in a two-way stream, waste substances which must be eliminated by the fetus and substances from the mother which are needed for respiration, growth, fluid balance, and immunological defense of the fetus (Fig. 7-28).

The basis of placental function is the pattern of circulation of maternal and fetal blood in relation to the villi and the placental barrier. On the maternal side, blood enters into the intervillous space through the open ends of about 30 spiral arteries of the uterus at a pressure of about 70 to 80 mm (Ramsay, 1965; Wilkin, 1965). The arterial blood, rich in oxygen and nutrients, passes over the villi in small fountainlike streams and then under reduced pressure settles back to the maternal base of the placental compartment where it is removed via open-ended uterine veins (Fig. 7-25). The intervillous space occupied by blood in the mature placenta is about 150 ml, and near term this volume of blood is replaced about

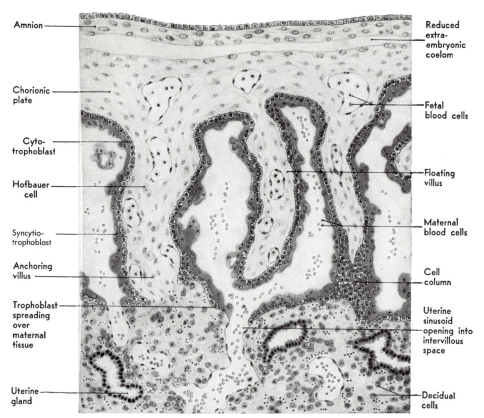

Amnion

Reduced extra-embryonic coelom

Chorionic plate

Fetal blood cells

Cyto-trophoblast

Hofbauer cell

Floating villus

Syncytio-trophoblast

Maternal blood cells

Anchoring villus

Cell column

Trophoblast spreading over maternal tissue

Uterine sinusoid opening into intervillous space

Uterine gland

Decidual cells

Fig. 7-26 Semischematic drawing to show the relations of the chorionic villi and the trophoblast to the maternal tissues of the placenta. (Redrawn, with modifications, from J.P. Hill, 1931, Phil. Trans. *Roy. Soc. London*, ser. B, vol. 22.)

three times per minute. On the fetal side, blood enters the placental villi through branches of the umbilical arteries. Despite the fact that anatomically the blood is arterial, it is the physiological equivalent of venous blood—poor in oxygen and high in CO_2 and waste products. In the terminal branches of the villi the fetal vessels are in the form of capillary networks, and in this region the bulk of placental exchange occurs. The now replenished blood returns to the fetus through the drainage system of the umbilical vein.

The major functions of the placenta consist of transport and synthesis. Transport functions in both directions across the placental barrier. The surface area available for exchange is greatly increased not only by the branches of the chorionic villi but also by vast numbers of microvilli which project from the surface of the syntrophoblast (Fig. 7-27). From the maternal side several classes of substances are transported (Fig. 7-28). One consists of easily diffusible substances, such as oxygen, water, and inorganic ions. Another class is composed of low-molecular-weight compounds, such as sugars, amino acids, and

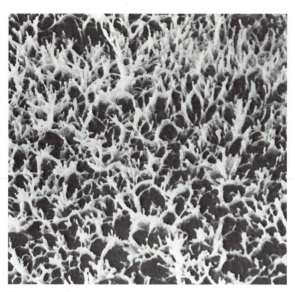

Fig. 7-27 Scanning electron micrograph of the syncytial trophoblastic surface of the human placenta in the twelfth week of pregnancy. Numerous microvilli increase the absorptive surface of the placenta. X9000. (Courtesy of Dr. Staffan Bergström.)

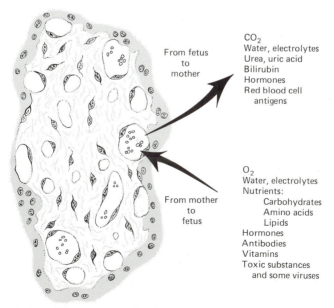

From fetus
to
mother

CO_2
Water, electrolytes
Urea, uric acid
Bilirubin
Hormones
Red blood cell
    antigens

O_2
Water, electrolytes
Nutrients:
    Carbohydrates
    Amino acids
    Lipids
Hormones
Antibodies
Vitamins
Toxic substances
    and some viruses

From mother
to
fetus

Fig. 7-28 Diagram illustrating the major forms of exchange between a fetus and its mother across a placental villus.

lipids, which serve as substrates for anabolic processes within the embryo. Their transfer requires active transport across the membranes of the placenta. Larger molecules, such as protein hormones and antibodies, require a means of transport involving pinocytosis as well as diffusion. An important class of transported macromolecules is maternal antibodies, which protect the newborn infant from disease until the immune system of the infant becomes functional. In many other mammals, such as cattle, the placental barrier does not allow the passage of maternal antibodies into the fetus. The newborn calf must obtain antibodies from its mother's milk during the first 36 hours after birth (while its intestinal villi are still capable of absorbing undigested proteins) or it will become an immunological cripple. From the fetal side, the principal substances transported across the placenta are CO_2, water, electrolytes, urea, and other waste products of fetal metabolism.

The placenta is known to synthesize four hormones in the syntrophoblastic layer. Two are protein hormones (*human chorionic gonadotropin* and *human placental lactogen*[3]) and two are steroids (*progesterone* and *estrogens*). Chorionic gonadotropin, the hormone responsible for maintaining the corpus luteum, is produced early by the trophoblastic tissues, even before implantation. The presence of this hormone in maternal urine has been the basis for many of the common tests for pregnancy. Under the sustaining influence of chorionic gonadotropin, the corpus luteum continues to produce progesterone and estrogens, which, in turn, act upon the endometrium so that it continues to provide adequate support for the growth of the embryo. Within a couple of months the placenta synthesizes estrogens and progesterone in such quantities that pregnancy can be maintained even if the corpus luteum is surgically removed.

The part of the chorion not involved in the formation of the placenta also undergoes interesting changes. During the last half of pregnancy, the chorion laeve is pushed by the growing embryo tight against the uterine walls and actually fuses with the opposite decidua parietalis, thus obliterating much of the uterine cavity (Fig. 7-31). Adherent to the inner face of the chorion laeve is the amnion, which during the third month expands to fill the entire chorionic sac (Fig. 7-29) and soon thereafter becomes loosely attached to its inner face. At term, the amniotic sac contains almost a liter of amniotic fluid. A section passing through the tissue between the amniotic cavity and the muscular layer of the wall of the uterus, in a region clear of the placenta, will show a merging of three originally separate structures. From the embryo toward the uterus these are, in order, the amnion, the chorion laeve, and the decidua parietalis (Fig. 7-31).

Birth and the Afterbirth The placental attachment normally occurs relatively high up in the body of the uterus. This results in the much-thinned decidua capsularis, the chorion laeve, and the adherent amnion being the only

---

[3]Also called *chorionic somatomammotropin*, this poorly understood hormone has both somatotropic and prolactinlike activity.

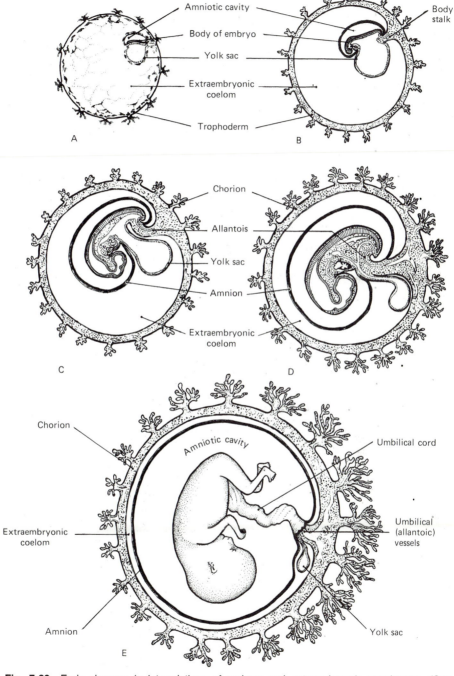

Fig. 7-29 Early changes in interrelations of embryo and extraembryonic membranes. (See colored insert.)

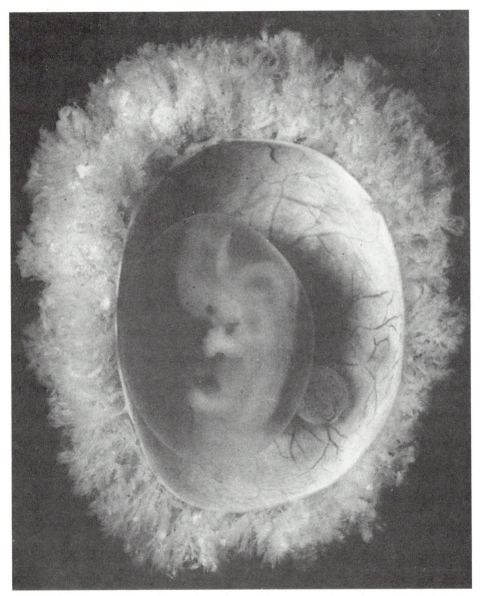

Fig. 7-30 Chorionic vesicle of seventh week opened to expose the embryo within its intact amnion. The small sphere to the right of the amnion is the yolk sac. (Photograph by Chester Reather of Carnegie. Embryo No. 8537A.)

structures lying over the cervical outlet (Fig. 7-31). Together they form one composite fibrous membrane. With the beginning of the muscular contractions which mark the onset of labor, the amniotic fluid is squeezed into this thin part of the chorionic sac and the sac acts as a preliminary dilator of the cervical canal. As the periodic contractions become more frequent and more powerful, the investing membranes rupture at this region, freeing the embryo from its fetal envelopes but leaving the placenta still attached within the uterus. The retention of the placenta is of vital importance, for the process of birth ordinarily extends over several hours, and were the fetus to be prematurely cut off from its uterine associations it could not survive the resulting interruption of its oxygen supply.

Continued uterine contractions, aided by voluntary contractions of the abdominal muscles, force the fetus into the slowly enlarging cervical canal until it is dilated sufficiently to permit the fetus to begin to move out of the uterus. When this has been accomplished, the obstetrician speaks of the first stage of

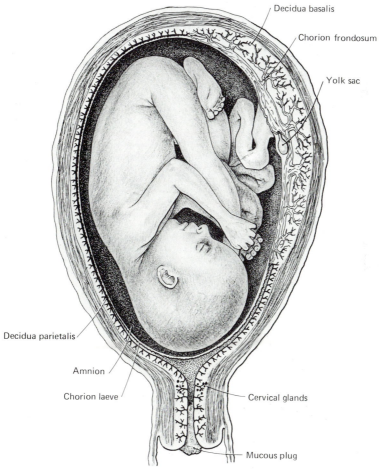

Fig. 7-31 Diagram showing the relations to the uterus of a 5-month fetus and its membranes. The amniotic cavity is darkly stippled; the uterine cavity is lightly stippled.

labor as having been passed. The second stage of labor is much briefer than the first. Once the fetus passes the cervical canal it moves promptly through the vagina to "present itself" at the perineum. Dilation of the vulval orifice of the vagina progresses much more rapidly than did dilation of the cervix, and once the presenting part of the body—usually the head—passes this outlet, the rest of the body emerges rapidly. With the tying off and cutting of the umbilical cord the relations with the uterus and placenta are ended and the newborn infant is for the first time an independently living individual.

In the usual course of events, some 15 or 20 minutes after the delivery of the fetus the uterus begins again to go into a series of contractions which loosen the placenta and the decidua from its walls and finally expel them. This is called the third stage of labor. Associated with the placenta are the torn remnants of the ruptured amnion, chorion laeve, and umbilical cord. This entire mass constitutes the "afterbirth."

Chapter 8

Basic Body Plan of Young Mammalian Embryos

As the embryo gets past the stage of gastrulation, the period of organogenesis begins. So many important events are occurring simultaneously that it is very difficult to obtain a clear picture of the details necessary to understanding the development of individual components by studying whole embryos at a given stage. Thus the remaining chapters of this text will describe the differentiation and morphogenesis of specific tissues and organs. The bulk of the descriptive material will be based upon mammalian embryos, principally the pig and human, but so much of the experimental work elucidating many developmental mechanisms has been conducted on embryos of amphibians and birds that references to this work will be integrated into the text. The main purpose of this chapter is to provide an overall structural plan for the mammalian embryo at the start of the period of organogenesis.

CHANGES IN BODY CONFIGURATION FROM PRIMITIVE STREAK TO EARLY ORGANOGENESIS

Young mammalian embryos begin to form their neural plate shortly after the early primitive-streak stage, at about the same time that the notochord commences its rapid elongation (Fig. 8-1). During the early stages of somite formation there is a striking similarity in the general organization of the body of a young chick (Fig. A-11) and a young pig (Fig. 8-1) or a young human embryo

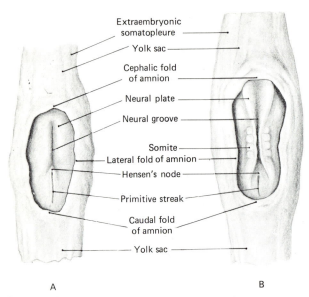

Extraembryonic somatopleure

Yolk sac

Cephalic fold of amnion

Neural plate

Neural groove

Somite

Lateral fold of amnion

Hensen's node

Primitive streak

Caudal fold of amnion

Yolk sac

A   B

Fig. 8-1 Drawings (X15) of pig embryos at the first appearance of the neural groove (*left*, from Carnegie Collection, C160-68) and at the time of the first somites (*right*, from Carnegie Collection, C190-2 and C196-1).

(Fig. 8-2). As if to anticipate its ultimate cerebral dominance, the human embryo, however, soon begins to show a relatively larger neural plate in its future forebrain region (cf. Figs. A-15 and 8-3). The cephalic part of the neural plate of the young human embryo is so large, and expanding laterally so rapidly, that its closure is delayed as compared with the early closure in other embryos. In the midbody region, however, the general arrangement of structures and layers is essentially similar (cf. Fig. A-25D with Fig. 8-3E, and Fig. A-25E with Fig. 8-3F). Toward the end of the third week a young human embryo loses its originally straight body axis (Fig. 8-4A). and begins to show a flexure in its cranial region (Fig. 8-4B). During the fourth week this cranial flexure rapidly becomes more marked (Fig. 8-5).

The basic body plan in a pig embryo of 5 mm (Fig. 8-6) or a human embryo of just over a month (Fig. 8-7) is much the same as that of a chick of 4 to 4½ days (Fig. A-38). There are, of course, minor differences. In birds the eyes and the midbrain, which will contain the optic centers, are relatively larger and further developed. The heart, liver, and mesonephros are more conspicuous and more advanced in mammalian embryos (cf. Figs. A-38 and 8-9).

EXTERNAL FEATURES OF 4- TO 6-MILLIMETER EMBRYOS

The Cephalic Region   In the cephalic region the thin superficial ectoderm leaves the contours of the brain clearly suggested (Fig. 8-6). The nasal (olfactory) placodes have appeared as a pair of local thickenings in the ectoderm at the rostral end of the head (Fig. 8-14E), and the bulging caused by the

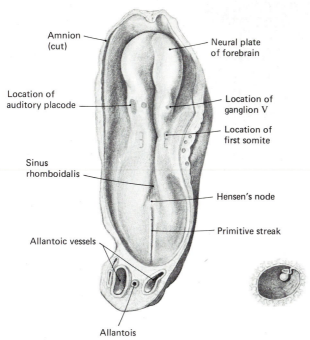

Amnion (cut)

Neural plate of forebrain

Location of auditory placode

Location of ganglion V

Location of first somite

Sinus rhomboidalis

Hensen's node

Primitive streak

Allantoic vessels

Allantois

Fig. 8-2 A human embryo at the beginning of somite formation (X45). (After Ingalls, 1920, *Carnegie Cont. to Emb.*, vol. 11.)

growing optic cups clearly marks the position of the eyes (Fig. 8-7). In the more advanced embryos in this general range the nasal placodes have been depressed below the general surface to lie in the floor of the nasal pits (Fig. 8-7). Cleared and stained preparations show the location of the auditory vesicles well, and the brain walls appear sharply outlined (Fig. 8-9).

The Oral Region and the Branchial Arches and Clefts Flanking the future oral region and caudal to it, in the region which will be under the chin, the branchial arches appear as strongly marked local elevations. This entire region is, at this stage, so compressed against the thorax that one gets a very incomplete view of its structure unless an embryo is decapitated and the head viewed in ventral aspects (Figs. 8-12 and 13-1A and B). It is then seen that the *maxillary processes* form the lateral parts of the upper jaw and that the two mandibular elevations merge with each other in the midventral line to form the arch (*mandibular arch*) of the lower jaw. Posterior to the mandibular are three similar arches, the *hyoid and the unnamed third and fourth postoral arches*, all of which appear clearly in lateral views (Figs. 8-6 and 8-7).

Between the branchial arches are deep furrows which mark the position of ancestral gill clefts. Although in mammalian embryos these furrows do not ordinarily break through into the pharynx, they are commonly called *clefts* because of their phylogenetic significance. Only the most cephalic of these clefts is named (*hyomandibular cleft*; Fig. 8-6); the others are designated by their

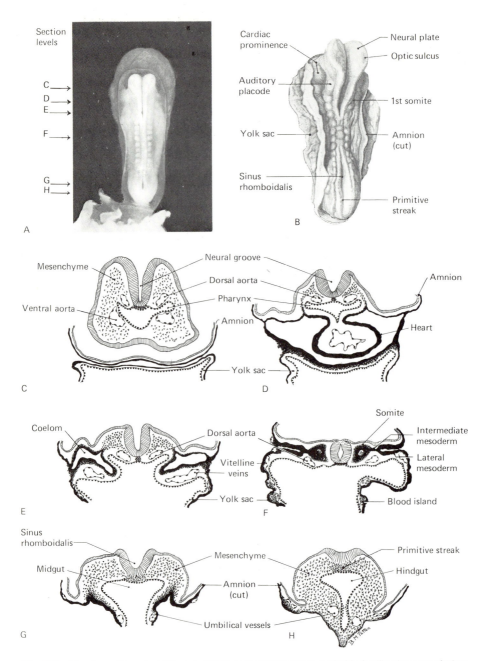

Fig. 8-3 Structure of human embryo in the 7- to 8-somite stage, probable fertilization age of 18 to 19 days. (A) Bartelmez 8-somite embryo (University of Chicago, H 1404), photographed (X12½) before sectioning. (B) Reconstruction of the Payne 7-somite embryo (X22). C-H Sections of Bartelmez embryo at levels indicated in (A). Projection outlines (X60) schematically represented with ectoderm hatched; endoderm, a beaded line; mesenchyme, angular stippling; and the more solid parts of the mesoderm in black.

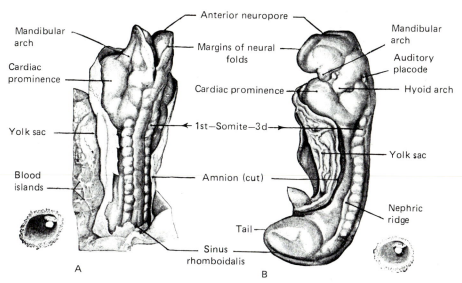

Fig. 8-4 Two human embryos of about 3 weeks' fertilization age. (A) Corner 10-somite embryo; probable age about 20 days (X25). (After Corner, 1929, *Carnegie Cont. to Emb.*, vol. 20.) (B) Heuser 14-somite embryo; probable age about 22 days (X30). (After Heuser, 1930, *Carnegie Cont. to Emb.*, vol. 22.) Sketches in lower corners show actual size of respective embryos and their chorionic vesicles.

postoral numbers. The entire region about the third and fourth postoral clefts becomes especially deeply depressed and is known as the *cervical sinus* (Fig. 8-7).

The Thorax At this stage of development the neck is not yet elongated and the head seems to be set directly on the thorax. The precociously large heart is the most massive thoracic structure, and the surface bulge it causes is known as the *cardiac prominence* (Fig. 8-7). The lungs are as yet but minute primordial cell clusters that do not cause any recognizable surface landmarks. Somites are serially arranged in pairs, beginning in the cephalic region not far caudal to the auditory vesicle and extending into the caudal region (Fig. 8-9).

The Abdomen Just caudal to the cardiac prominence is a shallow, superficial groove (shown, but not labeled, in Fig. 8-7). Beneath this groove lies a shelf of concentrated mesenchyme known as the *septum transversum* (Fig. 8-13). As is clearly suggested by its position just caudal to the heart and just cephalic to the liver, the septum transversum is the primordium of the ventral portion of the diaphragm. It already shows where the line of demarcation between thorax and abdomen is being established.

The Belly Stalk Conspicuous in the midventral region is the belly stalk. In younger embryos of the age range we are considering the ventral body walls and their amniotic extensions have not yet tightly clothed the yolk stalk (Fig. 8-6). At

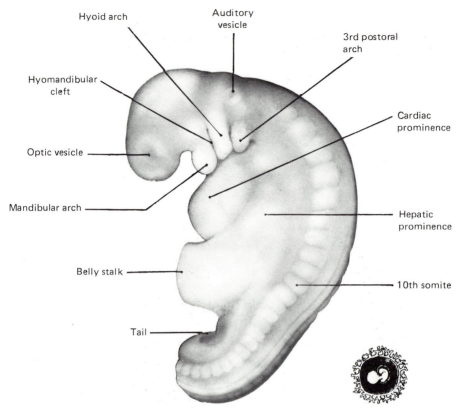

Fig. 8-5 Human embryo toward end of fourth week. Retouched photograph (X26) of embryo 6097 in the Carnegie Collection; crown-rump length 3.6 mm; 25 pairs of somites. Sketch, lower right, shows actual size of embryo and its chorionic vesicle.

this stage, they are about in relation to each other as shown diagrammatically in Fig. 7-29C. A little later the belly stalk will be more sharply delineated (Figs. 8-7 and 7-29D).

The Appendage Buds The appendage buds first appear as flangelike projections from the body wall extending over the territory occupied by about 6 somites (Fig. 8-6). A little later they become paddle-shaped (Fig. 8-8). The arm bud slightly leads the leg bud in developmental progress.

The Tail At this stage the human embryo has every bit as well developed a tail as the pig (cf. Figs. 8-6 and 8-7). Later in development the human tail normally undergoes regressive changes (Fig. 16-28) that leave us with only our symbolic coccyx. Regression of the human tail is apparently brought about by intrinsic cell death (Fallon and Simandl, 1978). Occasionally this regression fails to occur, and a human infant is born with a sizable and unmistakable tail.

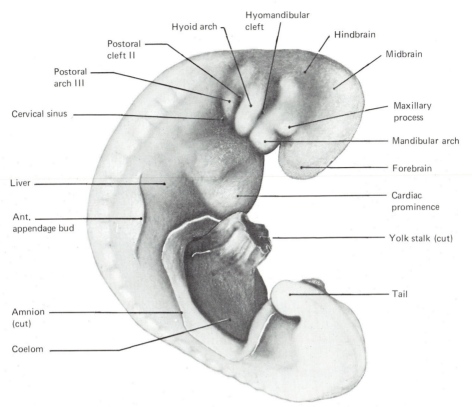

Fig. 8-6 Photograph (X20) of a pig embryo of 5 mm. Compare with Fig. 8-7 showing human embryo of closely comparable developmental stage.

THE NERVOUS SYSTEM

The Brain The central nervous system has progressed from a simple neural tube to one in which the fundamental subdivisions of the brain are taking shape. At first, three subdivisions (vesicles) are apparent. These are the forebrain (*prosencephalon*), the midbrain (*mesencephalon*), and the hindbrain (*rhombencephalon*) (Fig. 11-13A). At the stage under consideration the brain is in the phase of transition from the three- to the five-vesicle condition. There is a clear indication of separation of the forebrain into a rostral *telencephalon* and a *diencephalon*, and the thickening of the walls of the more rostral part of the hindbrain presages the differentiation of the *metencephalon* from the *myelencephalon* (Fig. 11-13C).

Projecting from the ventrolateral walls of the *diencephalon* are the *optic vesicles*, which are beginning to invaginate to form the optic cups (Fig. 8-14D). In response to an inductive signal from the optic vesicle, the ectoderm overlying the optic cup has thickened and begun to invaginate to form the lens vesicle (Figs. 8-14D and 12-2D).

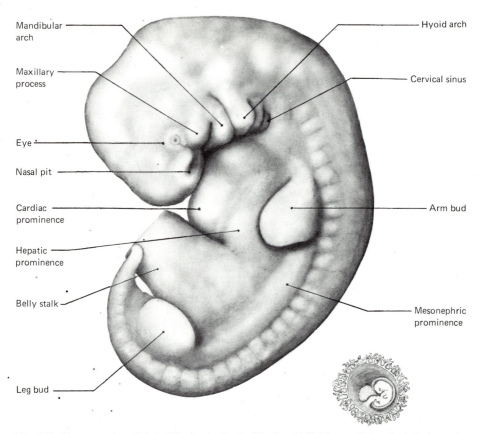

Fig. 8-7 Human embryo late in fifth week after fertilization; C-R 6.5 mm. Retouched photograph (X17) of embryo 6502 in the Carnegie Collection. Small sketch, lower right, shows actual size of embryo and its chorionic vesicle.

In the change from the three- to five-vesicle condition of the brain, the mesencephalon remains undivided. Its extent is indicated by constrictions between itself and the adjacent diencephalon and metencephalon (Figs. 11-13B and 8-13). The point of transition from metencephalon to myelencephalon is marked by little more than the thinning out of the roof of the neural tube where the myelencephalon begins (Fig. 8-13).

The Cranial Nerves Of the 12 pairs of cranial nerves characteristically present in mammals the first nerve (olfactory) and the second nerve (optic) are associated with organs of special sense, the primordia of which are already well established. The actual fibers, however, are not as yet readily recognizable. The third, fourth, and sixth nerves to the eye muscles, the eleventh to muscles of the neck, and the twelfth to the muscles of the tongue are primarily motor in function. The semilunar ganglion of the fifth cranial (trigeminal) nerve lies opposite the thickest part of the metencephalon (Figs. 8-9 and 8-14A). The still

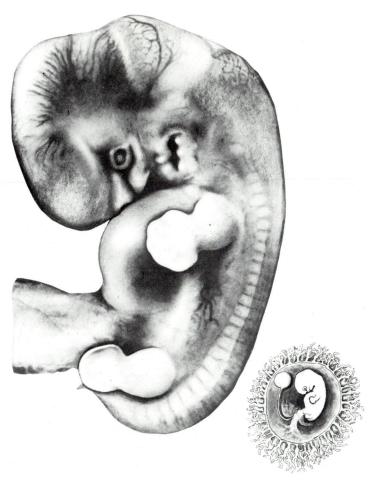

Fig. 8-8 Human embryo a little over 6 weeks after fertilization, C-R 14 mm. [Retouched photograph (X8) of embryo No. 1267A in the Carnegie Collection.]

unseparated mass of cells which will later be regrouped to form the ganglia of the seventh (facial) nerve and the eighth (acoustic) nerve appears just rostral to the auditory vesicles (Figs. 8-9 and 8-14A). Just caudal to the auditory vesicle are the ganglia of the ninth (glossopharyngeal) nerve and the tenth (vagus) nerve.

The Spinal Cord and Spinal Nerves Caudal to the myelencephalon the neural tube is more slender and gives rise to the spinal cord. Later the cord undergoes complex histogenetic changes which will be discussed in Chap. 11. During these early stages it is a relatively simple-appearing tube with a slitlike central canal and walls of closely packed cells of ectodermal origin. Alongside the spinal cord the ribbonlike neural crest has just broken up and formed the spinal (and also cranial) ganglia.

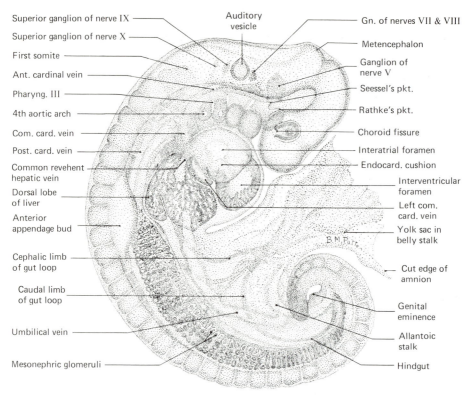

Superior ganglion of nerve IX

Superior ganglion of nerve X

First somite

Ant. cardinal vein

Pharyng. III

4th aortic arch

Com. card. vein

Post. card. vein

Common revehent hepatic vein

Dorsal lobe of liver

Anterior appendage bud

Cephalic limb of gut loop

Caudal limb of gut loop

Umbilical vein

Mesonephric glomeruli

Auditory vesicle

Gn. of nerves VII & VIII

Metencephalon

Ganglion of nerve V

Seessel's pkt.

Rathke's pkt.

Choroid fissure

Interatrial foramen

Endocard. cushion

Interventricular foramen

Left com. card. vein

Yolk sac in belly stalk

Cut edge of amnion

Genital eminence

Allantoic stalk

Hindgut

B.M.Patten

Fig. 8-9 Projection drawing (X17) of a lightly stained and cleared 5-mm pig embryo.
Abbreviations: *Endocard . cushion*, endocardial cushion of atrioventricular canal; *gn .*, ganglion; *pkt .*, pocket.

THE DIGESTIVE AND RESPIRATORY SYSTEMS

The Oral Region The intraembryonic gut tract of young mammalian embryos at first ends blindly at both its cephalic and its caudal ends (Fig. 8-10B and C). Where the oral opening is destined to be, there is a depression called the *stomodeum*. As the stomodeum deepens, the ectoderm of its floor comes into direct contact with the endoderm of the foregut to form the *stomodeal*, or *oral*, *plate*. During the stage under consideration the oral plate breaks through and establishes the oral opening into the foregut (Figs. 8-10D, 8-11, and 8-12). With this change we speak of the old stomodeal depression as the *oral cavity*. At this stage it is very shallow, but as development progresses, the oral cavity becomes much deeper because of the growth of surrounding structures (Chap. 13).

Arising medially as a slender ectodermal diverticulum from the rostral part of the stomodeum is *Rathke's pocket* (Figs. 8-13 and 15-4). From its first appearance Rathke's pocket is in close relationship to the *infundibular process* from the floor of the diencephalon. Together they will form the *hypophysis*.

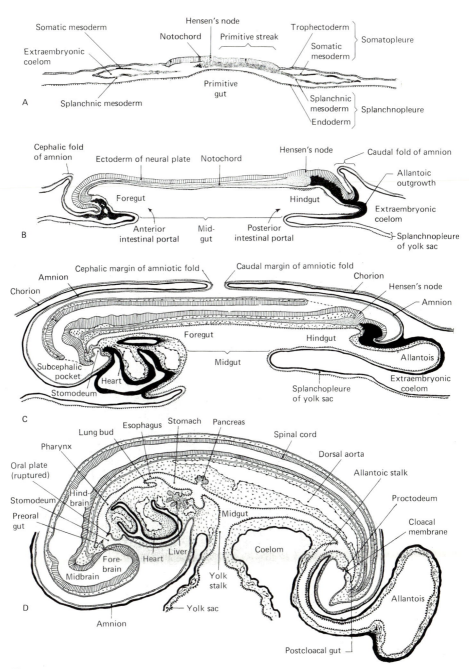

Fig. 8-10 Sagittal sections of pig embryos to show establishment and early regional differentiation of the gut. The drawings indicate, schematically, conditions (A) in the primitive-streak stage; (B) at the beginning of somite formation; (C) in embryos having about 15 somites; (D) in embryos having about 25 somites, or about 5 mm.

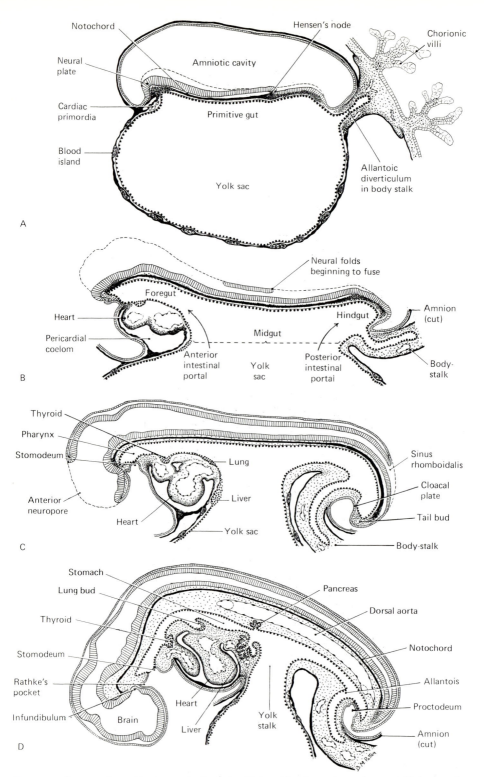

Fig. 8-11 Sagittal plans of human embryos in the third and fourth weeks to show establishment of the digestive system. (A) At the beginning of somite formation, about 16 days. (B) Seven somites, about 18 days. (C) Fourteen somites, about 22 days. (D) Toward the end of the first month.

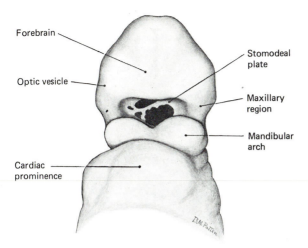

Forebrain

Optic vesicle

Cardiac prominence

Stomodeal plate

Maxillary region

Mandibular arch

Fig. 8-12 Breaking through of the stomodeal plate to establish oral opening into foregut as seen in a face view of a human embryo of the fourth week. (Drawn X30, from stereophotographs of embryo 6097 in the Carnegie Collection.)

The Pharynx   The breaking through of the oral plate does not occur at the extreme cephalic end of the foregut, so there remains a bay known as the *preoral gut* (Fig. 8-10D), or *Seessel's pocket* (Fig. 8-13). Except as a landmark in early stages Seessel's pocket is of no great importance because it gradually disappears without giving rise to any specific adult structure.

The cephalic portion of the foregut is mainly involved in the formation of the pharynx. Shallow dorsoventrally, the pharynx is broad laterally. There are four pouches on either side, each one extending out toward the corresponding external gill furrow (Fig. 8-14B). Where a pharyngeal pouch comes close to an external gill groove, the tissue between the pharyngeal lumen and the outside world is reduced to a very thin membrane consisting of the endoderm of the pharyngeal lining and the ectoderm in the floor of the gill-furrow with no intervening mesenchyme. (Fig. 8-14B). Although in mammalian embryos this membrane does not ordinarily break through to form an open gill cleft similar to that in our water-living ancestors, the phylogenetic significance of the topography is so unmistakable that the furrows are often referred to as *gill clefts*. The significance of the relations in this region is further emphasized by the location of the *aortic arches*, which lie in closely packed mesenchymal tissue between the gill clefts. In the embryos of birds and mammals the aortic arches do not form capillary beds as they do in gill-breathing animals. Nevertheless, the major ancestral pattern is there in the way the aortic arches pass from the ventrally located heart to the dorsal aorta by way of the gill arches flanking the pharynx (Figs. 8-14B and 8-17).

In older embryos a number of important structures arise from the endodermal lining of the pharynx. Of these only the small cluster of cells which constitute the primordium of the *thyroid gland* has emerged at this stage. The thyroid gland arises from the floor of the pharynx in the midline at about the

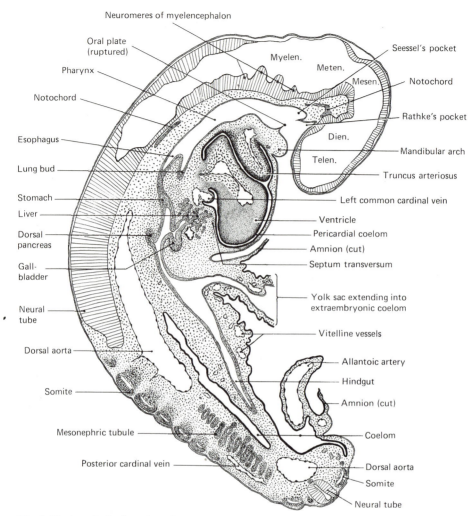

Fig. 8-13 Longitudinal section of 5-mm pig embryo (X25). The caudal end of an embryo in this stage of development is usually somewhat twisted to one side (see Fig. 8-6). For this reason, sections which cut the cephalic region in the sagittal plane pass diagonally through the posterior part of the body. For a schematic plan of a completely sagittal section of an embryo of about this age see figure 8-10, D.

level of the second gill arch (Figs. 8-14C and 15-4). The further development of this primordium and others that appear later in the pharyngeal region will be discussed in Chap. 15.

The Trachea and Lung Buds In the floor of the pharynx, at the level of the most posterior pair of pharyngeal pouches, a median ventral groove is rapidly converted into a tubular outgrowth parallel to the digestive tract. This groove is the *laryngotracheal groove*, and the tubular outgrowth which is formed by its prolongation caudad is the *trachea*. In 4- to 6-mm embryos the caudal end of the trachea has become enlarged and is just beginning to bifurcate to form the *bronchial* or *lung buds* (Figs. 8-13, 8-14D and E, and 14-9A and B).

The Esophagus and Stomach   Caudal to the pharynx the digestive tube is distinctly narrowed to form the *esophagus*. At this stage the esophagus is very short (Fig. 8-10D). It becomes elongated later as the stomach moves farther back in the body. A slight dilation of the foregut dorsal to the heart is all that presently indicates the location of the stomach (Fig. 8-10D).

The Liver and Pancreas   Immediately caudal to the stomach are the outgrowths of the gut which constitute the primordia of the *pancreas*, the *liver*, and the *gallbladder*. The pancreas at this stage consists of two independent parts, a well-defined dorsal bud and a smaller ventral bud (Fig. 8-13). The original hepatic diverticulum arises ventrally from the gut almost directly opposite the dorsal pancreatic bud (Fig. 8-10D). By the 5-mm stage it has given rise to a considerable mass of branching cords of endodermal cells which are the primordia of the glandular tubules of the liver (Figs. 8-13 and 14-6). These cell cords tend to push ventrad and a little cephalad to constitute the main mass of the growing liver. They soon spread out against the dorsocaudal face of the septum transversum (Fig. 8-13). The narrowed proximal portion of the original evagination from the gut persists as the duct draining the liver, and a diverticulum of it becomes enlarged to form the gallbladder (Fig. 14-6).

The Intestines   The elongation of the intestines which later results in their characteristic coiling has not yet commenced in 5-mm embryos. The entire length of the gut tract still lies essentially in the sagittal plane of the body. In the region of the narrowing midgut, where there is still a fairly broad communication with the yolk sac, a slight ventral bend is beginning to appear (Fig. 8-10D). This is the start of a U-shaped bend that rapidly deepens and becomes a highly characteristic feature of the gut tract in slightly older embryos (Fig. 14-14).

The lengthening of the foregut and the hindgut, as they are "floored in" by closure of the ventral body walls, soon reduces the primitive midgut to a small opening into the yolk stalk. The yolk stalk, therefore, serves as an excellent landmark indicating what part of the developing intestinal tract came from the foregut and what part was derived from the hindgut.

Within foregut territory we have just noted the beginning of many specialized structures—the pharyngeal pouches and their derivatives, the primordia of the respiratory organs, the stomach, the pancreas, and the liver. In contrast, the hindgut is relatively late in declaring its definitive regions. At this stage the only special feature of the hindgut is the allantoic stalk arising as a diverticulum from its caudal portion (Fig. 8-10D). The slightly dilated region of the hindgut from which the allantoic stalk emerges is the *cloaca*.

Near the caudal end of the cloaca is a depression in the ventral body wall called the *proctodeum*. It deepens in a manner similar to that exhibited by the

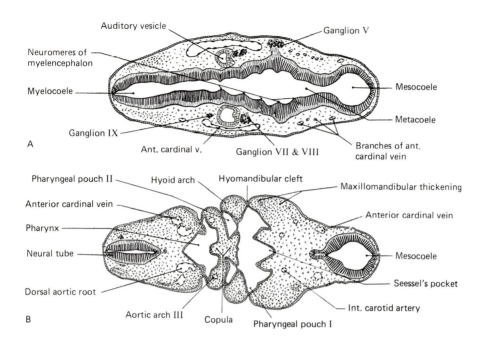

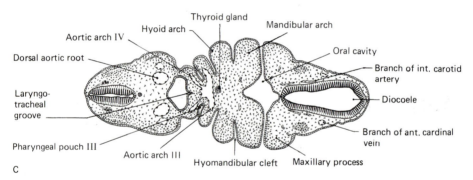

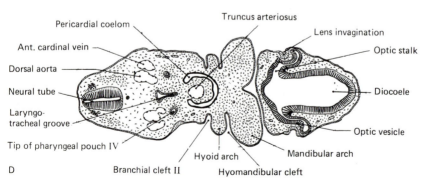

Fig. 8-14 Transverse sections of 5-mm pig embryo (X18). Compare with entire embryo (Fig. 8-6) and with longitudinal section of embryo of same age (Fig. 8-13).

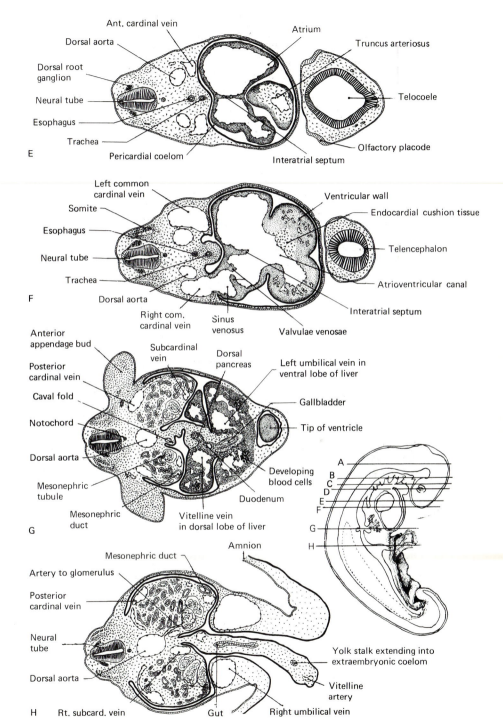

E — Ant. cardinal vein, Dorsal aorta, Dorsal root ganglion, Neural tube, Esophagus, Trachea, Pericardial coelom, Atrium, Truncus arteriosus, Telocoele, Olfactory placode, Interatrial septum

F — Left common cardinal vein, Somite, Esophagus, Neural tube, Trachea, Dorsal aorta, Right com. cardinal vein, Sinus venosus, Valvulae venosae, Ventricular wall, Endocardial cushion tissue, Telencephalon, Atrioventricular canal, Interatrial septum

G — Anterior appendage bud, Posterior cardinal vein, Caval fold, Notochord, Dorsal aorta, Mesonephric tubule, Mesonephric duct, Subcardinal vein, Dorsal pancreas, Vitelline vein in dorsal lobe of liver, Duodenum, Developing blood cells, Tip of ventricle, Gallbladder, Left umbilical vein in ventral lobe of liver

H — Mesonephric duct, Artery to glomerulus, Posterior cardinal vein, Neural tube, Dorsal aorta, Rt. subcard. vein, Gut, Right umbilical vein, Vitelline artery, Yolk stalk extending into extraembryonic coelom, Amnion stomodeum at the oral end of the foregut, leaving only a thin separating membrane of proctodeal ectoderm and cloacal endoderm. This membrane is known as the *cloacal plate*, or *cloacal membrane* (Fig. 8-10D). With its rupture, which occurs considerably later than the rupture of the oral plate, the originally blind hindgut has established an outlet.

THE MESODERM

From the simple condition in which it was divided into somitic (paraxial), intermediate, and lateral plate components, the mesoderm has begun to undergo some specialization. The somites have developed past the purely epithelial stage (Fig. 6-16B). The cells of the sclerotomal region have broken away from the main body of the somite, leaving the dermatomal and myotomal layers (Fig. 6-16D).

The intermediate mesoderm has differentiated into the tubules and ducts of pronephric and mesonephric kidneys (Fig. 8-14G and H), transient stages in the ontogenesis of the definitive urogenital system. By looking ahead to the schematic drawings of Figs. 16-4 and 16-5, one can better appreciate the relationship between the mesonephros and the metanephros, the permanent kidney. The urogenital (intermediate) mesoderm is concentrated in a pair of ridges, which run on the dorsal mesentery (Figs. 8-14G and 16-14). At this stage there is still no trace of gonadal tissue.

At a very early stage, the lateral plate mesoderm splits into two layers, one associated with ectoderm and one associated with endoderm. By convention, a layer of mesoderm plus ectoderm is called a *somatopleure* and a layer of mesoderm plus endoderm is called a *splanchnopleure*.

THE CIRCULATORY SYSTEM

Basic Plan of the Embryonic Circulatory System   The embryonic circulatory system can be analyzed in terms of three major circulatory arcs, with the heart as the common center and pumping station. One arc is entirely intraembryonic in its distribution. Its vessels bring food materials and oxygen to all parts of the growing body and from them they return waste materials. The other two circulatory arcs have both intra- and extraembryonic components. The vitelline arc carries blood to and from the yolk sac. The other arc carries blood to and from the allantois for gaseous interchange (Fig. A-44). As the blood from the three arcs is returned to the heart for recirculation, it is constantly being mixed so that its food material, oxygen and accumulated waste products are maintained at serviceable levels.

In placental mammals radical changes in the source of food supply and basic living conditions, as compared with reptiles and birds, alter the way the two extraembryonic circulatory arcs operate. The yolk sac of higher mammalian embryos, although formed in characteristic relations to other structures, is small and empty; its circulatory arc, therefore, has lost its significance as a purveyor of food. Nevertheless, the vessels of this arc still form and are for a time quite conspicuous. In what may be called a "phylogenetic hangover," the vitelline vessels, although deprived of their primary function, still bring the first blood cells into the embryonic circulation from their place of formation in the yolk-sac splanchnopleure, just as they did in ancestral forms with food-laden yolk sacs.

In mammalian embryos the allantoic arc takes over the functions abandoned by the vitelline arc as well as continuing to carry its own earlier responsibilities. The change in the embryo's immediately surrounding environment does, however, alter the manner in which the interchanges are carried out. Living within the uterus of its mother, a mammalian embryo cannot, as does the chick, bring its allantois close to the outside atmosphere, with only a porous shell intervening (Fig. A-44). Instead its allantois is either closely applied to the uterine lining, as in the pig (Fig. 8-17A), or sends into the uterine mucosa little rootlets of its own tissue (called *chorionic villi*), as in the human (Fig. 8-17B). In either situation, maternal blood and fetal blood are brought so close together that the fetal blood can absorb food and oxygen from the maternal blood and pass its own waste materials back to the maternal circulatory system. Thus, the allantoic circulation of a mammalian embryo serves as a provisional mechanism for food-getting, for respiration, and for excretion.

If these basic functional considerations are kept in mind, there is such inescapable logic to the way the circulation of an embryo operates that its organization is easy to remember. We shall now follow the early establishment of the heart and blood vessels, using mammalian development as a descriptive base, up to the pattern seen in 4- to 6-mm pig embryos or 4-week human embryos.

The Establishing of the Heart   In mammalian embryos, the heart arises from paired primordia situated ventrolaterally beneath the pharynx (Fig. 8-15). The cardiac primordia are composed of two layers as well as paired right and left. The inner layer is called the *endocardium*, because it is destined to form the internal lining of the heart. The outer layer, derived from thickened splanchnic mesoderm, is known as the *epimyocardium*, because it will give rise both to the heavy muscular layer of the heart wall (*myocardium*) and to its outer covering (*epicardium*).

While these changes have been occurring in the heart, folding-off of the embryonic body has been going on with concomitant progress in the closure of the foregut at the level of the heart (cf. Fig. 8-15A and B). As a result the paired endocardial tubes are brought progressively closer together. Finally they are approximated to each other and fused into a single tube lying in the midline (Figs. 8-15 and 8-16).

In the same process the epimyocardial layers are bent toward the midline enwrapping the endocardium. Ventral to the endocardial tubes the epimyocardial layers of opposite sides come into contact with each other. Where this contact occurs the limbs of the mesodermal folds next to the endocardium fuse with each other, forming an outer layer of the heart no longer interrupted

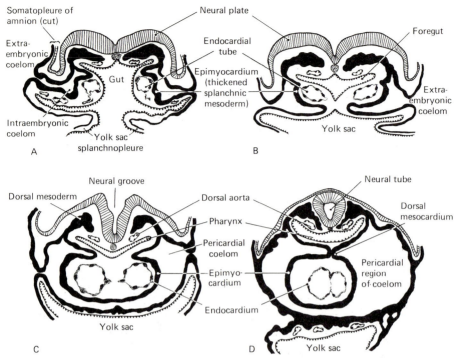

Fig. 8-15   Sections cut transversely through the cardiac region of pig embryos of various ages to show the origin of the heart from paired primordia. (Projection diagrams X65, from series in the Carnegie Collection.) (A) 5-somite embryo; (B) 7-somite embryo; (C) 10-somite embryo; (D) 13-somite embryo.

ventrally (Fig. 8-15C). Thus the originally paired right and left coelomic chambers become confluent to form a median unpaired pericardial cavity in the same process which establishes the heart as a median structure. Dorsally the right and left epimyocardial layers become contiguous, but here they do not fuse immediately as happens ventral to the heart. They persist for a time as a double-layered supporting membrane called the *dorsal mesocardium*. In this manner the heart is established as a nearly straight, double-walled tube suspended mesially in the most cephalic part of the coelom.

The early heart soon undergoes a change in shape from a straight tubular form to an S-shaped configuration. The morphology of these changes is similar to that in the chick embryo (Figs. A-48, A-49, A-50). During this phase the heart functions like a simple, tubular peristaltic pump. The veins converging to enter the heart become confluent in a thin-walled chamber called the *sinus venosus* (Figs. 8-14F and 17-22C and D). The sinus venosus opens through a slitlike orifice into the atrial portion of the heart. Backflow of blood is prevented by well-developed flaps known as the *valvulae venosae* (Figs. 8-14F and 8-19). From the atrial region, which is just beginning to bulge out into right and left chambers, blood enters the muscular ventricles. Internal subdivisions of the

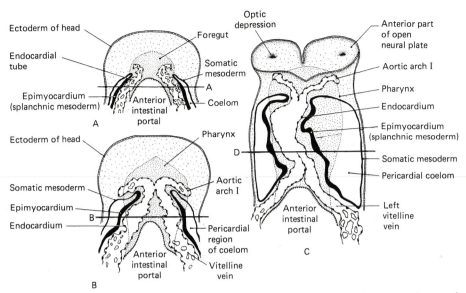

Fig. 8-16 Diagrams showing progress of fusion of cardiac primordia in the pig as seen in ventral views. (A) 5-somite embryo; (B) 7-somite embryo; (C) 13-somite embryo. The embryos are supposed to be viewed as transparent objects with the outlines of the cardiac primordia showing through. The lines *A*, *B*, and *D* indicate the levels of sections A, B, and D in Figure 8-15.

early atrial and ventricular regions is just beginning. Details of cardiac partitioning will be covered in Chap. 17. From the ventricle, blood passes into the *truncus arteriosus* and then on to the body by way of the ventral aortic roots. Often the transitional area between ventricle and truncus is called the *conus* (Fig. 17-21). Despite the early internal modifications of the structure of the heart, the blood entering the caudal end of the heart through the sinus venosus is still pumped out of the heart into the truncus as an undivided stream, just as was the case in younger embryos containing a straight tubular heart.

The Formation of Blood and Blood Vessels   While the heart is becoming established, the main vascular channels characteristic of young embryos are also making their appearance. In a manner similar to the genesis of the endocardial tubes themselves, cords and knots of mesodermal cells become aggregated along the future course of a developing vessel. These strands of cells then form hollowed-out tubes, lined by a layer of thin, flattened *endothelial cells*. In later stages some vessels also become extended by the formation of budlike outgrowths from their walls. Typically, a meshwork of small vascular channels is formed first. Gradually, some of these primitive channels become enlarged to form the main vessels. These channels, at least in the chick, are made up of two layers of flattened cells—an inner presumptive endothelium and another layer of flattened mesodermal cells surrounding it. A middle layer of collagen fibers and basal lamina forms between the two layers of flattened cells, following which the inner cellular layer differentiates into true endothelium and the outer layer into smooth muscle. Murphy and Carlson (1978) have suggested that the cellular differentiation of the layers of the vascular wall might be triggered by the intervening extracellular matrix material. Eventually, their walls are reinforced by the addition of circularly disposed connective-tissue fibers and smooth-muscle cells.

Blood cells and early blood vessels form as aggregates of splanchnic mesodermal cells lining the yolk sac (Figs. 17-2 and 17-3). These aggregates are called *blood islands*. The differentiation of the central cells of the blood islands into blood cells is described in Chap. 17. The cells around the early blood cells become flattened and differentiate into vascular endothelial cells. As the blood islands develop, they coalesce and become incorporated into the vitelline circulatory arc. These vitelline channels feed blood cells into the early beating heart, which distributes them throughout the vascular channels (Fig. 8-17). In human embryos the first blood cells are drawn from the yolk sac into the vascular system toward the close of the third week of development.

The Aortic Arches Vertebrate embryos pump their blood from the ventrally located heart around the pharynx to the dorsal aorta by way of a series of blood vessels called *aortic arches*. In most of the higher forms, six pairs of aortic arches are formed during the course of development. They appear in cephalocaudal sequence, the aortic arch that passes around the pharynx in the mandibular arch being the first one formed (Figs. A-26 and 17-7A and B). At no stage of development are all six pairs of aortic arches present simultaneously, for the most cephalic arches undergo regression before the sixth arch is formed. In mammalian embryos of the stage with which we are concerned here, one usually finds the first three pairs of arches well developed and the fourth pair starting to form (Figs. 8-17B and 17-7E). In a few of the relatively more advanced specimens the fourth arch has become a sizable vessel and the first arch has begun to regress (Fig. 17-7F). The ultimate fate of each of the aortic arches will be taken up in detail in Chap. 17.

The Dorsal Aorta and Its Major Branches When first formed, the dorsal aorta is a paired vessel throughout its entire length (Fig. A-20). This condition persists cephalic to the level of the arm buds (Fig. 17-9). In the region where they receive blood from the aortic arches, the paired aortae are usually called the *dorsal aortic roots*. Caudal to the level of the arm buds the right and left primitive aortae move toward the midline and fuse with each other to establish the unpaired condition of the dorsal aorta familiar in the adult (Fig. 17-11).

There are three conspicuous pairs of arterial trunks leading off the dorsal aorta in young embryos. The main *vitelline artery* arises at midbody level and extends ventrally to supply the vitelline vascular plexus (Fig. 8-17). Toward the caudal end of the aortae, major arterial trunks grow out along either side of the allantoic stalk (Fig. 8-17). These vessels remain paired and are known either as the *allantoic arteries*, because of their primary association with the allantois, or

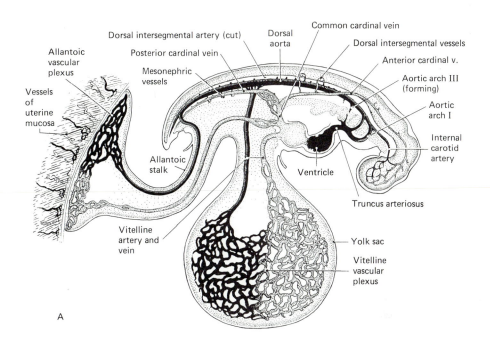

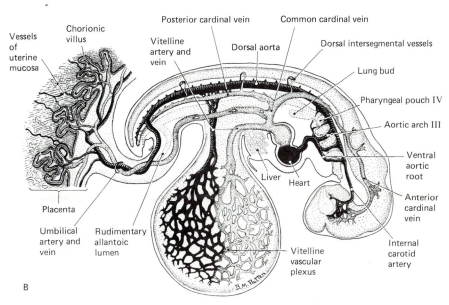

Fig. 8-17 Semischematic diagrams showing the basic plan of the circulatory system of young mammalian embryos. At this stage all the major blood vessels are paired right and left. For the sake of simplicity only the vessels on the side toward the observer are shown. (A) Pig embryo of about 4 mm (age about 16 to 17 days); (B) human embryo of about 4.5 mm (fertilization age about 4 weeks).

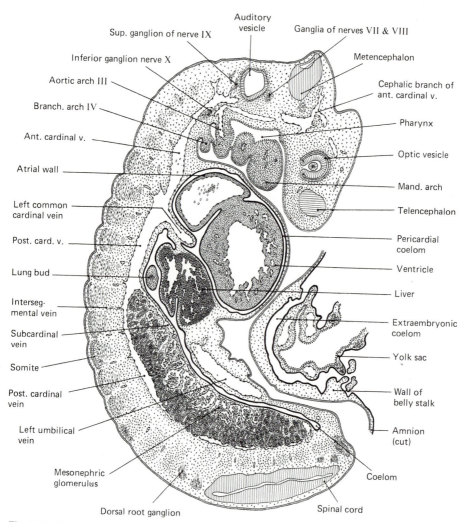

Fig. 8-18 Drawing of a parasagittal section of a 5.5-mm pig embryo. (Projection outlines X22.) The section was taken from the series to the left of the midline, at a plane especially favorable for showing the cardinal veins.

as the *umbilical arteries*, because they ultimately travel a considerable distance of their course within the umbilical cord.

The third pair of conspicuous branches from the aorta in young embryos are the *internal carotid arteries* which grow to the developing brain from the points where the first aortic arches join the dorsal aortic roots (Figs. A-32 and 8-17). The fact that these arteries become conspicuous so early is part of the general picture of the precocious growth of the cephalic region which has already been emphasized.

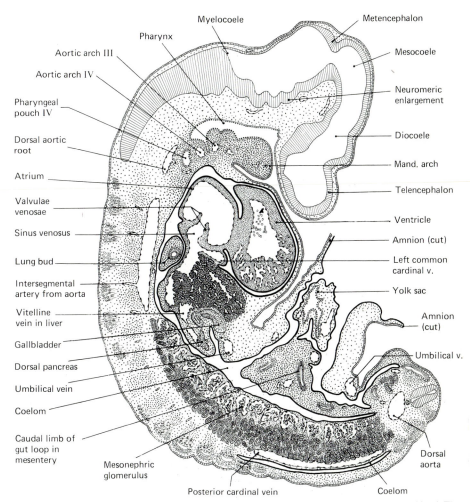

Fig. 8-19 Drawing of parasagittal section of 5.5-mm pig embryo. (Projection outlines X22.) The section was taken from the series to the right of the midline, at a plane favorable for showing the entrance of the sinus venosus into the right atrium.

The Cardinal Veins   The main return channels for the intraembryonic circulatory arc are the cardinal veins. The *anterior cardinal veins* arise as small channels collecting the blood from the head. These channels become confluent to form main vessels on either side of the developing brain (Fig. 8-17). Caudal to the level of the auditory vesicles, they dip ventrally a little, running in the dorsolateral body walls (Fig. 8-18). At the level of the heart they bend sharply downward, joining the posterior cardinals to form the *common cardinal veins* (Fig. 8-18). Continuing in the ventrolateral body walls, the common cardinals turn abruptly mesiad to enter the sinus venosus.

The *posterior cardinal veins* are the primary drainage channels of the caudal half of the body (Fig. 8-17). They originate through the confluence of small tributaries in the region of the tail and the posterior appendage buds. Their characteristic location in the caudal part of the body is along the dorsal border of the mesonephros, along which they receive numerous small vessels from the mesonephroi and the body wall (Figs. 8-14G and H, and 8-18). As they pass beyond the region of the mesonephros they course cephalad in the body wall dorsal to the liver to join the anterior cardinals in the formation of the common cardinals (Fig. 8-18).

In embryos of 4 to 6 mm the *subcardinal veins* are rather inconspicuous vessels. Later in development, however, they play an important role in the formation of the adult venous pattern in the abdominal region. At this stage they pursue a somewhat irregular course along the ventral to ventromedial border of the mesonephros (Fig. 8-14G and H). At the cephalic pole of the mesonephros the subcardinal veins bend dorsally and empty into the posterior cardinal veins (Fig. 8-18). Throughout the mesonephric region there are many small vessels interconnecting the subcardinal veins with the posterior cardinal veins.

The Vitelline Veins   Arising as collecting channels in the vitelline vascular plexus, the main vitelline veins (sometimes called the *omphalomesenteric veins*) pass toward the embryonic body, one along either side of the yolk stalk (Fig. 8-17). As they enter the body level they turn sharply cephalad, and then at the cardiac level, the right and left veins converge to empty into the sinus venosus (Fig. 17-18A). The growing cell cords of the liver soon break up the proximal portion of the vitelline veins into irregular channels, called *hepatic sinusoids* (Fig. 17-18B).

The Allantoic, or Umbilical, Veins   As was the case with the corresponding arteries, the return channels from the allantoic arc may be referred to as the *allantoic veins* because of their primary relations to the allantois or as the *umbilical veins* because, in older stages, a considerable part of their course is in the umbilical cord. Both terms are commonly used; which one is employed is usually determined by what relationship one wishes to emphasize at the moment. These veins have their origin in the capillary vessels of the allantois. Their basic relations are the same whether the allantois remains saccular, as it does in the pig (Fig. 8-17A), or forms only a rudimentary lumen, as it does in the human being (Fig. 8-17B). When they are first formed, the allantoic veins are paired throughout their entire length. They lie either side of the allantoic stalk up to the body level (Fig. 8-17) and then course cephalad in the lateral body walls (Fig. 8-14H). When they are first formed they run in the lateral body wall all the way to the cardiac level, where they join the common cardinal veins just as they empty into the sinus venosus (Fig. 8-17). This simple, direct route, however, does not long persist. Chap. 17 will outline how a series of fusions and obliterations of much of the right umbilical vein results in a single left umbilical vein entering the heart via the liver.

Chapter 9

The Differentiation of Connective Tissue, Skeletal Tissue, and Muscle

The connective tissues (which include cartilage and bone) and muscle constitute a family of tissues which arise from an embryonic tissue called *mesenchyme* (Fig. 9-1A). Regardless of the path along which they differentiate, mesenchymal cells are homogeneous in appearance and give no morphological clues which would enable one to predict their developmental fate. For many years mesenchyme was considered to be a mesodermal derivative, but there is considerable evidence that much of the mesenchyme of the head arises from the ectodermal germ layer—more specifically, the neural crest. Mesenchyme now seems to be a morphological least common denominator which tantalizingly masks both the origin and future fate of a given cell.

Differentiation of the tissues covered in this chapter follows two main patterns. In connective tissue proper, and in cartilage and bone, one of the main signs of cellular differentiation is the secretion of extracellular products—fibers and matrix material. Differentiation of muscle, on the other hand, is dominated by the intracellular accumulation of contractile proteins in highly ordered arrays. Muscle tends to be highly stable in its final differentiated state, but there is greater plasticity among the connective tissues. Under both experimental and pathological conditions the cellular elements of fibrous connective tissue in mature individuals can be stimulated to form nodules of cartilage or bone.

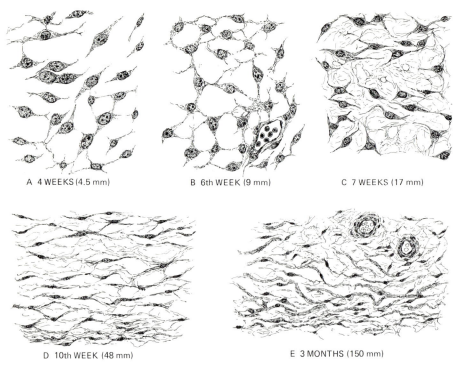

A 4 WEEKS (4.5 mm)   B 6th WEEK (9 mm)   C 7 WEEKS (17 mm)

D 10th WEEK (48 mm)   E 3 MONTHS (150 mm)

Fig. 9-1 Five stages in histogenesis of loose fibroelastic connective tissue. (All drawings, by projection, X500.)

HISTOGENESIS OF FIBROUS CONNECTIVE TISSUE

The Development of Fibroelastic Connective Tissue In the adult we recognize various types of connective tissues according to the kind and arrangement of the fibers present. The most abundant and widespread of the connective tissues is the "fibroelastic" type. This tissue is composed of an interlacing system of tough bundles of collagenous fibers and elastic fibers which give to the system resilience and adaptability to positional changes. It is this type of tissue which, with the emphasis on its collagenous fiber component, forms the connective-tissue investment of a bone known as its *periosteum* or the covering layer of cartilage called *perichondrium*. With a more generous admixture of elastic fibers this same general type of connective tissue holds muscle elements together as their peri- and epimysial sheaths. With a still more liberal elastic component, a similar connective tissue holds nerves and blood vessels in their mobile relations to muscle groups or supports the clusters of fat cells that lie in the subcutaneous zones over the muscles.

The specialized characteristics of a connective tissue, such as its particular mixture of collagenous and elastic fibers, are established relatively late, and their development can be considered more profitably in a course in histology than in the present connection. It is nevertheless pertinent at this time to see the general steps by which connective tissues take shape. A large proportion of connective tissue in early embryos arises from the somites or, in the case of the head, from the neural crest. The earliest mesenchymal cells tend to be stellate or spindle-shaped (Fig. 9-1A). Not infrequently processes from one cell appear to be connected with those of another (Fig. 9-1B). Although at one time it was believed that the processes of adjacent mesenchymal cells join to form a syncytial network, electron microscopic research has shown that the mesenchymal cells retain their identity as single cells. However, the tips of some processes from adjoining cells may be in extremely close contact with one another, forming specialized *gap junctions*. Gap junctions are thought to facilitate intercellular communication.

Soon delicate fibrils make their appearance among the mesenchymal cells (Fig. 9-1C). When fibers begin to become a conspicuous component of the young connective tissue, the term *fibroblast* is used in referring to the cells of mesenchymal origin which are producing the fibers. In the earliest stages, the cells and fibers of the developing connective tissue often possess no particular orientation, but later the young connective tissue takes on a quite characteristic appearance. One begins to see a definite pattern in the direction of the fibers, correlated with the mechanical conditions under which the tissue is commencing to function (Fig. 9-1D). Later still, the fibers begin to run in definite bundles (Fig. 9-1E).

In developing fibroelastic connective tissue, collagenous fibers are the first to appear in the extracellular spaces. Only later are elastic fibers formed. The origin of collagenous fibers from fibroblasts is known in considerable detail, but considerable uncertainty still shrouds the synthesis of elastic fibers. It is likely that elastic fibers are also secreted by fibroblasts and/or smooth muscle cells.

FORMATION OF THE SKELETON

The skeleton is a remarkably complex structure. In contrast to the static image conjured up by a skeletal preparation made of dried bones, skeletal tissue is highly reactive in both its prenatal development and postnatal state. To a large extent, the location and morphology of skeletal elements in the body depend upon interactions with the soft tissues associated with them. In postnatal life the morphology of the skeleton adapts to changing patterns of mechanical function. These changes are mediated by some of the same cellular processes of bone formation and destruction that occur in the embryo.

There are two major divisions of the skeleton. The *axial skeleton* includes the bones of the head, the vertebral column, and the ribs. The bones of the axial skeleton surround important soft tissues, principally the brain and spinal cord. So intimate is their relationship to the structures which they protect that their initial formation is dependent upon inductive influences from the central nervous system and their final size and morphology are the result of intrinsic potential combined with growth pressures from the underlying soft tissues (Hall, 1978). The *appendicular skeleton* consists of the bones of the limbs and the limb girdles. The bones of the limb differ in several respects from those of the axial skeleton. In contrast to the bones of the head and vertebral column, they are the central structures and are surrounded by the soft tissues with which they are associated. The evidence for the induction of limb bones by ectodermal structures is much more tenuous than that for induction of axial bones.

There are two major modes of bone formation. Many bones (e.g., long bones) first form in the embryo as tiny cartilaginous models of the bones. The cartilage then calcifies and becomes resorbed. Remnants of the cartilaginous model serve as a framework upon which true bone is deposited. This mode of osteogenesis, starting with a cartilaginous phase, is called *endochondral bone formation*. The other mode of bone formation, seen for instance in the flat bones of the skull, consists of the direct deposition of bone matrix by bone-forming cells without the preexistence of a cartilaginous model of the bone. This direct mode of bone formation is called *intramembranous bone formation*. The histological structure of mature bone is determined more by the mechanical conditions imposed upon it than by its mode of formation. Although there appears to be some degree of intrinsic control governing the morphogenesis of bones, forces of compression and tension are major secondary determinants of form.

This section on formation of the skeleton will begin with the differentiation and histogenesis of cartilage and bone at the cell and tissue level. Then formation of some of the characteristic skeletal elements will be covered. In dealing with bone formation it is difficult to separate those processes that are "pure" embryology from those that are commonly considered to be within the scope of histology textbooks. The essential elements of osteogenesis and endochondral bone formation will be outlined here, but for details the reader is referred to any of the standard histology texts.

Cartilage Formation

Cartilage forms from mesenchyme in many areas of the embryo, such as the limbs, vertebral column, respiratory tract, and skull. In some cases, for example, the cartilaginous precursors of vertebrae, chondrogenesis is known to be initiated by an inductive process. In other cases, for example, limb bones, the chondrogenic stimulus has not been definitely determined. Regardless of the initiatory mechanism, the sequence of chondrogenesis is remarkably similar wherever it occurs.

One of the most characteristic features of cartilage differentiation is the secretion of collagen and extracellular matrix materials, such as *chondroitin sulfate*. There appears to be no key moment when synthesis and secretion of these substances is switched on, for even in cartilage that is known to be induced, small amounts of matrix material is secreted before or in the absence of the inductive event (Lash, 1968). Thus in the case of cartilage, induction greatly accelerates rather than initiates specific pathways of molecular synthesis.

Morphologically, chondrogenesis begins with the proliferation and aggregation of mesenchymal cells of either mesodermal or neural crest origin into tight clusters of *chondroblasts*. The relative significance of local proliferation versus aggregation in the formation of the precartilage condensations remains to be determined. This early stage of chondrogenesis is associated with high levels of *hyaluronic acid*. The aggregated chondroblasts then begin to secrete sufficient quantities of collagen and mucopolysaccharide matrix materials to push the cells away from one another. During this stage of differentiation, the levels of hyaluronic acid decrease. Conversely, the levels of hyaluronidase and chondroitin sulfate increase. Hall (1978) has stressed the importance of hyaluronic acid in the proliferation and migration of several varieties of cells and the removal of hyaluronic acid (probably through the action of hyaluronidase) in their differentiation.

As the cartilage matrix is increased in amount, the cartilage cells embedded in it both multiply and become more widely separated from each other (Fig. 9-2B and C). This method of increasing mass by mitosis and the continued secretion of matrix is known as *interstitial growth*. Cartilage, and other hard materials, can also expand by *appositional growth*, the laying down of matrix by chondrogenic cells on the outer edge of the mass of cartilage. In time, the matrix becomes more rigid, with a resultant checking of interstitial growth.

The formation of a rigid matrix first occurs centrally in an area of developing cartilage. Then appositional growth assumes increasing importance. While the cartilage has been increasing in mass, it has been acquiring a peripheral investment of compacted mesenchyme. This layer soon becomes specialized into a connective-tissue covering called the *perichondrium* (Fig. 9-2). The layer of perichondrium next to the cartilage is less fibrous and more richly cellular than the outer layer; the cells in it continue to proliferate and become active in the secretion of cartilage matrix. For this reason it is known as the *chondrogenic layer* of the perichondrium. The perichondrial cells that synthesize matrix ultimately become enveloped in it and are incorporated into the mass of cartilage. Appositional growth of cartilage continues long after interstitial growth has ceased in the earliest formed matrix.

Histogenesis of Bone

Much remains to be learned about the nature of the cells that give rise to bone. From the early stages of osteogenesis several kinds of cells are involved in the formation of bone. *Osteoblasts*, which lay down the matrix and ultimately become entrapped in the matrix as *osteocytes*, differentiate from mesenchyme which, as in cartilage, may originate from mesodermal or neural-crest ectoderm, depending upon the location of bone formation. Closely associated with bone formation is localized bone removal. This is accomplished by multinucleated *osteoclasts*. For many decades the origin of osteoclasts has been the subject of considerable controversy. According to one viewpoint, they arise from the same population of cells that produces the osteoblasts. According to another interpretation, osteoclasts arise from a separate, hematogenous population of cells. The accumulated evidence favors the latter view. A third population of

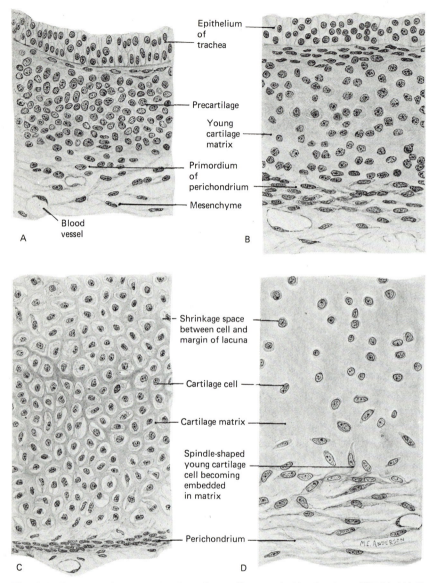

Epithelium of trachea

Precartilage

Young cartilage matrix

Primordium of perichondrium

Mesenchyme

Blood vessel

A

B

Shrinkage space between cell and margin of lacuna

Cartilage cell

Cartilage matrix

Spindle-shaped young cartilage cell becoming embedded in matrix

Perichondrium

C

D

Fig. 9-2 Stages in histogenesis of cartilage. (Camera lucida drawings, X450.) (A) Tracheal cartilage from embryo of eighth week (C-R 23 mm). (B) Tracheal cartilage from embryo of ninth week (C-R 33 mm). (C) Cricoid cartilage from embryo of tenth week (C-R 48 mm). (D) Tracheal cartilage from fetus of twenty-fifth week (C-R 240 mm).

cells associated with developing bone constitutes the *marrow*, an extremely complex group of cells which will not be dealt with in this chapter.

The differentiation of bone involves the production of an abundant intercellular matrix of specialized collagenous fibers and ground substance which has a strong tendency to calcify rather than take up water like cartilage matrix. Throughout both embryogenesis and postnatal life the growth and maintenance of bone is based upon a delicate balance between the deposition of new bone and the resorption of previously deposited bone. These two opposed processes often take place within a few hundred microns of one another. Particularly in later growth and remodeling the internal architecture of a bone is highly responsive to changes in its mechanical environment. An exciting hypothesis regarding growth control in bone is that through the piezoelectric[1] properties of the osseous matrix, mechanical deformation is translated into differences in electrical potential. These differences are then sensed by bone-forming cells, which react in predictable ways. According to Bassett (1971), a net negative charge stimulates osteoblastic activity and the deposition of new bone, whereas osteoclastic activity and bone resorption occur in areas with a net positive charge. This concept has recently been put to medical use in the stimulation of fracture healing in bones. Some idea of the state of this field can be gained by reading the reports given in a conference on bioelectricity (Liboff and Rinaldi, 1974).

Intramembranous Bone Formation In an area where intramembranous bone formation is about to begin one finds mesenchymal cells clustered together in strung-out groups and numerous blood vessels (Fig. 9-3A, lower part). These cells secrete a delicate axis of collagen fibers, which becomes flanked on either side by a row of cells (Fig. 9-3A). Accompanying the secretion of collagen fibers is the secretion of mucopolysaccharide matrix material. Many of these matrix molecules bind to the collagen fibers, completing the formation of the organic *osteoid framework*.

The osteoid framework has a strong attraction for calcium salts. In an early spicule of membraneous bone one end is usually broader and more mature than the other. In this area *calcification* (the deposition of calcium salts) of the matrix is taking place. Calcium and phosphate ions in soluble form are brought to areas of developing bone by the bloodstream. The local conditions which promote the deposition of inorganic bone salts remain poorly defined. Accompanying this change in the matrix is a change in the cells which are producing it. At the youngest end of a spicule of forming bone, the cells still resemble mesenchymal cells, with prominent processes of pale-staining cytoplasm. In the areas of calcification these cells, now properly called *osteoblasts*, have lost their cytoplasmic processes and are more deeply staining. The osteoblasts become aligned along surfaces of forming bone in regular rows.

When ossification has progressed to such a point that the original strand is completely invested by bone matrix, we say a *trabecula* (Latin, meaning *little beam*) has been formed (Fig. 9-3B). As the osteoblasts continue to secrete and thereby thicken the trabecula, the new matrix added is not laid down uniformly. The osteoblasts work more or less in cycles, depositing a succession of thin layers of matrix. Each of these layers of the matrix is called a *lamella* (Fig. 9-4). As the

---

[1]*Piezoelectricity* is a term borrowed from the physical sciences and refers to electrical potential produced by the mechanical deformation of certain types of nonconducting crystals. Some biological structures possess similar properties. In the case of bone, the collagen fibers in the matrix are piezoelectric.

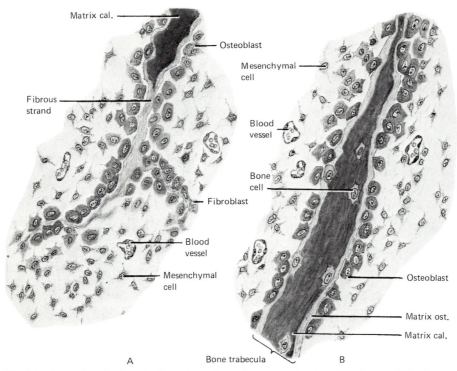

Fig. 9-3 Formation of trabecula of membrane bone. Projection drawings from the mandible of a pig embryo 130 mm in length. The two parts of this illustration cover a single trabecula which was growing from either end. The areas shown in A and B were directly continuous in the actual material, the bottom of B fitting on the top of A. The areas were separated in drawing merely as a matter of economy in the space occupied by the illustration.

Abbreviations: *Matrix cal.,* osteoid matrix impregnated with calcium salts; *Matrix ost.,* osteoid matrix not yet impregnated with calcium salts.

row of osteoblasts is forced back with the deposit of each succeeding lamella, not all the cells free themselves from their secretion. Here and there a cell is left behind and, as its former neighbors continue to pile up new matrix, becomes completely buried. An osteoblast so caught and buried is called a *bone cell*, or *osteocyte* (Fig. 9-3B), and the space in the matrix which it occupies is called a *lacuna*. The bone cells, thus entrapped, of necessity cease to be active bone formers, but they play a vital part in the maintenance of the bone already formed. They have delicate cytoplasmic processes radiating into the surrounding matrix through minute channels called *canaliculi*. The processes of one cell are in communication with the processes of its neighbors (Fig. 9-4). Thus the bone cells nearer to blood vessels absorb and hand on materials to those farther away, which in turn utilize these materials in maintaining a healthy condition in the organic part of the bone matrix.

As the various trabeculae in an area of developing bone grow, they inevitably come in contact with each other and fuse. Thus trabeculae, at first

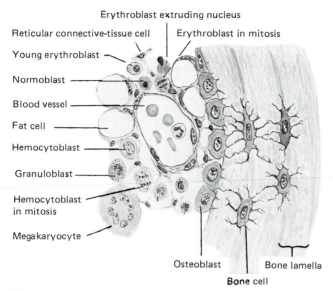

Erythroblast extruding nucleus

Reticular connective-tissue cell | Erythroblast in mitosis

Young erythroblast

Normoblast

Blood vessel

Fat cell

Hemocytoblast

Granuloblast

Hemocytoblast
in mitosis

Megakaryocyte

Osteoblast   Bone lamella

Bone cell

Fig. 9-4   A small area of bone and adjacent marrow as seen in highly magnified decalcified sections. The drawing has been schematized somewhat to emphasize the relations of the cytoplasmic processes of the osteoblasts and the bone cells so important in nutrition. In the adjacent marrow, developmental stages of various types of blood cells have been suggested.

isolated, soon come to constitute a continuous system. Because of its resemblance to a latticework (Latin, *cancellus*), bone in this condition, where the trabeculae are slender and the spaces between them extensive, is known as *primary cancellous bone*. The spaces between the trabeculae become occupied by blood-forming cells and are known as *marrow spaces*. (Fig. 9-4).

Endochondral Bone Formation   In endochondral bone formation a model of the bone is first formed from hyaline cartilage. Later in development the cartilage becomes removed and bone is deposited in its place. The actual process of bone formation appears to differ little from that seen in intramembranous bone formation.

When a mass of cartilage is about to be replaced by bone, very striking changes in its structure take place. The cells which have been hitherto secreting cartilage matrix begin to hypertrophy. Almost concurrently, calcium salts are deposited in the matrix surrounding them. The calcified matrix does not permit an adequate exchange of oxygen and metabolites between blood vessels and the cartilage cells. Consequently the cells die and are not able to maintain the integrity of the matrix surrounding them. The matrix becomes eroded, and this process of destruction continues until the cartilage is extensively honeycombed. Meanwhile the tissue of the perichondrium overlying the area of cartilage erosion becomes exceedingly active. There is rapid cell proliferation, and the new cells, carrying blood vessels and young connective tissue with them, begin to invade the honeycombed cartilage (Fig. 9-5).

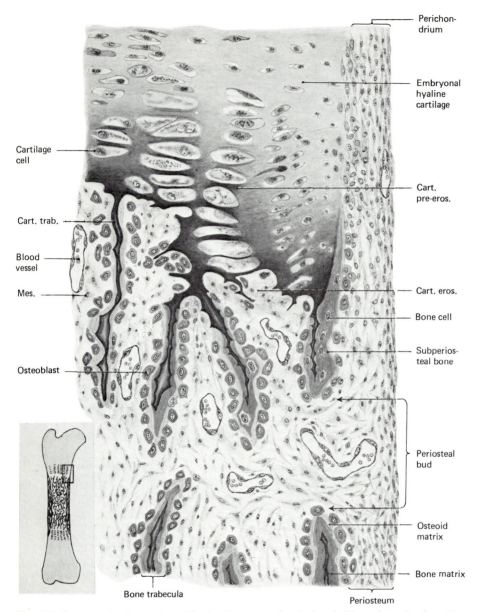

Fig. 9-5 Drawing showing periosteal bud and an area of endochondral bone formation from the radius of a 125-mm sheep embryo. The small sketch indicates the location of the area drawn in detail.

Abbreviations: *Cart. eros.,* area from which cartilage has recently been eroded; *Cart. preeros.,* area with cartilage cells enlarged and arranged in rows presaging erosion; *Cart. trab.,* remnant of cartilage matrix which has become calcified and serves as an axis or core about which bone lamellae are deposited to form a bone trabecula; *Mes.,* mesenchymal cell.

It is a striking fact that during its formation cartilage is devoid of blood vessels, the nearest vessels to it being those in the perichondrium. The invasion of cartilage by blood vessels definitely announces its impending disintegration as cartilage and at the same time is the initial step in the formation of bone. For this reason the enveloping layer of connective tissue, previously called *periochondrium* because of its relation to the cartilage, is now called *periosteum* because of the relations it will directly acquire to the bone about to be formed. A mass of periosteal tissue (*periosteal bud*, Fig. 9-5) grows into an area of honeycombed cartilage, carrying in potential bone-forming cells. These cells come to lie along the strandlike remnants of cartilage, just as osteoblasts arranged themselves along fibrous strands in membranous bone formation. The actual deposition of bone proceeds in the same manner endochondrally as intramembranously, the only difference being that strandlike remnants of cartilage serve as axes for the trabeculae in the former, whereas the deposit began on fibrous strands in the latter. Extensions and fusions of the growing trabeculae soon result in the establishment of typical cancellous bone similar to that formed intramembranously.

Development of Characteristic Skeletal Elements

Details of the formation of many of the characteristic skeletal elements vary from region to region and even from bone to bone. It is beyond the scope of this book to deal with the formation of all of the different bones or groups of bones. For a review of the mechanisms involved in the development of a number of specific bones, the reader is referred to the recent book by Hall (1978). In this section we shall deal with the formation of long bones in the appendicular skeleton and the formation of the major elements of the axial skeleton in both the head and the trunk.

Long Bones   The long bones are characteristically of endochondral origin. The cartilage in which they are preformed is a temporary miniature of the adult bone. Ordinarily there are several ossification centers involved in the formation of long bones. The first to appear is the one in the shaft, or *diaphysis*. The location of this center is shown schematically in Fig. 9-6A. Such details as the cartilage erosion which precedes its appearance and the manner in which the deposit of bone was initiated have already been considered (Fig. 9-5). Of interest now is the relation of such an endochondral ossification center to other centers and to the formation of the bone as a whole. Almost coincidentally with the beginning of bone formation within the cartilage, the overlying periosteum begins to add bone externally (Fig. 9-6B). In view of the fact that the bone-forming tissue carried into the eroded cartilage arose from the periosteum, this activity of the periosteum itself is not surprising.

The ossification which starts at about the middle of the shaft soon extends toward either end until the entire shaft is involved (Fig. 9-6C), leaving the two ends (*epiphyses*) still cartilage. Toward the end of fetal life, ossification centers

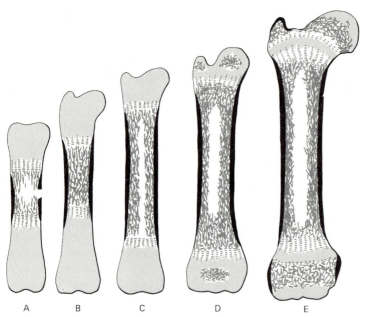

Fig. 9-6 Diagrams showing ossification in a long bone. Light gray areas represent cartilage; dark gray and black areas indicate bone. (A) Primary ossification center in shaft. (B) Primary center plus shell of subperiosteal bone. (C) Entire shaft ossified. (D) Ossification centers have appeared in the epiphyses. (E) Entire bone ossified except for epiphyseal cartilage plates and articular surfaces.

appear in the epiphyses. The numbers and locations of these epiphyseal centers vary in different long bones. In most cases there is at least one center in each epiphysis. Some of the phalanges are exceptions to this general rule, showing an epiphyseal center at only one end. In some of the larger long bones there may be two or three centers in one epiphysis.

Between the bone formed in the diaphysis and that formed in the epiphysis there persists a mass of cartilage, known as the *epiphyseal plate*, which is of vital importance for the growth in length of the bone. We should expect from the rigidity of bone matrix that interstitial growth could not account for its increase in length. This was long ago demonstrated experimentally by exposing a developing bone and driving into it three small silver pegs, two in the shaft and one in the epiphysis. The distance between the pegs was recorded, the incision was closed, and development was allowed to proceed until a marked increase had occurred in the length of the bone. On again exposing the pegs, the two in the shaft were found to be exactly the same distance apart as when they were driven in, but the distance between the pins in the shaft and that in the epiphysis had increased in proportion to the increase in the length of the bone. This clearly indicates that the epiphyseal plates constitute a sort of temporary, plastic union between the parts of the growing bone. Continued increase in the length of the shaft is accomplished by the formation of new cartilage and its replacement by bone at the epiphyseal plate. These epiphyseal plates persist during the entire postnatal growth period. Only when the skeleton has acquired its adult size do they finally become replaced by bone which joins the epiphyses permanently to the diaphysis.

As the bone increases in length there is a corresponding increase in its diameter. That the increase in diameter is accomplished by the deposition of new bone by the periosteum can be demonstrated by feeding a growing animal alizarin compounds or the antibiotic tetracycline. Both of these compounds have an affinity for growing bone. If they are administered and withheld in alternating cycles, the bones show alternating rings of stained or nonstained osseous material and demonstrate that growth in diameter is due to continued appositional growth beneath the periosteum. As new bone is added peripherally, there is a corresponding resorption centrally, resulting in the formation of a central cavity, called the *marrow canal*, within the bone (Fig. 9-6C). Recent studies on chick-quail chimeras have shown that osteoclasts and the cells of the *bone marrow* itself, which fills the newly formed marrow cavity, arise from blood-borne cells which become secondarily associated with the developing bone (Kahn and Simmons, 1975).

Formation of Joints   In an area where a freely movable joint (*diarthrosis*) is to be formed between two skeletal members, there is at first only a vaguely outlined precartilage concentration of mesenchyme. Gradually the mesenchyme becomes perceptibly more densely aggregated in areas where cartilage formation is about to start. As the cartilage miniatures of the future bone become better defined, the joint becomes localized as an area between them, where the mesenchyme is less concentrated (Fig. 9-7A). When the perichondrium takes shape, it extends around the ends of the skeletal members so that where the joint is to be formed there is for a time an area of loosely woven young fibrous connective tissue (Fig. 9-7B).

Meanwhile, ossification starts in the shafts of the bones, but their epiphyseal ends remain cartilaginous. The thinning out (Fig. 9-7C) and final disappearance (Fig. 9-7D) of the connective tissue from around the epiphyses establish the joint cavity. Even after epiphyseal ossification centers have

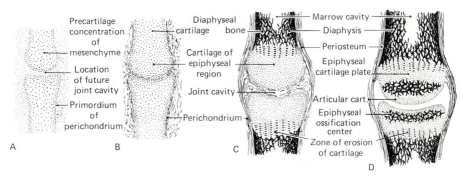

Fig. 9-7   Schematic diagrams showing four stages in development of a joint.

appeared, the articulating ends of the bones at a diarthrodial joint continue to be capped by cartilage, which provides a smooth bearing surface lubricated by the synovial fluid in the joint cavity.

The ligaments of the joint are formed by further development of the investing connective tissue which was retracted and concentrated peripherally to form the joint capsule. Eventually this young connective tissue of the ligamentous capsule becomes strengthened by the addition of more and more heavy bundles of collagen fibers. The ends of certain of these bundles become buried in the expanding tissues of the heads of the bones adjacent to the joint, thus holding the articulating members in proper relation to each other.

The initial development of a joint is accomplished on the basis of intrinsic factors, but maintenance of a joint requires some form of mechanical function. If an embryonic joint is subjected to prolonged culture in vitro or if the limb is paralyzed, fusion of the tissues of the joint is the rule.

Formation of Vertebrae and Ribs   The segmental nature of the mature vertebral column reflects its origin from the somites of the early embryo. The vertebral column arises from mesodermal cells originating in the sclerotomal portion of the somite (Fig. 6-16C and D). Although the cells of the sclerotome possess a chondrogenic bias and produce small numbers of molecules characteristic of the skeletal matrix, their participation in formation of axial skeletal elements depends upon inductive influences from the notochord and the ventral half of the spinal cord (Holtzer and Detwiler, 1953). Specific details of the inductions vary among the classes of vertebrates, but there is increasing evidence that elements of the extracellular matrix produced by the notochord and spinal chord are the effective inductive agents when they come in contact with migrating cells of the sclerotome. In birds the centra of the vertebrae are induced by the notochord and the neural arches by the spinal cord.

The earliest morphological step in the formation of the centrum of a vertebra is a migration of cells from the sclerotomal portions of the somites on either side toward the midline, where they become clustered about the notochord (Fig. 9-8A). The sclerotomal cells from each somite pair are densely packed in the caudal part and loosely packed in the cranial part (Fig. 9-8B). In the human embryo some cells from the condensed caudal portion migrate cranially and begin to differentiate into the intervertebral disk (Prader, 1947). The remainder of the condensed portion of the sclerotome then migrates caudad, whereas the loosely packed cells of the following somite migrate craniad (Fig. 9-8B). These migrating masses of cells derived from two somites then join to form the primordium of the centrum of a vertebra in a position interdigitating between two myotomes (Fig. 9-8C). Soon, paired concentrations of mesenchymal cells extend dorsally and laterally from the centrum to establish the primordia of the neural arches and the ribs.

The derivation of the vertebrae from mesenchymal contributions of adjacent somites places them between somites rather than opposite them. This means that when muscle tissue develops from the myotomal part of the somites,

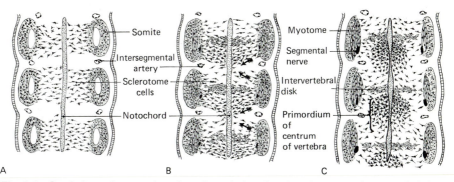

Fig. 9-8 Semischematic coronal sections through dorsal region of young embryos to show how vertebrae become intermyotomal in position. Note that the primordium of a centrum is formed by cells originating from sclerotomes of both adjacent pairs of somites.

it lies across the intervertebral joints, a location which effectively sets it up in proper mechanical relation to the skeletal units which it will move.

In the chick, both the spinal ganglia and the notochord are separately involved in the morphogenesis of certain segments of the vertebral column. If the spinal ganglia are removed, neural-arch cartilage forms, but it forms an unsegmented rod even though individual centra form (Fig. 9-9B). Conversely,

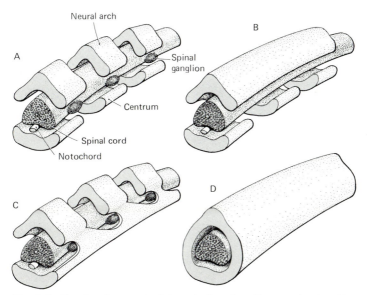

Fig. 9-9 Results of removal of the spinal cord and/or notochord on the morphogenesis of the vertebral column of the chick embryo. (A) Normal embryo showing segmental arrangement of the neural arches, centra, and spinal ganglia. (B) Excision of the spinal ganglia results in unsegmented neural arches but does not affect segmentation of the centra. (C) Excision of the notochord results in unsegmented centra but does not influence the morphogenesis of the neural arches. (D) Excision of both notochord and spinal ganglia results in a totally unsegmented vertebral column. (After Hall, 1977, Adv. Anat. Embryol. Cell Biol. **53**(IV):1.)

removal of the notochord results in the absence of segmented centra (Fig. 9-9C). If both notochord and spinal ganglia are removed, an unsegmented cylinder of cartilage forms around the spinal cord (Fig. 9-9D). The effect of the spinal ganglia upon segmentation of the neural arches may be largely mechanical, for they are present before the vertebrae begin to form and may act as barriers to the migration of sclerotomal cells (Hall, 1977). Little is known about the role of the notochord in morphogenesis of the centra of the vertebrae.

The stage in which the earliest recognizable parts of the skeleton are sketched in mesenchymal concentrations is frequently spoken of as the *blastemal stage* (Fig. 9-10). It is rapidly followed by the *cartilaginous stage*. In the development of the spinal column, conversion to cartilage begins first in the blastemal mass in the region of the centrum, and then chondrification centers appear in each neural and each costal process. These spread rapidly until all the centers fuse and the entire mass is involved. The cartilage miniature of the vertebra thus formed is at first a single piece showing no lines of demarcation where the original centers of cartilage formation become confluent and no

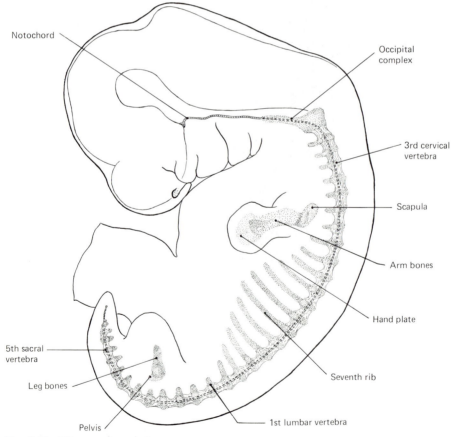

Fig. 9-10 Diagram of precartilage primordia in a 9-mm human embryo. (Adapted from several sources, chiefly the work of Bardeen.)

foreshadowing of the separate parts of which it will be made up after the cartilage has been replaced by bone. By the time ossification begins, the rib cartilages become separated from the vertebra (Fig. 9-11A), but the cartilaginous primordium of the vertebra itself remains in one piece. The locations of the endochondral ossification centers which appear in a vertebral cartilage are indicated schematically in Fig. 9-11A. These centers of ossification spread until the entire vertebra is ossified. Figure 9-11B to 9-11E shows the homologous components in vertebrae from different levels. One can see that all of the ribs contain components that are developmentally homologous with the ribs.

During the formation of the vertebral column those regions of the notochord which are within the developing vertebrae themselves eventually disappear. Between vertebral bodies mesenchymal cells surrounding the notochord form the *intervertebral disks*. Within the intervertebral disks the notochord persists as a mucoid structure known as the *nucleus pulposus*.

Development of the Skull The skull consists of two major subdivisions— the *neurocranium*, which surrounds the brain, and the *viscerocranium*, which

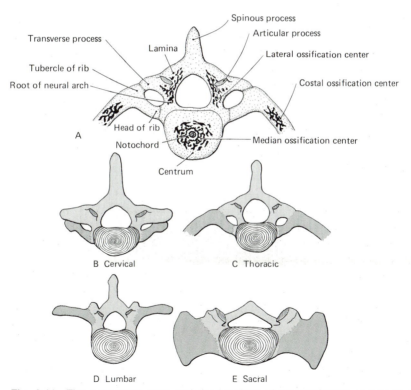

Fig. 9-11 The component parts and the locations of the ossification centers in developing vertebrae. (A) Diagram showing locations of various ossification centers in a thoracic vertebra and the associated ribs. (*B–E*) Drawings showing characteristic components of vertebrae from different levels. The neural-arch component is represented by light gray, the costal component by darker gray and the centrum by concentric lines.

surrounds the oral cavity, pharynx, and upper respiratory tract. Both of these subdivisions have components which arise as cartilaginous models and are later replaced by endochondral ossification. Both also have membrane bones which arise by direct ossification from mesenchyme. The vertebrate skull is so complex, from both the ontogenetic and the phylogenetic standpoint, that only a brief outline of the important aspects of its development will be given here. Those interested in greater detail are referred to the monograph by DeBeer (1937).

The bones of the skull arise from mesenchyme, but in contrast to the rest of the skeleton many of the mesenchymal precursor cells appear to have originated from the neural crest. Like the vertebral column, virtually all the bones of the skull are formed as the result of inductions between epithelial structures and skeletogenic mesenchyme. The *chondrocranium* (the cartilaginous base of the neurocranium) is induced by the notochord, whereas the membranous bones of the neurocranium are induced by the parts of the brain which they ultimately protect (Fig. 9-12). The elements of the viscerocranium (Fig. 15-5), on the other hand, require an inductive stimulus from the pharyngeal endoderm in order to take shape.

The *cartilaginous neurocranium*, which forms the base of the skull, consists of several masses of cartilage. Around the cranial end of the notochord is a cartilaginous plate called the *parachordal cartilage*, which is derived from the sclerotomal portions of the four occipital somites (Figs. 9-13A). This plate forms

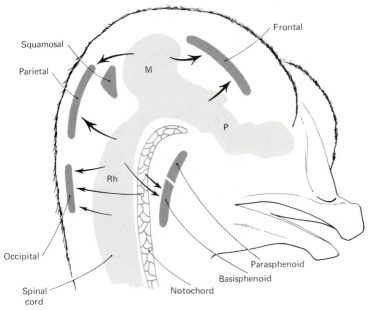

Fig. 9-12 Formation of bones in the developing avian skull as the result of inductive interactions between specific regions of the central nervous system and/or notochord and the surrounding ectomesenchyme of the head. (After Schowing, 1968, *J. Embryol. Exp. Mroph.* **19**:83.)

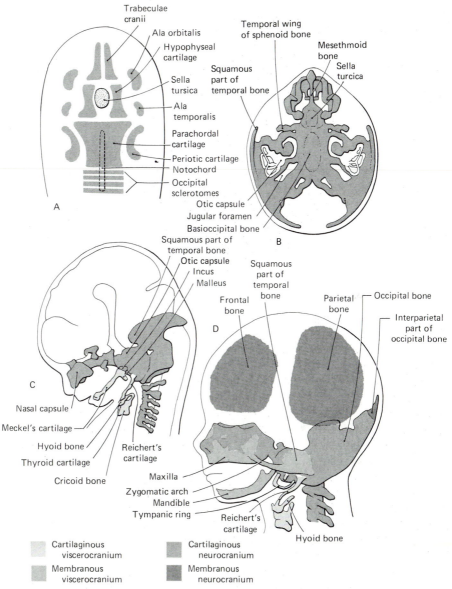

Fig. 9-13 Diagrams illustrating the origins and development of the major bones of the skull. (A) Embryo of about 6 weeks (viewed from above), showing the primordial cartilages that will form the chondrocranium. (B) Embryo of about 8 weeks (viewed from above), showing the chondrocranium. (C) Lateral view of the embryo illustrated in B. (D) Skull of 3-month embryo. (Adapted from several sources.)

the base of the occipital bone at the base of the skull. Rostral to the parachordal cartilage are the *prechordal cartilages* (constituting the paired *hypophyseal cartilages*), which form the bone surrounding the pituitary gland, and the *trabeculae cranii*, which form the ethmoid bone in the nasal region. Lateral to this axis are other pairs of cartilaginous elements which are associated with the sense organs (Fig. 9-13A). These cartilaginous structures are eventually replaced by bone.

The *membranous neurocranium* includes the large flat platelike bones of the cranial vault. After induction by specific parts of the brain, these bones remain separated by fibrous areas called *cranial sutures* as well as by larger soft areas called *fontanelles*. Throughout the period of fetal and postnatal growth, these bones adapt to the changing size and growth patterns of the brain.

The *cartilaginous viscerocranium*, principally of neural-crest origin (Fig. 6-13), is an integral part of the branchial arch system. Details of its further development will be presented in Chap. 15. The bones of the *membranous viscerocranium* form in association with the cartilaginous core of the first branchial arch. They ultimately form the adult bones of the upper and lower jaws, as well as part of the temporal bone, which becomes incorporated into the neurocranium.

Progress of Ossification in the Skeleton as a Whole Each of the more than 200 bones of the human body has its own developmental history involving the formation of the connective tissue or the cartilaginous mass which precedes it; appearance of local erosion centers if the bone is preformed in cartilage; number, location, and time of appearance of ossification centers; growth in length and diameter; development of epiphyses; time of fusion of epiphyses and diaphysis; and finally the development of muscle ridges and articular facets. It would be neither possible nor desirable in a book of this sort to attempt any systematic survey of the development of all, or even the majority, of the bones.

In addition to information on specific bones it is sometimes desirable to have information on the state of development exhibited by the skeleton as a whole. In the early stages of embryonic development the best way to obtain such information is to study an embryo that has been treated so that it becomes transparent. If the embryo is stained with alizarin, the ossification centers will stand out particularly clearly. The human embryos in Figs. 9-14, 9-15, and 9-16 will provide a good visual summary of early ossification in the skeleton as a whole. It is not profitable use of a student's time to attempt to memorize the centers of ossification of all the major bones, but these figures will provide a readily available source of reference for following up points of interest which may arise.

THE MUSCULAR SYSTEM

Recent work has shown that many varieties of nonmuscle cells in both embryos and adults contain filamentous proteins which possess varying degrees of contractile ability. This enables them to move or to undergo certain changes in shape. The contractile proteins are related to the actin and myosin in muscle cells, but typically they constitute only a small portion of the cytoplasmic contents of the cells. True muscle cells are cells which during the course of their

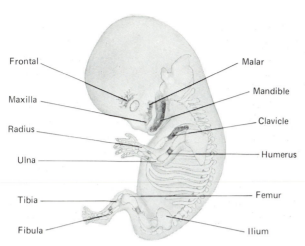

Fig. 9-14   Human embryo of 23½ mm (eighth week) stained with alizarin and cleared to show ossification centers. (Projection drawing, X2.5, by Jane Schaefer. University of Michigan Collection, EH 105.) (See colored insert.)

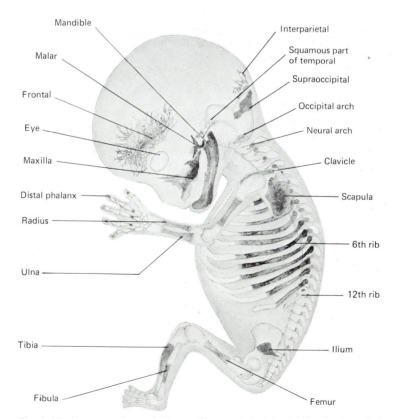

Fig. 9-15   Human embryo of 39 mm (nine weeks) stained with alizarin and cleared to show the progress of ossification. (Projection drawing, X2.5, by Jane Schaefer. University of Michigan Collection, EH 149.) (See colored insert.)

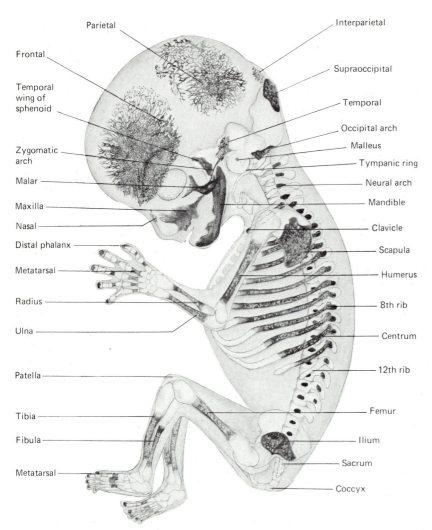

Fig. 9-16 Human embryo of 49 mm (tenth week) stained with alizarin and cleared to show the developing skeletal system. (Projection drawing, X2.5 by Jane Schaefer, University of Michigan Collection, EH 70.) (See colored insert.)

differentiation have produced massive amounts of contractile proteins, to the extent that these proteins overshadow the other constituents of the cell. Contractility is the dominant function of muscle cells, but the important role of muscle in the overall energy metabolism of the body must not be overlooked. At the cellular level, three types of muscle are recognized: skeletal muscle, cardiac muscle, and smooth muscle.

Grossly, most of the muscles of the body, particularly the skeletal muscles, fall into logical groups on the basis of their embryologic and phylogenetic origins. These groupings and their relationship to the major divisions of the motor (efferent) component of the nervous system are summarized in Table 9-1.

Table 9-1   Embryologic Origins of the Major Classes of Muscle

Embryologic origin	Muscle	Functional type	Innervation type*
Preotic myotomes	Extrinsic eye muscles	Striated, voluntary	Somatic efferent (cranial nerves III, IV, VI)
Somites (occipital myotomes)	Tongue (hypobranchial) muscle	Striated, voluntary	Somatic efferent (cranial nerve XII)
Somites (myotomes)	Trunk muscles; diaphragm; limb muscles	Striated, voluntary	Somatic efferent (spinal nerves)
Branchial arches	Branchiomeric (visceral) muscles; muscles of mastication; muscles of facial expression; pharyngeal and laryngeal muscles	Striated, voluntary	Special visceral efferent (cranial nerves V, VII, IX, X)
Splanchnic mesoderm	Cardiac muscle	Striated, involuntary	General visceral efferent
Splanchnic mesoderm	Muscles of gut and respiratory tracts	Smooth, involuntary	General visceral efferent
Local mesenchyme	Other types of smooth muscle (e.g., vascular)	Smooth, involuntary	General visceral efferent

*See Figs. 11-1 and 11-24 for designations of functional classes of nerves.

Skeletal Musculature

The skeletal musculature, comprising some 40 to 45 percent of the total weight of the adult human body, includes over 300 named muscles. Despite the anatomical diversity of the skeletal musculature, the development and relations of the fundamental structural unit, the muscle fiber, are remarkably similar throughout the body. In this section we shall first deal with the differentiation of individual muscle fibers; then the development of some of the major muscle groups in the body will be outlined.

Myogenesis   The cytodifferentiation of a multinucleated skeletal muscle fiber represents a classical case of progressive specialization during which the precursor cells sacrifice certain general functions, such as the ability to undergo mitotic division, as they are acquiring characterisitc new properties, in this instance the ability to contract. Muscle cells are derived from mesodermal mesenchyme, in most cases from the myotomal regions of the somites. Initially, these mesenchymal cells are presumably capable of forming the cells of cartilage, bone, and connective tissue, but in some poorly understood manner the developmental fate of certain mesenchymal cells becomes restricted to forming muscle even though morphologically these cells, now called *presumptive myoblasts*, cannot be distinguished from other mesenchymal cells at this time. Holtzer (1970) calls cell divisions that result in the restriction of or cause a change in the developmental pathways of a cell *quantal mitoses*, in contrast to *proliferative mitoses*, after which there is no change in the differentiative potential of the cell.

At some critical phase in their development the presumptive myoblasts lose the ability to undergo mitosis and begin to acquire the capability of synthesizing myosin and other myofibrillar proteins (Fig. 9-17). At this stage they are called *myoblasts*. Postmitotic myoblasts are arrested in the G_1 phase of the cell cycle and normally never again undergo DNA synthesis. These cells then prepare for fusion, the next major step in myogenesis. Fusing myoblasts come together side by side and rapidly lose the individual cell membranes which separate them. The biological events surrounding myoblastic fusion and their developmental significance have been the subject of intense inquiry and considerable controversy in recent years. This research has been carried out mainly under in vitro conditions and is beyond the scope of this book. For an introduction to the voluminous literature in this area, the reader is referred to reviews by Doering and Fischman (1974), Holtzer et al. (1972), and Konigsberg and Buckley (1974).

The formation of multinucleated muscle fibers by the fusion of separate cells was, itself, the subject of considerable controversy for many decades. According to another interpretation, repeated division of the nuclei of myoblasts without cytokinesis was the mechanism of multinucleation. Several lines of experimentation determined that the fusion mechanism was the correct one. One of the most convincing experiments involved the use of allophenic mouse embryos (Mintz and Baker, 1967). In this experiment (Fig. 9-18) mouse embryos homozygous for different *isozymes* (different molecular forms of the same enzyme) of the enzyme isocitrate dehydrogenase were fused (see Fig. 4-13). These isozymes (aa and bb) have different rates of electrophoretic migration. According to the internal division model, only enzyme forms aa and bb would be expected to be found, but if myoblasts bearing different genotypes fused, an intermediate ab form of the enzyme would also be formed. The latter case held true, thus confirming the validity of the fusion·model.

A newly fused, multinucleated skeletal muscle fiber is called a *myotube* (Fig. 9-17C). These syncytial structures are extremely actively engaged in the synthesis of *actin* and *myosin*, the principal contractile proteins of muscle. Myosin, a large filamentous protein composed of several subunits, is synthesized on characteristic helical polyribosomes, containing 50 to 60 ribosomes. The newly synthesized contractile proteins undergo *self-assembly* into thick and thin filaments, which in turn become assembled into functional units around the peripheral regions of the myotube. As the myotube matures, the contractile filaments occupy a greater share of the cytoplasm, and the nuclei migrate from their central location to positions just beneath the plasma membrane. By this stage the structure is properly known as a *skeletal muscle fiber*. Further growth of the myotube, or the muscle fiber, is accomplished by the cytoplasmic fusion of additional myoblasts with the muscle fiber. Some myoblasts do not immediately fuse with the muscle fiber but instead remain as unspecialized mononucleated cells in a position between the muscle fiber and its surrounding basal lamina. These cells are known as *satellite cells* (Fig. 9-17D). Later in postnatal life

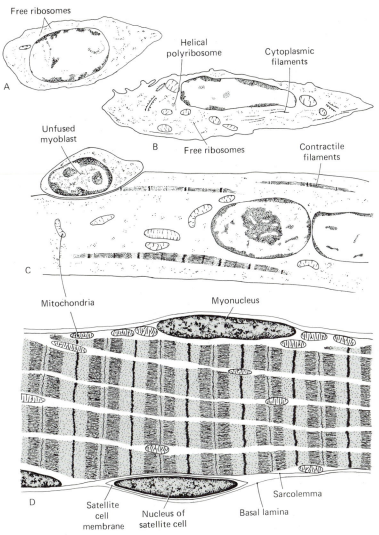

Fig. 9-17 Major steps in the embryonic differentiation of a skeletal muscle. (A) Unspecialized mesenchymal cell. (B) Myoblast. This spindle-shaped cell possesses large numbers of free ribosomes, including helical polyribosomes upon which the myosin molecules are formed. Cytoplasmic filaments and microtubules are present, but identifiable contractile proteins are not found. (C) Myotube. This is a long multinucleated cell formed by the fusion of mononucleated myoblasts. The nuclei are arranged in long central chains. Myofilamentogenesis is actively occurring, and bundles of well-ordered contractile filaments are present in the periphery. The large numbers of ribosomes attest to the continued protein synthetic activity. The small mononucleated cell alongside the myotube will eventually fuse with the myotube during further maturation. (D) Cross-striated muscle fiber. The nuclei have moved to a peripheral location, and the bulk of the cytoplasm (sarcoplasm) is filled with bundles of contractile filaments demonstrating the characteristic banding pattern of skeletal muscle.

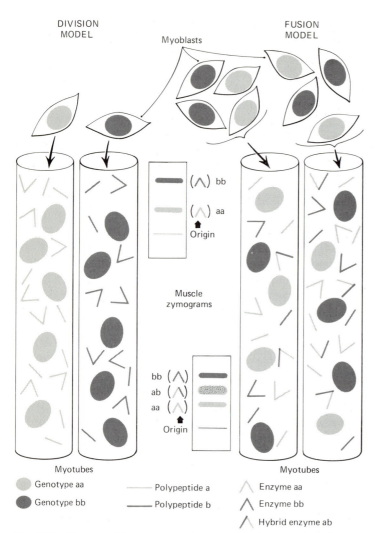

Fig. 9-18 The use of isozyme differences and allophenic mice in testing the "division" vs. the "fusion" model of myogenesis. The enzyme, isocitrate dehydrogenase, is a dimere composed of two polypeptide chains, aa or bb, depending upon the isozyme type. In allophenic embryos made from embryos of genotypes aa and bb, the internal division model would produce muscle fibers having only aa or bb isozyme bands after electrophoresis (upper zymogram). According to the fusion model, myoblasts of different genetic backgrounds would fuse into hybrid forms (*ab*) of the enzyme (lower zymogram). The latter result was obtained. (Adapted from Mintz and Baker, 1967, *Proc. Nat. Acad. Sci.* **58**:592.)

satellite cells or their progeny can fuse with the growing muscle fibers. After damage to a muscle, satellite cells become activated and become the source of new muscle fibers during regeneration (Mauro, 1979).

Differentiation of a skeletal muscle fiber is not completed when it becomes filled with myofilaments and peripheral nuclei. Another major step, requiring interaction with a motor nerve, results in the final enzymatic and functional differentiation of the muscle fiber into one of several types (fast, slow, and intermediate). From a functionally indifferent state, information supplied by the nerve causes changes in the mitochondria and the contractile proteins themselves which result in a muscle fiber being fast or slow contracting, or fatigable or fatigue-resistant. This last stage of neurally induced differentiation occurs after birth and represents a constant interaction throughout life. The nature of the neural influence is not known. Arguments in favor of the two main options, chemical trophic effects or electrical stimuli, are given in a recent symposium report (Drachman, 1974).

Primary Arrangement of Myotomes in Young Embryos   By the sixth week of development the human embryo has formed about 39 pairs of somites in a cephalocaudal progression. The exact count is uncertain because of the difficulty in assessing the segmental value of the primordia of the eye muscles and also the variable number of poorly defined somites in the rudimentary caudal region. Surprisingly constant are the 8 cervical, 12 thoracic, 5 lumbar, and 5 sacral somites. Adding to these 30 the usual 4 occipital and 5 caudal myotomes gives the 39 pairs of somites indicated in the conventionalized diagram of Fig. 9-19. Such a count leaves out of consideration the problematical eye-muscle primordia.

The stippled areas in Fig. 9-19 indicate approximately the extent of the myotomes when they first differentiate (cf. Fig. 6-16C). The unstippled ventral extensions from these areas suggest schematically the general territory of the embryonic body into which each myotome extends. It is now well established that the limb muscles are of somitic origin (Chap. 10). The developing muscle fibers in the myotomes at first run in a craniocaudal direction. In certain areas the primordial muscle masses undergo changes in their fiber direction and extensive secondary shifting out of their areas of origin; but many of the trunk muscles, especially the intercostal muscles and the muscles of the vertebral column, retain essentially their original segmental relations. Even in regions where the changes in the relations of the muscles themselves are difficult to follow, the cutaneous nerves still exhibit a distribution which gives a striking demonstration of the areas of the adult body which are derived from the several segmental zones of young embryos (cf. Figs. 9-19 and 9-20). The innervation of the various primordial muscle masses is acquired very early from nerves arising at the same segmental level in the body, so that when secondary changes in the position of the muscle occur the already attached nerve is pulled along with the muscle. The level of origin of a muscle is, therefore, indicated by the segmental level at which the nerve supplying it arises. Moreover, the path of the nerve in reaching the muscle suggests the path followed by the muscle in arriving at its adult location. A good example of this sequence of events is the association of the phrenic nerve with the muscle of the diaphragm. Its almost direct course from the fourth and fifth cervical nerves of the young embryo to the developing diaphragm is clearly shown in Fig. 11-15. In striking contrast is its long course in

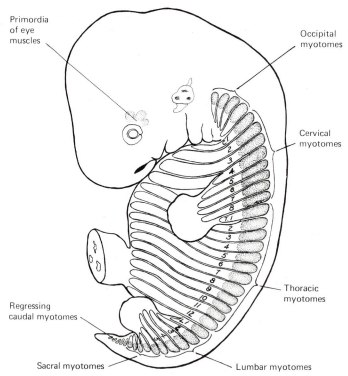

Primordia of eye muscles

Occipital myotomes

Cervical myotomes

Thoracic myotomes

Regressing caudal myotomes

Sacral myotomes

Lumbar myotomes

Fig. 9-19 Schematic diagram showing the regions into which the embryonic myotomes extend. The stippled area indicates the approximate size of the original mesodermal somite; the unstippled ventrolateral areas outlined opposite each somite suggest the territory into which the myotomes tend to extend. Compare with Fig. 9-20. (See colored insert.)

the adult from the neck, along the pleuropericardial folds, to the diaphragm, where it takes its final position in the body far caudal to the segmental level of its origin.

Muscles of Trunk and Body Wall The more dorsal and deeper parts of myotomes of the trunk tend to retain their segmental character with least modification. Here they lie closely interdigitated with the developing axial skeleton and become attached to it as the intervertebral muscles. In the deeper parts of the ventrolateral extensions of the myotomes at thoracic levels, also, the original segmental arrangement is kept in the form of the intercostal muscles. At the same time tangential splitting of the myotomes tends to separate off some of their more superficial parts, which become much more modified than the deeper layers. These superficial muscle masses give rise to the long muscles of the neck and back. One of the best examples of the end-to-end fusion of myotomal fibers to form massive suprasegmental muscles is furnished by the development of the sacrospinalis muscle group (Fig. 9-22). From their position dorsal to the axial skeleton these muscles are frequently characterized as *epaxial trunk muscles*

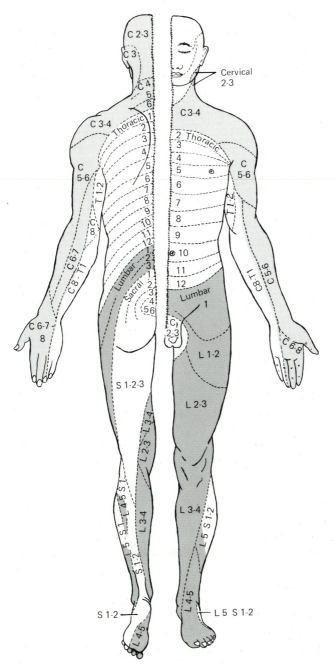

Fig. 9-20 Segmental areas of the adult body as indicated by distribution of cutaneous nerves. (Adapted from various sources.) Compare with Fig. 9-19. (See colored insert.)

(Fig. 9-21). Ventral to the vertebral column other long muscles are formed by a similar process. Such muscles (e.g., the *psoas major* and *quadratus lumborum*) are spoken of as *hypaxial trunk muscles*.

Muscles of Head and Neck   In young embryos the head is broadly attached to the trunk with the limits of the future cervical region ill defined. Moreover, some of the muscles which make their first appearance in the zone of transition from head to trunk are carried caudad from their level of origin as they develop, whereas others move farther rostrad. It is therefore more logical and convenient in a brief account such as this to deal with the development of the musculature in the cervicocephalic region as a whole.

Typical mesodermal somites are not found cephalic to the future occipital region. The mesoderm of the head is made up mostly of migrating cells which collectively are designated as *mesenchyme*. Conspicuous masses of these mesenchymal cells are packed in beneath the ectodermal covering of the facial region and the branchial arches. It is from such cells that the major part of the musculature of the neck and face arises. Because of this derivation from the mesoderm of the series of branchial arches, such muscles are described as being of *branchiomeric origin*. In this part of the body where somites are absent the branchial arches make excellent descriptive landmarks. The statement that a nerve supplies the second branchial arch, or that a muscle arises in the third, has the same exactitude that designation of the somitic level of origin gives in other parts of the body.

In general the muscles of branchiomeric origin tend to retain the innervation characteristic of the gill-arch stages in phylogeny and ontogeny. Thus we find the muscles which arise from a primordial mesenchymal mass in the

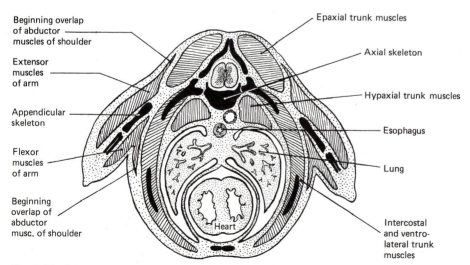

Fig. 9-21 Schematic diagram of some of the major developing muscle masses to show their relations to the axial and to the appendicular skeleton.

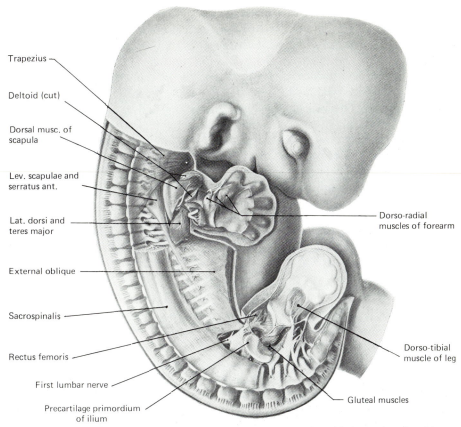

Trapezius

Deltoid (cut)

Dorsal musc. of scapula

Lev. scapulae and serratus ant.

Lat. dorsi and teres major

External oblique

Sacrospinalis

Rectus femoris

First lumbar nerve

Precartilage primordium of ilium

Dorso-radial muscles of forearm

Dorso-tibial muscle of leg

Gluteal muscles

Fig. 9-22 Superficial dissection (X10) to show the developing trunk and limb muscles of an 11-mm human embryo. (After Bardeen and Lewis, 1901, *Am. J. Anat.,* vol. 1.)

mandibular arch supplied by motor fibers in the mandibular division of the fifth (trigeminal) nerve. This group includes the *muscles of mastication* and others involved in swallowing (see Fig. 15-3). The utilization of the muscle masses of the mandibular arch region for chewing and swallowing seems quite natural in view of their primary relations.

The *muscles of facial expression* have a much less direct and simple origin. With no muscular tissue between the skin and the bones of the rostral part of the skull, a fish's change of facial expression is limited to the opening and closing of the mouth. Since the muscle arising in the mandibular arch has been preempted for chewing and swallowing, the muscles of facial expression must be derived by the migration of other primordial muscle masses into facial territory. With the functional regression of the gills, which occurs with the assumption of air breathing, the primordial muscle arising in the region of the hyoid arch is the nearest available source. The muscle in this region has started to increase in conspicuousness in amphibia, with the addition of an outer superficial layer which acts as a primitive constrictor of the neck. In mammals both the deep and the superficial layers of this hyoid-arch musculature extend into the facial region. Human embryos of the sixth and seventh weeks show this evolutionary process sketched out in recapitulation. The primordial superficial muscle sheet spreads out into the front and sides of the head, and to the thin, superficial layer (the *platysma*) in the cervical region. The muscles controlling the movements of the nose and lips are derived largely by subgrouping and rearrangement of the primordial deep layer of muscle from the hyoid-arch region. The foregoing brief sketch serves to emphasize the fact that the facial muscles migrate much more extensively during development than is the case with most of the rest of the cephalic musculature. In their migration they carry along with them branches of the seventh nerve and the external carotid artery which originally furnished the nerve and vascular supply to the hyoid arch. It is only when one sees the early stages in the development of these muscles that there seems to be any logic whatever in their arrangement, their innervation, or their blood supply in the adult.

Cardiac Musculature

The heart is formed from the splanchnic mesoderm of the early embryo (Fig. A-22). The cardiac muscle cells themselves arise from mesenchymal cells located in the inner layer of the epimyocardium. Early cardiac myoblasts, like those of skeletal muscle, are spindle-shaped mononuclear cells, but they do exhibit certain unique features. Perhaps foremost among these is the presence of comparatively large numbers of myofibrils in the cytoplasm and the consequent ability of cardiac myoblasts to undergo pronounced contractions. Another feature which distinguishes the developing cardiac muscle cell from its counterpart in skeletal muscle is the ability of the former to undergo mitotic divisions even though the cytoplasm contains numerous bundles of contractile filaments (Rumyantsev, 1967; Manasek, 1968). Many types of cells in the body seem to have lost the ability to divide if they have already produced specialized cytoplasmic structures characteristic of their fully differentiated state. In view of the requirement for early and continuous functioning of the heart during embryonic development, it is not too surprising that the cells of the heart deviate from this general rule by producing contractile filaments while they increase in number. Cardiac muscle cells do not fuse to form a syncytium as do skeletal muscle myoblasts. Instead adjoining cells develop specialized intercellular connections that in mature muscle account for the appearance of the *intercalated disks*, which are found between the cells. Specialized structures of the intercalated disks facilitate intercellular communication.

Smooth Musculature

The smooth musculature arises from mesoderm that in the course of development applies itself as an outer coat around the primary epithelial lining of hollow internal organs. Much of the smooth muscle, for example that surrounding the digestive and respiratory tracts, arises from splanchnic mesoderm, but as a general rule, smooth muscle seems to differentiate from whatever type of mesenchyme surrounds the epithelial component of a structure. There is evidence that the smooth muscle in many of the blood vessels arises from somatic mesoderm and that the smooth muscle of the iris (the *sphincter pupillae* muscle) is of ectodermal origin. In general, smooth muscle cells are organized into bands or sheets held in place within the organ by connective tissue. The smooth musculature operates under the regulatory control of the autonomic nervous system but cannot be made to respond directly to our will. Very little is yet known of the developmental mechanisms underlying the cytodifferentiation and histogenesis of smooth muscle.

Chapter 10

Limb Development

The appendages of vertebrates have undergone a remarkable degree of diversification during the unfolding of the vertebrate classes. Paleontological evidence suggests that the earliest appendages consisted of a pair of fin folds that ran along the entire length of the body of some of the most primitive fishes. Soon this arrangement gave way to discrete outgrowths, the pelvic and pectoral fins, which may represent the localized persistence of portions of the original fin folds. With the spreading of vertebrates into terrestrial habitats, their limbs took on a considerably different configuration from that of their aquatic forebears. Since then, the fundamental arrangement and the pattern of development of the tetrapod limb have been preserved despite the evolutionary radiation that has led to the appearance of structures as diverse as flippers, wings, hooves, and the human hand.

Initiation of Limb Development   The limb arises as a condensation of cells from the lateral plate mesoderm and its ectodermal covering (Fig. 10-1). The limb primordia in amniotes appear along the *Wolffian ridges* (Fig. 10-2), which run along the lateral surface of the body. The limb does not begin to form until the primordia of most other organs are laid down and the musculature of the body wall beneath the undifferentiated limb bud is already well differentiated.

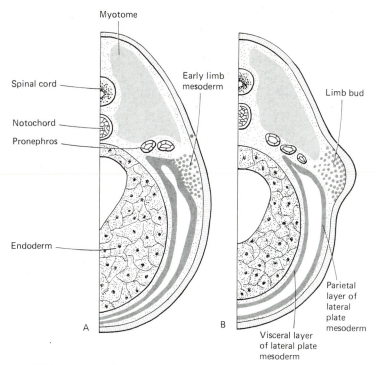

Fig. 10-1 Diagrams illustrating the origin of the amphibian limb bud. *(After Balinsky, 1975,* An Introduction to Embryology, *4th ed. W. B. Saunders Co., Philadelphia.)*

This is reminiscent of the phylogenesis of the chordates, in which appendages arose only after the chordate line was well established. The cephalocaudal rule of development applies to the limbs, and at any given time the anterior limbs are more advanced than the posterior ones.

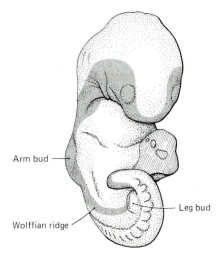

Fig. 10-2 Ventrolateral view of a 30-somite (4.2 mm) human embryo, showing the thickened ectodermal ring (gray). The portion of the ring between the upper and lower limb buds is known as the Wolffian ridge. *(After O'Rahilly and Gardner, 1975,* Anat. Embryol. **148**:1.)

Some form of activation of the lateral mesodermal cells seems to be required for the initiation of limb development, but the nature of the stimulus remains poorly defined. The early limb primordium is located close to the somites, and there is evidence that the presence of somites is necessary for limb development to occur. The proximity of the pronephros to the limb primordium is intriguing, but definitive evidence for a major causal relationship between pronephros and limb is lacking.

After the mesodermal cells of the early limb primordium are activated, they next act upon the overlying ectoderm. In most classes of vertebrates the ectoderm responds to the influence of the future limb mesoderm by thickening.

Experiments have demonstrated that the mesoderm rather than the ectoderm is the prime mover in early limb development. If limb mesoderm is combined with early flank ectoderm, the ectoderm thickens and participates in the formation of a limb. In contrast, a combination of limb ectoderm with flank mesoderm does not result in limb development. These experiments not only illustrate the primacy of the limb mesoderm, but they show that at this early stage the ectoderm of the body is still not completely determined, because ectoderm that would normally form flank skin can still assume the specializations that are involved in promoting outgrowth of the limb.

Regulative Properties of the Early Limb Primordium   The limb primordium is a *self-differentiating system*; i.e., once established, it contains all of the information required to attain its normal form even if removed from the body of the embryo. The early *limb disk*[1] is also a highly regulative system. This has been shown by some simple but highly instructive experiments (Fig. 10-3): (1) When part of a limb disk is removed, the part that is left continues to develop into a perfectly normal limb. (2) When a limb disk is split into two halves which are prevented from fusing, each half gives rise to a complete limb. (3) When two equivalent halves of a limb disk are placed together, a single limb forms. (4) When two harmonious limb disks are superimposed, the cells become reorganized so that they form but a single limb. (5) Disaggregated limb mesoderm, grafted to the flank, reorganizes and often forms a normal limb.

These experiments demonstrate that the limb primordium possesses regulative properties very similar to those of entire early embryos, such as the sea urchin (on which Hans Driesch's experiments defining regulative characteristics were first made) and the mammalian blastocyst, which has received careful study only in recent years.

Axial Determination of the Limbs   The limb is an asymmetric structure which is commonly defined in terms of three axes—the proximodistal (P-D), the anteroposterior (A-P), and the dorsoventral (D-V). Experiments conducted by Ross Harrison early in this century demonstrated clearly that the individual axes

---

[1]A *limb disk* is the name given to the group of limb-forming cells before they begin to project beyond the surface of the body wall as the limb bud.

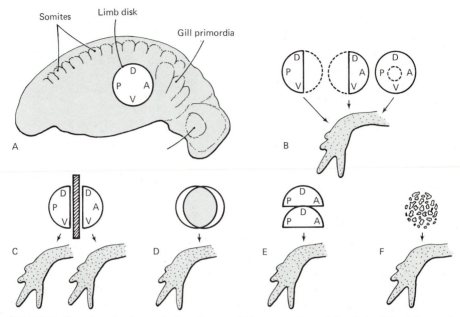

Fig. 10-3 Experiments demonstrating the regulative properties of the limb disk of the early amphibian embryo. (A) Normal embryo. (B) Normal limb development after removal of either half or the central region of the limb disk. (C) When halves of the limb disk are separated by a barrier, each half develops into a complete limb of the same polarity. (D) Superimposition of two limb disks of the same polarity results in the formation of one normal limb. (E) Combining two identical halves of a limb disk results in a single limb. (F) Mechanical disruption of the limb disk followed by its reaggregation is followed by normal development. *(After Harrison and Swett.)*

of the limb are fixed (determined) at different times of development. This was done by rotating the limb disks of *Ambystoma* embryos (Fig. 10-4). When a right forelimb disk was rotated 180° around its center (Fig. 10-4A), the limb that formed looked like a normal left arm growing from the right side of the body. A left limb disk grafted in place of the right disk with only its A-P axis reversed with respect to that of the body of the host also developed into a normal looking left arm, whereas a left arm disk grafted to the right side with only the D-V axis reversed developed into a normal right arm. In all three of these experiments development along the A-P axis of the early limb disk proceeded on the basis of information intrinsic to the rotated limb disk. On the other hand, development along the D-V axis was dictated by the body of the host and not by axial information residing within the disk. Relations along the P-D axis were not altered in these experiments. These results showed that the A-P axis of the early limb disk was determined but that the D-V axis was not. Other work showed that the P-D axis was also undetermined at this time.

At a later stage of embryonic development the same operations as those outlined above were performed (Fig. 10-4B). The types of limbs that grew out from these rotated primordia differed from those in the previously described

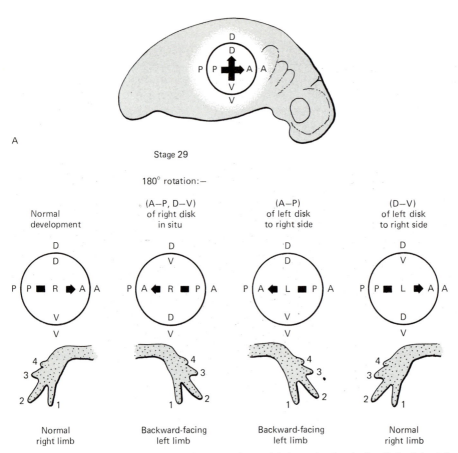

Fig. 10-4 Diagrams of experiments demonstrating axial determination in the limb disk of the amphibian (*Ambystoma*) embryo by means of grafting and rotation of the limb disk with respect to the surrounding tissues. (A) Experiments performed on early embryos showing that only the anteroposterior axis is determined. (B) Experiments performed on older embryos, showing that both the anteroposterior and dorsoventral axes and determined. For details, see the text. *(After Harrison and Swett.)* Abbreviations: A, anterior; D, dorsal; P, posterior; V, ventral; L, left, R, right.

series. Now both the A-P and the D-V axis of the outgrowing limbs corresponded entirely with the original axes of the rotated limb primordia and were independent of any influence by the body.

Additional experiments of a similar nature showed that the P-D axis was determined last. Thus it was established that in the limb the sequence of axial determination is A-P→D-V→P-D (Swett, 1937). This sequence in the limb may represent a general one in the major external organs, for a similar sequence of axial determination occurs in the developing ear and in the retina of the eye.

Outgrowth of the Limb Bud As soon as the limb primordium begins to project from the surface of the embryo, it may be properly called a *limb bud*.

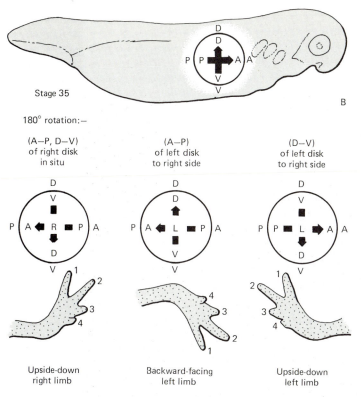

Stage 35

B

180° rotation:—

| (A—P, D—V) of right disk in situ | (A—P) of left disk to right side | (D—V) of left disk to right side |

Upside-down right limb

Backward-facing left limb

Upside-down left limb

Structurally, the limb bud consists of a mesodermal core of homogeneous-appearing mesenchymal cells, which are supplied with a capillary network. Nerve fibers have not yet grown into the early limb bud. Covering the mesodermal core of the limb bud is a sheet of ectoderm which, in most vertebrates, is thickened at the apex of the limb bud.

In 1948, Saunders conducted an illuminating study that opened the way for a vast amount of subsequent experimentation on the mechanism of limb development. He was interested in discovering the role, if any, of the band of thickened ectoderm (*apical ectodermal ridge*) that covers the convex outer margin of the wing bud in chick embryos. He removed the apical ectodermal ridge from the wing buds of chick embryos and found that shortly thereafter outgrowth of the limb ceased (Fig. 10-5). This experiment demonstrated a critical interaction between the apical ectodermal ridge and the underlying mesodermal core during growth of the wing bud.

Apical Ectodermal Ridge

Structure of the Ridge   The appendage buds of most vertebrates possess a terminal thickening. Typically, the thickening is in the form of an apical ecto-dermal ridge, running along the anteroposterior margin of the bud. There is remarkable uniformity in the shape and appearance of the apical ridge

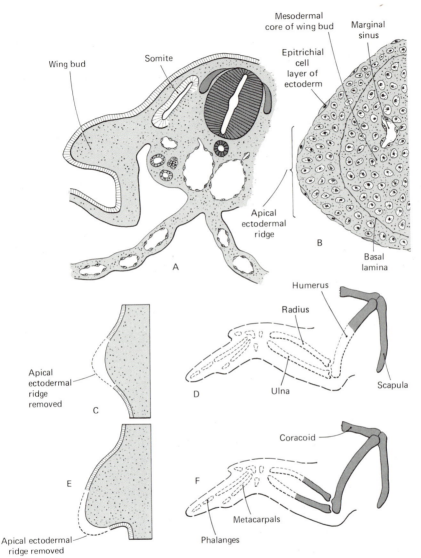

Fig. 10-5 The effect of the removal of the apical ectodermal ridge on the development of the wing bud. (A) Diagram showing location of the center of mesodermal proliferation involved in forming the core of the wing bud. (B) Cell detail drawing of tip of wing bud of 3-day chick to show the apical ridge. (C) Ridge removal in 3-day chick. (D) Skeletal deficiency resulting from third day removal of apical ridge. (E) Ridge removal at 4 days. (F) Skeletal deficiencies resulting from fourth day removal of apical ridge. In D and F the parts of the skeleton which develop are stippled; the parts failing to develop are shown in outline only *(C–F, showing the results of apical ridge removal on skeletal development, based on the work of Saunders, 1948, J. Exptl. Zool., 108.)*

throughout the vertebrates. In birds and mammals it commonly stands out as a nipple-shaped structure in cross section (Figs. 10-5 and 10-6), whereas in fishes and some reptiles the apical epidermis appears to be thrown into a tight fold. The amphibians show an unusual dichotomy in the morphology of the limb ectoderm. Anuran amphibian embryos have apical ectodermal ridges that are of the same general form as those of birds and mammals. Urodeles, on the other hand, do not have apical thickenings on the limb buds (Fig. 10-6C). Why this latter group does not have apical ridges and how limb outgrowth occurs in their absence have not yet been explained.

The apical ridge begins to take shape shortly after the limb bud begins to project from the lateral surface of the embryos. It is oriented along the anteroposterior (pre- to postaxial) plane of the limb bud and is somewhat asymmetrical in shape, i.e., it is thicker along its posterior portion in birds. The apical ectodermal ridge typically attains its greatest degree of development when the limb is at the paddle-shaped stage of development. Thereafter, as the digits begin to differentiate, the ridge begins to regress.

There are two principal histological configurations of the apical ectodermal ridge in higher vertebrates. In birds it is arranged as a pseudostratified columnar epithelium (Fig. 10-5B), whereas in mammals it is a stratified cuboidal or squamous epithelium. Kelley and Fallon (1976) have described in detail the structure of the human apical ectodermal ridge during its major stages of development and decline (Fig. 10-7). One of the major structural characteristics of the apical ectodermal ridge is the presence of numerous gap junctions connecting the cells of the ridge with one another (Fig. 1-10). Fallon and Kelley (1977) have postulated that through the gap junctions the cells of the apical ectodermal ridge are electrically and metabolically coupled. Histochemical studies (Milaire, 1969) have shown that the cells of the apical ridge contain large amounts of glycogen and RNA, as well as high concentrations of both alkaline and acid phosphatases. These chemical properties are characteristic of groups of embryonic cells that are actively involved in growth and differentiation.

Properties of the Apical Ectodermal Ridge   Although there is still considerable discussion concerning its exact role in development of the limb, the apical ectodermal ridge is considered by most researchers to act as a stimulator of outgrowth of the limb bud. Other hypotheses cast the apical ridge in a permissive or mechanical role with respect to growth and morphogenesis of the limb (Amprino, 1965). Several lines of evidence favor the former viewpoint. Shortly after surgical removal of the apical ectodermal ridge, outgrowth of the limb bud ceases (Saunders, 1948). A genetic wingless mutation of chickens provides a natural experiment that produces similar results (Zwilling, 1949). In this mutant, early wing development is normal, but then it comes to a halt. Correlated in time is the regression and disappearance of the apical ectodermal ridge. Conversely, additional apical ectodermal ridges, whether applied surgically in the laboratory or provided through the action of a genetic mutant (*eudiplopodia*), result in the formation of supernumerary wing tips in chick embryos.

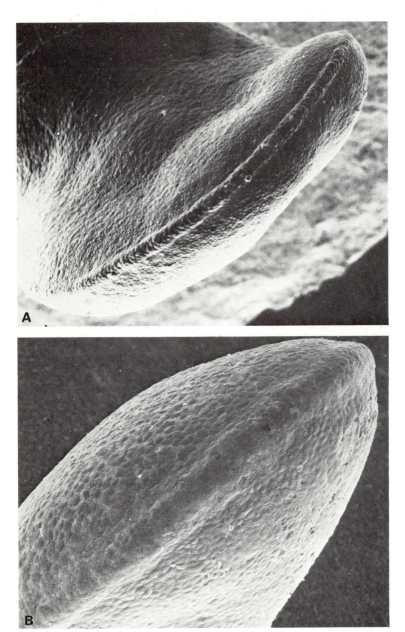

Fig. 10-6 Scanning electron micrographs of limb buds in the embryos of (A) the chick (stage 26); (B) the human (stage 15); (C) the axolotl (stage 39). Note the pronounced apical ectodermal ridge in the limb buds of the chick and human embryos and the absence of a ridge in the axolotl. (*A, courtesy of Dr. John Fallon; B, from Kelley and Fallon, 1976,* Devel. Biol. **51***:241; courtesy of the authors; C, from Tank et al., 1977,* J. Exp. Zool. **291***:417.)*

Fig. 10-6 Continued

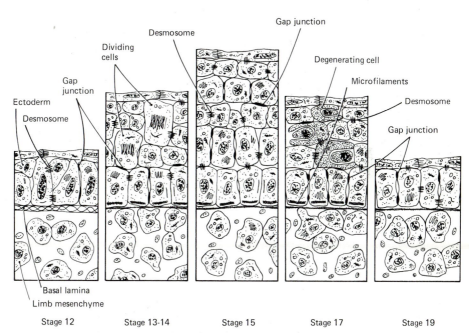

Fig. 10-7 Schematic diagram illustrating the changes in structure during the building-up and reduction of the human apical ectodermal ridge. At stage 12, the basal lamina between ectoderm and mesoderm is double-layered, with cross-links. At later stages it is a single-layered structure. *(Adapted from Kelley and Fallon, 1976, Devel. Biol., 51:241.)*

Whatever its exact role in limb development, the apical ectodermal ridge seems to be remarkably nonspecific in its actions. Surgical exchange experiments have shown that the apical ridge of a chick embryo can promote outgrowth of limb mesoderm not only from other species of birds but also from mouse embryos. The apical ectodermal ridge does not specify the proximodistal level of morphogenesis of the limb but rather acts as a nonspecific stimulator of outgrowth. Rubin and Saunders (1972) grafted apical ridges from older chick leg buds (and consequently at a more distal level) onto young leg buds and found that normal development ensued, despite the difference in age and proximodistal level between the apical ectodermal ridge and the underlying mesoderm. When the apical ridge begins to regress, however, it gradually loses its ability to foster growth of the limb bud. Even when the apical ectodermal ridge is dissociated or turned inside out it can reorganize and fulfill its usual role. Although there is some indirect evidence that the cells of the apical ectodermal ridge secrete diffusible substances into the underlying mesoderm of the limb bud, further evidence must be accumulated before this function can be considered to be established.

Mesoderm of the Early Limb Bud

Structure   The mesoderm of the early limb bud consists of a homogeneous-appearing mound of mesenchymal cells embedded in a loose matrix of collagen fibers and mucopolysaccharide (glycosaminoglycan) ground substance. The cells are undifferentiated in appearance, with large nuclei, prominent nucleoli, and scanty basophilic cytoplasm. Mitotic activity of these cells is prominent throughout the limb bud. Beneath the apical ectodermal ridge is a zone, several hundred microns thick, which is rich in ground substance and in which the cells have a high rate of division. This region has been called the *progress zone* (Summerbell, Lewis, and Wolpert, 1973), and it may play an important role in growth and morphogenesis of the limb. The limb bud is well vascularized and contains a prominent marginal sinus beneath the apical ectodermal ridge. Nerves are not present within the early limb bud, but nerve fibers grow in as development progresses. Differentiation of the mesoderm will be dealt with later in this chapter.

Role of the Mesoderm   The interaction between ectoderm and mesoderm in the limb bud is not entirely a one-way process. There is evidence that the mesoderm of the early limb bud stimulates the early ectoderm to form the apical ectodermal ridge and that once the apical ridge is established a continuing influence of the mesoderm is required for maintaining its integrity.

Mutants for winglessness and polydactyly in chickens have furthered our understanding of the important role of the mesoderm. In the wingless mutant early development of the wing bud is normal, but soon the apical ridge degenerates and further outgrowth of the wing bud ceases. Recombination experiments have shown that if "wingless" mesoderm is covered with normal wing ectoderm, the "normal" apical ridge undergoes degeneration and development of the wing is arrested (Zwilling, 1956). Similarly, if limb ectoderm is placed over nonlimb mesoderm, the apical ectodermal ridge regresses. In polydactyly, combinations of normal mesoderm and mutant ectoderm give rise to normal limbs, whereas mutant mesoderm plus normal ectoderm produces polydactylous appendages. Such experiments indicate that the mesoderm continuously stimulates the overlying ectoderm to maintain an active apical ectodermal ridge which, in turn, stimulates further outgrowth of the mesoderm of the limb bud. Zwilling (1961) has proposed that the mesodermal influence is due to the existence of a mesodermal "maintenance factor," but to date this factor has not yet been isolated or defined.

It has been repeatedly demonstrated that the character of the morphogenesis of the appendage is governed by the mesodermal component rather than by the ectoderm. If wing mesoderm is combined with ectoderm of the foot, a wing covered with feathers develops, whereas if distal hindlimb mesoderm is covered with wing ectoderm, a leg and foot covered with scales results. The mutual interactions between limb mesoderm and ectoderm even operate across species barriers. Limb mesoderm of a duck combined with ectoderm of a chick gives rise to a webbed foot, and early limb bud mesoderm of a mouse forms toes when it is placed in approximation to ectoderm of the wing bud of a chick (Cairns, 1965).

Cell Death in the Developing Limb   Among the many processes involved in the development of the limb bud, cell death is prominent (Saunders et al., 1962). Appearing in the forelimb first in the region of the future axilla (armpit), patches of dying cells are later seen in the elbow region and between the developing digits; in the hindlimb a similar pattern is seen (Fig. 10-8). The area of cell death in the future axilla of the chick is particularly well defined in the chick embryo, and it is called the *posterior necrotic zone.* More circumscribed areas of cell death are associated with differentiation of the skeleton, particularly the joints, and with the histogenesis of certain groups of muscles. Even in the development of mammalian digits, the abbreviation of the first digit, containing only two phalangeal segments as opposed to three for the other digits, may be due to the establishment of a localized zone of cell necrosis at the end of the first digital primordium (Milaire, 1977). In cases of webbed digits, whether normal for the species (e.g., the duck) or as a developmental anomaly (*zygodactyly*), the persistence of webs of soft tissue between the digits is associated with reduced amounts of cell death.

Prominent areas of cell death, particularly in the interdigital areas, have been seen in all major groups of amniote embryos (reptiles, birds, and mammals, including humans), but interdigital cell death has not been seen in amphibian embryos. Cameron and Fallon (1977) speculate that cell necrosis, as a developmental mechanism in limb development, may have originated phylogenetically in conjunction with the branching of the amniotes from the anamniote line of vertebrates.

The cells comprising the necrotic areas of the limb buds appear to be genetically determined to die at a particular stage of development. At a certain

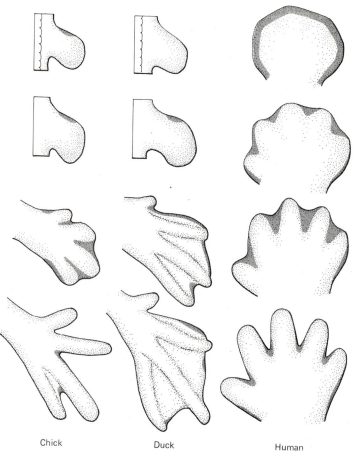

Chick   Duck   Human

Fig. 10-8 Zones of cell death at various stages of embryonic development of the chick and duck foot and the human hand. *(Chick and duck after Saunders and Fallon, 1966, in* Major Problems in Developmental Biology, *ed. M. Locke, Academic Press, New York; human after Menkes et al., 1965,* Rev. Roumaine Embryol. Cytol. **2***:161.)*

stage of development, what Saunders calls a *death clock* is set, and the affected cells, although apparently healthy for some time afterwards, are committed to dying at a critical stage of development (Saunders, 1969). The visible onset of cell death is preceded by a reduction in protein and nucleic acid synthesis and an increase in the activity of hydrolytic enzymes. Grafting experiments involving cells of the posterior necrotic zone of chick embryos have shown that although the death clock is set at approximately stage 17 of development, reversibility or a "reprieve of the death sentence" is possible up to stage 22 if the cells of the posterior necrotic zone are grafted to the dorsal side of the wing bud. After stage 22, however, the commitment to death is irreversible, even though obvious signs of cell necrosis are not visible until stage 24.

Zone of Polarizing Activity and Morphogenetic Control of the Developing Limb   Another discrete area of limb mesenchyme with characteristic properties is located near the posterior edge of the limb bud near its junction to the body wall. The cells of this area, known as the *zone of polarizing activity*, stimulate the formation of a supernumerary appendage if grafted to the anterior (preaxial) part of the limb bud (Fig. 10-9). The zone of polarizing activity was first identified in the wing bud of the chick embryo, very close to the posterior necrotic zone of cell death (Saunders and Gasseling, 1968). Since then a polarizing zone has been identified in the hind limb of birds and in limb buds of urodele and anuran amphibians, reptiles, and mammals, including human beings.

In the wing bud of the chick, on which the vast majority of the experimental work has been performed, polarizing activity, as assayed by grafting techniques, is only weakly detectable at the earliest stages of limb outgrowth (stages 15 and 16). Thereafter it is quite strongly developed until the late stages, when early digital differentiation begins (Fig. 10-10). MacCabe et al. (1977) have developed an in vitro assay system showing that the zone of polarizing activity is the source

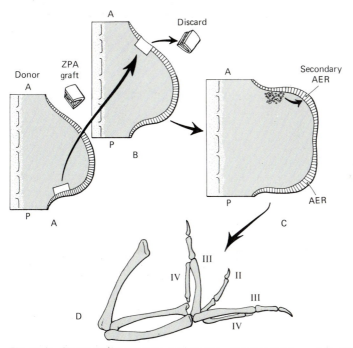

Fig. 10-9 Scheme illustrating the grafting of a zone of polarizing activity (ZPA) from one wing bud of a chick embryo into the anterior part of another wing bud. A secondary apical ectodermal ridge forms distal to the graft and a duplicated wing tip develops. The figure on the right shows the skeleton of such a wing and illustrates the mirror image symmetry of the two wing tips. *(After Saunders, 1972,* Ann. N. Y. Acad. Sci. **193***:29.)* Abbreviations: A, anterior; P, posterior.

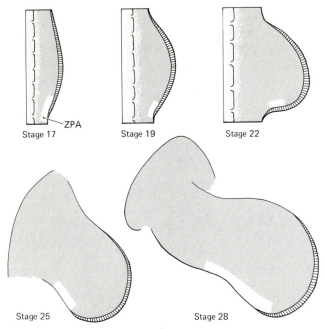

Fig. 10-10 Schematic drawings (at the same scale) showing the changing location of the zone of polarizing activity (ZPA) in the chick wing at different stages of development. The stages are those of Hamburger and Hamilton (1951). *(Adapted from Saunders, 1972, Ann. N.Y. Acad. Sci. 193:29.)*

of a diffusible factor (apparently not identical to Zwilling's maintenance factor) which can cause the thickening of the cells of the apical ectodermal ridge.

Polarizing activity is not only confined to the zone of polarizing activity. Recent work has shown that cells of the somites, mesodermal tissue medial to the wing bud, and even mesonephros of the chick possess polarizing activity. Despite the unquestioned activity of the cells of the zone of polarizing activity, their role in normal limb outgrowth and morphogenesis is uncertain, for if this zone is completely excised, development of the appendage proceeds entirely normally. According to one suggestion (Fallon and Crosby, 1975), polarizing activity may play a role in normal development at the earliest inductive stages, but not during subsequent stages of limb development. Other opinions either assign a much more important ongoing function during limb morphogenesis or suggest that this zone might not have any function in normal limb development.

Morphogenetic Control of the Developing Limb   Despite the considerable experimental progress that has been made in the analysis of certain aspects of limb development, our understanding of the overall control of the process is still far from complete. This has led investigators to propose a number of widely differing hypotheses of limb morphogenesis. Detailed consideration of these is beyond the scope of this book, but some of the more influential ideas will be summarized below. Typically, investigators have viewed morphogenetic control of the developing limb in terms of three axes, as defined by the Cartesian coordinate system, and most hypothetical mechanisms have been confined to explaining development along a particular axis.

Outgrowth along the proximodistal axis has received by far the most attention. According to the Saunders-Zwilling hypothesis (Zwilling, 1961), which centers on the interactions between the apical ectodermal ridge and the underlying mesoderm, the apical ectodermal ridge exerts an inductive action upon the mesoderm, causing the latter to increase in mass by cell proliferation. The continued activity of the apical ridge is, in turn, controlled by the influence of the mesoderm upon the apical ridge, in the form of the yet undefined mesodermal maintenance factor.

The role of the ectoderm is viewed in a completely different light by Amprino (1965), who has postulated that the epithelial hull of the limb bud exerts its morphogenetic effect by mechanical means. He feels that the growth of the ectoderm results in a proximodistal shifting of that layer with respect to the mesoderm of the limb bud and that the ectodermal shifting creates new space that is quickly filled in by the proliferating mesodermal cells.

A more recent hypothesis is known as the *progress zone model* (Summerbell et al., 1973; Wolpert et al., 1975). The progress zone is the distal 300 μm of mesoderm beneath the apical ectoderm of the wing bud in the chick. Under the apparent influence of the apical ectodermal ridge, the cells in the progress zone are said to remain both undifferentiated and in a continuous state of mitosis. As these cells divide, some of the daughter cells leave the progress zone and are then free to differentiate into discrete components of the appendage. What the cells will become (i.e., their proximodistal level of differentiation) depends upon how long the cells have resided in the progress zone. Cells that leave the progress zone early during the course of limb development are supposed to be endowed with positional information enabling them to develop into the humerus or femur, for instance, whereas cells that leave the progress zone at a later time would contain positional information compatible with development of tissues of the next segment of the limb (i.e., radius or ulna). The last of the cells to emerge from the progress zone would form the terminal phalangeal bones.

Morphogenetic control along the anteroposterior and dorsoventral axes of the limb has received relatively little attention. Several investigators have suggested that the zone of polarizing activity, which now appears to release a morphogenetically active, diffusible substance into the mesoderm of the limb bud, may generate by this means a morphogenetic gradient that is used as a reference point by which cells can localize their positions along the anteroposterior axis of the limb. Regarding establishment of the dorsoventral axis, even less is known except that in the chick the dorsal and ventral non-apical-ridge ectoderm possesses the ability to establish the dorsoventral axis of distal, nondifferentiated segments of the limb bud (Pautou, 1977).

Morphology of Later Limb Development   Up to this point, our consideration of limb development has been confined to the early stages, in which the limb bud consists of a mass of homogeneous-appearing mesodermal cells covered by a sheet of ectoderm. Despite our inability to detect differences among cells or groups of cells of the limb bud, even with the most sophisticated morphological or chemical techniques presently available, experimental analysis has revealed that during these early developmental stages axial determination and the pattern for the morphogenesis of specific components of the limb have been set up. On the basis of this morphogenetic blueprint the differentiation of individual tissue components and cell types and later growth of the limb occur. This sequence of events is becoming so clear that it is almost axiomatic to say "morphogenesis precedes differentiation."

When viewed grossly, limb development proceeds from a flattened disk of determined cells to an early mound-shaped outgrowth, the limb bud (Fig. 10-11). As the limb bud grows out, it flattens distally, forming a paddle-shaped

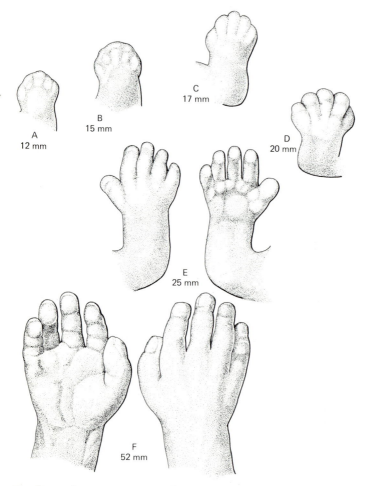

Fig. 10-11 Stages in the development of the human arm and hand. The lengths indicated refer to the crown-rump length. *(After Scammon, from Retzius.)*

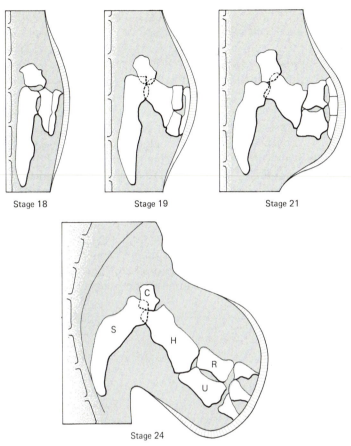

Stage 18   Stage 19   Stage 21

Stage 24

Fig. 10-12 Maps of prospective bone-forming areas of the developing wing bud. Based upon reconstructions of limb buds bearing radioisotopically labelled grafts of tissues. Stages based on Hamburger and Hamilton (1951). *(Adapted from Stark and Searls, 1973. Devel. Biol., 33:138.)* Abbreviations: C, coracoid; H, humerus; R, radius; S, scapula; U, ulna.

structure. Early digital primordia appear like rays within the thin paddle, and the elbow or knee becomes readily recognizable. By this time essentially all of the major components of the limb have been laid down, and further development consists principally of growth and the rotation of the limb to its final definitive configuration.

The differentiation of limb components follows a pronounced proximodistal gradient and a lesser gradient along the anteroposterior axis, the direction of the latter gradient varying among the classes of vertebrates. This means that even though all of the prospective parts of the limb are represented in the early limb bud (Fig. 10-12), the *stylopodium* (upper arm or thigh) differentiates first, followed by the *zeugopodium* (forearm or leg) and finally by the *autopodium* (hand or foot).

Differentiation of the Skeleton The skeletal tissues arise from cells originally present within the limb bud. Why some cells of the limb bud differentiate into

skeletal tissues is not known, although there are indications that local environmental factors, under the influence of the vascular bed of the limb bud, play a
role in determining which mesenchymal cells will form skeletal structures and
which will not (Caplan, 1977).

The first morphological indication of formation of the skeleton is the
condensation of mesenchymal cells in the central core of the proximal part of the
limb bud. These cells begin to secrete matrix material and soon hyaline cartilage
becomes recognizable. In the mammalian forelimb, the scapula and humerus
form first (Forsthoefel, 1963). Then the postaxial components of the zeugopodium and autopodium (ulna, and digits IV and V) appear. Finally, the preaxial
skeletal elements (radius, and digits III, II, and I) become established (Fig.
10-13). In amphibian limbs the gradient of differentiation along the anteroposterior axis is reversed, and the radius is established before the ulna. The skeleton
of the limb remains cartilaginous until the gross form of the limb is well
established. Only at a later period in development does the replacement of
cartilage by bone begin. The formation of joints is discussed in Chap. 9.

Differentiation of Muscles   For many years the origin of the musculature in
the limb has been debated. According to one opinion, the muscles arise in situ
from the somatopleural cells that originally gave rise to the limb bud. According
to a second point of view, cells derived from the somites secondarily enter the
limb bud and subsequently form muscle. The controversy now seems decided in

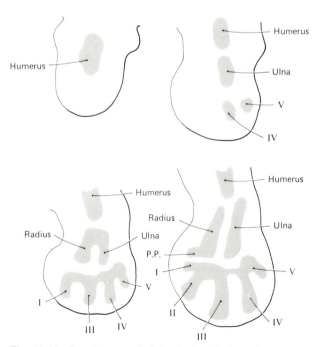

Fig. 10-13   Development of skeletal primordia in embryonic forelimbs of the mouse. *(Drawings
based upon photomicrographs in Forsthoefel, 1963,* Anat. Rec. **147:29**.*)* Abbrev: p.p. prepollex

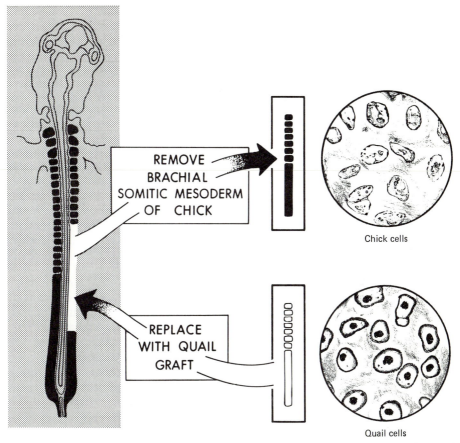

Fig. 10-14 Diagram of an operation in which a row of somites is removed from a chick embryo and replaced by somites from a quail embryo. The dense masses of chromatin in the nuclei of quail cells serve as natural cellular markers. *(Adapted from Christ et al., 1977, Anat. Embryol. 150:171.)*

favor of the latter viewpoint by the use of cellular marking experiments involving the removal of somites adjacent to the limb region in chick embryos and replacing them with a graft of somites from the Japanese quail (Christ et al., 1977; Chevallier et al., 1977) (Fig. 10-14). Quail cells can be easily distinguished from cells of the host chick embryo because the nuclei of quail cells contain large masses of dense chromatin (DNA) associated with the nucleus (Fig. 10-14). The grafted quail tissue becomes readily integrated with the body of the host, and quail cells undertake surprisingly normal patterns of migration within the chick tissues.

In these experiments, cells from the quail somite grafts migrated into the limb buds of the chick, and their final location within the differentiated tissues of the limb was determined on the basis of their nuclear staining characteristics. The muscle fibers of the limbs were shown to be of quail somitic origin, whereas the cells making up the tendons and the intramuscular connective tissue came from the limb bud of the chick embryo.

The formation of muscular elements in the limb takes place shortly after the skeletal elements begin to take shape. Čihák (1972) has described four fundamental phases in the ontogenesis of muscle pattern:

1   The condensation of mesenchymatous cells to form common flexor and extensor muscle masses. In these masses, layers corresponding to the fundamental muscular layers of the adult limbs take shape.

2   The formation of individual muscle primordia within the common muscle masses. Within the wing bud of the chick embryo, Shellswell and Wolpert (1977) have described a binary pattern of splitting of the common muscle masses first into groups and then into individual muscles. By the time the muscle primordia have taken shape, the muscle cells themselves are in the myoblastic or early myotube stage. Almost nothing is known about the mechanisms involved in splitting of the common muscle masses into individual primordia.

3   The phase of reconstruction of the muscle primordia. When first formed, the muscle primordia are in a phylogenetically primitive arrangement. During the phase of reconstruction, many of the primordia become rearranged to a configuration characteristic of the species. For example, in the human hand, some muscle primordia from different layers fuse to form single muscles (e.g., dorsal interosseous muscles). In contrast, other primordia (e.g., parts of the contrahentes muscle layer of the palm) disappear through cell death, despite the fact that the cells within them have differentiated to the point of containing myofilaments (Grim, 1972).

4   The development of the definitive form of the muscle. Prominent in this phase is the formation of connective-tissue elements and their integration with the muscle fibers. It is now known that the bodies of the muscles first take shape apart from the tendons and that the muscles and their tendons become joined together only at a later time. Because of their highly independent modes of origin, it appears likely that when the primordia of the skeletal elements, muscles, and tendons are being established, the cells comprising them are taking their cues from a common set of morphogenetic guidelines that were established when their original pattern was set up.

After the muscles in the limb have been fully formed, they must still undergo considerable growth to keep pace with the rest of the body, both during the fetal period and after birth. In many species new muscle fibers are added to the muscles up to about the time of birth. Shortly after birth the formation of new muscle fibers ceases and the number of muscle fibers remains more or less constant throughout postnatal life. Muscle fibers are syncytial structures, and their nuclei can no longer divide. Growth of the individual muscle fiber is accomplished by the incorporation of satellite cells (see page 280) into the fiber and the addition of new units of contractile proteins (sarcomeres), usually at the ends of the muscle fibers.

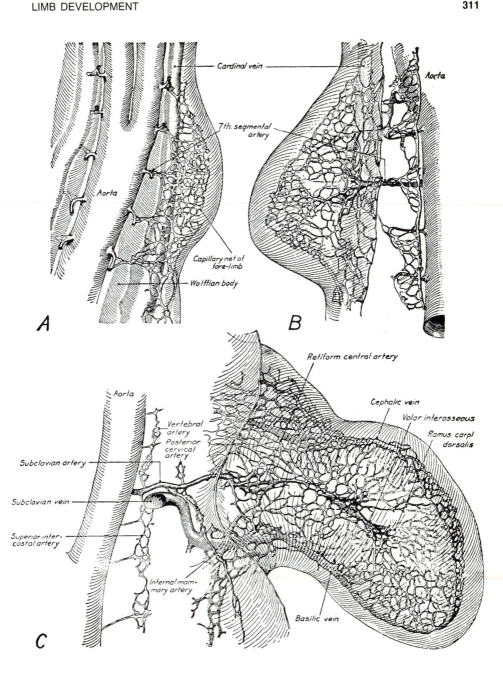

Fig. 10-15 Diagrams from injected pig embryos showing three stages in the early development of the anterior limb bud. (A) Embryo of 4.5 mm (equivalent to 4-week human embryo). (B) Embryo of 7.5 mm (equivalent to 5-week human embryo). (C) Embryo of 12 mm (equivalent of 6-week human embryo). *(After Wollard, 1922, Carnegie Cont. to Embryol. **14**:139.)*

Development of the Vascular Supply   The earliest limb bud is supplied by a
fine capillary network arising from several segmental branches of the aorta (Fig.
10-15). Soon, some channels within the network are favored over others. One of
the first to take shape is a primary central artery, which distributes blood to the
limb bud. Blood from the central artery travels through a capillary network and
then collects in a *marginal sinus* that extends around the apex of the limb bud,
just beneath the apical ectodermal ridge, and returns the blood to the body
through venous channels coursing along both the pre- and postaxial borders of
the limb.

The capillary networks and the marginal sinus itself respond to growth of
the limb bud by sending out new vascular sprouts. A careful study by Seichert
and Rychter (1971) has shown that the marginal sinus in the developing wing of
the chick is constantly being reformed by the coalescence of capillary sprouts
which grow out apically from the already existing marginal sinus. Blood first
drains from the marginal venous channels into superficial venous plexuses of the
body, but as development continues, a progressively greater proportion of the
blood is shunted into deeper channels. As the digital rays become established,
the apical marginal sinus begins to break up, but the proximal portions of the
marginal channels persist in mammals as the *basilic* and *cephalic veins*. The
fundamental arrangement of deeply situated arteries and superficial veins
persists in the adult limb.

The formation of the major arteries of the extremities is a complex process
in which several major arterial trunks successively take ascendency and then
regress, to be replaced by still other major vessels. As an example of this, the
major steps in the development of the definitive arterial pattern of the human
arm are illustrated in Fig. 10-16.

The role of the vasculature in limb development has been viewed several
different ways over the years. Some of the earliest opinions suggested that
vascular patterns at specific stages of development were phylogenetically fixed
and, in a sense, almost independent of the tissues developing alongside them. A
more widespread viewpoint is that the vascular pattern to a large extent responds
to the growth and development of other tissues in the limb bud and that
morphogenesis of the vascular system is almost entirely a secondary phenomenon, being dependent upon the differentiation of other tissues within the limb.
In recent years increasing attention has been paid to the possibility that the
pattern of the vasculature may, of itself, act as a primary agent in morphogenesis
and differentiation. The close relationship between the marginal sinus and the
apical ectodermal ridge is looked upon by some as a functionally important
association instead of as merely an anatomical juxtaposition. According to
others (e.g., Caplan and Kautroupas, 1973), local differences in the metabolic
environment, brought about by the patterns of the embryonic arteries and veins,
may be a determining factor in the course of differentiation followed by some
groups of cells. More information is required before the role of blood vessels in
limb development can be firmly established, but it is likely that the blood vessels
will be shown to be prime movers in some aspects of development and secondary
responders in other aspects.

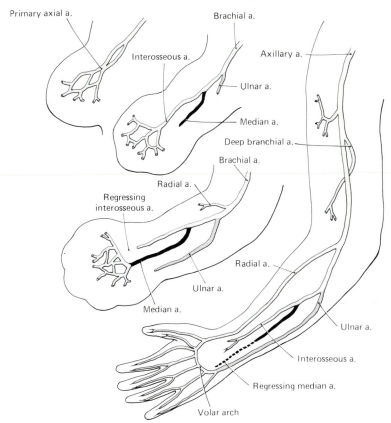

Fig. 10-16 Development of the arteries of the human arm.

Innervation of the Embryonic Limb The early limb bud forms in the absence of nerves, and it remains uninnervated during much of the early period of outgrowth. In higher vertebrates, early skeletal primordia and common muscle masses have already been laid down before nerves grow into a given segment of a limb. The morphological delineation of individual muscles and establishment of nerve patterns often seems to go hand in hand.

The unique interrelationships between nerves and many components of the limb have stimulated a number of questions concerning morphogenetic influences of nerves upon the limb and factors that determine the pattern of innervation of the limb. It is by now quite well established that overall morphogenesis of limbs is independent of the nerve supply. This has been shown in a number of grafting experiments in which limb buds, placed in ectopic locations either on the body or, in birds, on the chorioallantoic membrane, undergo normal development.

In another type of experiment, *aneurogenic limbs* can be created by surgically removing the neural tube of early embryos. Under these circumstances the limbs grow in the absence of nerves. Both gross morphology of the limbs and the patterns of internal tissues are nearly normal. The nerves do, however, appear to influence the rate of mitosis of mesenchymal and epithelial tissues, and some minor deviations from normal morphology in aneurogenic limbs are probably due to deficiencies in the number of cells available to form the primordia of certain structures. In this way, nerves may indirectly affect morphogenesis.

The factors that determine the pattern of nerves within a limb are still being debated. Extreme views are (1) that the pattern of innervation is determined entirely by the configuration of the other tissue components of the limb, the ingrowing nerves passively following environmental cues, and (2) that the ingrowing nerves themselves are uniquely specified, allowing them to home in upon specific, predetermined structures within the limb. Both of these postulates have been subjected to experimental analysis, and it now appears that the answer to this question may lie somewhere between these two extremes. If limbs are grafted to ectopic sites or if pieces of the embryonic nervous system are grafted so that the limb is innervated by foreign nerves, the main pattern of nerve trunks is often recognizable, but the overall pattern of innervation is usually abnormal (Piatt, 1956, 1957). Yet other work on fishes and amphibians has strongly suggested that motor nerve fibers have a preference for a particular muscle, even to the point of functionally displacing other nerve fibers that have been allowed secondarily to innervate the muscle.

Rotation of the Limbs Both the anterior and posterior limbs undergo pronounced changes of position during embryonic life. It is easiest to use the human limb as the model for describing these changes. At the stage of early differentiation of the digits the arm bud projects outward at right angles to the body. (This is equivalent to holding your arm horizontally, with the thumb upward and palm forward when standing.) As the elbow joint forms, the forearm and hand bend ventrally so that the palm faces the trunk. (Without dropping the upper arm at the shoulder, flex your elbows, bringing the palms of your hands against your chest.) Then the arm undergoes a 90° rotation about its long axis so that the elbow points in a caudal direction. (Drop your elbow to your side.)

The leg undergoes a similar set of axial changes. Early in development the leg bud projects straight out from the body parallel to the arm, with the future great toe cranial (like the thumb) and the sole facing the same direction as the palm. As the knee forms, the lower leg and foot bend so that the sole faces the trunk. The leg also undergoes a 90° rotation, but in an opposite direction to that of the arm. This brings the knee to point in a cranial direction.

Despite the changes in configuration of the limbs, their homologous structures and early embryonic relations can be determined by the segmental pattern of their innervation. In the arm, the thumb and radial side of the forearm (preaxial structures) receive the most cranial innervation of the limb, as do the great toe and tibial side of the leg (Fig. 9-20).

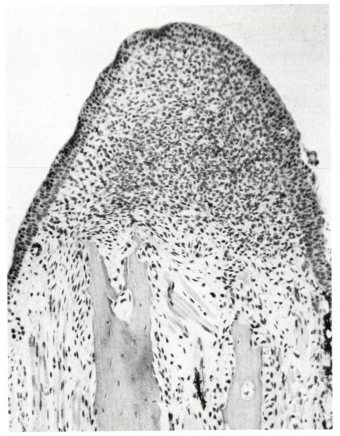

Fig. 10-17 Regeneration blastema in an adult newt. Twenty-five days previously the forearm was amputated through the radius and ulna. The blastema is the dense aggregation of cells in the top half of the figure. In the lower half one can see the ends of the amputated radius and ulna and some of the forearm muscles. (From B.M. Carlson, in G.V. Sherbet, ed. *Neoplasia and Cell Differentiation,* S. Karger, Basel.)

Limb Regeneration   The limbs of some vertebrates (e.g., salamanders) are endowed with the capacity to regenerate after they have been amputated. Amputation is followed by epidermal wound healing, which covers the amputation surface, and by the removal of internal debris created by the original wound. Within a few days, the limb stump enters the period of *dedifferentiation*, during which the differentiated tissues of the distal part of the limb break up and are replaced by a population of primitive-appearing cells. These cells aggregate and proliferate at the end of the limb stump to form a regeneration blastema (Fig. 10-17), which is similar in many respects to an embryonic limb bud. As the blastema grows it forms a new limb in a manner that parallels very closely the sequence of events that occurs in the embryonic limb bud.

Although there are many parallels between the embryonic development and regeneration of a limb, there are also some differences, reflecting the tighter integration of the regenerating limb to the body. At least three conditions—a wound epidermis, mesodermal damage, and an adequate nerve supply—are normally required for the initiation of limb regeneration. The role of other factors, such as hormones and bioelectric currents, remains unclear.

Both the dedifferentiative phase and subsequent morphogenesis involve the reading of positional information and response to it by the cells involved in regeneration. In contrast to the embryonic limb bud, which must set up a system of positional values, the regenerating limb develops in association with mature cells which have retained a record of their position and which, under appropriate conditions, can express this information.

The regeneration of extremities is most successful in fish, the urodelan amphibians, and the tadpoles of anurans. During metamorphosis, most frogs and toads lose the ability to regenerate limbs. Lizards are renown for their ability to regenerate tails, but their limb regenerative capacity is poor. The limbs of birds and mammals do not regenerate. A major question in regeneration is whether the limbs of higher forms have lost the capacity to regenerate entirely or whether they retain a latent ability to regenerate if missing factors are supplied. Numerous attempts have been made to stimulate the regeneration of limbs in frogs and in higher vertebrates. Some degree of success has been reported by such means as changes in the hormonal environment, increasing the nerve supply, or the application of electrical currents to the amputated limbs. A number of factors are likely to be involved in the loss of regenerative power, and the stimulation of partial regeneration of limbs in higher vertebrates suggests that the loss of ability to regenerate may not be irreversible.

Chapter 11

The Development of the Nervous System

ORGANIZATION OF THE MAMMALIAN NERVOUS SYSTEM

Most students using this book have not taken detailed courses in neuroanatomy or neurobiology, and many have been exposed only briefly to the functional organization of the nervous system. Without some background knowledge of the adult nervous system, the student is likely to find many aspects of its development barren and discouraging. This section is designed to provide an adequate background for understanding the elements of neuroembryology presented in this chapter.

Neurons and Synapses The fundamental functional unit of the nervous system is the *neuron*, a highly specialized cell consisting of a large cell body and one or more processes called *axons* and *dendrites*, which extend from the cell bodies of the neurons. These processes may be very long (up to a meter in humans), and they are often highly branched, making functional connections with processes of many other neurons or other end organs, such as muscle fibers. The specialized points at which one neuronal process contacts another is called a *synapse*, and it is at this point where an electrical impulse from one neuron is transmitted to the other through the mediation of a chemical transmitter substance, such as *acetylcholine* or *epinephrine*. Groups of interconnected

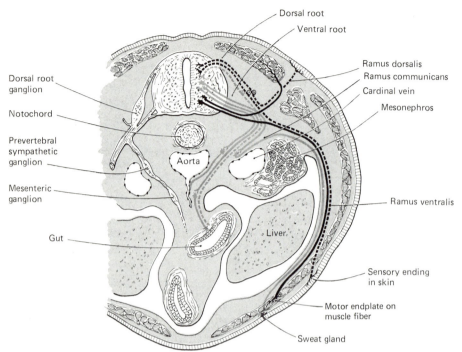

Fig. 11-1 Schematic diagram indicating the various connections made by the neurons which develop in a typical spinal nerve. (Modified from Froriep.)

The neurologist classifies the fibers in a spinal nerve according to their relations and functions. The components of a typical spinal nerve on this basis are:

I. Afferent
   A. General Somatic Afferent
      1. *Exteroceptive,* i.e., fibers conducting impulses from the external surface of the body such as touch, pain, temperature. (Represented in this figure by dark broken lines.)
      2. *Proprioceptive,* i.e., fibers carrying impulses of position sense from joints, tendons, and muscles. (Not represented in this diagram.)
   B. General Visceral Afferent
      Fibers from viscera (interoceptive) by way of sympathetic chain ganglion, white ramus communicans, and dorsal root; cell bodies in dorsal root ganglion; no synapse before reaching cord. (Illustrated in this figure by the light broken line.)
II. Efferent
   A. General Somatic Efferent
      Motor neurons to skeletal muscle; cell bodies in ventral columns of gray matter; fibers emerge by central roots. (Illustrated in this figure by dark solid lines.)
   B. General Visceral Efferent
      Two-neuron chains from cord to glands and to smooth muscle of viscera and blood vessels. The first neurons (preganglionic) have their cells of origin in the lateral column of the gray matter of cord from first thoracic to third lumbar level. Fibers leave cord by ventral root, turn off in white ramus communicans to end in synapse with the second-order neurons (postganglionic) of the two-neuron chain. (Illustrated by the light solid lines.)
      Note the various destinations of the visceral efferent paths, e.g.:
      1. Fibers to smooth muscle of gut wall; impulse relayed by second-order motor neurons from synapses in mesenteric ganglion.
      2. Fibers (vasomotor) to smooth muscle of blood-vessel wall; impulses relayed by second-order motor neurons from synapses in prevertebral ganglia.
      3. Fibers (pilomotor) to muscles about hair follicles, and fibers (sudomotor) to sweat glands in skin. Impulses in both these cases relayed from synapses in prevertebral ganglia by

neurons are organized into numerous networks or pathways, which are responsible for the transmission and processing of discrete functional messages throughout the body.

The nervous system does not consist of neurons alone. In the central nervous system (brain and spinal cord) the nerve cell bodies and their processes are closely associated with large numbers of smaller cells, called *glial cells*, which are said to fulfill functions as diverse as nutrition, support, insulation, and even memory, according to some researchers. The nervous system is richly vascularized, and blood vessels of different dimensions abound on and within nervous tissue.

Neuronal processes are either *myelinated* or *unmyelinated*. In the central nervous system, myelinated nerve processes comprise much of the *white matter* of the brain and spinal cord; unmyelinated nerve fibers are found within the *gray matter*. Proper function of nerves in both the central and the peripheral nervous system requires that the nerve fibers be well insulated. In myelinated nerves insulation is provided by a multilayered sheath of a lipid material, called *myelin*, wrapped around the neuronal processes (Fig. 11-21). Myelin sheaths prevent the short-circuiting of electrical signals within the nervous system much like the insulation around individual electrical wires in a large cable. In the central nervous system *oligodendroglial cells* (one of the types of glial cells) are the myelinating agents. In the peripheral nervous system myelination of nerve fibers is accomplished by the *Schwann cells*, derived from the neural crest (Chap. 6).

Functional Classes of Neurons Functionally, neurons can be divided into three main groups. Some neurons carry impulses from sense organs (*receptors*) and sensory nerve endings toward the spinal cord and brain. These are called *afferent* neurons. Others conduct motor impulses away from the brain and spinal cord to muscles or glands (*effectors*), which respond by appropriate activity. Such neurons are said to be *efferent*. In the spinal cord, and especially in the brain, there are countless neurons having many relatively short processes which can transfer an incoming sensory impulse to any of a number of efferent neurons with which their various processes connect. These are *association* neurons. These three types of neurons plus the receptors and effectors, by which the nervous system makes contact with and influences the rest of the body, constitute the cellular basis for the integrated function of the organism.

The Organization of a Spinal Nerve The structures that we call *nerves* in the dissecting room are actually bundles of neuronal processes which connect

---

second-order motor fibers passing back over the ramus communicans and thence to periphery via branches of spinal nerve.

Most of the fibers in the spinal nerve are medullated and therefore whitish in appearance. The visceral second-order motor ("postganglionic") fibers, however, lack a myelin sheath and are grayish in appearance. Certain of such second-order visceral motor fibers (e.g., pilomotor and sudomotor) which run back along the ramus communicans after synapse in a prevertebral ganglion account for the so-called "gray bundle of the ramus communicans."

peripheral structures with the brain or spinal cord. The cell bodies are either buried in the central nervous system, where they are called *nuclear masses*, or they are massed at some point along the nerve to form a ganglion. The nerve itself consists only of the long slender neuronal processes and the sheaths which protect them.

Figure 11-1 will serve as a general introduction to the organization of a typical nerve. The components of a nerve can be subdivided two ways. The first would be into afferent and efferent nerve fibers, which have been defined in the previous section. Within the *territory*[1] of a nerve the scattered processes of afferent neurons converge toward the main trunk of the nerve. Close to the spinal cord, all of the afferent nerve fibers collect into a single branch known as the *dorsal root* (Fig. 11-1). A grossly visible expansion along the dorsal root marks the location of the *dorsal root ganglion*, where the cell bodies of the afferent neurons are located. Afferent (sensory) neurons are called *unipolar neurons* because one prominent process leads from the cell body. From this process one major branch leads to the periphery. The other major branch travels down the remainder of the dorsal root into the dorsal part of the spinal cord, where it makes synaptic connections with processes of other neurons.

The efferent neurons arise in the ventral part of the gray matter of the spinal cord, where the cell bodies are located. Efferent neurons are large cells with numerous smaller processes (dendrites) and one large process (axon) extending from the cell body. Because of their numerous processes, these are called *multipolar neurons*. The dendrites synapse with and receive signals from other neuronal processes within the spinal cord. The information received from the dendritic processes may be translated into an electrical signal (action potential), which goes from the cell body down the axon toward the effector organ for that neuron. As they leave the spinal cord, the axons of the efferent neurons are collected into the *ventral root* (Fig. 11-1), which shortly joins with the dorsal root to form a common nerve trunk containing a mixture of afferent and efferent nerve fibers.

The other major way in which the components of a peripheral nerve can be subdivided is into *somatic* and *visceral* divisions. Somatic neurons supply the skin, muscles, and tendons, whereas visceral neurons carry signals to and from the gut and other visceral organs. In both the somatic and visceral afferent parts of a nerve, the processes of only one neuron make the complete connection between the receptor and the spinal cord. Somatic efferent nerves are comprised of only the axons of cell bodies located in the spinal cord. On the other hand, visceral efferent nerves are composed of two-neuron chains, connecting the spinal cord to the effector organ (Fig. 11-1). The first-order neurons originate within the gray matter of the spinal cord and send out axons toward the periphery along the ventral roots. Soon these processes lead off into separate autonomic nerve branches (*ramus communicans*, singular) and head toward the

---

[1]A nerve territory is that part of the body which is supplied by processes from a single nerve. The territories of the cutaneous nerves are outlined in Fig. 9-20.

viscera. Here they ramify into complex networks of autonomic nerves, along which enlargements (ganglia) are seen. These ganglia contain the cell bodies of the second-order autonomic neurons, which in turn send axons directly into the visceral organs to be innervated. For further details of the organization of the autonomic nervous system a textbook of anatomy or neuroanatomy should be consulted. The legend of Fig. 11-1 presents a functional classification of the components of a typical spinal nerve.

Ganglia   The cell bodies of the somatic efferent and the first-order visceral efferent nerves are located within the central nervous system, but the cell bodies of the other components of peripheral nerves are found in aggregations at specific sites along the nerves. The dorsal root (sensory) ganglia of the spinal nerves are segmental aggregates of neural-crest cells (Fig. 6-14), whereas the corresponding ganglia of the cranial nerves (V, VII, VIII, IX, X, and XI) arise from ectodermal placodes (thickenings) in the head region (Fig. 12-1). The ganglia containing the cell bodies of second-order autonomic neurons also arise from aggregations of cells originating in the neural crest (Fig. 6-14).

Reflexes   The least common denominator of integrated neural function is the *reflex arc*, which constitutes the neural connections between sensory input and a motor response. The simplest type of reflex is one in which a local center in the spinal cord receives sensory impulses coming to it by way of afferent neurons and sends out motor impulses along efferent neurons which activate effectors in the same segmental level that was stimulated. Such an automatic local response is known as a simple *intrasegmental reflex* and represents the most primitive type of mechanism in the vertebrate action system (Fig. 11-2, arc 1).

By transmission of a sensory impulse longitudinally along the spinal cord through the mediation of association neurons, the outgoing impulse for the same stimulus may be imparted to efferent neurons in several adjacent metameres. Such a mechanism is termed an *intersegmental reflex* (Fig. 11-2, arc 2), and it calls into play the concerted response on the part of a group of effectors. The part of the cord principally involved in relaying segmental and intersegmental responses is the centrally located gray matter. The gray matter is composed chiefly of association neurons and of the cell bodies of motor neurons which send their processes into the spinal nerves.

A sensory stimulus can also be carried to other groups of nerve fibers which run longitudinally up the peripheral white matter of the spinal cord toward the brain in the form of well-defined pathways or tracts. Phylogenetically, the conduction pathways in the peripheral part of the spinal cord increase in conspicuousness concomitantly with the increasing extent to which the brain assumes a coordinating control over the basic reflexes which constitute the primary function of the cord. The sensory (afferent) conduction paths are for the most part grouped in the dorsal portion, whereas the motor (efferent) tracts are found in the lateral and, to a lesser extent, in the ventral regions of the cord.

Higher in the central nervous system the reflex arcs become more complex,

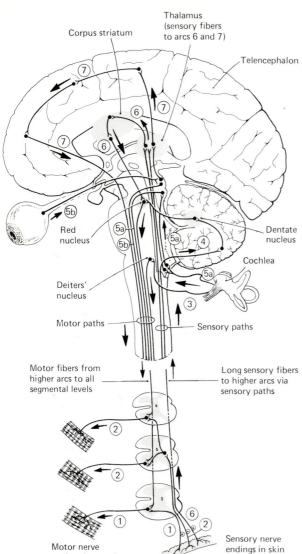

Arc 7

VOLUNTARY AND INHIBITORY
CONTROL
Choice of response based on memory
of past experiences. (Via pyramidal
tract.)

Arc 6

AUTOMATIC ASSOCIATED
CONTROL
of complex muscular actions. (Via
striatorubrospinal tract.)

Arc 5

a. AUDITORY REFLEXES
E.g., Automatic response to sudden
noise.

b. VISUAL REFLEXES
E.g., Automatic response to blinding
flash of light.
(Both a and b via tectospinal tract.)

Arc 4

SYNERGIC CONTROL
Automatic coordinating control of
muscular actions. (Via rubrospinal
tract.)

Arc 3

Automatic coordinating control of
muscular actions. (Via rubrospinal
tract.)

Arc 3

EQUILIBRATORY CONTROL
Automatic balancing reactions. (Via
vestibulo-spinal tract.)

Arc 2

INTERSEGMENTAL REFLEX
Impulse carried by association neurones to neighboring segments causing
coordinated response of muscles in
several segments.

Arc 1

INTRASEGMENTAL REFLEX
Response limited to segment stimulated.

Fig. 11-2 Schematic diagram showing the nature of the activities carried out in various parts of the central nervous system.

as the input from the special sensory systems of the head are added to the total neural input and vast numbers of association neurons connect discrete fiber tracts to the various coordinating and association centers of the brain. The varieties of response to a given afferent stimulus are immense, not only because of the number of involuntary reflex centers but also because of the ability of the cerebral cortex to direct facultative responses on the basis of past experience.

EMBRYOGENESIS OF THE NERVOUS SYSTEM

A number of fundamental processes are involved in the embryogenesis of the nervous system. Some dominate the embryo during their occurrence; others are limited to certain parts of the nervous system. The major processes are (1) induction, both the primary induction of the neural plate and the secondary inductions emanating from the early brain and spinal cord; (2) proliferation, both as a response to primary induction and as a prelude to the morphogenesis and growth of specific portions of the nervous system; (3) cell migration, of which there are many striking examples at many stages and in many regions; (4) the differentiation of neurons and glial cells, including both structural and functional maturation; (5) the formation of specific connections between groups of neurons; (6) the stabilization or elimination of interneural connections, resulting in the death of unconnected cells; and (7) the development of integrated neural function, allowing the embryo and newborn to undertake coordinated movements.

EARLY STAGES IN ESTABLISHING THE NERVOUS SYSTEM

Recent experiments conducted on amphibian embryos (Hirose and Jacobson, 1979) suggest that even in the early cleavage stages individual blastomeres give rise to consistent populations of cells within the nervous system (and within other areas of the body). Each blastomere is said to serve as the source of a clone of cells, which ultimately constitute a *clonal domain*. This is determined by injecting a single blastomere with horseradish peroxidase as an intracellular label. After varying numbers of cell divisions, the embryos are fixed and histochemically stained for peroxidase activity. Only those cells which are the progeny of the originally labeled blastomere are stained. In this way maps of clonal domains derived from identifiable blastomeres can be constructed. An example of a map of a clonal domain in the central nervous system of a *Xenopus* embryo is shown in Fig. 11-3. This illustrates the remarkable precision in the fates of specific cells of early embryos even as they are swept up in the massive morphogenetic movements and are involved in the inductive processes and the wave of proliferation that follows.

The initial induction of the central nervous system from ectoderm by the notochord has already been covered in Chap. 6. In addition to bringing about a change in the shape of the cells of the future neural plate and neural tube (Figs. 6-8 and 6-11), the primary inductive event stimulates the proliferation of the neuroepithelial cells. Increased proliferation by the responding tissue is a common consequence of induction in other systems, as well. We have also seen the origin of the neural groove by the infolding of the thickened neural plate ectoderm, the closure of the neural groove to form the neural tube, and the separation of the tube from the overlying parent ectoderm (Figs. 8-1, 8-2, and 8-3). Coincident with the separation of the neural tube from the cutaneous

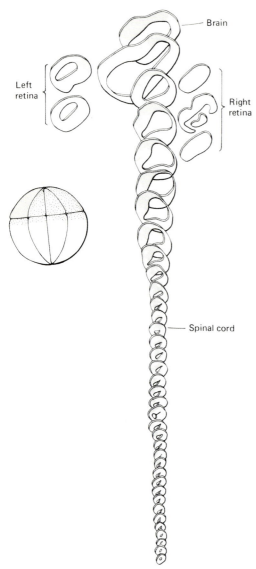

Fig. 11-3 Demonstration of cell lineages in the formation of the central nervous system in *Xenopus* (Anura). Horseradish peroxidase was injected as a marker into one blastomere (LD 1.2) of a cleaving embryo (gray cell in sketch of embryo). The animal was allowed to develop into a larva (stage 32) and was then fixed and serially sectioned. Cells descended from the original labelled blastomere, retain the peroxidase label and are stained gray on the representative sections shown in this figure. *(Adapted from Hirose and Jacobson, 1979, Devel. Biol. 71:191.)*

ectoderm, the cells of the neural crest migrate out from the junction between neural and cutaneous ectoderm and either become widely distributed throughout the body or remain near the spinal cord as the segmentally arranged spinal ganglia (Fig. 6-14).

Almost as soon as it is independently established, the neural tube becomes markedly enlarged cephalically. This expanded portion is the primordium of the brain. Caudally the neural tube (forerunner of the spinal cord) remains of relatively uniform diameter.

In its enlargement the brain at first exhibits three regional divisions—the primary forebrain, midbrain, and hindbrain, or, to use their more technical synonyms, the *prosencephalon, mesencephalon,* and *rhombencephalon* (Fig. 11-13A). The three-vesicle stage of the brain is short-lived. The prosencephalon becomes subdivided into two regions, *telencephalon* and *diencephalon;* the mesencephalon remains undivided; and the rhombencephalic region becomes differentiated into *metencephalon* and *myelencephalon.* Thus in place of three vesicles five are established. These stages in the development of the brain are clearly visible in embryos between 8 and 17 mm in length (Figs. 11-11, and 11-13C and D). Using these familiar conditions as a basis, we are ready to trace the later differentiation of some of the more important parts of the nervous system.

HISTOGENESIS OF THE CENTRAL NERVOUS SYSTEM

Establishment of the Neuroepithelium The ectoderm of the open neural groove and early neural tube is a thick pseudostratified epithelium. An epithelium of this type appears to contain several layers of cells because the nuclei are found in several different planes, but in reality the nuclei are located within very long and somewhat irregularly shaped cells, the cytoplasm of which extends throughout the entire thickness of the epithelium. Thus, in contrast to commonly held viewpoints of former years, the early neuroepithelium is not a multilayered structure. The *neuroepithelial cells* possess a high degree of mitotic activity, and in confirmation of the early work of Sauer (1935), autoradiographic research, with tritiated thymidine used as a marker, has indicated that during different phases of their life history, the nuclei of the neuroepithelial cells occupy different positions within the neural epithelium (Sauer and Walker, 1959; Langman et al., 1966). The synthesis of DNA occurs in nuclei located close to the external limiting membrane (Fig. 11-4). Following this phase, the nuclei move within the cytoplasm of the neuroepithelial cells toward the lumen, and then the cells undergo mitotic division.

Before closure of the neural tube, the nuclei of the daughter cells again move toward the external limiting membrane, where they may again synthesize DNA and repeat the germinative cycle. In this manner the number of neuroepithelial cells is greatly increased. Following closure of the neural tube some of the neuroepithelial daughter cells migrate away from the lumen past the DNA-synthesizing neuroepithelial cells to positions immediately beneath the

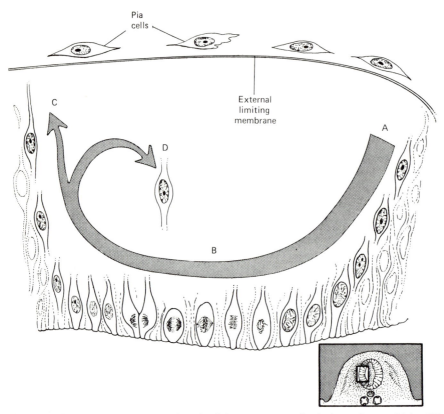

Fig. 11-4 Diagrammatic representation of cellular events occurring in the early neural tube. When it first forms, the neural tube is a pseudostratified epithelium with the external limiting membrane serving as its base and the boundary of the central canal serving as its apex. The nuclei of neuroepithelial cells begin synthesizing DNA in the basal areas (A) and then migrate within the cytoplasm toward the apical part of the cells where they undergo mitosis (B). Nuclei of the daughter cells then migrate back toward the external limiting membrane where they either begin to differentiate as neuroblasts (C) or return to the pool of proliferating neuroepithelial cells (D). *(Adapted from several sources, chiefly Sauer and Langman.)*

external limiting membrane (Fig. 11-4C). These cells, called *neuroblasts*, begin to produce processes which are the forerunners of axons and dendrites.

As the primitive neuroepithelium matures, it is possible to subdivide it into layers (Fig. 11-5).[2] The innermost layer, called the *ventricular* or the *ependymal layer*, contains cells which are still in the mitotic cycle. Ultimately this layer becomes the *ependyma*, a columnar epithelium which lines the central canal and

---

[2]The terminology of the layers of the developing neural tube is in a state of flux. Most textbooks still refer to three layers, the ependymal layer, the mantle layer, and the marginal layer. The Boulder Committee (1970) has advocated replacing these names with *ventricular layer, intermediate layer,* and *marginal layer.* In addition, this committee designates a fourth layer, the *subventricular layer,* between the ventricular and the intermediate layer. The latter terminology is becoming increasingly used in the research literature.

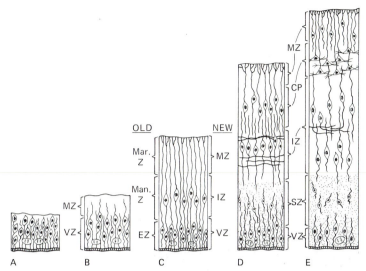

Fig. 11-5 Diagram illustrating successive stages in the layering of the wall of the developing brain (after Rakic). The designation of the layers is that recommended by the Boulder Committee *(Anat. Rec.* **166***:257).* At the earliest stage (A) the wall of the brain is a pseudostratified epithelium. Section "C" compares the traditional with the Boulder terminology. Abbreviations: CP, cortical plate; EZ, ependymal zone; IZ, intermediate zone; MZ (Mar. Z), marginal zone; Man. Z, mantle zone; SZ, subventricular zone; VZ, ventricular zone. Cell proliferation occurs in the ventricular zone and, to a lesser extent, the subventricular zone. As the brain matures, the layers of the cerebral cortex develop in the region of the cortical plate.

the ventricular system of the central nervous system. The peripheral part of the neural tube now consists of an expanding zone containing numerous cell processes but few cell bodies. This outer, cell-poor layer is called the *marginal layer*; it ultimately becomes the white matter of the central nervous system (see page 329). The postmitotic neuroblasts which leave the inner ventricular layer form an intermediate layer of densely packed cells, called the *mantle layer*. This layer will become the gray matter.

Neuroblasts Some of the daughter cells generated in the neuroepithelium lose the ability to undergo mitosis and migrate toward the outer wall of the neuroepithelium (Fig. 11-4C). These cells, called *neuroblasts*, are initially bipolar, having slender processes which connect with both the central luminal border and the external limiting membrane of the neural tube. The *bipolar cells* soon become detached from the inner luminal border by a retraction of the inner process, thus becoming *unipolar cells*. The unipolar cells then accumulate large amounts of rough endoplasmic reticulum (*Nissl substance*) in their cytoplasm and send out several dendrites. At this stage the *multipolar neuroblasts* are expending much of their developmental energy toward the production of the cytoplasmic processes which will make connections with other parts of the nervous system. The way in which axons and dendrites grow out will be described on page 344. There are many specific patterns of differentiation of individual neurons throughout the nervous system, ranging from the early formation of large macroneurons to the later differentiation of microneurons. The morphology of both the nerve cell bodies and their processes varies greatly from region to region within the nervous system. Details are left to more specialized texts.

Neuroglia   After the first wave of neuroblast formation, other neuroepithelial cells follow a course of differentiation leading to the formation of neuroglial cells. Surprisingly little is yet known about either the functions of mature neuroglial cells or their development in the embryo. Of the three classes of neuroglial cells, two, the *astrocytes* and the *oligodendroglial cells*, arise from neurectoderm. The third type, the *microglia*, are active in phagocytosis after damage to the nervous system and are presumably of mesenchymal origin (possible a form of macrophage), although some researchers feel that microglia arise from neurectoderm.

The astrocytes and oligodendroglial cells are said to arise from a common intermediate cell type known as a *glioblast* (*spongioblast*) (Fig. 11-6). Some glioblasts develop long, fine processes and differentiate into astrocytes. The processes of astrocytes become closely associated with capillaries, and it has been suspected that among other functions they may serve as a transit system for metabolites within the substance of the central nervous system. The oligodendroglial cells are smaller and structurally more simple than astrocytes, and they become recognizable later in development. They appear as satellites around the

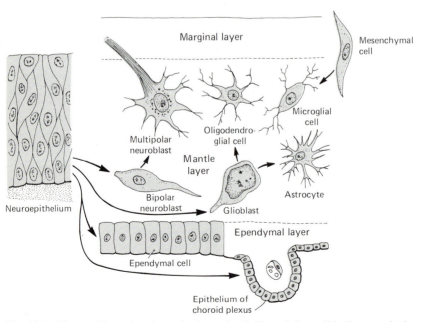

Fig. 11-6 Diagram illustrating the major lines of cell differentiation within the neural tube.

cell bodies of neurons and are also involved in myelination of nerve fibers within the white matter of the central nervous system. In contrast to neuroblasts and neurons, neuroglial cells retain the ability to divide, even in the adult.

Ependymal Cells   Other cells of the original neuroepithelium retain their epithelial character and differentiate into the columnar *ependymal cells* that line the central canal of the brain and spinal cord (Fig. 11-6). Much remains to be learned about the functions and potentialities of ependymal cells. In lower vertebrates the ependymal cells retain the potential for differentiating into neurons during regeneration of the central nervous system.

The Meninges   The early neural tube is surrounded by a loose mesenchyme which forms a thin membrane (*primitive meninx*). In the early fetal period the primitive meninx becomes subdivided into two layers, a relatively thick outer layer (*ectomeninx*) of mesodermal origin, which forms the *dura mater*, and a thin inner layer (*endomeninx*) of presumed neural crest origin, which ultimately becomes further subdivided into the thin *pia mater* and the middle *arachnoid layer*. Soon spaces which form within the pia-arachnoid layer become filled with the specialized *cerebrospinal fluid*, which bathes the spinal cord.

Development of Gray and White Matter in the Spinal Cord and Brain
During the period of development when the neurons differentiate, the appearance of the spinal cord undergoes very marked changes. Some of the neuroblasts in the mantle layer of the cord send out processes very early in development. Others remain undifferentiated and continue to proliferate for a time, causing continued growth in the mantle layer. As it grows in mass the mantle layer takes on a very characteristic configuration, becoming butterfly-shaped in cross section. With this change in shape, and with the transformation of its glioblasts into neuroglia and its neuroblasts into characteristic nerve cells, the mantle layer becomes the so-called gray matter of the spinal cord (Fig. 11-7).

During the growth of the mantle layer the originally extensive lumen of the neural tube is reduced, by obliteration of its dorsal portion, to form the small central canal characteristic of the adult cord (Fig. 11-7). The central canal is now lined by the epithelioid ependymal layer.

Meanwhile the outer (marginal) layer of the cord has been increasing extensively in mass. This increase is due to the secondary ingrowth of longitudinally disposed neuronal processes which constitute the conduction paths between the various levels of the spinal cord and the brain (Fig. 11-2). Because each of these fibers is enveloped in a sheath rich in myelin, the region of the cord in which they lie has a characteristic whitish appearance, which contrasts strongly with the gray color of the richly cellular portion of the cord derived from the mantle layer. For this reason the fibers in the marginal layer of the cord are said to constitute its white matter. The main groups of these fibers are more or less marked off from each other by the dorsal and ventral horns of the gray matter. They are known as the *dorsal, lateral,* and *ventral columns* of the

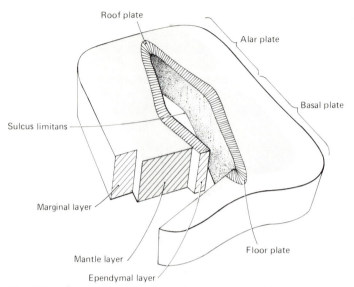

Fig. 11-7   Diagram illustrating the major subdivisions of the neural tube.

white matter of the cord (Fig. 11-8). The dorsal columns contain the main tactile and proprioceptive paths to the brain; the ventral columns are primarily motor; and the lateral columns contain important ascending fiber tracts to the brain and also some of the main motor paths from brain to cord and thence to spinal nerves (Fig. 11-2).

In dealing with the topography of the neural tube in cross section, it is customary to designate its thickened side walls as the *lateral plates*, its thin dorsal wall as the *roof plate*, and its thin ventral wall as the *floor plate* (Fig. 11-7). Extending along the inner surface of each lateral plate is a longitudinal sulcus (*sulcus limitans*), which divides the lateral plate into a dorsal afferent part (*alar plate*) and a ventral efferent part (*basal plate*, Fig. 11-24).

Histogenesis of the developing brain involves some unique problems and some unique solutions. A readily apparent problem involves the location of the gray and white matter. As we have just seen for the spinal cord and brainstem, the gray matter, containing the nerve cell bodies, remains in the center and becomes surrounded by the meylinated neural processes which constitute the white matter. In the forebrain and cerebellum, on the other hand, the gray matter is located at the periphery, often in several discrete layers, and the white matter is on the inside.

One cannot properly speak of the brain as a homogeneous entity because almost every part has a unique pattern of histogenesis. Isolated examples will be presented here for illustrative purposes. Developmental differences are not only present from one region to another in the embryonic brain, but they are also related to time. This can be illustrated by patterns of cellular proliferation. In almost all areas of the spinal cord or brain the proliferation of cells that will

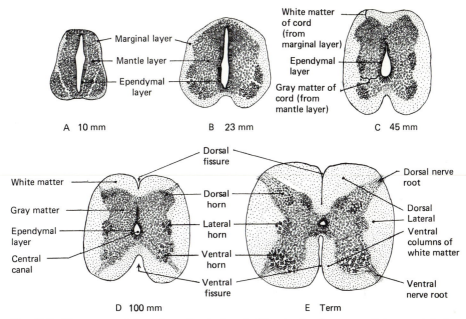

Fig. 11-8 Transverse sections through spinal cord of the pig at various ages. Note especially the parts of the adult cord derived from the ependymal, mantle, and marginal layers of the embryonic neural tube.

become neurons occurs near the central canal. Except for a few specialized areas, such as the cerebellum, the young neurons are postmitotic cells, that is, they will not divide again, although they may undertake relatively extensive migrations. If tritiated thymidine is injected into an embryo and the animal is not killed until it is mature, the radioactive label, which is incorporated into the DNA, remains concentrated in the neuroblasts that were undergoing their last round of DNA synthesis at the time the isotope was available. Older, postmitotic nerve cells do not incorporate the isotope at all, and other cells which remain in the mitotic cycle gradually lose the label by progressive dilution with each cell division. Radioautographs of brains prepared from mature mice which were exposed to [H³]thymidine at various times during pregnancy reveal strikingly different patterns of proliferation (Fig. 11-9).

One of the most important processes in histogenesis of the brain is cell migration. From their site of origin near the ventricles of the brain, young neurons migrate out toward the periphery, where they often settle down in several discrete layers. According to Rakic (1975) young postmitotic neurons, usually fairly simple bipolar cells, use long processes of specialized cells, called *radial glial cells*, as guides in their journey to the periphery (Fig. 11-10). The radial glial cells, which ultimately become a form of astrocyte, are very prominent during the period of active neuronal cell migration. Their cell bodies are located close to the ventricular lumen, but a long process from each cell extends nearly to the surface of the brain; the migrating neurons are guided

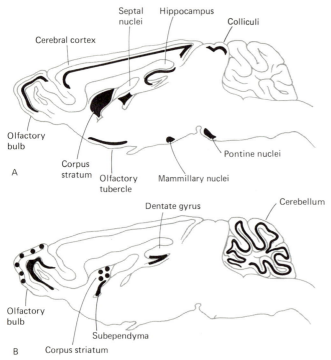

Fig. 11-9 Sagittal sections through the brains of adult mice which were exposed to ³H-thymidine at 15 days (A) and 19 days (B) of gestation. Autoradiographic preparations were made from the sectioned brains. The location of labeled cells (gray) illustrates the differing patterns of cell division during development. *(Modified from Langman, 1975, Birth Defects XI(7):85.)*

along their way by these radial processes. In multilayered areas of brain cortex, large neurons, constituting the innermost layer, migrate out first. Subsequent layers of gray matter are formed by smaller neurons migrating through the first and other previously formed layers to form a new layer of neurons at the periphery, so that whatever the number of layers of cortical gray matter, the innermost layer is first formed and the outermost layer is the last (Angevine and Sidman, 1961).

REGIONAL DIFFERENTIATION OF THE SPINAL CORD AND BRAIN

Spinal Cord After the earliest period of histogenetic activity, in which precursor cells begin to differentiate into neuronal and glial cells, development of the spinal cord is dominated externally by the development of the spinal nerves, one pair for each body segment.

During the period of development when the neurons are differentiating and acquiring their sheaths, the spinal cord undergoes marked changes in its relations within the body. In young embryos the neural tube extends the entire length of the body and into the tail (Fig. 11-11). As the spinal column is formed,

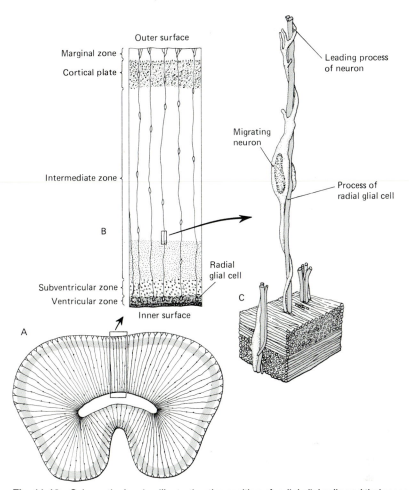

Fig. 11-10 Schematic drawing illustrating the position of radial glial cells and their association with peripherally migrating neurons during development of the brain (*After Rakic, 1975*, Birth Defects **XI**(7):100.)

the growth of the neural arches encloses the spinal cord in the "neural canal." Up to about 3 months the neural canal and the spinal cord are coextensive and the segmentally arranged nerves pass outward through intervertebral spaces directly opposite their point of origin. After this period, differential growth is such that neither the vertebral column nor the neural tube keeps pace with the expansion of the caudal part of the body (Fig. 11-12). The spinal cord lags much farther behind than does the vertebral column. Because the nerves are already established before these changes in relations occur, they appear to be dragged out caudally and pass back through the neural canal until they arrive at the intervertebral space which was originally opposite their point of origin (cf. Fig. 11-12A to 11-12D). Since it is the cephalic parts of the two systems which are

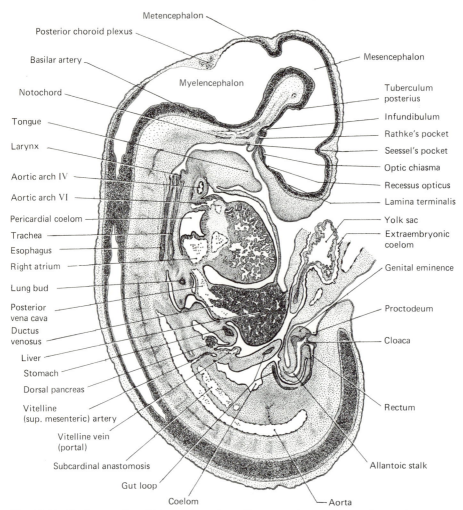

Fig. 11-11 Sagittal section of 10-mm pig embryo. The general arrangement of internal structures is essentially similar to that in human embryos of the sixth week.

fixed with reference to each other, the extent of displacement is progressively greater in the more caudal regions. In a human fetus at term the spinal cord ends at about the level of the third lumbar vertebra, except for a small vestigial strand (*filum terminale*) representing the regressing terminal portion of the primitive neural tube (Fig. 11-12D). Postnatally, this differential growth continues until adulthood, when the end of the cord usually lies near the level of the first lumbar vertebra. Thus the sacral and coccygeal nerves emerging from the cord course almost directly downward for a considerable distance. The group of nerves thus pulled out in the lower portion of the spinal canal constitutes the *cauda equina*, so called because of its fancied resemblance to a horse's tail.

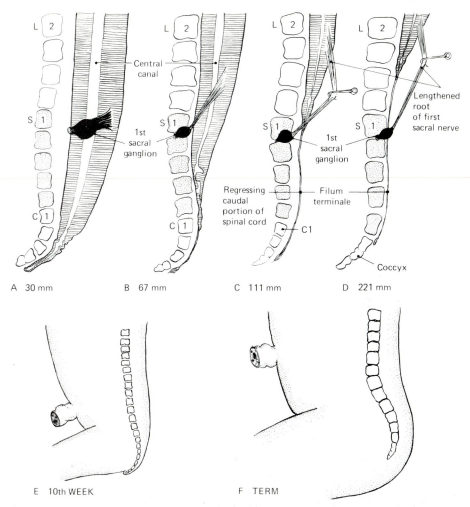

Fig. 11-12 Diagrams showing changes in relations of caudal end of spinal column and spinal cord due to differential growth. (A)-(D) Relations of the first sacral nerve and ganglion at different ages used as an indicator of the changing position of the spinal cord within the spinal canal. *(After Streeter, 1919, Am. J. Anat., vol. 25.)* (E) Silhouette of embryo of 10 weeks; (F) at term, showing the shift cephalad of caudal part of spinal column. *(Redrawn, with some modification, after Schultz.)*

Myelencephalon   The myelencephalon of the embryo becomes the *medulla* of the adult brain (Figs. 11-13 and 11-18G). In many respects its organization is quite similar to that of the spinal cord. A major difference is the central canal and the roof plate. Very early in development the lumen of this part of the neural tube becomes dilated, foreshadowing its ultimate fate as the large cavity in the medulla known as the *fourth ventricle* (Fig. 11-14C). At the same time its roof becomes very thin (Fig. 11-14A). Small blood vessels from the pia mater press into the membranous ependymal roof and push it ahead of them into the

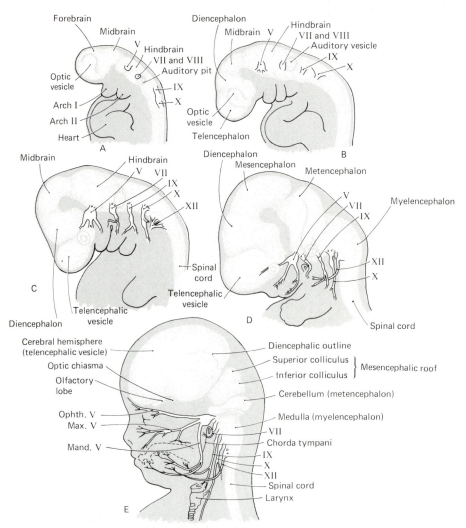

Fig. 11-13 Five stages in early development of brain and cranial nerves. *(Adapted from various sources, primarily figures by Streeter and reconstructions in the Carnegie Collection.)* (A) At 20 somites; probable fertilization age of 3½ weeks. *(Based on the Davis embryo.)* (B) At 4 mm; fertilization age of about 4 weeks. (C) At 8 mm; fertilization age of about 5⅓ weeks. (D) At 17 mm; fertilization age of about 7 weeks. (E) At 50 to 60 mm; fertilization age of about 11 weeks.

The cranial nerves shown are indicated by the appropriate Roman numerals: V, trigeminal; VII, facial; VIII, acoustic; IX, glossopharyngeal; X, vagus; XI, spinal accessory; XII, hypoglossal.

Abbreviations: Ch. T., chorda tympani branch of seventh nerve; *Hy.,* hyoid arch; *Md.,* mandibular arch; *Mand. V,* mandibular division of trigeminal nerve; *Max. V.,* maxillary division; *Ophth. V,* ophthalmic division of fifth cranial nerve.

lumen of the fourth ventricle. This freely branching group of vessels and their overlying ependymal epithelium are known as the *choroid plexus of the fourth ventricle* or the *posterior choroid plexus* (Figs. 11-11 and 11-18G). This plexus, plus others which differentiate within the roof of the third ventricle and in the walls of the lateral ventricles of the telencephalon, secrete *cerebrospinal fluid,* a

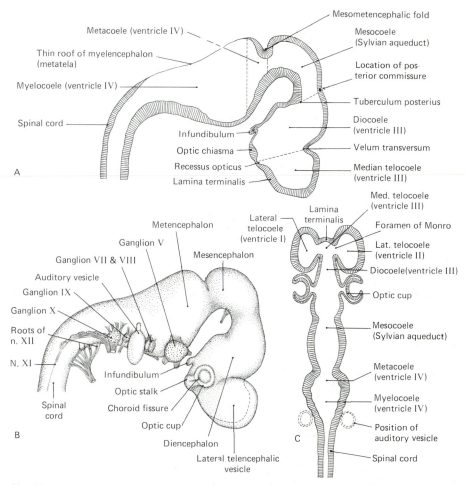

Fig. 11-14 Diagrams to show the topography of the brain shortly after the transition from the 3-to the 5-vesicle stage. (A) Sagittal section. The conventional lines of demarcation between the adjacent brain vesicles are indicated by broken lines. (B) Surface view of brain with position of cranial ganglia and nerve roots indicated. (C) Schematic frontal plan of brain as it would appear if the flexures had all been straightened out before cutting.

clear fluid with a special composition which bathes the tissues of the brain and spinal cord.

The walls of the neural tube in the brain region show early histological changes similar to those which occur in the walls of the spinal cord, with the resulting establishment of ependymal, mantle, and marginal layers. The ependymal layer of the myelencephalon becomes the epithelioid lining of the fourth ventricle. The mantle layer gives rise in part to columns of gray matter continuous with those in the cord and also forms more or less distinct cell masses (nuclei) associated with the roots of the more posterior cranial nerves (Fig. 11-24). The marginal layer receives an ingrowth of longitudinally disposed myelinated fibers constituting the conduction pathways between the spinal cord and nerves and the more rostral parts of the brain (Fig. 11-2). The sulcus limitans is especially strongly marked during the early stages of the development of the myelencephalic region. Later it becomes masked in certain regions by the growth of underlying nuclei, but wherever it persists it is a valuable landmark in dealing with the location of nuclei and fiber tracts. In the brain, as in the cord, afferent centers develop dorsal, and efferent centers ventral, to the sulcus limitans (Figs. 11-1 and 11-24).

Metencephalon   The dorsolateral walls of the neural tube in the metencephalic region undergo very extensive growth and give rise to the *cerebellum* of the adult brain. Histogenesis of the cerebellar cortex occurs relatively late in embryonic development and in many species continues actively after birth. Although the complexity of cerebellular histogenesis does not permit it to be covered here, it remains an important and widely studied system for understanding how neural interconnections are made within the central nervous system. Reflecting these internal changes, the early smooth contours of this region (Figs. 11-13C and 11-15) become broken up by the development of a complex series of folds (Figs. 11-13D and E, and 11-18D and E) which impart a very characteristic appearance to the cerebellum. In this part of the brain develop the synaptic centers concerned with the coordination of complex muscular movements (Fig. 11-2, arc 4).

Relatively late in development great groups of fibers which form the paths of intercommunication between the cerebellum and other parts of the nervous system appear superficially in the walls of the metencephalon. These form the ventral prominence known as the *pons* (Fig. 11-18G) and the *cerebellar peduncles* which extend over the lateral walls of the metencephalon. Deep to, and partly intermingled with, these superficial groups of fibers lie continuations of the same longitudinal fiber tracts which, on their way to and from the brain, traverse the marginal layer of the medulla. Still deeper are the masses of cells which originate from the mantle layer of this part of the neural tube. These cells are clustered in definite centers (nuclei) which are associated with the cranial nerves of the metencephalic level (Fig. 11-24).

The original lumen of the neural tube in the metencephalic region remains of considerable size. Since there is no line of demarcation between it and the lumen of the medulla, it is regarded as the anterior part of the fourth ventricle (Fig. 11-14).

Mesencephalon   The dorsolateral walls of the mesencephalon give rise to two pairs of rounded elevations known as the *corpora quadrigemina* (Fig. 11-18G). The two more rostral prominences, called the *superior colliculi*, are the synaptic center for visual reflexes (Fig. 11-2, arc 5b); and the two more caudal prominences, called the *inferior colliculi*, are the synaptic centers for auditory reflexes (Fig. 11-2, arc 5a).

In nonmammalian vertebrates the superior colliculi are greatly enlarged and are known as the *optic lobes*. In lower vertebrates the optic lobes are the

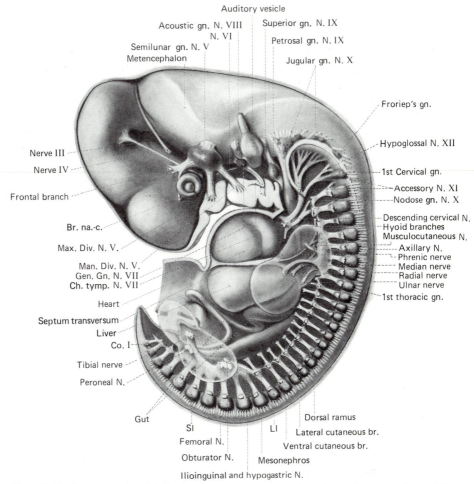

Fig. 11-15 Reconstruction (X13) of the nervous system of a 10-mm human embryo. *(After Streeter, 1908, Am. J. Anat., vol. 8.)*
   Abbreviations: Br. fr., frontal branch of ophthalmic division of trigeminal nerve; *Br. na.-c.,* nasociliary branch of ophthalmic division of trigeminal nerve; *Ch. tymp.,* chorda tympani branch of facial nerve; *Co. 1,* ganglion of first coccygeal nerve; *Gn.,* ganglion; *L. 1,* ganglion of first lumbar nerve; *N.,* nerve; *S. 1,* ganglion of first sacral nerve.

primary visual centers. With the great development of the forebrain in mammals, visual image processing is taken over by the thalamus and visual centers of the cerebrum, but the visual reflexes remain in the superior colliculi, the primitive visual centers.

   The ventrolateral parts of the mesencephalic walls constitute the main pathway over which fibers pass to and from the more rostral parts of the brain (Fig. 11-2, arcs 6 and 7). The fact that these fiber tracts in the mesencephalon are continuations of the longitudinal tracts encountered in the myelencephalon and in the floor of the metencephalon should be especially emphasized. These tracts in the mesencephalic floor are designated as the *cerebral peduncles*.

With the great thickening of its walls, the lumen of the mesencephalon becomes relatively reduced to form a narrow canal joining the lumen of the metencephalon and myelencephalon (fourth ventricle) with the lumen of the diencephalon (third ventricle). This canal is known as the *cerebral aqueduct*, or *aqueduct of Sylvius* (Figs. 11-18G).

Diencephalon The diencephalon undergoes a number of striking local modifications during its development. Its roof becomes thin, and vessels developing on its outer surface force it ahead of them in fingerlike processes which project into the third ventricle as the *anterior choroid plexus*, or *choroid plexus of the third ventricle* (Fig. 11-18G). In the median part of the diencephalic roof, caudal to the point of origin of the choroid plexus, the *epiphysis*, or *pineal body*, appears as a small local evagination (Fig. 11-18G).

In the floor of the diencephalon a median diverticulum called the *infundibulum* is formed. The distal portion of Rathke's pocket loses its original connection with the stomodeal ectoderm and becomes closely applied to the infundibulum (cf. Figs. 11-11 and 17-33). Somewhat later these two structures become intimately fused to form an endocrine gland known as the *hypophysis* (Figs. 11-18G and 15-1).

Very early in development the optic vesicles arise as outgrowths from the ventrolateral walls of the prosencephalon (Figs. 11-13A and 12-2B). When the prosencephalon is divided into telencephalon and diencephalon the optic stalks open into the brain very near the new boundary (Fig. 11-16C). In fact the median depression in the floor of the brain opposite their point of entrance is regarded as the ventral landmark which establishes the demarcation between telencephalon and diencephalon (*recessus opticus*; Figs. 11-14A and 11-11). Immediately caudal to the optic recess there is a marked thickening in the floor of the diencephalon where part of the fibers of each optic nerve cross to the other side of the midline. This point of crossing of optic nerve fibers is known as the *optic chiasma* (Figs. 11-14A and 11-11). Beyond the chiasma the crossed and uncrossed fibers on either side run together as the *optic tracts*. The optic tracts pass along the lateral walls of the diencephalon where some of their fibers end. Others pass to the visual reflex centers in the superior colliculi (Fig. 11-2, arc 5b). Still others concerned with the interpretation and memory of visual impulses pass by way of the geniculate nuclei to the visual areas of the cerebral cortex.

The dorsal parts of the lateral walls of the diencephalon become greatly thickened by multiplication of neuroblasts in the mantle layer. These thickened regions are known as the *thalami*. The dorsal portion of the thalamus is the gateway of fibers passing from the cord and the *brainstem*[3] to the cerebral

---

[3]*Brainstem* is a commonly used term for designating those portions of the brain other than the telencephalon, diencephalon, and cerebellum.

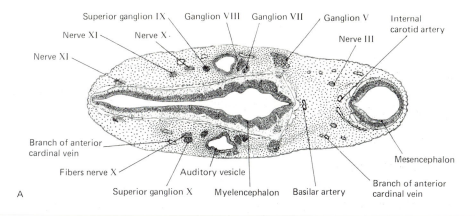

A

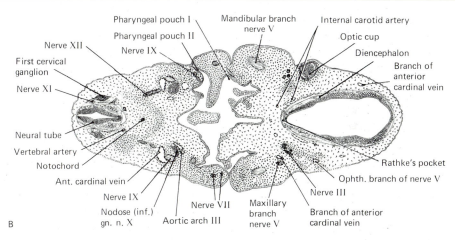

B

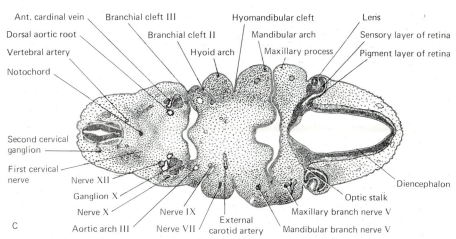

C

Fig. 11-16 Three transverse sections from the series of the 9.4-mm pig embryo used in making the reconstruction illustrated in Fig. 17-13. (Projection drawings, X15.) By laying a straight-edge across either of these reconstructions at the level of the marginal line bearing the serial number of a cross section, its relations within the body as a whole are precisely indicated. (A) Section 96 passing through the myelencephalic region; (B) section 142 passing through the pharynx; (C) section 159 at the level of the optic cups.

hemispheres (Fig. 11-2, arcs 6 and 7). In it are large synaptic centers which act as relay stations.

The thickening of the lateral walls of the diencephalon greatly reduces the width of its lumen. In its central portion the two walls come in contact and fuse, forming across the third ventricle a conspicuous connection known as the *massa intermedia*.

Telencephalon   The telencephalon consists of the most rostral part of the neural tube including paired dorsolateral outgrowths. These outgrowths first appear as roughly hemispherical evaginations called the *lateral telencephalic vesicles* (Fig. 11-14C). Although the division of the lateral walls of the neural tube into alar and basal plates is not clearly marked this far forward in the brain, the telencephalic evaginations, because of their general relations, are regarded as involving the alar plates.

At first the cavities within the two telencephalic vesicles are broadly continuous with the primary lumen of the neural tube (Fig. 11-14C). Later in development these openings into the lateral vesicles appear relatively much smaller; nevertheless, they persist, even in the adult, as the "foramina of Monro" (Fig. 11-18G). Vascular plexuses push through each foramen of Monro into the lateral ventricles of the telencephalic lobes to form the *lateral choroid plexuses*.

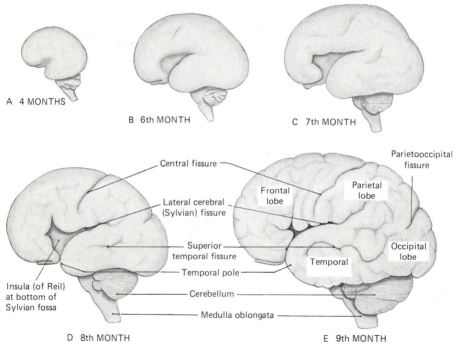

A 4 MONTHS

B 6th MONTH

C 7th MONTH

Central fissure

Parietooccipital fissure

Frontal lobe

Parietal lobe

Lateral cerebral (Sylvian) fissure

Superior temporal fissure

Occipital lobe

Temporal

Temporal pole

Insula (of Reil) at bottom of Sylvian fossa

Cerebellum

Medulla oblongata

D 8th MONTH

E 9th MONTH

Fig. 11-17   Lateral views of fetal brains at various stages in development. *(Modified from Retzius.)*

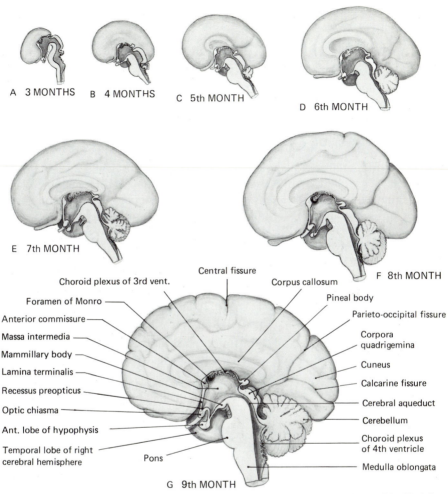

Fig. 11-18 Sagittal sections of fetal brains at various stages of development *(Modified from Retzius.)*

Once established, the lateral lobes of the telencephalon undergo exceedingly rapid growth (Fig. 11-17). At first the telencephalic lobes are smooth in contour and without striking local differentiations. Relatively late in development they become much convoluted, and certain regional divisions become clearly marked.

A conspicuous fissure called the *sulcus rhinalis* divides the ventral part of the telencephalic lobes from the dorsal. The region ventral to the sulcus rhinalis is chiefly concerned with the olfactory sense and is therefore often called the *rhinencephalon*. It includes the olfactory bulb and the olfactory tract. The rhinencephalon reaches its maximum relative development in lower forms, particularly those which rely heavily upon their sense of smell. In higher mammals it becomes largely overshadowed by the tremendous growth of the more dorsal portions of the cerebral cortex.

Dorsal to the sulcus rhinalis the outer walls (*pallium*) of the telencephalic vesicles constitute the nonolfactory portions of the cerebral cortex. These cortical areas are phylogenetically the newest portions of the brain. In them are located the suprasegmental centers concerned with memory, voluntary action, and inhibitory control (Fig. 11-2, arc 7).

THE DEVELOPMENT OF A PERIPHERAL NERVE

A typical peripheral nerve consists of a motor, a sensory, and an autonomic component (Fig. 11-1). With the exception of the dorsal root ganglion, which contains the cell bodies of the sensory neurons, a peripheral nerve is composed of nerve processes and their cellular and extracellular investments.

Axonal Outgrowth   The formation of a nerve begins with the outgrowth of axonal processes from some of the large motor neurons within the basal plate of the spinal cord or brainstem (Fig. 11-19). The demonstration by Harrison (1910) that embryonic nerves grow by the progressive extension of processes from the nerve cell body put to rest decades of debate regarding not only the mechanism of nerve growth but also the structural nature of the nervous system itself. In this experiment Harrison explanted cells from the neural tubes of frog embryos into serum in a hollow in a glass slide and watched the processes grow out (Fig.

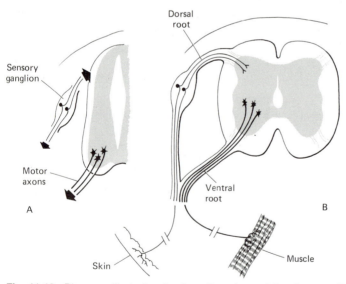

Fig. 11-19 Diagrams illustrating the formation of a peripheral nerve. (A) Early stage, showing outgrowth of axons (arrows) from developing motor neurons in the basal plate of the spinal cord and the outgrowth of sensory processes toward both the spinal cord and the periphery. (B) Outgrowth completed, with nerve processes extending between peripheral end organs to the spinal cord.

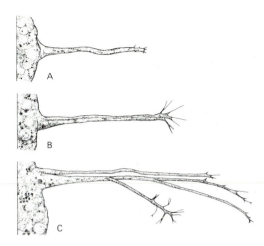

Fig. 11-20 Outgrowth of a group of embryonic nerve fibers in culture. The growing tip of the nerve sends out pseudopodial projections which are in a constant state of flux. (A) 24 hours; (B) 25½ hours; (C) 34 hours. *(After Harrison, 1910, J. Exp. Zool. 9:787.)*

11-20). An important problem was solved, and in the process a new technique, tissue culture, was established.

The first processes which grow out in the formation of a nerve are called *pioneering fibers*, and they typically arise from large neurons. The tip of the slender process of an outgrowing nerve fiber is expanded to form a *growth cone*, from which numerous fine pseudopodial processes (*filopodia*) actively extend and retract (Fig. 11-20). If this outgrowth is followed, it can be seen to advance by amoeboid movements, tending to progress along anything that furnishes a favorable substratum along which it may move. The growth cone appears to pull out behind itself a long slender cytoplasmic process (the young nerve fiber), but in reality much of the motive force for outgrowth lies in the cell body, which actively produces and sends down the growing nerve fibers the materials for further growth.

The continued production and transport of cytoplasmic materials down the nerve fibers is called *axonal transport* (Weiss and Hiscoe, 1948), and it is a major means by which a neuron is able to maintain continued and adaptive contact with the periphery. Even in the adult, if a nerve is severed, the part of the nerve distal to the cut degenerates, but new growth cones form at the ends of the proximal processes and renewed outgrowth, in this case called *regeneration*, brings the nerve back into contact with its end organ.

How does an outgrowing process of a peripheral nerve find its proper place in the body? Over the years many hypotheses have been put forth, but we still do not have a definitive answer. Most peripheral nerves seem to grow out by a continued testing of the microenvironment in the region of the growth cone by the filopodia. Direct observation has shown that the filopodia retract from unfavorable substrates, but that they remain fixed to a favorable substratum and direct the entire nerve processes in that direction. In vitro experimentation (Weiss, 1934) has shown that the physical nature of the substratum is an important determinant of the direction and nature of outgrowth. This has been called *contact guidance*. Still unresolved in peripheral nerves are the possibility of specific chemical guidance and the role of local electrical fields in determining the direction of growth.

The Components of a Developing Peripheral Nerve   The development of a spinal nerve begins with the outgrowth of axons from the large motor neurons in the ventral horn of the spinal cord. Meanwhile, the neurons in the sensory ganglia alongside the spinal cord send out toward the cord processes which ultimately enter the dorsal horn of the gray matter (Fig. 11-19). Sensory processes also grow toward the periphery, where they soon join the outgrowing motor nerve fibers in a common nerve trunk.

The outgrowing motor and sensory nerve fibers typically reach the area to be innervated when histodifferentiation of the latter is still incomplete. Axons from the large motor neurons eventually make contact with developing myotubes in the early muscles. After a period of exploration and tentative functional contacts, branches of the motor nerve fiber ultimately establish definitive *neuromuscular junctions* (*motor end plates*) with a group of several to several hundred muscle fibers, forming the basis of a *motor unit* within the muscle.

Late in development certain sensory processes appear to act upon some small groups of muscle fibers in an almost inductive fashion, causing the muscle fibers to form *muscle spindles*, specialized *proprioceptive units* that lie among the other muscle fibers within a muscle (Zelená, 1964). Muscle spindles are supplied with small motor nerve fibers, as well. Other sensory nerve processes go to the developing skin and some deeper tissues, and they undergo differentiation into a wide variety of specialized sensory terminations.

With these changes at the periphery of the nerve, the proximal processes in the spinal cord make several varieties of connections with developing association neurons or tracts within the spinal cord. When the central connections are complete, the essential neural circuitry for a reflex arc is established and reflex responses to peripheral sensory stimuli begin to appear.

In addition to making the proper connections, the developing nerve fibers must become insulated from one another. The myelination of peripheral nerve fibers is accomplished by the wrapping of portions of the cells many times around the nerve processes (Fig. 11-21). Some peripheral nerve fibers are considered to be unmyelinated, but even these are partially embedded within the cytoplasm of Schwann cells.

Autonomic Nerves   Accompanying the motor and sensory fibers within the proximal parts of spinal nerves are efferent and afferent fibers of the involuntary, or *autonomic, nerves* (Fig. 11-1). The efferent nerve fibers leave the spinal cord through the ventral roots along with the motor axons, but then they leave the nerve through a separate branch and make connection with the cell bodies of second-order neurons of neural-crest origin, located in chains of ganglia that run on each side of the body ventral to the vertebral column or in other, isolated

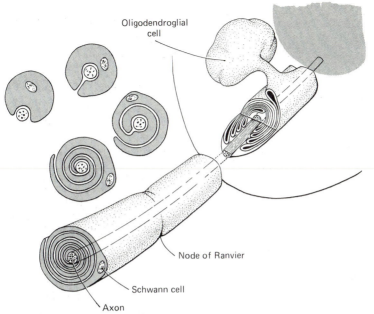

Oligodendroglial cell

Node of Ranvier

Schwann cell

Axon

Fig. 11-21 Myelination in the central and peripheral nervous systems. In a peripheral nerve myelination is accomplished by Schwann cells wrapping around the axon many times like a jelly roll (*upper left*). In the central nervous system oligodendroglial cells are the myelinating agents.

ganglia scattered throughout the thorax and abdomen. These second-order neurons, part of the sympathetic division of the autonomic nervous system, send out processes to the visceral structures, including blood vessels and skin glands. Afferent autonomic fibers return from these structures to their cell bodies, which are also located in the dorsal root ganglia. Central connections are then made with the spinal cord.

The differentiation of autonomic neurons involves at least two major steps (Bunge et al., 1978). The first occurs early in the development of the neural crest, when some cells become determined to be components of the autonomic nervous system. These cells then migrate down the appropriate pathway and settle down in relation to a specific end organ. The second major stage in the differentiation of an autonomic nerve involves the choice of the neurotransmitter that the nerves will use. Although it is becoming increasingly apparent that perhaps more than 10 different neurotransmitters are employed in various regions of the nervous system, the bulk of the autonomic nerves are either *cholinergic* (i.e., use acetylcholine as the transmitter) or *adrenergic* (i.e., use norepinephrine). The second-order (terminal) neurons of the parasympathetic division are cholinergic, whereas those of the sympathetic division are adrenergic. When autonomic neurons first arrive at their destination in the body, they are still capable of differentiation into either cholinergic or adrenergic neurons. Through the processing of microenvironmental signals from their end organs and possibly also the pattern of neural stimuli impinging upon them from the central nervous system, the neurons finally respond by committing themselves to becoming adrenergic or cholinergic.

Effect of the Periphery upon Development of the Spinal Cord   Later development of the spinal cord is greatly influenced by its relationship with peripheral structures. Mere inspection of a mature spinal cord reveals swelling in the cervical and lumbar regions, from which the nerves to the upper and lower limbs arise. These differences are also reflected in the nerves themselves. The diameter of the nerves supplying the limbs is much greater than that of nerves found in other regions of the trunk.

The relationship between peripheral load and the degree of development of the nervous system has been amply demonstrated in experimental studies. Reduction of the mass of peripheral tissue supplied by a given nerve is followed not only by a decrease in diameter of the nerve that would have supplied the tissue but also by a reduction in size of the corresponding sensory ganglion and ventral (motor) horn. The reduction in size is largely due to a greater amount of neuronal cell death than in controls. Even in normal development, large numbers of neurons die within the spinal cord (Fig. 11-22). Hughes (1961) found that in the lumbar region of the spinal cord in normal *Xenopus* almost 90 percent of the original motor neurons die by the time of metamorphosis. After removal of a limb bud, the effect is still more striking. Hamburger (1958) removed a limb bud in 2½-day chick embryos and by 6 to 7 days found a massive wave of cell death which eliminated the entire population of 20,000 motor neurons normally destined to supply the limb. In contrast, the addition of peripheral tissue into the area supplied by a nerve (such as the transplantation of a limb bud onto the flank of an early embryo) results in a dramatic increase in the size and number of nerve cells going to that area (Fig. 11-23; Detwiler, 1920).

It has also been shown that specific chemical influences profoundly affect the development of some components of the nervous system. Nerve growth factor, a protein that has been isolated from several normal and abnormal mammalian tissues, exerts a profound effect upon the growth of sensory, and particularly sympathetic, neurons in a wide variety of vertebrate embryos (Levi-Montalcini, 1958). The demonstration of nerve growth factor has stimulated a search for other specific chemical factors which might exert an effect upon the developing nervous system as well as upon other components of the body.

THE CRANIAL NERVES

In dealing with the spinal nerves we dealt with four functional types of neurons—somatic afferent, somatic efferent, visceral afferent, and visceral efferent (Fig. 11-1). In the cranial nerves we find not only these same types of neurons but also subtypes with more restricted distribution and more specialized function (Fig. 11-24). The eye and the ear, for example, are very highly

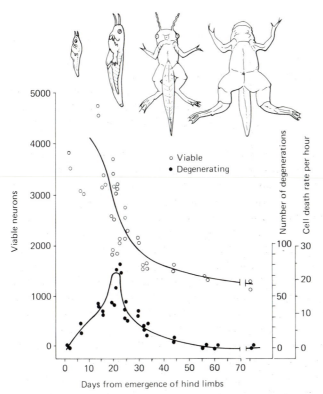

Fig. 11-22 Numbers of viable and degenerating neurons in the ventral horn of the spinal cord of *Xenopus* at different stages of development. *(Adapted from Hughes, 1961,* J. Embryol. Exp. Morph. **9**:269.)

differentiated and sharply localized somatic sense organs. Therefore their fibers are set apart from the general somatic afferent category as *special somatic afferent fibers*. From the taste buds sensory nerve fibers carry gustatory impulses to the brain. Because these fibers are associated with organs of special sense, they are called *special visceral afferent fibers*, concerned with such functions as touch, temperature, and pain. The musculature in the pharyngeal region differs from other visceral musculature in that it is striated, so the motor fibers to it are distinguished from other visceral efferent fibers by calling them *special visceral efferent neurons*.

The spinal nerves are segmentally arranged, and all of them are built on the same general plan. The cranial nerves, although most likely derived from segmental structures in the phylogenetic past, have lost some of their regular segmental arrangement and have become very highly specialized. One of the major phylogenetic changes has been the lack of union of the dorsal and ventral roots of the cranial nerves. Instead, the cranial nerves of mammals represent, for the most part, either the primitive dorsal or the primitive ventral roots. Two

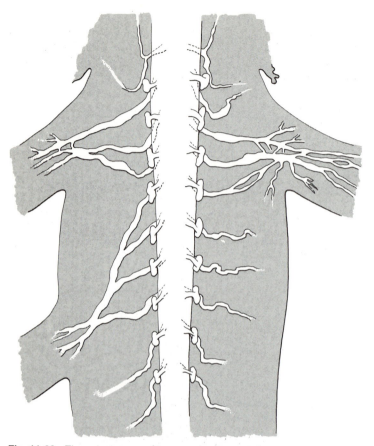

Fig. 11-23 The spinal nerves of a normal salamander (*right*) and one to which an extra limb has been grafted (*lower limb on left*). The nerves (and corresponding sensory ganglia) to both normal and grafted limbs are larger than those supplying only the body wall (*After Detwiler*).

of the cranial nerves (I and II—the olfactory and the optic nerves) could be more appropriately considered to be extensions of brain tracts rather than true nerves.

Several of the cranial nerves (III, IV, VI, and XII) appear to have evolved from primitive ventral roots. This plus the myotomally derived muscles which they innervate (extraocular muscles and tongue muscles) place them in the somatic motor category. (For a schematic representation of the functional classes of cranial nerve fibers and their representation in the brain, see Fig. 11-24.) Many of the remaining cranial nerves (V, VII, IX, X, and XI) can be traced back to the primitive dorsal roots, through which run not only afferent (sensory) nerve fibers but also the motor nerve fibers (special visceral efferent) leading to the branchial arches and the striated muscles arising from the branchial arch mesenchyme (see Fig. 15-3). The motor neurons running through these nerves

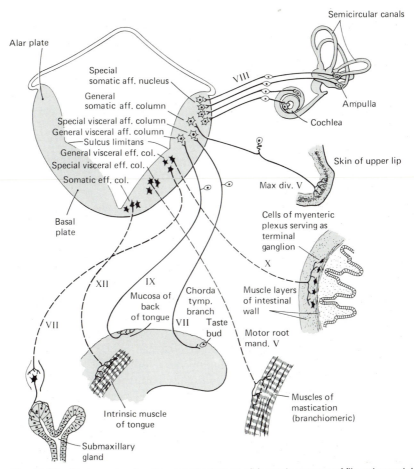

Fig. 11-24 Diagram showing the central relations of the various types of fibers in cranial nerves. It should be emphasized that the diagram is a schematic composite and that no one cranial nerve contains all the types of fibers shown. As was the case with the spinal nerves (Fig. 11-1), the efferent nuclei or clusters of efferent nerve cells are located ventrolaterally, in the basal plate of the neural tube wall. The afferent fibers of the cranial nerves have their cell bodies located outside the neural tube in ganglia (cf. Fig. 11-13). For further explanation see text.

arise within the substance of the brain, but the sensory neurons and their ganglia are derived from ectodermal placodes, or the neural crest. Overall, there are greater similarities between cranial and spinal nerves than one would expect after superficial examination.

Olfactory Nerve (I)   Unlike other sensory nerves the olfactory nerve lacks a ganglion. Another unique feature is that all its fibers are nonmyelinated. These fibers arise from cells in the epithelial layer lining the olfactory pits (Fig. 11-26). From there they grow centripetally into the olfactory bulbs. In the olfactory bulbs (Fig. 13-4D) the fibers of the nerves terminate in synapses with other neurons which relay the impulses along the olfactory tracts to centers in the rhinencephalon.

Optic Nerve (II)   As is the case with the olfactory nerve, the optic nerve fibers arise from peripherally located cells and grow centripetally. Neuroblasts situated in the sensory layer of the retina in close association with its photosensitive cells send out processes which leave the optic cup through the choroid fissure. These fibers then traverse the optic stalk and enter the brain in the diencephalic floor. At their point of entrance the two optic nerves[4] intersect. At the intersection, some of the fibers from each nerve cross over to the opposite side, so that each eye has central connections with both sides of the brain.

Oculomotor Nerve (III)   As its name implies, the oculomotor nerve contains efferent fibers to muscles moving the eye. The cluster of neuroblasts from which it arises is located in the basal plate of the mesencephalon. Emerging from the floor of the mesencephalon (Figs. 11-25 and 11-26), their fibers pass directly to the orbital region and innervate the inferior oblique and the superior, inferior, and internal rectus muscles of the eyeball.

Trochlear Nerve (IV)   The *trochlear nerve* is a motor nerve to the superior oblique muscle of the eye. Its nucleus of origin, like that of the third nerve, is located in the basal plate of the mesencephalon. Its fibers do not leave directly from the ventrolateral walls of the brain, as usually happens in a motor nerve. Instead they pass to the dorsal wall of the mesencephalon (Figs. 11-25 and 11-26) and cross before emerging.

Trigeminal Nerve (V)   The *trigeminal nerve* takes its name from the fact that it has three main divisions—the ophthalmic, the maxillary, and the mandibular (Figs. 11-15 and 11-25). As is indicated by its large semilunar ganglion, the fifth nerve has a large number of sensory fibers. It has, nevertheless, enough motor fibers associated with its mandibular branch to be regarded as a mixed nerve (Fig. 11-27). The names of its branches clearly indicate its distribution to the facial region.

Abducens Nerve (VI)   The *abducens nerve* takes its name from the fact that it controls the external rectus muscle, contraction of which makes the eyeball rotate outward. Its nucleus lies in the basal plate of the myelencephalon, from which its fibers emerge ventrally just caudal to the pons and pass toward the orbit (Figs. 11-25 and 11-27).

---

[4]Since both the photosensitive cells themselves and the retinal ganglion cells sending out their associated nerve fibers arise in the walls of the optic cup, which is in turn an evagination of the embryonic forebrain, what we commonly call the *optic nerve* is, strictly speaking, not a nerve but a fiber tract arising within a modified portion of the brain wall.

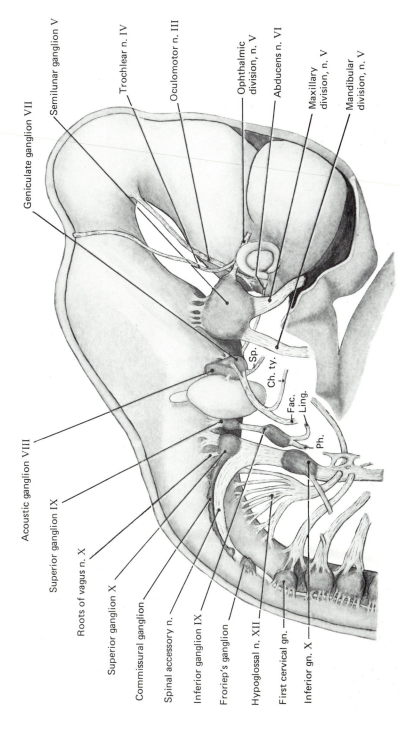

Fig. 11-25 Reconstruction of the brain and cranial nerves of a 12-mm pig embryo. (*After F. T. Lewis, 1902, Am. Jour. Anat. Vol. 2.*) *Abbreviations:* *Ch. ty.,* chorda tympani branch of the seventh (facial) nerve; *fac.,* facial nerve; *Ling.,* lingual branch of ninth nerve; *Ph.,* pharyngeal branch of ninth nerve; *Sp.,* greater superficial petrosal nerve.

Geniculate ganglion VII

Semilunar ganglion V

Trochlear n. IV

Oculomotor n. III

Ophthalmic division, n. V

Abducens n. VI

Maxillary division, n. V

Mandibular division, n. V

Acoustic ganglion VIII

Superior ganglion IX

Roots of vagus n. X

Superior ganglion X

Commissural ganglion

Spinal accessory n.

Inferior ganglion IX

Froriep's ganglion

Hypoglossal n. XII

First cervical gn.

Inferior gn. X

Sp.

Ch. ty.

Fac.

Ling.

Ph.

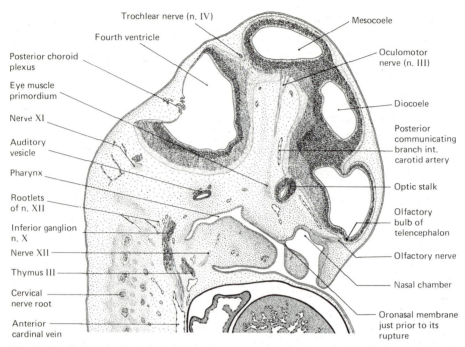

Fig. 11-26 Projection drawing of parasagittal section of head of 15-mm pig embryo. The section is to the right of the midline, in a plane especially favorable for showing the relations of the nasal pits and of the olfactory and oculomotor nerves.

Facial Nerve (VII)   The *facial nerve* is primarily motor, but the presence of the geniculate ganglion on its root shows that it also carries some sensory fibers. The majority of its sensory fibers pass by way of the chorda tympani branch (Figs. 11-24 and 11-25) to join the mandibular branch of the fifth nerve. These fibers are concerned with the sense of taste. The nerve's motor fibers arise from a nucleus situated in the basal plate of the myelencephalon and innervate the muscles of facial expression.

Auditory Nerve (VIII)   At first the ganglionic mass from which the fibers of the eighth nerve arise is closely associated with the geniculate ganglion of the seventh nerve (Fig. 11-25). Gradually these ganglia become entirely distinct. Still later the ganglion of the eighth nerve divides into two parts, a *vestibular ganglion* and a *spiral ganglion* (Fig. 12-13F and G). With the division of the ganglion, the nerve fibers arising from its cells become grouped into two main bundles, one associated with each ganglion. Meanwhile the otic vesicle has differentiated into two distinct parts, the *cochlea*, which is the organ of hearing, and the group of *semicircular canals*, which, together with the *utriculus* and *sacculus*, constitute an organ of equilibrium. The spiral ganglion and the cochlear branch of the eighth nerve become associated with the auditory part of

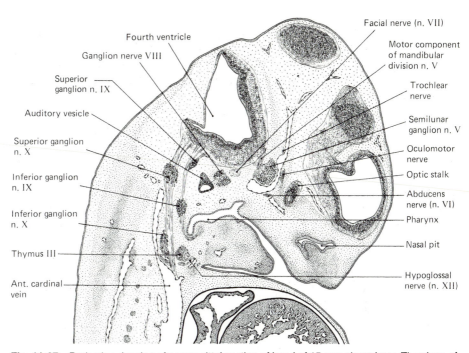

Fig. 11-27 Projection drawing of parasagittal section of head of 15-mm pig embryo. The plane of section is slightly farther to the right than that shown in the preceding figure. It is particularly favorable for showing the position of origin and the ganglia of the trigeminal, glossopharyngeal, and vagus nerves.

the mechanism. The vestibular ganglion and its branch of the eighth nerve become associated with the semicircular canals.

Glossopharyngeal Nerve (IX) The *glossopharyngeal* is a mixed nerve, but by far most of its fibers are sensory. The ganglion cells from which these sensory fibers arise are grouped in two clusters, one near the root of the nerve (*superior ganglion*) and one farther peripherally on its course (*inferior* or *petrosal ganglion*) (Figs. 11-25 and 11-27). The cell bodies in the superior ganglion give rise to fibers which innervate a small cutaneous area of the external ear. The inferior ganglion contains cell bodies which give rise to fibers concerned with general sensibility in the region about the root of the tongue or which supply the taste buds. The efferent fibers arise from nuclei in the basal plate of the myelencephalon. Some of these fibers innervate the stylopharyngeus muscle. Others are secretory fibers to the parotid gland.

Vagus Nerve (X) The *vagus* is a mixed nerve carrying five different types of fibers. Fibers from cells in its superior (jugular) ganglion bring in sensory impulses from the skin in the region of the external ear. Fibers from cells in the inferior (nodose) ganglion bring in sensory impulses from the pharynx, larynx, trachea, esophagus, and the thoracic abdominal viscera. Afferent fibers having cells of origin in the inferior ganglion carry gustatory impulses from scattered taste buds in the region of the epiglottis. From nuclei of origin in the myelencephalon efferent fibers extend to striated muscles of the pharynx and larynx. Other efferent fibers run to various terminal ganglia; from there the impulses are relayed to the visceral musculature over second-order motor neurons (Fig. 11-24).

The Accessory Nerve (XI)   The commissural ganglion (Fig. 11-25) which appears so closely associated with the accessory nerve is really a continuation of the superior (jugular) ganglion of the vagus. Froriep's ganglion usually disappears in the adult, so that the accessory nerve is left without ganglia. Its fibers, practically all efferent, originate not only from the posterior part of the myelencephalon but also from the first five or six segments of the spinal cord (Fig. 11-25). A large number of the fibers of the accessory nerve run with those of the vagus nerve to sympathetic ganglia, from which the motor impulses are relayed to the smooth muscle of the viscera. Other fibers from the accessory nerve join similar vagus fibers to striated muscles in the pharynx and larynx. Most of the fibers arising from the cervical part of the cord turn off in the external ramus to end in the trapezius and sternocleidomastoid muscles.

Hypoglossal Nerve (XII)   The *hypoglossal nerve* is composed almost entirely of motor fibers. They arise from an elongated nucleus in the posterior part of the myelencephalon and emerge in several separate roots, which join to form a single main trunk (Fig. 11-25). Peripherally they are distributed to the muscles of the tongue.

THE DEVELOPMENT OF NEURAL FUNCTION IN THE EMBRYO

When dealing with the intricacies of morphology and mechanisms in the developing nervous system, one must not lose sight of the main reason for having a nervous system, namely, the generation and coordination of most of the functional activities of the body. In many developing systems form and function are closely linked. This is particularly true in the nervous system, where both the appropriate neural pathways and the end organs must be sufficiently mature to permit function. Early studies on neural function involved the gross examination of movement patterns and reflexes in embryos and fetuses. More recently, behavioral testing has been correlated with anatomical and physiological studies of the segments of the nervous system involved in the function under study.

The Development of Motor Function in the Human Embryo   Many aspects of development of the neuromuscular system are reflected in the movements of the embryo as a whole. Until about 6 weeks, the embryo merely rocks back and forth passively in its bath of amniotic fluid, with the only intrinsic movements being the regular beating of the heart. Toward the end of the second month the embryo makes its first spontaneous movements. These are simple side to side twitchings which serve to indicate the functional maturation of the musculature in the body wall. Actually, about a week earlier the embryo is capable of making weak twitches in the neck in response to striking the lips or nose with a fine bristle (Hooker, 1952). This behavioral pattern signifies that the first functional reflex arcs have been laid down. In succeeding days, the twitching movements and simple reflex capabilities extend progressively caudad, in conformance with the general cephalocaudad rule of development.

By the start of the third month reflex movements of the hands can be elicited; they are followed a week later by foot reflexes. The beginnings of limb function are accompanied by rapid development of facial movements. At 10 or 11 weeks early swallowing and the first movements that foreshadow the rhythmical breathing motions of the chest are seen. The fetus seems to anticipate its major postnatal requirements by developing swallowing, breathing, and grasping movements early. By the end of the third month of pregnancy almost the entire skin of the fetus is sensitive to touch, and although its movements are still feeble, the fetus is very active at this time. Yet only rarely can the mother detect these movements, mainly because the fetus is still so small it could comfortably fit into the palm of one's hand.

Until the twelfth week the movement patterns of the fetus are highly irregular, and the magnitude of a response often far exceeds the intensity of the stimulus. Soon, however, the pattern of largely undirected responses to a stimulus ceases and more predictable reflex responses begin to occur.

Toward the end of the fourth month the mother can often feel fetal movements ("quickening"). The grip of the fetus becomes stronger, and weak but nonsustained breathing movements are possible. The fetus soon begins to alternate periods of rest with periods of activity. Very premature fetuses, born late in the fifth month of pregnancy, are able to breath spontaneously for short periods of time, but these breathing movements still cannot be sustained. The remainder of pregnancy is occupied to a great extent by the maturation of reflex and behavioral patterns which have already been set up. During this period the development of sensory function proceeds rapidly and in a fairly regular sequence. After the early appearance of general cutaneous sensation, the functions of taste, balance (vestibular system), hearing, and vision mature, often in that sequence (Gottlieb, 1976; Bradley and Mistretta, 1975). The maturation of function continues for several years into postnatal life and the sequence of functional changes often follows the morphological sequence of myelination of the nerves involved.

Chapter 12

The Sense Organs

The sense organs are an individual's windows to the surrounding world. It should not be surprising, therefore, that they are derived principally from the outer, ectodermal germ layer. In anamniotes, the sense organs start out as thickened ectodermal placodes (Fig. 12-1), each one being the result of a secondary inductive stimulus emanating from some part of the developing central nervous system. The vertebrate sensory organs that are functionally most important and structurally most complex are the eyes and ears. Their development will be described in detail in this chapter. The ears are composite structures in which much of the sound collecting apparatus is derived from the branchial arch system and the pharynx. The organs of smell and taste are morphologically so intimately bound with the overall development of the face and oral region that their embryogenesis will be treated together in Chap. 13. The cutaneous senses are bound to the development of the peripheral nerves, which was described in the previous chapter.

THE EYE

The vertebrate eye is a very complex organ, the constituents of which are derived from several different primordial sources, both ectodermal and meso-dermal, in the cephalic region of the embryo. Of necessity, its proper function

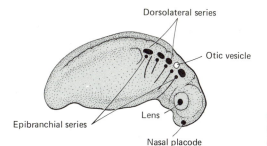

Dorsolateral series

Otic vesicle

Epibranchial series

Lens

Nasal placode

Fig. 12-1 Ectodermal placodes in the head of an amphibian *(Ambystoma)* larva. *(After Yntema, 1937, J. Exp. Zool. **92**:93.)*

requires almost perfect alignment of the components directly involved in vision, as well as a high degree of transparency of those through which light rays pass. The normal embryology of the eye demonstrates beautifully some of the fundamental processes and concepts of development. Early development of many of the components of the eye depend upon inductive interactions between one component and another. This is followed by a phase of intracellular differentiation, starting with a burst of mitosis and then RNA synthesis leading to the formation of specific classes of intracellular proteins in some cases or extracellular fibers and matrix in others. In some aspects of eye development the influence of extracellular materials and the migration of cells play an important role. The development of the retina and its central connections involves not only the differentiation of a very complex neural tissue but also one of the best-documented cases of integration between one part of the embryo and another. Finally, mechanical influences are important in obtaining and maintaining proper alignment of the tissues in the visual pathway. Even later in development, after many of the components of the eye have been laid down, they may continue to exert sustained influences upon one another.

Primary Optic Vesicle   After the initial induction of the neural ectoderm by the underlying archenteron, the first morphological indication of eye formation is an outpocketing of the wall of the diencephalon. In human embryos the evagination of the optic vesicles begins very early. By the middle of the third week (embryos of 7 to 9 somites) depressions appear in the still open neural plate where it widens out in the future forebrain region. These early depressions are called the *optic sulci* (Fig. 12-2A). They are the initial shallow concavities that later become deepened to form the *primary optic vesicles*. After the neural plate in the brain region has been closed, the optic vesicles appear in external views of reconstructions of the central nervous system as rounded protuberances from the lateral walls of the forebrain (Fig. 11-13A).

Formation of the Optic Cup from the Primary Optic Vesicle   Toward the end of the fourth week, the lumen of the primary optic vesicle is still broadly continuous with the lumen of the forebrain, and the walls of the vesicle show little differentiation from the condition of the parent forebrain walls (Fig. 12-2B). At the start of the fifth week, the distal portion of the optic vesicle

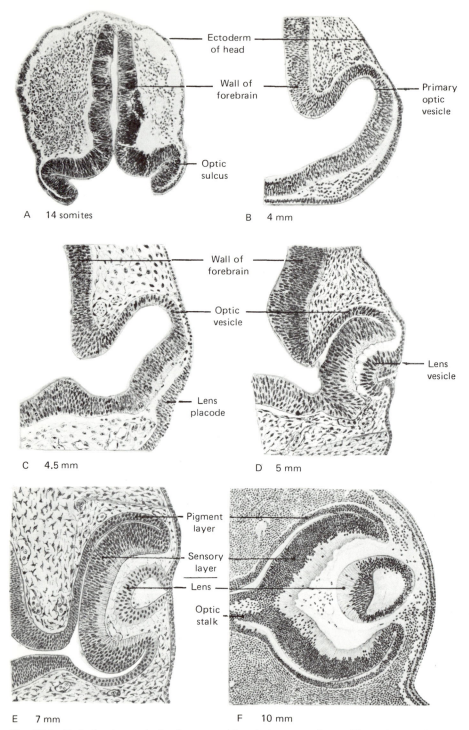

Fig. 12-2 Each development of optic cup and lens in human embryos. Photomicrographs and drawings from various sources placed in corresponding orientation and brought to same magnification (X100). *(A, from Heuser; B, from Fischel; C-E, from Ida Mann; F, from Prentiss.)*

begins to flatten (Fig. 12-2C) and very soon thereafter begins to become invaginated so that the single-walled primary vesicle becomes transformed into the double-walled optic cup (Fig. 12-2D). As the invagination becomes more complete, the original lumen of the optic vesicle is reduced to a vestigial slit between the inner and outer layers of the newly formed cup (Fig. 12-2C to 12-2F). At the same time there is rapid differentiation of the two layers of the cup. The outer layer becomes much thinner, and by the sixth week, in human embryos, it begins to show melanin granules which foreshadow its ultimate conversion into the *pigment layer of the retina*. The inner "collapsed" layer of the cup becomes much thickened, an indication that it has begun the elaborate series of changes by which it will become the *sensory layer of the retina*. This is the layer that will receive the visual images and convert them to signals which will be transmitted to other regions of the brain via the optic nerve.

The invagination which forms the optic cup occurs not at the center of the optic vesicle but eccentrically toward its ventral margin. This makes a gap in the continuity of the wall of the optic cup which is known as the *choroid fissure* (Fig. 12-3). As development of the eye progresses, the choroid fissure partially envelops the *hyaloid artery,* or *central artery of the retina* (Fig. 12-3E), a branch of the internal carotid artery which in the embryo supplies many of the structures within the eyeball. The *optic stalk* (Fig. 12-3B and C), along part of which the choroid fissure extends, becomes invaded by processes of the nerve cells which differentiate in the sensory layer of the retina. After these processes have grown down the wall of the optic stalk, on their way toward making synaptic connections in the brain, the optic stalk is known as the *optic nerve.*

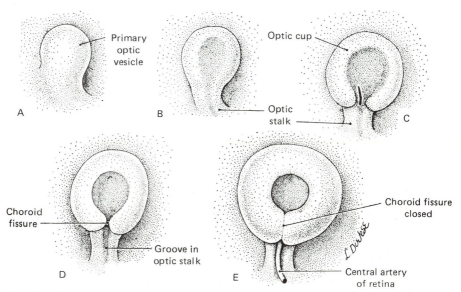

Fig. 12-3 Stages in the formation of the optic cup and the choroid fissure. *(Adapted from several sources, primarily Streeter.)*

Establishment of the Lens   Before it begins to invaginate, the primary optic vesicle comes into very close contact with the overlying head ectoderm (Fig. 12-2B). During this time the ectoderm, which has already been subjected to preparatory influences from other cephalic tissues, is induced by the optic vesicle to form lens (Spemann, 1938; Lewis, 1904). The effects of the inductive event are first made manifest at about the end of the fourth or the beginning of the fifth week (human embryos of 4 to 5 mm), when the superficial ectoderm immediately overlying the optic cup begins to develop a local thickening known as the *lens placode* (Fig. 12-2C).

The importance of close contact between the optic cup and the overlying ectoderm in lens induction is demonstrated by numerous experiments in which these structures have been separated either by increased distance or by the interposition of foreign objects placed between them. In most cases there is an interference with lens formation. These experimental results are supported by natural experiments in the form of genetic mutants in which lens formation fails to occur. In the *eyeless* mutants in mice, the optic cup does not come as close to the surface ectoderm as it should, and in the *fidget* mutant a reduction in mitosis in the optic cup interferes with proper contact with the ectoderm.

As the cavity in the optic cup deepens, the lens placode becomes invaginated into the cup to form an open *lens vesicle*, the lumen of which is sometimes called the *lens pit*. During the fifth week, the lens vesicle becomes closed (Fig. 12-2E) and then breaks away completely from the parent ectoderm to constitute a rounded epithelial body lying in the opening of the optic cup (Fig. 12-2F). Before the end of the sixth week, the cells on the deep pole of the lens are beginning to elongate, presaging their transformation into the long transparent elements known as *lens fibers* (Figs. 12-2F and 12-7A).

Differentiation of the Lens   The development of the lens into a transparent structure with the appropriate optical qualities involves a highly orchestrated sequence of intracellular differentiative events, culminating in the synthesis of a specific class of lens proteins called *crystallins*. As we have just seen, the early lens vesicle soon undergoes an asymmetrical development, with the deeper cells elongating into lens fibers and the cells of the outer pole retaining a low epithelial configuration. By the end of the seventh week, the lens fibers in human embryos have elongated sufficiently to make contact with the lens epithelium, thus reducing the original cavity in the lens vesicle to a potential slit. During the rest of the lifetime of the individual, there is a progressive buildup of the fibrous component of the lens. The outer epithelial portion remains, but its relative prominence decreases greatly. A broad view of the morphology of lens development can be gained by examining in sequence Figs. 12-2C to 12-2F, 12-7, and 12-8.

The essence of cytodifferentiation within the lens is the transformation of mitotically active epithelial cells from the outer pole of the lens to the elongated postmitotic cells, containing the crystallin proteins, in the main body of the lens. The transformation from epithelial cells to lens fibers takes place in an equatorial

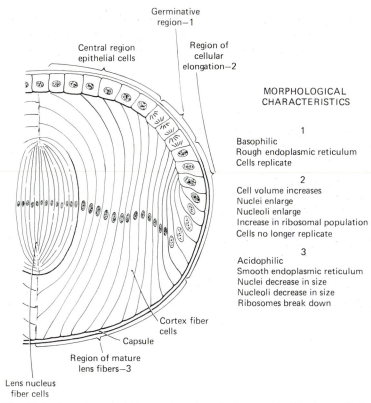

Germinative
region—1

Central region
epithelial cells

Region of
cellular
elongation—2

MORPHOLOGICAL
CHARACTERISTICS

1
Basophilic
Rough endoplasmic reticulum
Cells replicate

2
Cell volume increases
Nuclei enlarge
Nucleoli enlarge
Increase in ribosomal population
Cells no longer replicate

3
Acidophilic
Smooth endoplasmic reticulum
Nuclei decrease in size
Nucleoli decrease in size
Ribosomes break down

Cortex fiber
cells

Capsule

Region of mature
lens fibers—3

Lens nucleus
fiber cells

Fig. 12-4 Organization of the vertebrate lens. During growth of the lens, cells from the germinative region begin to elongate, stop dividing and begin to synthesize specific lens crystallin proteins. The elongated cells in the lens nucleus are the oldest of the lens fiber cells. Toward the edge of the lens, the fiber cells are successively younger. *(After Papaconstantinou, 1967,* Science **156**:338.)

region, which surrounds the entire lens (Fig. 12-4). The low lens epithelium just outside the equatorial region is mitotically active throughout in embryonic eyes, but postnatally the cells in the central epithelial region (Fig. 12-4) cease dividing. A germinative region of dividing cells remains. Daughter cells from this region move into the equatorial zone, where they lose the ability to divide and begin to elongate. At the same time, the cytological correlates of RNA synthesis (nuclear and nucleolar enlargement and increased numbers of ribosomes causing cytoplasmic basophilia) are becoming prominent. These changes are preparatory to the formation of massive amount of the lens crystallins within the elongated lens fibers.

From the equatorial zone the lens fibers continue to elongate. They lose the cytological characteristics of protein-synthesizing cells and become eosinophilic cells, with shrunken and condensed nuclei and a continuing reduction with age of the ribosomes and the rough endoplasmic reticulum on which the crystallins and other proteins are made. The body of the lens contains layers of lens fibers, organized much like an onion. Those at the very center of the lens (called the *nucleus*) are the first lens fibers that were formed in the embryo. The tips of the lens fibers grow toward the external and internal poles of the lens. At each pole of the lens there is a place, called the *lens suture*, where fibers arising at opposite points on the equator meet one another. Farther toward the periphery, in the cortex of the lens, the lens fibers are successively younger. New lens fibers are continually fed from the equatorial region onto the outer cortex throughout life.

The patterns of nucleic acid and protein synthesis correlate well with the cytological characteristics of the lens. DNA synthesis is found only in the low lens epithelium and is most concentrated in the germinative region (Fig. 12-5A). RNA synthesis is heavy in the lens epithelium, the equatorial region, and the cortex of the body, but it is absent in the core of the lens (Fig. 12-5B). Protein synthesis occurs throughout the lens, although in reduced amounts in the core (Fig. 12-5C), but experiments with actinomycin D have shown that protein synthesis in the core, in contrast to that in the epithelium, is due to a long-lived mRNA (Fig. 12-5D). The long-lived mRNA is an adaptation which allows some continued synthesis of crystallin proteins in cells whose nucleic acid synthesizing mechanisms have been turned off.

The crystallins comprise a family of lens proteins (α, β, γ, and δ types), which appear to be structural rather than enzymatic in nature. Different combinations of crystallins are found in different animal groups, for example, α, β, and δ crystallins are found in the avian lens, whereas the α, β, and γ varieties are found in the mammalian lens. The types of crystallins appear at different times during lens development. In the chick, δ crystallin can first be detected in the elongated cells of the invaginating lens placode at about 50 hours and the β crystallins a few hours later (Zwaan and Ikeda, 1968). The last to appear is α crystallin, which can be found in the early lens vesicle at about 80 hours. Although interpretations regarding the relationship of the crystallins to cytodifferentiation in the lens have varied rather widely over the years, there is increasing evidence that these proteins are products of differentiation more than causative factors influencing subsequent differentiation.

Later Development of the Lens   The influence of the retina upon the lens does not cease with the initial inductive event. A continuing retinal influence during the later embryonic life has been demonstrated by the Coulombres (1963), who surgically reversed the lens in chick embryos so that the low outer-surface epithelium faced the retina and the elongated cells of the deep layer faced the outside. Very rapidly, the low epithelial cells which were facing the retina began to elongate and formed an additional set of lens fibers (Fig. 12-6). On the corneal side of the rotated lens, a new lens epithelium formed. These structural adaptations are evidence of a mechanism which ensures a continuous alignment of the lens with the rest of the visual system during development.

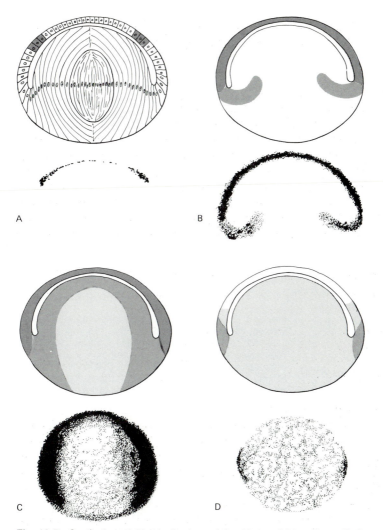

Fig. 12-5 Synthetic activities in the lens of the 12-day chick embryo. Embryos were exposed to specific isotopic precursors of proteins and nucleic acids, and autoradiographs were then prepared from tissue sections. In these figures drawings of lenses, with intensities of synthetic activities indicated by different levels of gray, are placed over tracings of autoradiographs, showing the densities of silver grains. (A) Pattern of [^{14}C]thymidine incorporation showing DNA synthesis concentrated in the germinative region. (B) Pattern of incorporation of [^{14}C]uridine, indicating most RNA synthesis occurs in the lens epithelial cells and in the outermost fiber cells of the lens cortex. (C) [^{14}C]leucine incorporation, showing protein synthesis is greatest in the lens epithelium and is progressively less toward the lens nucleus. (D) 8 hours after the administration of actinomycin D (an inhibitor of RNA synthesis) and later exposure to [^{14}C]leucine, cells of the lens epithelium no longer take the isotope, whereas those of the lens body continue to do so. This indicates the presence of long-lived mRNA in the lens fiber cells. (*Adapted from Reeder and Bell, 1965,* Science **150**:71.)

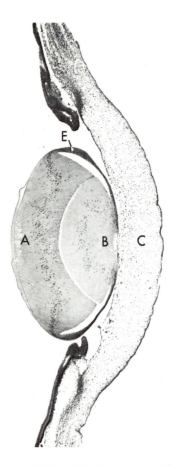

Fig. 12-6 Axial section through the lens of an 11-day chick embryo. At 5 days of incubation the lens was surgically reversed so that the interior lens epithelium faced the vitreous body and retina. The formerly low epithelial cells elongated to form new lens fibers (A). Due to the reversal of polarity of the equatorial zone of the lens, new epithelial cells (E) are added on the corneal face of the lens, covering the original mass of lens fibers (B). Abbreviations: *C.,* cornea. *(Courtesy of A. J. Coulombre, 1965, from* Organo-genesis, *De Haan and Ursprung, Holt, Rinehart and Winston, Inc., New York.)*

Iris and Ciliary Apparatus　As the lens increases in size, it settles back into the optic cup and the margins of the cup begin to overlap its edges. We can now recognize the thin overlapping part of the optic cup as the epithelial portion of the iris, and the reduced opening in front of the lens as the pupil. The iris contains muscles (*the dilator* and *sphincter pupillae*) which, as their names imply, regulate the size of the pupil. Embryologically, these muscles are unusual because they arise from neurectoderm instead of mesoderm.

In fetuses of the fifth month the ciliary region is readily identifiable by the marked folding which has involved this portion of the original optic cup (Fig. 12-8). Outside the epithelial layer is the loosely aggregated mesenchyme which will be organized into the muscular portion of the ciliary body. This ciliary muscle, by altering the tension on the suspensory ligament of the lens, controls the lens curvature and thereby helps the eye to change focus so that objects at different distances can be made to cast sharp images on the retina. Normal development of the ciliary body appears to depend upon the correct amount of intraocular pressure. If some of the fluid within the developing eyeball is allowed to escape, a defective ciliary body results (Coulombre and Coulombre, 1957).

Choroid Coat and Sclera Outside the optic cup, mesenchymal cells, largely of neural crest origin, early become massed in a concentrated zone. Reacting to an inductive influence from the optic cup, this mesenchymal coat becomes differentiated into an inner, highly vascular tunic known as the *choroid coat* (Fig. 12-7C) and an outer tunic composed of densely woven fibrous connective tissue known as the *scleroid coat*, or *sclera*. The tough sclera molds the eyeball and gives firm attachment to the muscles which move the eye in its socket.

Many inframammalian vertebrates possess a ring of cartilaginous or bony elements (*scleral ossicles*) which surround the outer margin of the cornea. Fourteen ossicles are present in the eye of the chick. Each ossicle arises as by the outgrowth of an epithelial papilla from the eye into the surrounding *ectomesenchyme* (mesenchyme derived from neural crest). By about the twelfth day in the chick embryo, ossification begins, and within 2 days it has expanded so that it forms a ring of overlapping bony elements. Deletion experiments suggest that the papillae may induce the ectomesenchyme to form skeletal tissue.

Cornea Continuous with the sclera in front, and forming the part of the eye overlying the lens and the iris, is the *cornea* (Fig. 12-8). In postnatal life the cornea is a multilayered structure which, like the lens, must be transparent and without optical aberrations in order for undistorted light rays to reach the retina. The mature cornea consists of an outer epithelium underlain by a basement membrane (*Bowman's membrane*) and an inner endothelium, also underlain by a basement membrane (*Descemet's membrane*). Between the outer and inner layers is a thick stroma, which consists of fibroblasts and layers of collagen fibers, arranged perpendicularly to one another (Fig. 12-10). In recent years the cornea has received considerable attention from developmental biologists because it is a very convenient system for studying relationships between cells and the extracellular matrix during development. For a detailed treatment of the morphology of corneal development, the reader is referred to the monograph by Hay and Revel (1969).

An inductive influence, emanating from the lens vesicle and the optic cup, stimulates the transformation of ordinary surface ectoderm to corneal epithelium (Lewis, 1904). This represents one of the terminal events of a long series of inductions involved in the formation of the eye. Some of the major inductive events are summarized in Fig. 12-9.

Shortly after the first inductive event, the future cornea is two cells thick and differs little from the general cutaneous ectoderm. Soon the basal layer of epithelial cells increases in height because of the elaboration of secretory organelles within the cells. In these cells the Golgi apparatus shifts toward the basal surface, where it is involved in the secretion of a collagenous extracellular matrix known as the *primary stroma* (Fig. 12-10).

The next major step in corneal development is the migration, over the inner surface of the acellular primary stroma, of the *corneal endothelial cells*, which arise from mesodermal mesenchyme associated with blood vessels around the lip

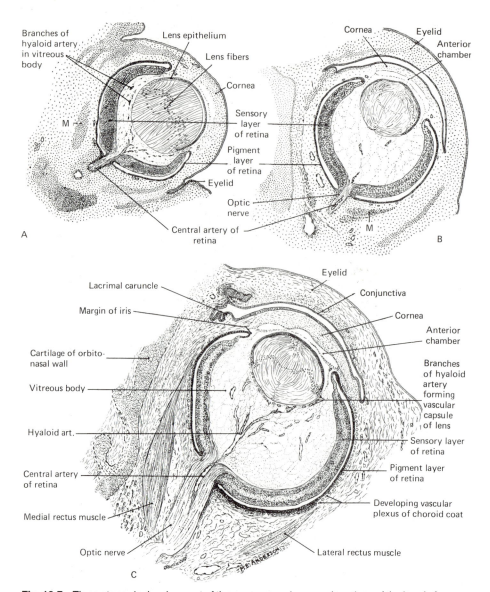

Fig. 12-7 Three stages in development of the eye as seen in coronal sections of the head of young human embryos. (A) From an embryo of 17 mm; about 7 weeks. (Projection drawing, X50, from University of Michigan Collection, EH 14.) (B) From an embryo of 33 mm; about middle of ninth week. (Projection drawing, X35, from University of Michigan Collection, EH 217.) (C) From an embryo of 48 mm; about middle of tenth week. *(Adapted from Ida Mann, X25.)* Abbreviation: *M.,* primordial mesenchymal concentration for eye muscle.

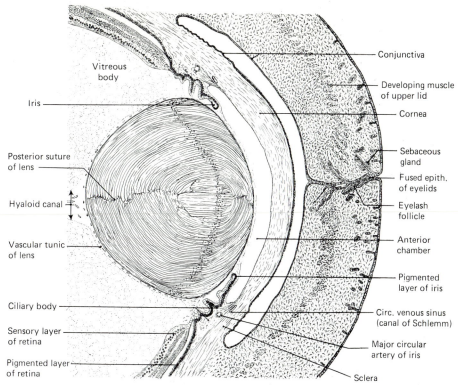

Fig. 12-8 Anterior part of the eye from a human fetus of about 19 weeks; 174 mm. Vertical section (X20) to show fused eyelids, developing ciliary region, and lens.

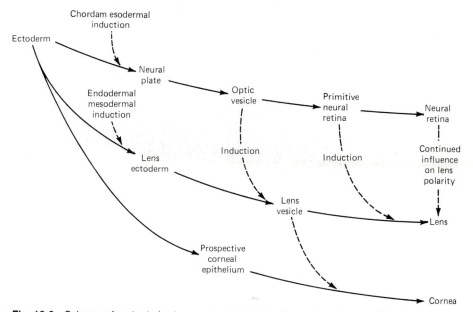

Fig. 12-9 Scheme of major inductive events occurring in the embryonic eye. Inductive events or tissue interactions are indicated by broken arrows.

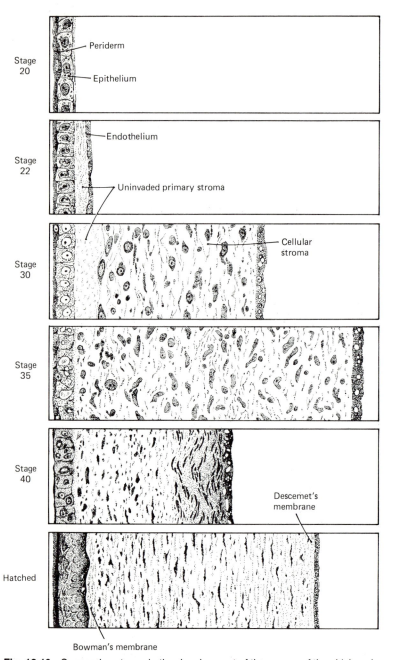

Fig. 12-10 Successive stages in the development of the cornea of the chick embryo. The stages indicated are those of Hamburger and Hamilton (1951). *(After Hay and Revel, 1969,* Fine Structure of the Developing Avian Cornea, Monogr. in Devel. Biol., *vol. 1, S. Karger, Basel.)*

of the optic cup. When the endothelial cells have ceased their migration, the cells change from a squamous to a cuboidal shape and form a complete inner lining of the cornea, which at this time consists of the outer epithelium and the inner endothelium bounding on either side the still acellular primary stroma (Fig. 12-10 stages 30 and 35).

After they have formed a complete layer, the corneal endothelial cells synthesize large amounts of hyaluronic acid and secrete it into the primary stroma. The water-binding properties of hyaluronic acid cause the stroma to swell greatly (Fig. 12-11). The swollen matrix seems to serve as a highly favorable substrate for cell migration, and fibroblasts of neural crest origin then invade the primary stroma and begin proliferating within it. The migratory phase of fibroblastic seeding of the primary stroma comes to an end when large amounts of hyaluronidase, probably synthesized by the fibroblasts themselves, are

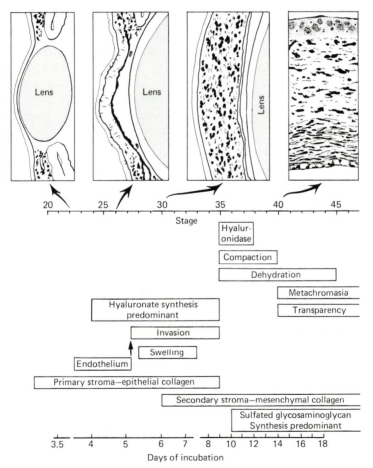

Fig. 12-11 Flow chart of significant events in corneal morphogenesis. *(Drawings after Hay and Revel, 1969,* Fine Structure of the Developing Avian Cornea, Monogr. in Devel. Biol., *vol. 1, S. Karger, Basel.) (Graph after Toole and Trelstad, 1971,* Devel. Biol. **26**:33.)

secreted into the stroma and break down much of the hyaluronic acid that is present. This signals a new phase, characterized by the settling in of the fibroblasts in their new location and their adding coarse collagen fibers to the substance of the stroma. (The correlation of high concentrations of hyaluronic acid with cell migration and its removal by hyaluronidase with subsequent stabilization of the cells is not confined to the cornea alone, but it may have widespread significance in developing systems.)

After being populated by fibroblasts, the primary stroma is called the *secondary stroma*. Only a thin layer of acellular primary stroma remains beneath the corneal epithelium. Both the outer epithelium and the inner endothelium of the cornea continue to secrete extracellular material beneath their basal surfaces. This results in the formation of Bowman's membrane beneath the epithelium (Fig. 12-10) and Descemet's membrane under the corneal endothelium.

Once its major components are established, the initially opaque cornea must become transparent and establish architectural properties appropriate for the transmission of a path of undistorted light into the eyeball. In the chick embryo the transparency of the cornea increases from 40 percent transmission in the 14-day embryo to 100 percent in the 19-day embryo. This is accomplished by the dehydration of the corneal stroma, a process which is stimulated by thyroxine secreted into the blood by the maturing thyroid gland. The role of thyroxine has been demonstrated by the premature dehydration of the cornea when thyroid hormone has been administered early and the retardation of dehydration by thyroid inhibitors. The clearing of the cornea results from the actions of thyroxine upon the corneal endothelium, which pumps out sodium from the corneal stroma into the anterior chamber of the eye. Water molecules follow the sodium ions, thus resulting in the dehydration of the stroma.

The curvature of the cornea is on a smaller radius than that of the remainder of the eyeball and thus appears to bulge out from the eye as a whole. The perfection of the corneal curvature is of great functional importance, for the cornea is what we might call the front lens of the eye and acts in conjunction with the crystalline lens in bringing light rays into focus on the retina.

Retina   After the formation of the optic cup, the inner and outer walls follow different pathways of differentiation. The outer wall becomes thin and highly pigmented, forming the *pigment layer of the retina* (Fig. 12-7). Cells of the inner wall of the optic cup continue to proliferate, causing this layer to increase in both thickness and circumference. These cells differentiate into neural elements, and the inner wall of the optic cup is then known as the *neural (sensory) layer of the retina*. Cells in the center of the retina mature first, leaving a zone of proliferating cells around the margins of the retina. As long as the retina grows, daughter cells from the marginal growth zone contribute to the substance of the retina. During its very early stages of formation, the polarity of the retina becomes fixed in a manner reminiscent of the determination of the limb axes. In the retina the nasotemporal (anteroposterior) axis is fixed first. This is followed by fixation of the dorsoventral axis, and finally radial polarity is established.

The mature neural retina is an extremely complex tissue, composed of three well-defined layers of neural cells. From the inside to the outside of the retina, these layers are called the *layer of ganglion cells*, the *layer of bipolar cells*, and the *layer of rods and cones*, the light-sensitive elements (receptors) of the eye. In the postnatal eye light striking the retina must first pass through the ganglion cell layer and then the bipolar cell layer before it strikes the rod and cone cells. These cells convert the light signal into neural signals, which are transmitted via elaborate networks of synapses through the bipolar cells to the ganglion cells. Long processes of the ganglion cells extend through the optic nerve to the primary visual centers of the brain (e.g., in the human; Fig. 11-2, arc 5b).

In the embryo, cells of the ganglion layer (i.e., the layer closest to the center of the eyeball) begin to differentiate first. Differentiation of remaining layers follows, with the rods and cones of the outermost layer of the neural retina taking shape last. The differentiation of cell layers within the retina is accompanied by waves of cell death, but the function of this phenomenon in the retina is still poorly understood. Although the development of synaptic connections within the retina is presently receiving considerable attention, the details are beyond the scope of this book. The neurons constituting the ganglion cell layer send out axons which grow out into the ventral wall of the optic stalk and thence toward the optic centers of the brain. In an extremely precise manner they make connections with the next order of neurons in the visual pathway.

Eyelids, Conjunctiva, and Associated Glands The eyelids start to develop during the seventh week as folds of skin growing back over the cornea (Fig. 12-7A). Once they have begun to form, the eyelids close over the eye quite rapidly, usually meeting and fusing with each other by the end of the ninth week. This fusion involves only the epithelial layers of the lids (Fig. 12-8), and the eyelashes and the glands that lie along the margins of the lids start to differentiate from this common epithelial lamina before the lids reopen. Signs of the loosening of the epithelial union can be seen in the sixth month, but it is ordinarily well into the seventh month before the eyelids actually reopen.

The space between the eyelids and the front of the eyeball is commonly referred to as the *conjunctival sac*. The most massive glands opening into the conjunctival sac are the *lacrimal glands*. They develop from multiple epithelial buds which make their first appearance during the ninth week. The lacrimal glands produce a thin, watery secretion which under normal conditions keeps the corneal surface cleaned and lubricated. When stimulated by local irritation, or under conditions of emotional disturbance, activated by way of the autonomic nervous system, they produce fluid in excess which overflows the lids as "tears." Under normal circumstances the fluid produced by the lacrimal glands, after bathing the conjunctival surfaces, passes into the nasal chamber by way of the *nasolacrimal duct*.

Changes in Position of the Eyes During development the eyes undergo a striking change in their relative position. In embryos of the sixth week they are far around on either side of the head (Fig. 13-1C). If, at this stage, a line is imagined passing through the optical axis of each eye and then prolonged to meet in the center of the head, the angle these two lines would form would be about 160°. In other words, the eyes look almost straight off to either side like the eyes of a fish. In such a position there can be no overlapping of their visual fields, which precludes the binocular type of vision so important to us in estimating distances. As the facial structures grow, the eyes are carried forward in the head, and as a result, their optical axes begin to converge. By the end of the seventh week (embryos of 17 to 19 mm) the ocular angle has been reduced to about 120°. By the eighth week the eyes are beginning to look quite definitely forward (Fig. 13-1F), and by the tenth week (embryos of 40 to 50 mm) the angle is approximately 70°—only about 10° wider than it is in the adult.

THE EAR

For convenience, the adult mammalian ear may be divided into three regions— external, middle, and internal. The external ear is essentially a sound-collecting funnel consisting of the *auricula*, or *pinna*, and the *external auditory canal*. The middle ear is a sound-transmitting mechanism involving a chain of three *auditory ossicles* which pick up the vibrations received by the eardrum and transmit them across the middle ear, or tympanic cavity, to the receptive mechanism of the internal ear. The internal ear is composed of an elaborate system of fluid-filled, epithelially lined chambers and canals constituting the so-called *membranous labyrinth*. This membranous labyrinth lies within the temporal bone in a similarly shaped, but larger, series of cavities constituting the *bony labyrinth*. The narrow space between the walls of the bony labyrinth and the membranous labyrinth is known as the *perilymphatic space* and is filled with the *perilymphatic fluid*. The sound-receiving portion of the membranous labyrinth is the *cochlea*, a curiously shaped structure spirally coiled in a manner suggestive of a snail shell. Closely associated with the cochlea is the vestibular complex concerned with equilibration. The vestibular portion of the membranous labyrinth is composed of the *sacculus*, the *utriculus*, and the three *semicircular ducts*, or *canals*. It is phylogenetically the most primitive part of the ear; in fact, it is the only part of the ear that has been differentiated in the fishes.

Formation of the Auditory Vesicle   The primordium of the membranous labyrinth, or inner ear, is the first part of the ear mechanism to make its appearance. Formation of the auditory vesicle is stimulated by an inductive action by the hindbrain upon the overlying ectoderm. In human embryos morphological changes are first noticeable early in the third week (2-somite stage; Fig. 8-2) when the superficial ectoderm on either side of the still-open neural plate becomes slightly thickened. This thickening is the start of the *auditory placode* which, by the middle of the third week (7-somite stage; Fig. 8-3B), becomes quite clearly marked. By the end of the third week the auditory placode has taken shape as a sharply circumscribed thickening in the ectoderm on either side of the developing myelencephalon (Fig. 12-12A). During the

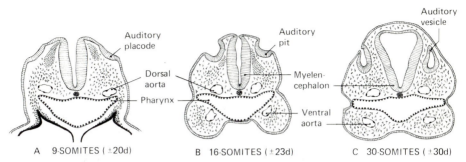

Auditory
vesicle

Auditory
placode

Auditory
pit

Dorsal
aorta

Myelen-
cephalon

Pharynx

Ventral
aorta

A   9-SOMITES (±20d)   B   16-SOMITES (±23d)   C   30-SOMITES (±30d)

Fig. 12-12 Formation of auditory vesicle as seen in cross sections of young human embryos.
(Modified from Arey.)

fourth week the placode is invaginated to form the *auditory pit* (Fig. 12-12B).
The pit becomes deepened, and finally its opening at the surface is closed. After
it has been closed off from the surface, the former auditory pit constitutes a sac
called the *auditory vesicle* (*otic vesicle*). Like the limb bud and retina, the major
axes of the auditory vesicle are sequentially determined, with the anteroposterior axis fixed first and the dorsoventral axis next. Grafting experiments have
suggested that in all three structures (eye, ear, and limb) axial determination is
accomplished by the recognition and interpretation of some general environmental cues that are present in the lateral walls of the embryo.

Many of the soft ectodermally derived tissues of the head induce the
mesenchyme surrounding them to form protective coverings of skeletal tissue.
For example, the early brain induces the formation of the flat bones of the
cranium; the eyeball induces the formation of the scleral cartilages; and the
olfactory placodes appear to stimulate the formation of hard tissues in the nasal
region. Similarly, the auditory vesicle induces the mesenchyme around it to form
the cartilaginous ear capsule (Balinsky, 1925). In contrast to the former
examples, in which the induced mesenchyme is of neural crest origin, the
mesenchyme forming the ear capsule is of mesodermal origin.

Differentiation of the Auditory Vesicle to Form the Inner Ear As the
auditory vesicle enlarges, it changes from its originally spheroidal shape and
becomes elongated dorsoventrally. About where the epithelium of the auditory
vesicle was separated from the superficial ectoderm, there develops a tubular
extension of the vesicle, known as the *endolymphatic duct* (Fig. 12-14A). As the
auditory vesicle expands laterally, the endolymphatic duct is left occupying a
progressively more median position in relation to the rest of the vesicle. Almost
from the outset of its differentiation the more expanded dorsal portion of the
auditory vesicle with which the endolymphatic duct is connected can be
identified as the primordium of the vestibular part of the membranous labyrinth,
and the more slender ventral extension is recognizable as the primordium of the
cochlea (Fig. 12-13).

By the close of the sixth week of development conspicuous flanges appear

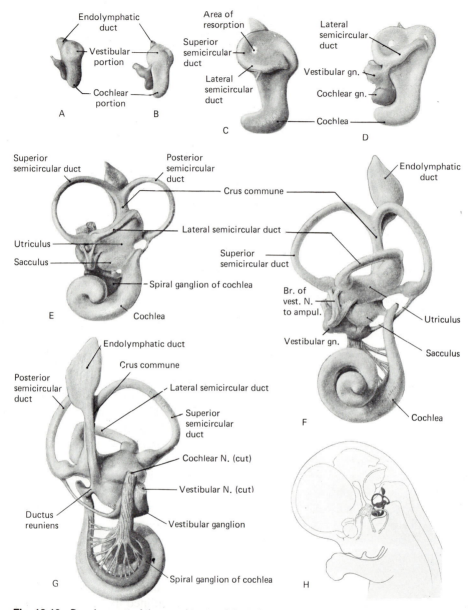

Fig. 12-13 Development of the membranous labyrinth in human embryos. *(After Streeter,* Am. J. Anat., *vol. 6, 1906.)* (A) 6 mm, lateral view. (B) 9 mm, lateral view. (C) 11 mm, lateral view. (D) 13 mm, lateral view. (E) 20 mm, lateral view. (F) 30 mm, lateral view. (G) 30 mm, medial aspect. (H) Outline of head of 30-mm embryo to show position and relations of developing inner ear.

on the vestibular portion of the auditory vesicle, foreshadowing the differentiation of the *semicircular ducts* (old term, semicircular canals). As the flanges push out from the main vesicle, their central portions become thin and finally undergo resorption so that the original semilunate flange becomes converted into a looplike duct (Fig. 12-13C to 12-13E). There are three such ducts formed, each occupying a plane in space approximately at right angles to the other two. While the semicircular ducts are taking shape, the vestibular portion of the auditory vesicle is becoming subdivided by a progressively deepening constriction into a more dorsal utricular portion and a more ventral saccular portion (Fig. 12-13E to 12-13G). When this division has occurred, the semicircular ducts open off the utriculus. Near one of their two points of communication with the utriculus, each semicircular canal forms a local enlargement known as an *ampulla*. Within the ampulla there develops a specialized area called a *crista*, containing neuroepithelial cells with hairlike processes projecting into the lumen of the ampulla. These specialized receptors are innervated by branches of the vestibular division of the eighth cranial nerve. Changes in the position of the head are accompanied by a lag in the movement of fluid within the semicircular ducts which results in mechanical stimulation of the neuroepithelial cells of the crista. The nerve impulses thus initiated pass over the appropriate central pathways (Fig. 11-2, arc 3) and make us aware of positional changes. In light of this function the significance of the arrangement of the three semicircular ducts in planes at right angles to each other is self-evident.

Specialized areas, called *maculae*, are developed in the sacculus and utriculus. The maculae contain neuroepithelial cells similar in general character to those in the cristae of the semicircular ducts and like them are supplied by branches of the vestibular division of the eighth cranial nerve. Impulses initiated in the maculae make us aware of static position in contrast to the sense of positional change mediated through the mechanism of the semicircular canals.

The *cochlear portion of the membranous labyrinth* is the sound-perceiving part of the ear mechanism. The cochlea (Fig. 12-13F and G) is a tiny snail-shaped structure which consists of three parallel ducts, a central *cochlear duct* and dorsal and ventral *perilymphatic ducts*. Running the entire length of the cochlear duct is the organ of Corti, a band of tissue containing very specialized auditory receptors, which are connected to terminal fibers of the auditory nerve in the *spiral ganglion* of the cochlea (Fig. 12-13G). Sound is perceived when sound waves in the air deform the *tympanic membrane* (ear drum; Fig. 12-15). The motions of the tympanic membrane are mechanically transmitted by the auditory ossicles of the middle ear to another membrane covering the *oval window* (Fig. 12-15), which in turn sets up waves in the fluid within the cochlea. The waves are then detected by the appropriate auditory receptors within the organ of Corti, which convert the mechanical stimulus into a neural signal. The neural signal travels along the auditory reflex arc outlined in Fig. 11-2, arc 5a.

Originating from the most ventral part of the auditory vesicle, the cochlea elongates rapidly during the sixth week and, in embryos of 11 to 13 mm, shows a sharp forward bend of its distal end (Fig. 12-13C and D). Elongation continues at an accelerated rate during the seventh and eighth weeks, and the initial bend rapidly develops into a spiral of 2½ turns (Fig. 12-13E and F). As the cochlea is thus differentiated, its originally broad connection with the vestibular portion of the membranous labyrinth becomes narrowed to the slender *ductus reuniens* (Fig. 12-13G). The cochlear division of the eighth nerve follows the cochlea in its growth changes, and its fibers fan out to be distributed along the entire length of the cochlear duct as the spiral ganglion of the cochlea (Fig. 12-13G).

Middle Ear   While the receiving mechanism of the ear is forming in the manner just described, the transmitting apparatus of the middle ear is also taking shape. Recall that in their initial relations the first pharyngeal pouches extend laterad so that their endodermal lining makes contact with the ectoderm at the bottom of the first gill furrow to form the gill plate. The distal portion of the pouch remains somewhat expanded to form the primordium of the middle-ear chamber, or *tympanic cavity*, but the proximal portion soon becomes

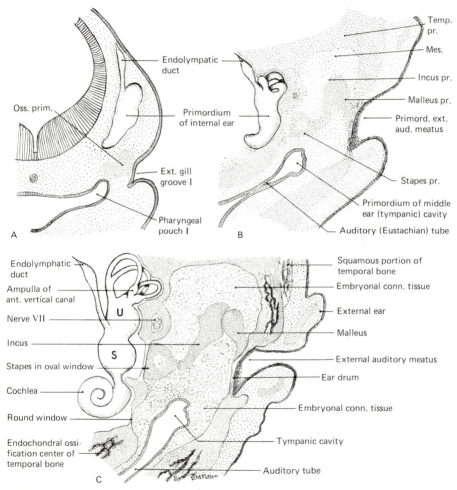

Fig. 12-14   Schematic diagrams showing three stages in the development of the middle-ear chamber and the auditory ossicles. Abbreviations: *Mes.*, mesenchyme; *Oss. prim.*, mesenchymal concentration which is the first indication of the primordia of the auditory ossicles; *Temp. pr.*, mesenchymal concentration where primordium of temporal bone is taking shape.

narrowed to form the *auditory (eustachian) tube* (Fig. 12-14A and B). The primary contact between the endoderm of the pharyngeal pouch and the ectoderm of the floor of the gill furrow does not last long. The blind outer end of the pouch, which constitutes the primordium of the tympanic cavity, pulls away from the surface, and a conspicuous concentration of mesenchyme appears adjacent to it (Fig. 12-14A). As development progresses, the mesenchymal cells of this primordial mass become organized into the cartilaginous precursors of the *auditory ossicles* lying between the developing inner ear and the retained portion of the first gill furrow, which may now be said to constitute the primordium of the external auditory meatus. At this stage the early ossicles lie above the primordial tympanic cavity, embedded in a very loose embryonic connective tissue (Fig. 12-14C).

Phylogenetically, the auditory ossicles are derived from bones involved in the suspension and articulation of the jaw in the lower vertebrates. The malleus and incus are derivatives of the first branchial arch and are the homologues of the articular (malleus) and quadrate (incus) bones, which are essential components of the articulation between the upper and lower jaws of lower vertebrates through the reptiles. The articular bone represents the ossified basal end of Meckel's cartilage (see Chap. 15), and it is the articular surface of the primitive lower jaw. The quadrate bone, a dermal bone, is the articular surface for the upper jaw. The third element (the stapes in mammals) is derived from the most dorsal portion of the hyoid (II) arch. In crossopterygian fishes this bone is called the *hyomandibula*, and it serves to anchor the jaw apparatus to the neurocranium in the vicinity of the inner ear. In amphibians, the hyomandibula is converted into a bone (*columella*) of the middle ear. The columella alone provides the mechanical connection between the tympanic membrane, located on the surface of the head, and the inner ear. In mammals, many of the dermal bones originally associated with the jaw in phylogeny have been eliminated or used for other purposes. In the latter category fall the articular and quadrate bones, which have been incorporated into the middle ear as the malleus and incus. The muscles and nerves which are associated with the middle-ear ossicles are appropriately derived from the first and second arches in accordance with their phylogenetic and ontogenetic origins (Fig. 15-3).

During the latter part of intrauterine life the connective tissue about the auditory ossicles begins to undergo rapid resorption, with a resultant expansion of the tympanic cavity. Eventually the ossicles come to lie suspended within the enlarged tympanic cavity with only a thin layer of epithelium overlying their periosteal investment. At the time of birth, however, there is still a residue of unresorbed embryonic connective tissue partially filling the tympanic space and more or less damping the free movement of the ossicles (Fig. 12-15). Full mobility of the ossicles is acquired within a few months after birth when the remaining loose connective tissue is resorbed. When this has occurred movement imparted by sound waves to the eardrum is freely transmitted by the ossicles to the membrane of the oval window to which the stapes is attached.

External Ear   The external ear is formed by the growth of the mesenchymal tissue flanking the first (hyomandibular) gill furrow of the young embryo. During the second month several nodular enlargements appear, some of them arising from mandibular arch tissue rostral to the first gill furrow, and others from the hyoid arch along the caudal border of the furrow. The coalescence of these tubercles and their further development mold the pinna of the ear (Fig. 12-16). In view of the number of separate growth centers involved, it is not surprising that the configuration of the fully formed external ear exhibits so wide a range of variations. These individual differences are easy to overlook until one begins to pay particular attention to the details of ear shapes. A little critical observation, however, will soon make it apparent why European police utilize the configuration of the ear as one of the very important features for the identification of persons in whom they are interested.

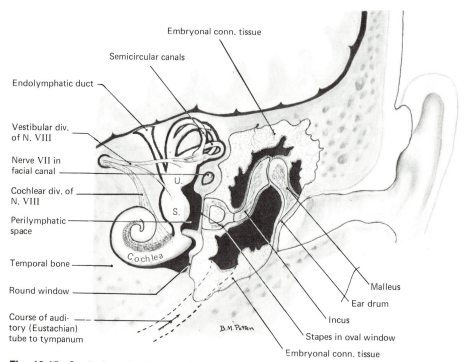

Fig. 12-15 Semischematic diagram showing ear mechanism at term. The cochlea has been turned mesiad to show its spiral course.

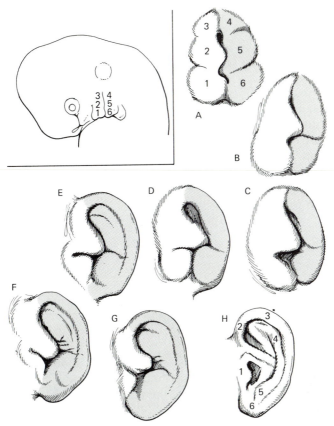

Fig. 12-16   Stages in development of the external ear. *(After Streeter, 1922,* Carnegie Contr. to Embryol. *14:111.)* The parts derived from the mandibular side of the cleft are unshaded; the parts from the hyoid side are shaded.

Chapter 13

The Face and Oral Region

In the early embryo of about 4 weeks the face, as such, consists of only a few primordial tissue masses partially surrounding the future oral region. The oral cavity is represented by an ectodermal depression, the *stomodeum*, which abuts upon the blindly ending anterior end of the foregut. A thin *oral (stomodeal) plate*, consisting of opposed layers of ectoderm and endoderm, maintains a temporary partition between the stomodeum and foregut. Rostral to the stomodeum, the area of the future face is occupied by the frontal prominence of the greatly overhanging forebrain. Laterally, the maxillary process of the first branchial arch is prominent, and caudal to the stomodeum, the mandibular arch and other emerging branchial arches are the dominant structures. Between the forebrain and the maxillary processes, the thickened nasal placodes are becoming visible.

The essence of facial development is the establishment of the primordial tissue masses, which act as building blocks, and their subsequent growth and rearrangement into recognizable facial structures. The stomodeal depression, even as late as when the oral plate ruptures and establishes communication between the cephalic end of the gut and the outside world, is very shallow (Fig. 8-12). The deep oral cavity characteristic of the adult is formed by the forward growth of structures about the margins of the stomodeum. Some idea of the extent of this forward growth can be gained by the fact that the tonsillar region of the adult is at about the level occupied by the oral plate before it ruptured and disappeared. The growth of the structures bordering the stomodeum, then, does not just give rise to superficial parts of the face and jaws but actually builds out the walls of the oral cavity itself. Formation of the midface region involves not only the relative displacement of the growing forebrain away from the mouth but also the filling in of the space between the oral cavity and the forebrain. This is accomplished by the building-up of a rim of mesenchymal cells around the pair of thickened ectodermal nasal placodes, by the medial migration of the eyes from their original positions at the side of the head, and by the ingrowth of the maxillary process into the area. Throughout much of the fetal period and the period of postnatal growth one of the most characteristic aspects of facial development is the continued increase in relative prominence of the midface region.

Cellular Origins of the Face   Most structures in the facial region originate from masses of mesenchyme covered by ectoderm and lined by either ectoderm or endoderm. In contrast to the rest of the body, most facial mesenchyme (except that forming the striated musculature) is of ectodermal rather than mesodermal origin. Early in the development the expanding forebrain and optic cups displace caudally the original head mesoderm. Then cells of the cephalic neural crest migrate into the interstices between the brain and the overlying ectoderm, replacing the original mesoderm (Ross and Johnston, 1972).

Once in place, the cephalic neural crest cells assume the appearance of ordinary mesenchymal cells. In the upper face the neural crest mesenchyme differentiates into connective tissue, cartilage, and bone, as well as tooth pulp. Most of the tissues of the mid- and lower face originate from the first branchial arch. The branchial arches are also invaded by neural crest cells, but here the original mesodermal cores of the arches are retained, later to differentiate into the muscles of the face and pharynx, whereas the neural crest cells that have surrounded the mesodermal cores again develop into skeletal and connective tissues. It is common for these cells, particularly the mesodermal precursors of the muscles, to undergo extensive migrations before settling down into their definitive locations. Posterior to the pharynx the skeleton and connective tissues arise from mesodermal mesenchyme.

The Face and Jaws   Because the front of the head of a young embryo is pressed against its thorax, it is not possible in the usual lateral views (Fig. 8-6 and 8-7) to see many of the interesting developmental changes going on in the facial region. In specially prepared mounts of the head which can be oriented so as to provide direct views of the face (Figs. 8-12 and 13-1A), the most conspicuous landmarks of a 4-week embryo are the stomodeal depressions and the mandibular arch, which constitutes its caudal boundary. Within the next week most of the structures which take part in the formation of the face and jaws are already clearly distinguishable (Fig. 13-1B). In the midline, cephalic to the oral cavity, is a rounded overhanging area known as the *frontal prominence*. On either side of the frontal prominence are horseshoe-shaped elevations surrounding the olfactory pits. The median limbs of these elevations are known as the *nasomedial processes*, and the lateral limbs are called the *nasolateral processes*.

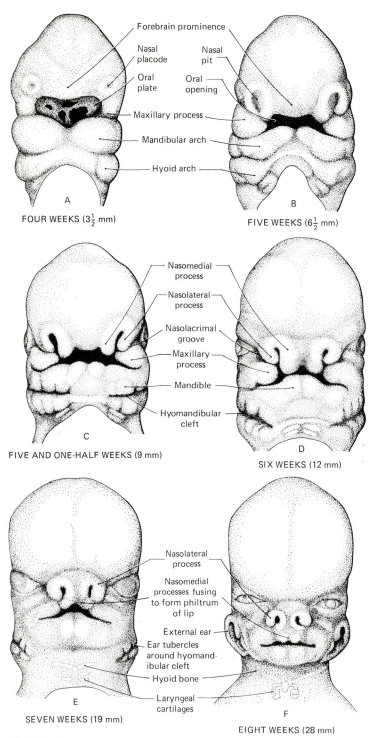

Forebrain prominence

Nasal placode

Nasal pit

Oral plate

Oral opening

Maxillary process

Mandibular arch

Hyoid arch

A

FOUR WEEKS ($3\frac{1}{2}$ mm)

B

FIVE WEEKS ($6\frac{1}{2}$ mm)

Nasomedial process

Nasolateral process

Nasolacrimal groove

Maxillary process

Mandible

Hyomandibular cleft

C

FIVE AND ONE-HALF WEEKS (9 mm)

D

SIX WEEKS (12 mm)

Nasolateral process

Nasomedial processes fusing to form philtrum of lip

External ear

Ear tubercles around hyomandibular cleft

Hyoid bone

Laryngeal cartilages

E

SEVEN WEEKS (19 mm)

F

EIGHT WEEKS (28 mm)

Fig. 13-1 Drawings showing, in frontal aspect, some of the important steps in the formation of the face. *(After William Patten, from Morris,* Human Anatomy, *McGraw-Hill Book Company, New York.)*

Growing toward the midline from the cephalolateral angles of the oral cavity are the *maxillary processes*. In lateral views of the head (Figs. 8-6, 8-7, and 13-2A) it will be seen that the maxillary process and the mandibular arch merge with each other at the angle of the mouth. Thus the structures which border the oral cavity cephalically are (1) the unpaired frontal prominence in the midline, (2) the paired nasomedial processes on either side of the frontal prominence, and (3) the paired maxillary processes at the extreme lateral angles (Fig. 13-1B). From these primitive tissue masses are derived the upper lip, the upper jaw, and the nose.

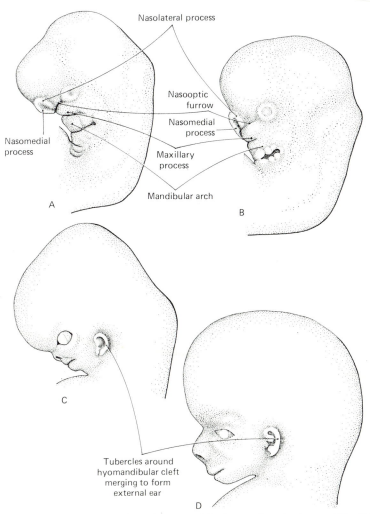

Fig. 13-2 Lateral views of development of the face and external ears. *(After William Patten.)* The embryos represented are the same as those drawn in face view in Fig. 13-1. (A) 5½ weeks; (B) 6 weeks; (C) 7 weeks; (D) 8 weeks.

The caudal boundary of the oral cavity is less complex, consisting of the paired primordia of the mandibular arch. Appearing first on either side of the midline are marked local thickenings resulting from the rapid proliferation of mesenchymal tissue. Until these thickenings have extended from either side to merge in the midline, there remains a conspicuous mesial notch; with their merging, the arch of the lower jaw is completed (Fig. 13-1B to 13-1F).

During the sixth week (Fig. 13-1C and D), marked progress is made in the development of the upper jaw. The maxillary processes become more prominent and grow toward the midline, crowding the nasal processes closer to each other. The nasal processes, meanwhile, have grown so extensively that the lower part of the frontal prominence between them is completely overshadowed (cf. Fig. 13-1B and D). The growth of the medial limbs of the nasal processes has been especially marked, and they appear almost in contact with the maxillary processes on either side. The groundwork for the completion of the upper jaw is now well laid down.

In the final steps in the formation of the upper jaw, the nasomedial processes move toward the midline and merge with each other (Fig. 13-1D). Soon afterward they fuse on either side with the maxillary processes to complete the arch of the upper jaw (Fig- 13-1E and F). The segment of the upper jaw which is of nasomedial origin gives rise externally to the upper lip in the region of the *philtrum* (Fig. 13-1E and F). A deeper triangular projection of the fused nasomedial processes becomes the premaxillary portion of the dental arch as well as the median (primary) part of the palate (Fig. 13-3).

Toward the close of the second month and the beginning of the third, when the molding of the soft parts is well under way, formation of the deeper-lying bony structures begins. The more medial portion of the maxillary bone, which carries the incisor teeth, arises from separate ossification centers formed in the part of the upper jaw which is of nasomedial origin. This origin of the incisive portion of the human maxilla emphasizes its homology with what is in lower forms a separate bone known as the *premaxillary*, or *intermaxillary*. In the skulls of human infants the sutures separating the incisive portion from the rest of the maxilla are likely to be still evident, and occasionally traces of them may be made out in the adult skull. The rest of the maxillary zone, carrying all the upper teeth behind the incisors, is developed in the part of the upper jaw which arises from the maxillary process. It is one of the first bones in the body to be calcified (Fig. 9-14, 9-15, and 9-16).

The Palate   Late in the second month, when the upper jaws have been established, the palatal shelves begin to make their appearance. These paired structures subdivide the more rostral portion of the original stomodeal chamber. Since the nasal pits break through above the level of the shelves, the formation of the palate in effect elongates the nasal chambers backward so that they open eventually into the region where the oral cavity becomes continuous with the pharynx (Fig. 13-4).

The palate as well as the arch of the upper jaw is contributed to by both the

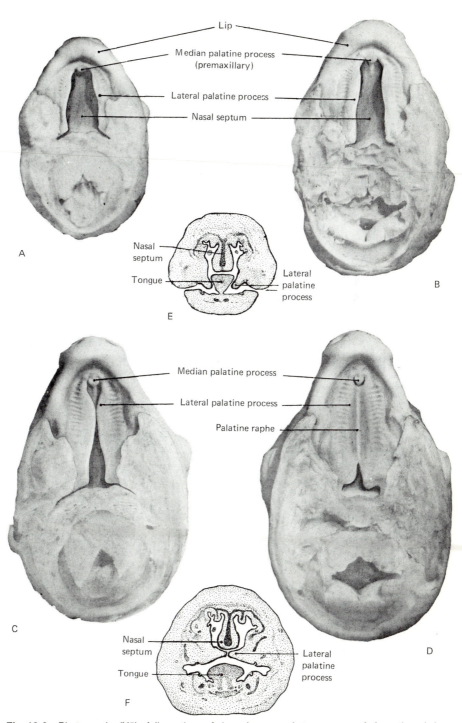

Lip

Median palatine process (premaxillary)

Lateral palatine process

Nasal septum

A

B

Nasal septum

Tongue

Lateral palatine process

E

Median palatine process

Lateral palatine process

Palatine raphe

C

D

Nasal septum

Tongue

Lateral palatine process

F

Fig. 13-3 Photographs (X5) of dissections of pig embryos made to expose roof of mouth and show development of palate. (A) 20.5 mm; (C) 26.5 mm. The diagrams of transverse sections are set in to show: (E) the relations before the retraction of the tongue from between the palatine processes; (F) after the retraction.

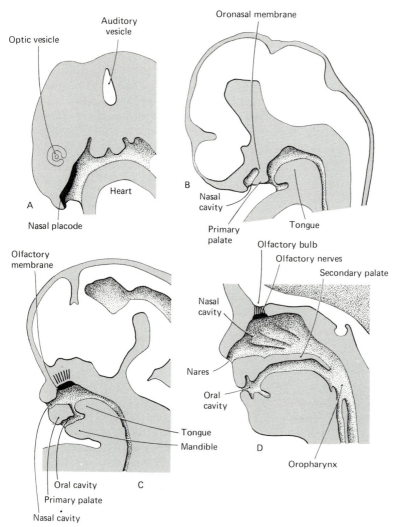

Fig. 13-4 Parasagittal sections of human heads showing the development of the nasal chambers. (A) 5 weeks; (B) 6 weeks; (C) 7 weeks; (D) 12 weeks.

nasomedial processes and the maxillary processes. From the intermaxillary (nasomedial) region a small, triangular, median portion of the palate *(primary palate)* is formed (Fig. 13-3A and B). The main part of the palate *(secondary palate)* is derived from that portion of the upper jaw which arises from the maxillary processes. Shelflike outgrowths arise on either side and grow toward the midline (Fig. 13-3A and B). When these palatal shelves first start to develop, the tongue lies between them, and they are directed obliquely downward so that their margins lie along the floor of the mouth on either side of the root of the tongue (Fig. 13-3E). As development progresses, the tongue moves down and the margins of the palatal shelves swing upward and toward the midline (Fig. 13-3F). Further growth brings them into contact with each other, and their fusion soon completes the main part of the palate (Fig. 13-3D). In the extreme rostral region the small, triangular, premaxillary (median palatine) process lies between the lateral palatine shelves, and they fuse with it instead of with each other. At the same time as the palate is being formed the nasal septum grows toward it and becomes fused to its cephalic face (Fig. 13-9). Thus the separation of right and left nasal chambers from each other is accomplished at the same time as the separation of the deeper portions of the nasal chambers from the oral cavity.

Sometimes the palatal shelves fail to fuse with one another, resulting in a congenital defect known as *cleft palate*. Because of the high incidence and clinical importance of cleft palate and related defects in humans, considerable effort has been expended in trying to understand the basis for the elevation and fusion of the palatal shelves. The many hypotheses, involving factors both intrinsic and extrinsic to the palatal shelves, have been reviewed by Ferguson (1978).

The Nasal Chambers   The first indication of the nose in human embryos is the formation of a pair of thickened areas of ectoderm (the *nasal*, or *olfactory*, *placodes*) on the frontal aspect of the head (Figs. 13-1A, and 13-4A). Almost as soon as they are formed, the nasal placodes sink below the general surface level so that the thickened epithelium constitutes the floor of the *nasal* (*olfactory*) *pits* (Fig. 13-4B). The mesenchymal tissue surrounding the nasal pits proliferates rapidly so that the pits are deepened both by their own progressive invagination and by the forward growth of the surrounding tissue. The bordering elevations become horseshoe-shaped, with their open ends toward the mouth. The two limbs of the nasal elevations are named, on the basis of their positions, the *nasomedial* and the *nasolateral processes* (Fig. 13-1C). At first the nasal pits are far apart, each being well to the side of the young facial region (Fig. 13-1B). As development progresses, the two nasal pits and their associated processes converge toward the midline (Fig. 13-1B to 13-1F). The nasomedial processes on either side eventually merge with each other to form the medial portion of the upper lip and the septum of the nose. The nasolateral processes become the alae (wings) of the nose.

While these external changes are taking place, the nasal pits are becoming progressively deeper and are extending backward and downward toward the oral cavity (Fig. 13-4B). During the seventh week the tissue separating the nasal pits from the oral cavity becomes thinned to merely a double layer of epithelium— the oronasal membrane. When this breaks through, as it soon does, the nasal pits open freely into the oral cavity just caudal to the arch of the upper jaw (Fig. 13-4C). The formation of the palate by the fusion of the palatal shelves greatly lengthens the original nasal chamber (Fig. 13-4D). The olfactory area differentiates within the roof of each nasal chamber. The olfactory receptor cells, which are actually very primitive bipolar neurons, differentiate within the epithelium itself among tall columnar cells called *sustentacular cells*. The apical process of a receptor cell forms a knob containing several highly modified cilia, which are assumed to be the chemical receptor sites (Frish, 1967). The basal process elongates and makes connections with other neurons in the olfactory bulb, through which olfactory stimuli are transmitted in the form of neural signals to the appropriate centers in the brain.

The Tongue   While the palate has been forming the roof of the mouth, the tongue has been taking shape in the floor. From a developmental standpoint the tongue may be conveniently described as a sac of mucous membrane which becomes filled with a mass of growing muscle. The reason for making this crude comparison is the fact that the covering of the tongue and the lingual muscles have different origins and undergo such striking changes in relative position that it is desirable to consider them separately. The primordial areas involved in forming the mucous covering of the tongue appear early in the second month of development. In 5-week embryos paired lateral thickenings, which arise as a result of the rapid proliferation of the mesenchyme beneath the overlying epithelium, are known as the *lateral lingual swellings* (Fig. 13-5A and B). Between them is a small median elevation known as the *tuberculum impar*. Behind the tuberculum impar is another median elevation appropriately called the *copula* (i.e., yoke), for it unites the second and third arches in a midventral prominence. The copula extends cephalocaudally from the tuberculum impar to the primordial swelling which marks the beginning of the epiglottis (Fig. 13-5A and B).

These areas, which are relatively distinct in young embryos, merge so early and so completely that any too dogmatic statement as to exactly what part of the surface of the adult tongue comes from each of them is unwise. There is retained, however, one unmistakable point of orientation which, for all practical purposes, gives us the general picture sufficiently clearly. This landmark is the *foramen cecum*, a small median pit in the dorsum of the tongue (Fig. 13-5D). Embryologically, the foramen cecum is a vestige of the invagination from the floor of the pharynx which gives rise to the thyroid primordium (Fig. 15-4). We know this invagination is formed at the cephalocaudal level where the first and second branchial arches merge with each other. When the lingual primordia begin to take shape, we find this pit between the tuberculum impar and the copula (Fig. 13-5C and D). In adult anatomy, the sulcus terminalis, with this same pit at its apex, is regarded as the boundary between the "body" and the "root" of the tongue.

Using the foramen cecum as a landmark, we can see that the mucosal covering of the body of the tongue arises from the first arch. Just how much of the mucosa of the body of the tongue comes from the tuberculum impar is a rather moot point of secondary interest. Certainly this area is soon overshadowed by the much more rapidly growing lateral lingual swellings and at most is responsible for only a small median area just distal to the foramen cecum. It is likewise difficult to state exactly the level of the adult tongue which represents

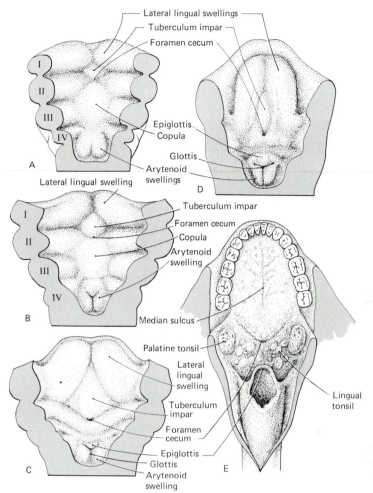

Fig. 13-5 Stages in the development of the human tongue. The upper part of the head has been cut away to permit viewing of the oropharyngeal region from above. The cut visceral arches are indicated by their numbers (*I-IV*), respectively. (A) Fourth week; (B) late in the fifth week; (C) early in the sixth week; (D) middle of the seventh week; (E) adult.

the place where ectoderm and endoderm become continuous when the oral plate ruptures. There can be no doubt, however, that the greater part of the body of the tongue is clothed with what was originally stomodeal ectoderm.

The innervation of the tongue reflects the multiple origins of its components. The musculature of the tongue is innervated by the hypoglossal (XII) nerve, in keeping with the presumed origin of the tongue muscle from the *occipital (postotic) myotomes*. Phylogenetically, the intrinsic tongue musculature is thought to be homologous with the *hypobranchial musculature* in lower vertebrates (fishes). This musculature consists of long projections from the occipitocervical myotomes which extend in a ventroanterior direction beneath the branchial apparatus. General sensation of the tongue correlates well with the branchial arch from which the mucosa was derived. Thus, the body of the tongue is innervated by the trigeminal (V) nerve. The root of the fully developed tongue is innervated by the sensory components of the glossopharyngeal (IX) and vagus (X) nerves. Overgrowth of the mucosa derived from the base of the second branchial arch by that of the third arch eliminates the second arch mucosa and its nerve supply from the root of the tongue.

The function of taste is accomplished by the taste buds, which form on the lingual papilla. The taste buds first appear during the seventh week of gestation as the result of an interaction between the fibers of special visceral afferent nerves (VII and IX) and the overlying epithelium. In many mammals taste buds are not confined to the tongue but are present in significant numbers on the mucosa of the soft palate, the nasoincisor duct, and the epiglottis (Mistretta, 1972). There is good evidence that the fetus is able to taste, and it has been hypothesized that the function of taste may be used by the fetus to monitor its intra-amniotic environment (Bradley and Mistretta, 1975).

Salivary Glands   The human salivary glands originate during the sixth and seventh weeks as ridgelike thickenings of the oral epithelium. Because of the relatively extensive epithelial shifts in the oral cavity, the germ-layer origins of the individual salivary glands are not definitely known, but most current opinions consider the parotid gland to be derived from ectoderm and the submandibular and sublingual glands to be endodermal structures.

The growth and morphogenesis of salivary glands is based upon continued interactions between the salivary epithelium and the surrounding mesenchyme. Surrounding the epithelial lobule in a developing salivary gland is a basal lamina containing glycosaminoglycans of different ages and compositions. Peripheral to the basal lamina in nonbranching parts of the lobule are bundles of collagen fibers (Fig. 13-6C). Branching of a primary lobule is accomplished by the contraction of well-ordered microfilaments within the epithelial cells and the deposition of collagen outside the basel lamina in the same area (Fig. 13-6D) Continued mitotic activity and the production of newly synthesized glycosamino-glycans at the tips of the secondary lobules ensure continued growth of the primordium of the gland (Bernfield et al., 1973).

DEVELOPMENT OF TEETH

In primitive vertebrates the teeth are smaller, more numerous, and distributed over much wider areas than is the case in mammals. In their simplest form they are plates with conical protruding tips consisting of a core of calcified material called *dentine* and an apical cap of much harder material called *enamel*. They are true dermal organs, as their dentine is formed by the connective-tissue layer of the skin, and their enamel by the epithelial layer. In the development of our own more highly specialized teeth it is interesting to see that the same dual origin (epithelium and underlying mesenchyme) is retained. Even though our teeth

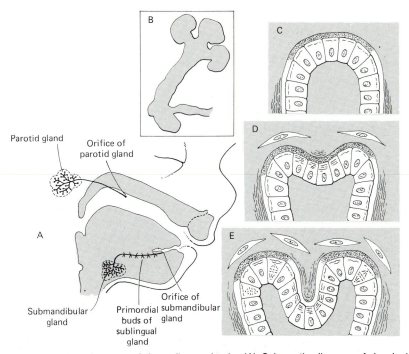

Fig. 13-6 Development of the salivary glands. (A) Schematic diagram of developing salivary glands in an 11-week human embryo. (B) Early salivary gland developing in culture. (C–E) Model depicting the relationship between cleft formation in the developing gland and the distribution of extracellular materials. *(After Bernfield, 1973, Am. Zool. 13:108.)* (C) Primary lobule showing the accumulation of newly synthesized glycosaminoglycans (stipple) within the basal lamina at the distal end of the lobule. Alongside the lobule collagen fibers (*wavy lines*) accumulate outside the basal lamina. (D) Early cleft formation within a developing lobule. Cleft formation is associated with contraction of bands of microfilaments within the cells at the apex of the lobule and the beginning of deposition of collagen fibers outside the basal lamina at the cleft. (E) Deepening of the cleft continues with an accentuation of the processes noted in D and a reduction of new glycosaminoglycan synthesis within the cleft.

start to form completely inside the gums instead of on a dermal surface, their enamel comes from specialized areas of epithelium which have grown down into the locations where the teeth are formed. Likewise their dentine comes from specialized mesenchymal cells, but of neural crest origin. When it is recalled that the epithelium which lines the tooth-forming part of the oral cavity is infolded stomodeal ectoderm, we can see that, highly specialized as they are as to both structure and development, our own teeth have retained fundamentally the same origin in ontogeny as they had in phylogeny.

Like so many other structures of dual origin, the development of teeth depends upon a reciprocal inductive interaction between epithelial and mesenchymal precursors (Koch, 1967). As a result of the inductive process, morphologically unspecialized ectodermal and mesenchymal cells of the oral region differentiate into highly organized complexes of secretory cells which themselves have only a transient existence. However, the products of their secretory activity, the teeth, are among the most morphologically stable elements of the body, and in many animals they persist throughout the entire lifetime of the individual.

With respect to morphogenetic control, the developing tooth shows a certain parallel to the limb bud, for in each system it is the mesenchymal component that determines form. For example, when the mesenchymal component of a molar tooth bud was combined in vitro with the epithelial component of an incisor, a molar tooth developed (Kollar and Baird, 1969). Conversely, after combining molar ectoderm with incisor mesenchyme, an incisor tooth took form. An intriguing experiment by Silbermann et al. (1977) demonstrated that the differentiative properties of dental mesenchyme are labile, whereas its morphogenetic properties are relatively stable. They combined mesenchyme from a molar tooth primordium of a mouse with ectoderm from the limb bud of a chick and then explanted the combined tissues onto the chorioallantoic membrane in a chicken egg. The explanted tissues grew into a structure having the shape of a tooth, but instead of being made of dentine, the normal differentiated end product of dental mesenchyme, the tooth was made of cartilage.

Dental Ledge   Local changes leading to tooth formation can be made out in the jaws of human embryos toward the close of the second month of development. By the seventh week a definite thickening of the oral epithelium can be seen on both the upper and lower jaw. This band of epithelial cells, which by the eighth week has begun to push into the underlying mesenchyme around the entire arc of each jaw, is known as the *dental ledge* (*dental lamina*) (Figs. 13-7 and 13-8A). Almost coincidently a cellular ingrowth slightly nearer the outside of the jaw begins to mark off the territory which is to become the lip from that which is to become the gum. This bandlike ingrowth of cells is known as the *labiogingival lamina* (Fig. 13-8A).

Enamel Organs   After the dental ledge is established, local buds arise from it at each point where a tooth is destined to be formed. Since these cells masses give rise to the enamel crown of the tooth, they are termed *enamel organs*. As would be expected, the enamel organs for the deciduous (milk) teeth are budded off from the dental ledge first, but the cell clusters which give rise to the enamel of the permanent teeth, although they arise later, are formed at a surprisingly early time (Fig. 13-11). They remain dormant, however, during the growth of the deciduous teeth and begin to develop actively only after the jaws have enlarged enough to accommodate the permanent dentition. The histogenetic processes involved in the formation of deciduous teeth and permanent teeth are essentially the same. The shape of the enamel organ suggests that of an inverted goblet, with the section of the dental ledge appearing like a somewhat distorted stem (Figs. 13-9 and 13-10). The epithelial cells lining the inside of the goblet early take on a columnar shape. Because they constitute the layer which secretes

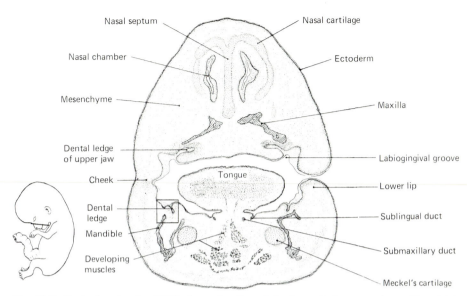

Fig. 13-7 Projection drawing (X20) of a section through the jaws of a human embryo of the eighth week; C–R 25 mm. *(University of Michigan Collection, EH 164.)* Outline, lower left, shows actual size of embryo, and the line across the jaws gives the location of the section. The rectangle about the dental ledge indicates the area represented at higher magnification in Figure 13-8, A.

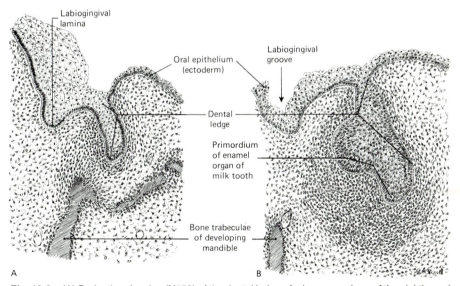

Fig. 13-8 (A) Projection drawing (X150) of the dental ledge of a human embryo of the eighth week. The drawing was made from the area indicated by the rectangle in Figure 13-7. (B) Projection drawing (X150) of a comparable area from a somewhat older embryo; C–R 30 mm. *(University of Michigan Collection, EH 15.)* Note appearance of primordium of enamel organ of the milk tooth as a local bud on the side of the dental ledge.

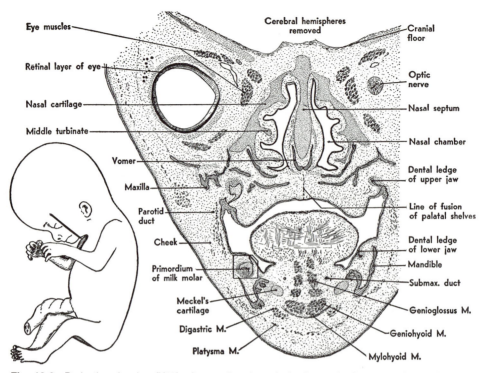

Fig. 13-9 Projection drawing (X10) of a section through the jaws of a human embryo of the eleventh week; C–R 58 mm. *(University of Michigan Collection, EH 198.)* Outline, left, shows actual size of embryo, and line across the jaws gives location of section. Small polygon enclosing primordium of milk molar indicates area represented at higher magnification in Figure 13-10.

the enamel cap of the tooth, they are called *ameloblasts* (enamel formers). The outer layer of the enamel organ is made up of closely packed cells which are at first polyhedral in shape but which soon, with the rapid growth of the enamel organ, become flattened. They constitute the so-called outer epithelium of the enamel organ. Between the outer epithelium and the ameloblast layer is a loosely aggregated mass of cells designated collectively, because of their characteristic appearance, the *enamel pulp*, or the *stellate reticulum* (Fig. 13-10).

Dental Papilla   Inside the goblet-shaped enamel organ there is a mass of mesenchymal cells of neural crest origin called the *dental papilla*. The dental papilla constitutes the primordium of the pulp of the tooth (Fig. 13-10). The cells of the dental papilla proliferate rapidly and soon form a very dense aggregation. A little later in development the enamel organ begins to assume the shape characteristic of the crown of the tooth it is to lay down (Fig. 13-11). At the same time the outer cells of the dental papilla take on a columnar form similar to that of the ameloblasts (Fig. 13-12). They are now called *odontoblasts* (dentine formers) because they are about to become active in the secretion of dentine. By

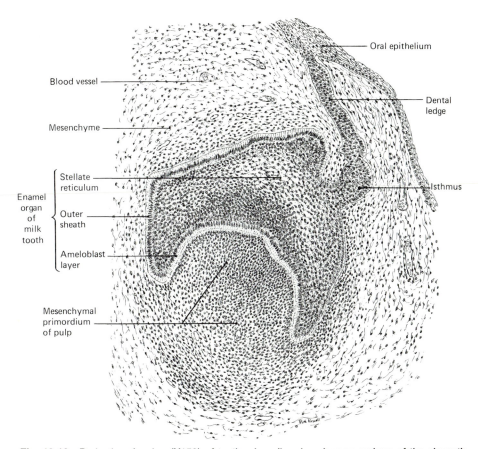

Fig. 13-10 Projection drawing (X150) of tooth primordium in a human embryo of the eleventh week. Location of area drawn is indicated by the rectangle in Fig. 13-9.

this time the dental ledge is losing its continuity with the oral epithelium, although traces of the original connection can still be identified in the mesenchyme at the lingual side of the tooth germ (Fig. 13-13).

Formation of Dentine With the preparatory developments complete, the tooth-forming structures are ready to go about the fabrication of dentine and enamel. As with bone, enamel and dentine are both composed of an organic matrix in which inorganic compounds are deposited. Although bone, dentine, and enamel are similar in having both organic and inorganic constituents in their matrix, with regard to composition and microscopic structure they are quite different in detail. Bone has approximately 45 percent organic material, dentine has only 28 to 30 percent, and enamel has less than 5 percent. Histologically they are totally unlike. Bone matrix has cells scattered through it in lacunae. Dentine has its cellular elements lying against one face, sending long processes into

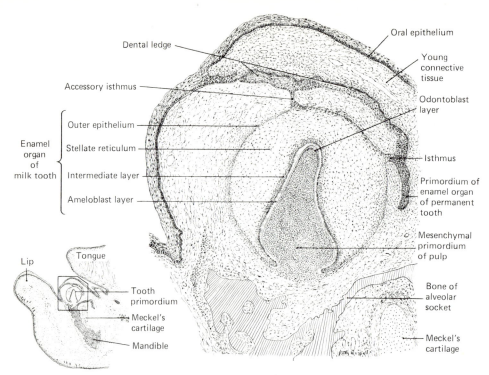

Figure. 13-11 Projection drawing (X50) of a parasagittal section of the lower jaw of a human embryo of the fourteenth week, passing through the primordium of a lower central incisor; C–R 104 mm. *(University of Michigan Collection, EH 145.)* The small sketch, lower left, indicates the relations of the area represented.

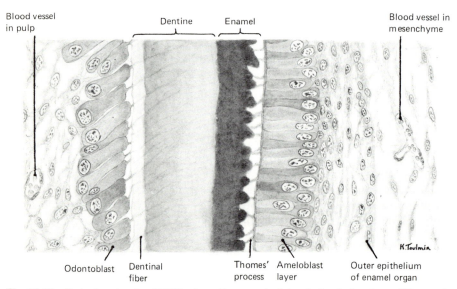

Fig. 13-12 Projection drawing (X425) of small segment of developing incisor from 130-mm pig embryo, showing formation of enamel and dentine. The conditions here illustrated are closely comparable with those seen in human embryos late in the fifth month.

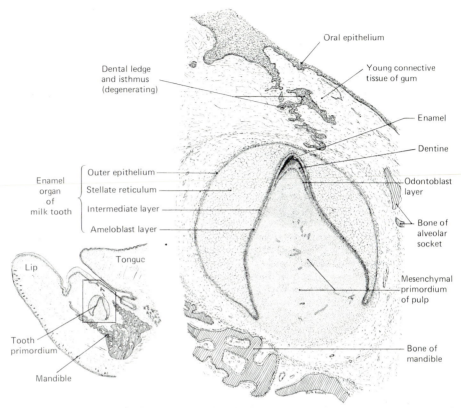

Fig. 13-13 Projection drawing (X40) of primordium of lower central milk incisor from a human embryo of 19 weeks' presumptive fertilization age; C–R 174 mm. *(University of Michigan Collection, EH 143.) (Orienting sketch, lower left, X5.)*

tubules in the matrix. Enamel is prismatic in structure, and the cells which formed the enamel are destroyed in the eruption of the tooth.

The first dentine is deposited against the inner face of the enamel organ, the odontoblasts drawing their raw materials from the small vessels in the pulp and secreting their finished product toward the enamel organ. As the odontoblasts continue to secrete additional dentine matrix, the accumulation of their own product inevitably forces the cell layer back, away from the material previously deposited. Strands of their cytoplasm become embedded in the material first laid down and are then pulled out to form the characteristic processes of the odontoblasts known as the *dentinal fibers* (Fig. 13-12).

Formation of the Enamel   While the dentine is being laid down by the cells of the odontoblast layer, the enamel cap of the tooth is being formed by the ameloblast layer of the enamel organ. As with the odontoblasts, the active cells of the ameloblast layer are columnar in shape (Fig. 13-12). The amount of organic material laid down as the framework of enamel is much less than that of

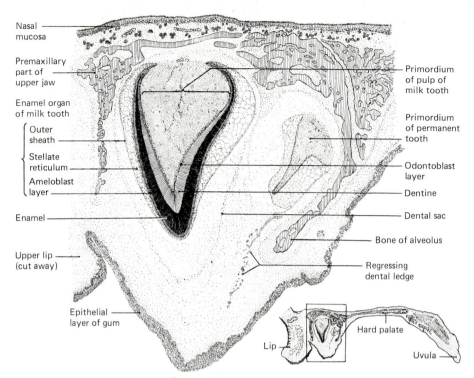

Fig. 13-14 Projection drawing (X8) of upper jaw of a human fetus at term, showing developing central incisor tooth. Orienting sketch in lower right-hand corner is actual size.

either bone or dentine; consequently, it is rather more difficult to make out its precise character and arrangement. Nevertheless, in decalcified sections it is possible to see delicate fibrous strands (*Thomes' fibers* or *processes*) projecting from the tips of the ameloblasts into the areas of newly formed enamel (Fig. 13-12). Enamel formation and dentine formation both begin at the tip of the crown (Figs. 13-13 and 13-14) and progress toward the root of the tooth. The entire crown is well formed before the root is completed, and it does not acquire its full length until the crown has entirely emerged.

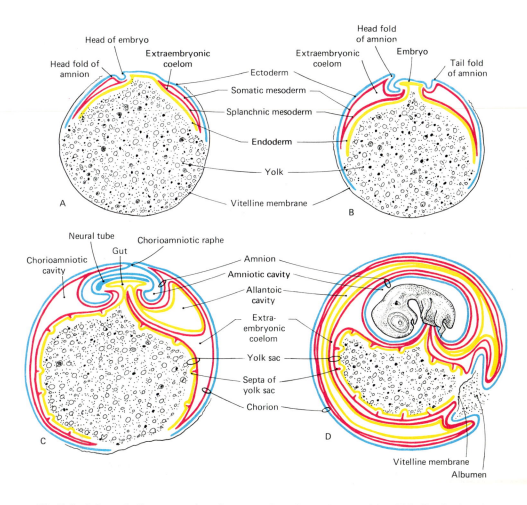

Fig. 7-4 Schematic diagrams to show the extraembryonic membranes of the chick. (D, after Lillie.) The embryo is cut longitudinally. The albumen, shell membranes, and shell are not shown; for their relations, see Fig. 7-1. (A) Embryo early in second day of incubation. (B) Embryo early in third day of incubation. (C) Embryo of 5 days. (D) Embryo of 9 days.

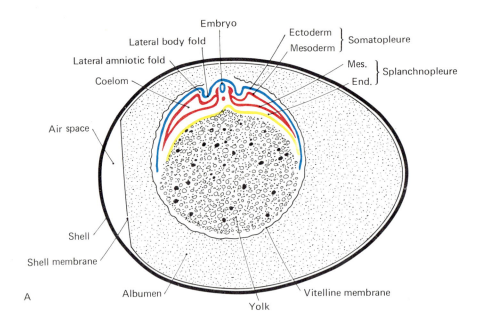

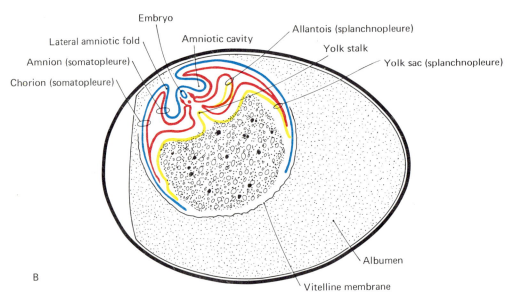

Fig. 7-1 Schematic diagrams to show the extraembryonic membranes of the chick. (After Duval.) The diagrams represent longitudinal sections through the entire egg. The body of the embryo, being oriented approximately at right angles to the long axis of the egg, is cut transversely. (A) Embryo of about 2 days' incubation. (B) Embryo of about 3 days' incubation. (C) Embryo of about 5 days' incubation. (D) Embryo of about 14 days' incubation.

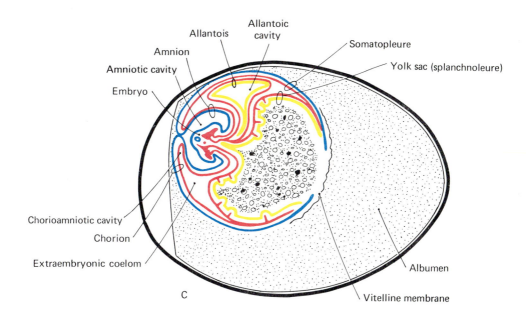

Allantoic cavity
Allantois
Amnion
Amniotic cavity
Embryo
Somatopleure
Yolk sac (splanchnoleure)
Chorioamniotic cavity
Chorion
Extraembryonic coelom
Albumen
Vitelline membrane

C

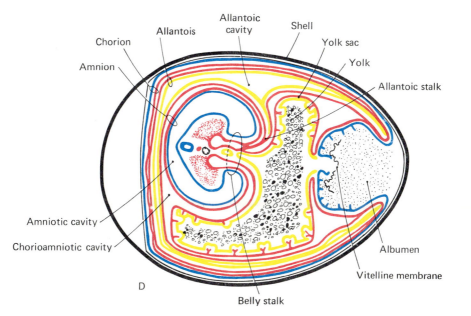

Allantoic cavity
Allantois
Chorion
Amnion
Shell
Yolk sac
Yolk
Allantoic stalk
Amniotic cavity
Chorioamniotic cavity
Albumen
Vitelline membrane
Belly stalk

D

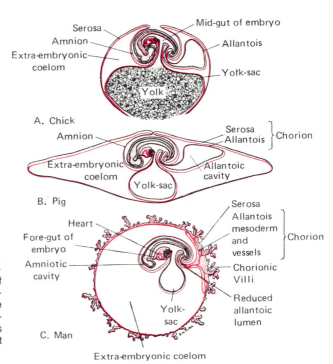

A. Chick

B. Pig

C. Man

Fig. 7-16 Diagrams showing inter-relations of embryo and extraembry-onic membranes characteristic of higher vertebrates. Neither the ab-sence of yolk from its yolk sac nor the reduction of its allantoic lumen radi-cally changes the human embryo's basic architectural scheme from that of more primitive types.

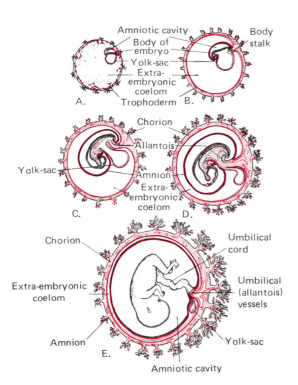

Fig. 7-29 Early changes in interre-lations of embryo and extraembryonic membranes.

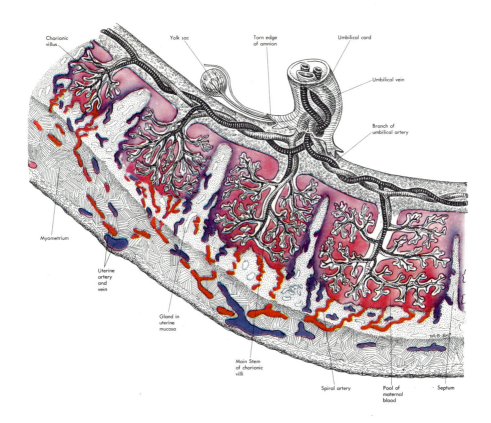

Fig. 7-25 Schematic diagram to show interrelations of fetal and maternal tissues in formation of placenta. Chorionic villi are represented as becoming progressively further developed from left to right across the illustration.

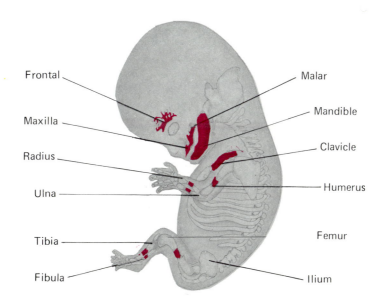

Fig. 9-14 Human embryo of 23½ mm (eighth week) stained with alizarin and cleared to show ossification centers. *(Projection drawing, X3, by Jane Schaefer. University of Michigan Collection, EH 105.)*

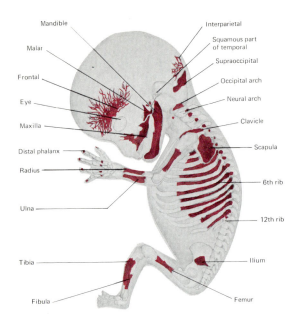

Fig. 9-15 Human embryo of 39 mm (nine weeks) stained with alizarin and cleared to show the progress of ossification. *(Projection drawing, X3, by Jane Schaefer. University of Michigan Collection, EH 149.)*

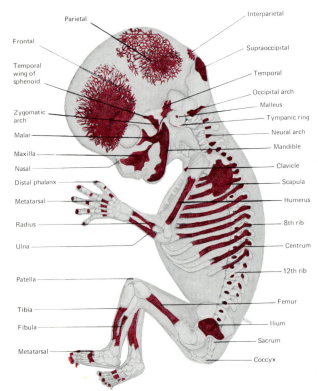

Fig. 9-16 Human embryo of 49 mm (tenth week) stained with alizarin and cleared to show the developing skeletal system. *(Projection drawing, X3 by Jane Schaefer, University of Michigan Collection, EH 70.)*

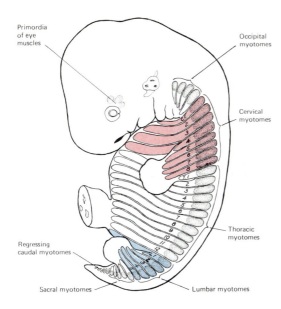

Fig. 9-19 Schematic diagram showing the regions into which the embryonic myotomes extend. The stippled area indicates the approximate size of the original mesodermic somite; the unstippled ventrolateral areas outlined opposite each somite suggest the territory into which the myotomes tend to extend. Compare with Fig. 9-20.

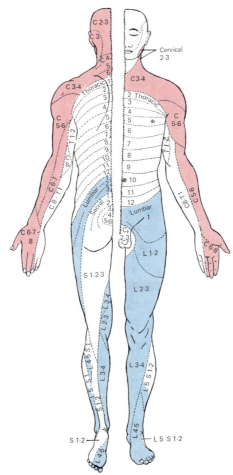

Fig. 9-20 Segmental areas of the adult body as indicated by distribution of cutaneous nerves. *(Adapted from various sources.)* Compare with Fig. 9-19.

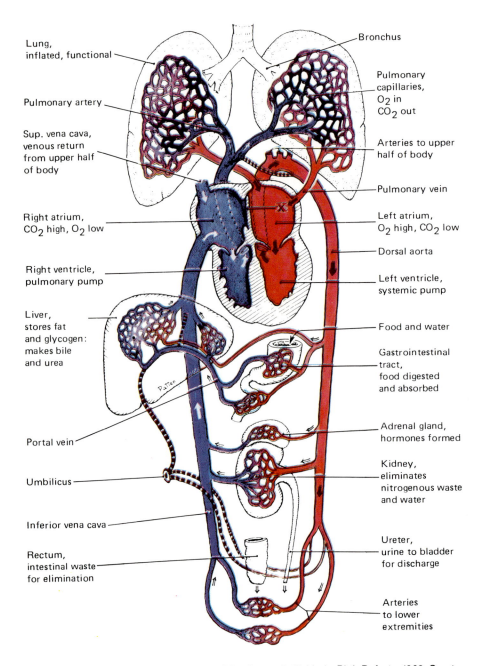

Lung,
inflated, functional

Bronchus

Pulmonary
capillaries,
O_2 in
CO_2 out

Pulmonary artery

Sup. vena cava,
venous return
from upper half
of body

Arteries to upper
half of body

Pulmonary vein

Right atrium,
CO_2 high, O_2 low

Left atrium,
O_2 high, CO_2 low

Dorsal aorta

Right ventricle,
pulmonary pump

Left ventricle,
systemic pump

Liver,
stores fat
and glycogen:
makes bile
and urea

Food and water

Gastrointestinal
tract,
food digested
and absorbed

Portal vein

Adrenal gland,
hormones formed

Umbilicus

Kidney,
eliminates
nitrogenous waste
and water

Inferior vena cava

Rectum,
intestinal waste
for elimination

Ureter,
urine to bladder
for discharge

Arteries
to lower
extremities

Fig. 17-1 Plan of the postnatal circulation. *(After Patten, in Fishbein,* Birth Defects, *1963. Courtesy of the National Foundation and the J.B. Lippincott Company, Philadelphia.)* The stippling has the same significance as in Fig. 17-35. The heavily cross-banded structures were important fetal vessels (cf. Fig. 17-35) which after birth ceased to carry blood and gradually became reduced to fibrous cords. The asterisk indicates the valvula foraminis ovalis in the closed position characteristic for postnatal life.

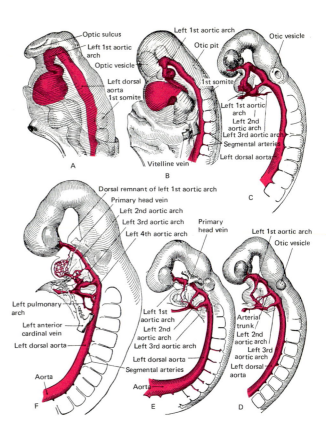

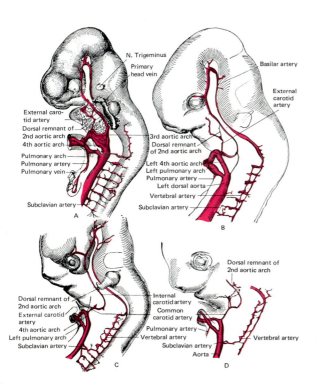

Fig. 17-7 Lateral aspect of vessels in the branchial region of pig embryos of various ages. (A) 10 somites; (B) 19 somites; (C) 26 somites; (D) 28 somites; (E) 30 somites; (F) 36 somites (6 mm). *(After Heuser,* Carnegie Cont. to Emb., *vol. 15, 1923.)* The drawings in this and the three following figures were made directly from injected specimens rendered transparent by treatment with oil of wintergreen.

Fig. 17-8 Continuation of the same series of lateral views of injected embryos as shown in figures 17-7. (A) 14 mm; (B) 17 mm; (C) 19.3 mm; (D) 20.7 mm.

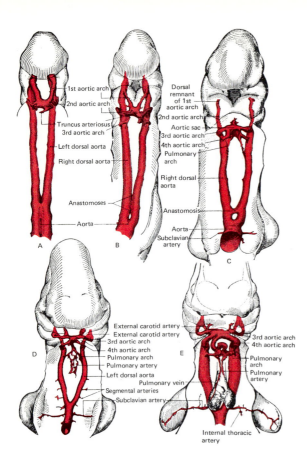

1st aortic arch
2nd aortic arch
Truncus arteriosus
3rd aortic arch
Left dorsal aorta
Right dorsal aorta
Anastomoses
Aorta
A

Dorsal remnant of 1st aortic arch
2nd aortic arch
3rd aortic arch
B

Aortic sac
2nd aortic arch
3rd aortic arch
4th aortic arch
Pulmonary arch
Right dorsal aorta
Anastomosis
Aorta
Subclavian artery
C

External carotid artery
External carotid artery
3rd aortic arch
4th aortic arch
Pulmonary arch
Pulmonary artery
Left dorsal aorta
Segmental arteries
Subclavian artery
D

3rd aortic arch
4th aortic arch
Pulmonary arch
Pulmonary vein
Pulmonary artery
Internal thoracic artery
E

Fig. 17-9 Ventral aspect of vessels in the branchial region of pig embryos of various ages. (A) 24 somites; (B) 4.3 mm; (C) 6 mm; (D) 8 mm; (E) 12 mm. *(After Heuser,* Carnegie Cont. to Emb., *vol. 15, 1923.)*

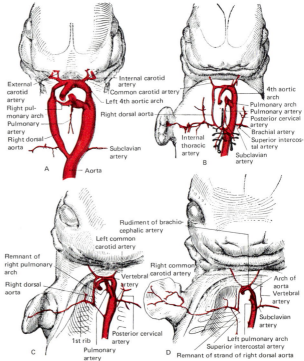

External carotid artery
Right pulmonary arch
Pulmonary artery
Right dorsal aorta
A

Internal carotid artery
Common carotid artery
Left 4th aortic arch
Right dorsal aorta
Subclavian artery
Aorta

4th aortic arch
Pulmonary arch
Pulmonary artery
Posterior cervical artery
Brachial artery
Superior intercostal artery
Subclavian artery
Internal thoracic artery
B

Remnant of right pulmonary arch
Right dorsal aorta
1st rib
Posterior cervical artery
Pulmonary artery
Vertebral artery
Left common carotid artery
Rudiment of brachiocephalic artery
C

Right common carotid artery
Arch of aorta
Vertebral artery
Subclavian artery
Left pulmonary arch
Superior intercostal artery
Remnant of strand of right dorsal aorta
D

Fig. 17-10 Continuation of the series of injected embryos shown in figure 17-9. (A) 14 mm; (B) 17 mm; (C) 19.3 mm; (D) 20.7 mm.

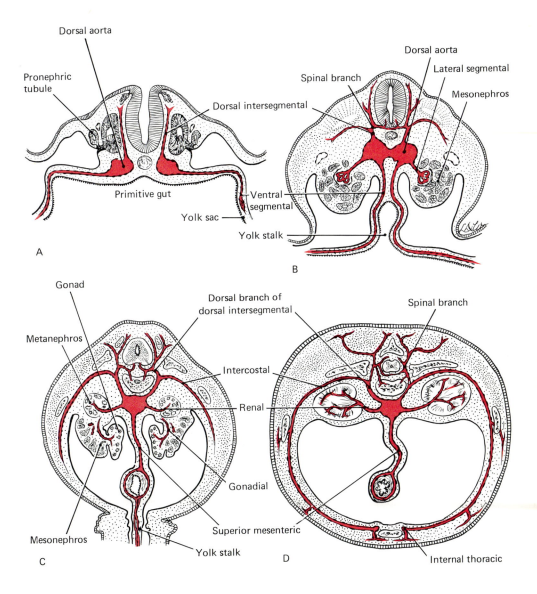

Fig. 17-11 Cross-sectional plans of the body showing relations of segmental branches of aorta at different stages of development.

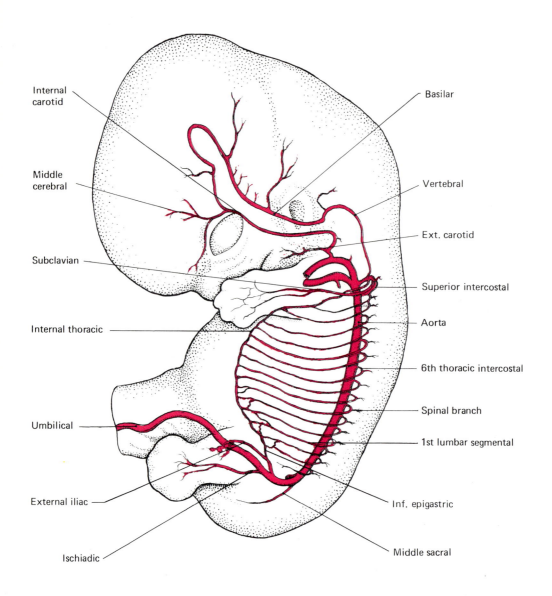

Internal carotid

Middle cerebral

Subclavian

Internal thoracic

Umbilical

External iliac

Ischiadic

Basilar

Vertebral

Ext. carotid

Superior intercostal

Aorta

6th thoracic intercostal

Spinal branch

1st lumbar segmental

Inf. epigastric

Middle sacral

Fig. 17-12 Arteries of the body wall in a human embryo of 7 weeks. *(Modified after Mall.)*

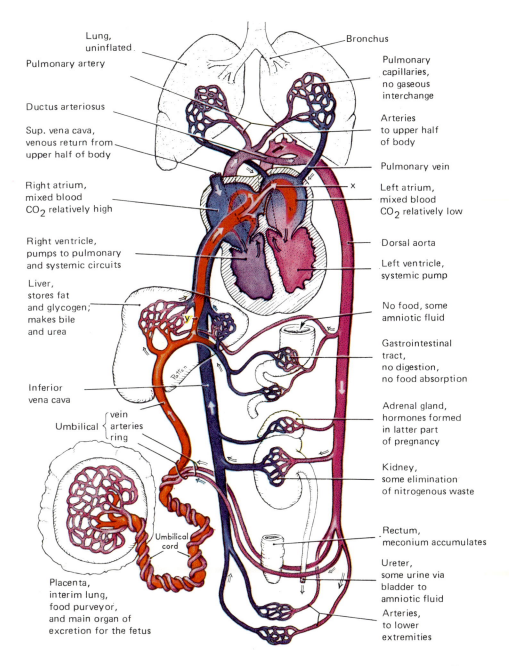

Lung, uninflated

Pulmonary artery

Ductus arteriosus

Sup. vena cava, venous return from upper half of body

Right atrium, mixed blood CO_2 relatively high

Right ventricle, pumps to pulmonary and systemic circuits

Liver, stores fat and glycogen; makes bile and urea

Inferior vena cava

Umbilical { vein, arteries, ring

Placenta, interim lung, food purveyor, and main organ of excretion for the fetus

Umbilical cord

Bronchus

Pulmonary capillaries, no gaseous interchange

Arteries to upper half of body

Pulmonary vein

Left atrium, mixed blood CO_2 relatively low

Dorsal aorta

Left ventricle, systemic pump

No food, some amniotic fluid

Gastrointestinal tract, no digestion, no food absorption

Adrenal gland, hormones formed in latter part of pregnancy

Kidney, some elimination of nitrogenous waste

Rectum, meconium accumulates

Ureter, some urine via bladder to amniotic fluid

Arteries, to lower extremities

Fig. 17-35 Plan of the fetal circulation at term. The ductus venosus in the liver is marked by †. The arrow in the heart marked by * indicates the passage of blood from the right atrium to the left as it occurs during atrial diastole. This flow pushes the valvula into the open positions here represented. When the atria contract, the valvula moves back against the septum, closing the foramen ovale against the return flow and thus forcing all the blood in the left atrium to enter the left ventricle. (*After Patten, in Fishbein, 1963,* Birth Defects, *Courtesy of the National Foundation and the J.B. Lippincott Company, Philadelphia.*)

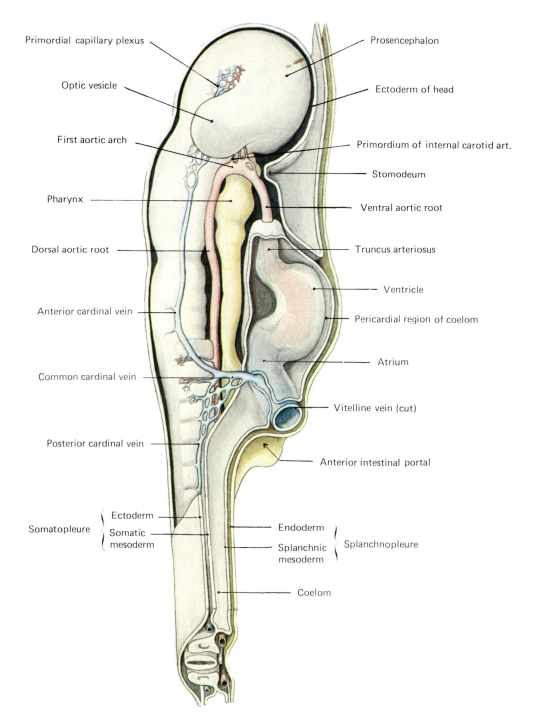

Primordial capillary plexus

Optic vesicle

First aortic arch

Pharynx

Dorsal aortic root

Anterior cardinal vein

Common cardinal vein

Posterior cardinal vein

Somatopleure
{ Ectoderm
{ Somatic
{ mesoderm

Prosencephalon

Ectoderm of head

Primordium of internal carotid art.

Stomodeum

Ventral aortic root

Truncus arteriosus

Ventricle

Pericardial region of coelom

Atrium

Vitelline vein (cut)

Anterior intestinal portal

Endoderm }
Splanchnic { Splanchnopleure
mesoderm }

Coelom

Fig. A-26 Diagrammatic lateral view of dissection of a 38-hour chick. The lateral body wall of the right side has been removed to show the internal structures. Note especially the relations of the pericardial region to that part of the coelom which lies farther caudally and the small anastomosing channels of the developing posterior cardinal vein from which a single main vessel is later derived.

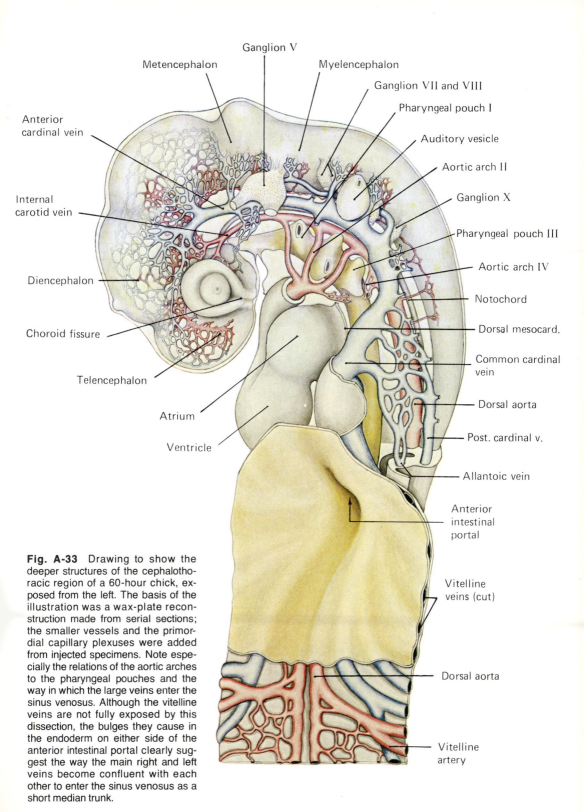

Ganglion V

Metencephalon

Myelencephalon

Ganglion VII and VIII

Pharyngeal pouch I

Auditory vesicle

Aortic arch II

Ganglion X

Pharyngeal pouch III

Aortic arch IV

Notochord

Dorsal mesocard.

Common cardinal vein

Dorsal aorta

Post. cardinal v.

Allantoic vein

Anterior intestinal portal

Anterior cardinal vein

Internal carotid vein

Diencephalon

Choroid fissure

Telencephalon

Atrium

Ventricle

Vitelline veins (cut)

Dorsal aorta

Vitelline artery

Fig. A-33 Drawing to show the deeper structures of the cephalothoracic region of a 60-hour chick, exposed from the left. The basis of the illustration was a wax-plate reconstruction made from serial sections; the smaller vessels and the primordial capillary plexuses were added from injected specimens. Note especially the relations of the aortic arches to the pharyngeal pouches and the way in which the large veins enter the sinus venosus. Although the vitelline veins are not fully exposed by this dissection, the bulges they cause in the endoderm on either side of the anterior intestinal portal clearly suggest the way the main right and left veins become confluent with each other to enter the sinus venosus as a short median trunk.

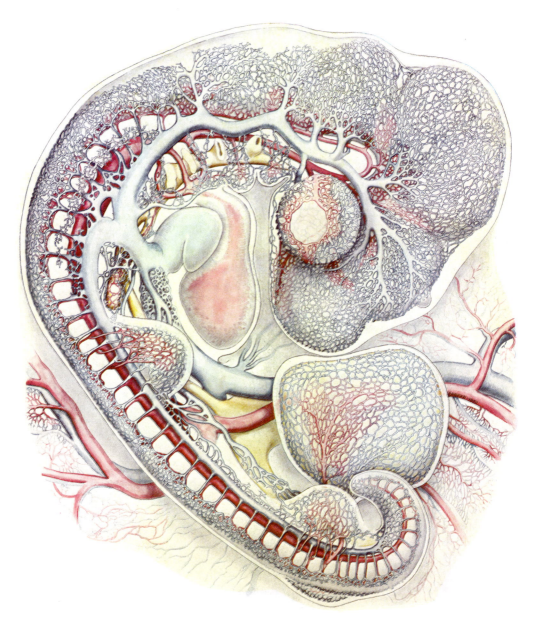

Fig. A-47 The blood vessels of a 4-day chick. The basis of the illustration was the same wax-plate reconstruction from which Figure A-46 was drawn. The smaller vessels and the primordial capillary plexuses were added from injected specimens. The labeled diagram of Figure A-46 will serve as a means of identifying the main vessels.

Chapter 14

Development of the Digestive and Respiratory Systems and the Body Cavities

THE DIGESTIVE SYSTEM

Early in the period of organogenesis (4-week human or 5-mm pig embryo) the digestive system consists of a relatively simple tube (Fig. 8-10D), bounded at the cranial end of the foregut by the degenerating stomodeal plate (Fig. 8-13 and 14-1A) and at the caudal end of the hindgut by the still-intact cloacal membrane. The midgut retains an open ventral connection with the yolk sac through the yolk stalk. At this stage the allantois is a prominent diverticulum from the ventral surface of the hindgut (Figs. 8-10D and 8-11D). Along the cranial half of the gut may be seen the earliest primordia of the glands associated with the digestive tract. So many new structures arise in the region of the pharynx that their embryology will be considered in a separate chapter (Chap. 15).

Further development of the gut proper involves the processes of elongation, herniation of part of the gut into the body stalk, rotation of several local regions of gut, and, finally, histogenesis and functional maturation. While these are taking place, the primordial digestive glands and respiratory system are growing out in complex branching patterns as the result of interactions between local gut endoderm and its enveloping mesoderm, with the deep involvement of extracellular matrix material produced by these tissues.

Esophagus   The pharynx becomes abruptly narrowed just caudal to the most posterior pouches. It is at this point that the primitive gut gives rise ventrally to the tracheal outgrowth (Figs. 8-13 and 11-11). The region of narrowing where the trachea becomes confluent with the gut tract may be regarded as the posterior limit of the pharynx. From this point to the dilation which marks the beginning of the stomach, the gut remains of relatively small and uniform diameter and becomes the esophagus (Fig. 14-1). The original

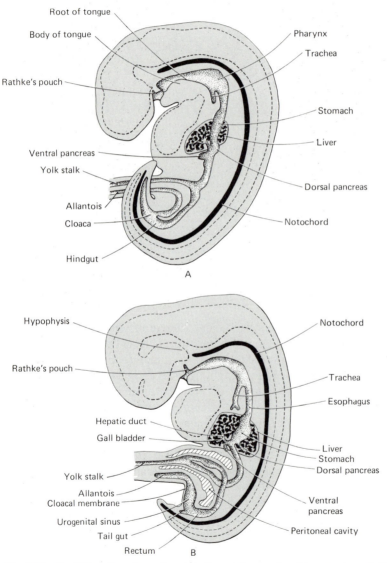

Fig. 14-1 Sagittal sections through early human embryos, showing early stages in the formation of the gut and associated structures. (A) Early in the fifth week. (B) Early in the sixth week.

endodermal lining of the primitive gut gives rise only to the epithelial lining of the esophagus and its glands. During the seventh and eighth embryonic weeks, the human esophageal epithelium proliferates and nearly occludes the lumen. In a number of species (e.g., the chick) the esophagus becomes completely occluded, only to reopen later as the result of cell degeneration within the epithelium. The connective tissue and muscle coats of the esophagus are derived from mesenchymal cells which gradually become concentrated about the original epithelial tube (Fig. 15-8D).

Stomach The region of the primitive gut destined to become the stomach is more or less clearly marked by a dilation (Fig. 14-1). Its shape, even at this early stage, is strikingly suggestive of that of the adult stomach. Its position, however, is quite different.

In young embryos the stomach is mesially placed with its cardiac (esophageal) end somewhat more dorsal than its pyloric (intestinal) end. It is slightly curved in shape, with the convexity facing dorsally and somewhat caudally and the concavity facing ventrally and somewhat cephalically. The positional changes by which it reaches its adult relations involve two principal phases: (1) the stomach is bodily shifted in position so that its long axis no longer lies in the sagittal plane of the embryo but diagonally across it; and (2) there is a concomitant rotation of the stomach about its own long axis so that its original dorsoventral relations are also altered. These changes in position are schematically indicated in Figs. 14-2 and 14-3. The shift in axis takes place in such a manner that the cardiac end of the stomach comes to lie to the left of the midline and the pyloric end to the right. Meanwhile rotation has been going on. In following the progress of rotation the best point of orientation is the line of attachment of the primary dorsal mesentery (for discussion of mesenteries, see page 419). While the stomach occupies its original mesial position, the mesentery is attached to it middorsally, along its convex curvature (Fig. 14-14). As the stomach continues to grow in size and depart from the sagittal plane of the body, it rotates about its own long axis. The convex surface to which the mesentery is attached, and which was at first directed dorsally, now swings to the left. Since the long axis of the stomach has in the meantime been acquiring an inclination, the greater curvature of the stomach comes to be directed somewhat caudally as well as to the left (Fig. 14-2E).

Histogenesis of the human gastric mucosa begins toward the end of the second month with the appearance of folds (*rugae*) and the first gastric pits. During the next few weeks the formation of gastric pits and the glands associated with them spreads throughout the wall of the stomach. Many of the specific cells involved in secretion begin to differentiate both morphologically and cytochemically early in the fetal period, but neither hydrochloric acid nor pepsin are present in the gastric contents until near term.

Omental Bursa The change in position of the stomach necessarily involves change in that part (*dorsal mesogastrium*) of the primary mesentery which

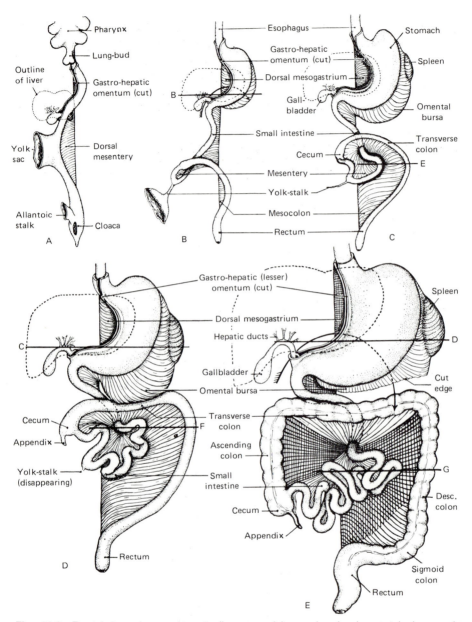

Fig. 14-2 Frontal-view plans, schematically summarizing major developmental changes in position of gut and in relations of mesenteries. *(Adapted from a number of sources.)* Crosshatched areas in E indicate the part of mesentery of the duodenum and the parts of the mesocolon which become fused to body wall. The heavy lines marked B, C, D, E, F, and G indicate the locations of the cross sections designated by the same letters in Fig. 14-3.

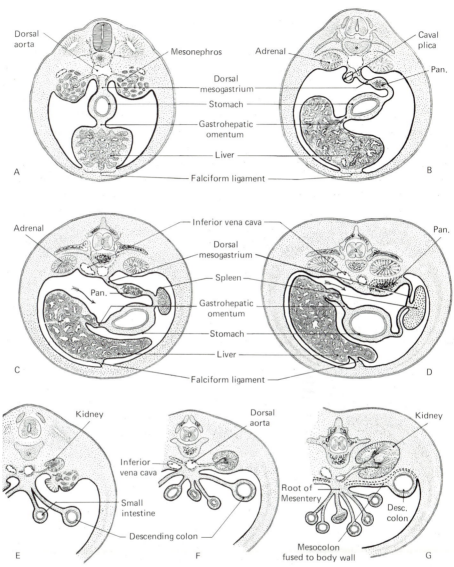

Fig. 14-3 Cross-sectional plans of developing body to show changes in relations of mesenteries. *(Adapted from several sources.)* (A)–(D) Sections at level of stomach and liver to show formation of omental bursa. (E)–(G) Sections at level of kidney to show fusion of parts of mesocolon to body wall. This illustration should be studied in comparison with Fig. 14-2, which shows corresponding stages in frontal plan. The arrows in C and D indicate the epiploic foramen.

suspends it in the body cavity. The dorsal mesogastrium is pulled after the stomach and forms a pouch, known as the *omental bursa* (Fig. 14-2C and D). The opening from the general peritoneal cavity into the bursa is known as the *epiploic foramen (foramen of Winslow).* (See arrow in Fig. 14-3C and D).

Intestines   The primitive gut is at first a fairly straight tube extending throughout the length of the body. Near its midpoint it opens ventrally into the yolk sac (Figs. 8-10D and 8-13). The first conspicuous departure from this condition is the formation of a hairpin-shaped loop in the future intestinal region. The closed end of this loop extends into the belly stalk (Fig. 14-14). The yolk stalk connects with the gut at the bend of the loop and forms an excellent point of orientation in following the series of foldings and kinkings by which the definitive configuration of the intestinal tract is established. The attachment of the yolk stalk is just cephalic to what will be the point of transition from small to large intestine. Thus all the gut between the yolk stalk and the stomach becomes small intestine, and except for about 2 feet of the terminal part of the small intestine, the gut caudal to the yolk stalk forms the large intestine.

The positional changes which bring about adult relationships are initiated by the throwing of a twist in the primary U-shaped bend of the gut which extends into the belly stalk. Viewed in ventral aspect, the twist in the gut tract is counterclockwise (Fig. 14-2B and C). The immediate result of this twist is to bring a considerable proportion of the original cephalic limb of the gut loop into a position in the belly posterior to the segment of the caudal limb which was twisted across it (Fig. 14-2D). This initial twist is the primary factor in establishing the fundamental positional relations of the large and small intestines. We can recognize immediately in the crossing segment of the caudal limb of the gut loop what we know in the adult as the *transverse colon*. We can see, also, just how it comes about that the adult jejunum and ileum lie in the abdomen below the level of the transverse colon (Fig. 14-2E).

The coiling so characteristic of the small intestine begins soon after the primary twist in the gut loop has occurred (Fig. 14-2D). That portion of the cephalic limb of the primary loop which emerges below the transverse colon is destined to become jejunum and ileum. This part of the intestine now begins to increase in length exceedingly rapidly and consequently becomes freely coiled on itself. The coiling begins while the twisted primary gut loop still projects out into the extraembryonic coelom of the belly stalk, giving the embryo at this age the appearance of having an umbilical hernia. By about the tenth week of human development the abdomen has enlarged sufficiently to accommodate the entire intestinal tract, and the protruding part of the intestinal loop moves back through the umbilical ring into its definitive position within the peritoneal cavity. In this retraction the coils of small intestine tend to slip into the abdominal cavity ahead of the protruding part of the colon. In doing so they crowd to the left the lower part of the colon which has remained from the first in the abdominal cavity. This establishes the descending colon in its characteristic position close against the body wall on the left (Figs. 14-2D and E, and 14-3F and G). When the upper part of the colon which projected into the belly stalk is finally drawn into the peritoneal cavity, its cecal end swings to the right and downward (Fig. 14-2E).

The duodenal epithelium begins a phase of rapid proliferation early in the second month, and by 6 to 7 weeks the exuberant epithelial growth temporarily occludes the lumen. By the end of the second month the continuity of the intestinal lumen is usually restored. At about this time aggregates of mesodermal cells begin to invade the stratified intestinal epithelium, which is developing small secondary lumina beneath its surface (Fig. 14-4). Coalescence of secondary lumina and continued mesodermal upgrowth results in the formation of the minute fingerlike intestinal villi, which greatly increase the absorptive surface of the intestine. Specific intestinal cell types begin to differentiate during villus formation. Histochemical studies have shown that many of the intestinal

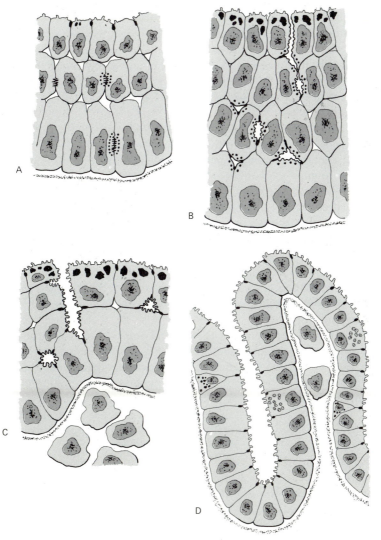

Fig. 14-4 Steps in the formation of duodenal villi in the fetal rat. (A) 15–16 days; (B) 17 days; (C) 18 days. *(After Mathan et al., 1976, Am. J. Anat. 146:73.)*

secretions are formed in small amounts within the mucosa during the fetal period, but secretion into the gut occurs later, and only in small amounts until birth. The intestines of mammalian fetuses contain a greenish material, called *meconium*, which consists of desquamated cells from the gut, hairs, and other materials swallowed along with the amniotic fluid and bile secretion.

Cecum and Appendix　The originally inconspicuous cecal dilation, which by the sixth week indicates the point of junction of small and large intestines, is destined to undergo marked local specialization. By 2 months changes in the position of the gut tract make the small intestine enter the large intestine almost at right angles instead of in alignment as was the case at first. At this angular junction the colon develops a diverticulum which is known as the *cecum* (Fig. 14-2C). For a time the cecum as a whole continues to enlarge. By the third month, however, its distal end begins to lag behind in growth, so that it appears as an extension which is definitely smaller in diameter. This slender extremity of the cecal diverticulum is the *vermiform appendix* (Fig. 14-2D and E).

Rectum and Anus　The attainment of adult conditions at the extreme caudal end of the digestive tract is so intimately associated with the development of the urogenital openings that changes in the cloacal region as a whole can more profitably be taken up later, in connection with the reproductive organs.

Liver　Very early in development the endodermal *hepatic diverticulum* from the floor of the foregut extends into the mesenchyme of the *septum transversum* (Fig. 14-13). This early hepatic outgrowth represents the first morphological evidence of a series of inductive processes which had already begun at an earlier embryonic stage (Fig. 14-5). The original diverticulum, in embryos of 5 to 6 mm, has become clearly differentiated into several parts (Figs. 8-13 and 14-6). A maze of branching and anastomosing cell cords grows out from it ventrally and cephalically.

In addition to the mesenchyme of the septum transversum, any mesenchyme derived from either the splanchnopleural or somatopleural components of the lateral plate mesoderm is capable of supporting continued hepatic outgrowth and differentiation. In contrast axial mesoderm allows only minimal survival and little progression of development of hepatic endoderm (Le Douarin, 1975). The distal portions of the hepatic cords give rise to the secretory tubules of the liver, and their proximal portions form the hepatic ducts. A dilation which is the primordium of the *gallbladder* originates from where the hepatic ducts become confluent. Closer to the gut tract is a separate outgrowth of cells which constitutes the ventral primordium of the pancreas.

The branching and anastomosing tubules which are distal continuations of the hepatic ducts constitute the actively secreting portion of the liver. Their position and extent in embryos of various ages are shown in Figs. 8-13, 11-11, 14-7C, and 14-14. The organization of these secreting units in the liver is quite characteristic. The *hepatic tubules* are not packed so closely together in a

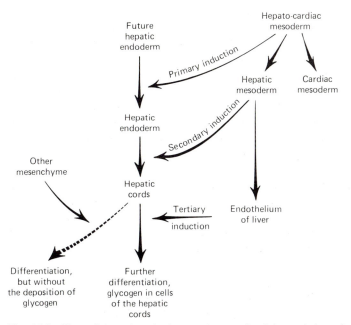

Fig. 14-5 Tissue interactions in the morphogenesis of the endodermal component of the liver. *(Modified from Deuchar, after Wolff.)*

framework of dense connective tissue as is usually the case in massive glands. Surprisingly little connective tissue is formed between them, and the intertubular spaces become pervaded by a maze of dilated and irregular capillaries known as *sinusoids*.

The development of the liver from early endodermal buds to its mature form involves not only an increase in mass and structural complexity but also a gradual acquisition of the metabolic pathways which enable it to carry out its manifold functions in postnatal life. A major function of the liver is the synthesis and storage of glycogen, which serves as a carbohydrate reserve for the entire body. The embryonic liver, particularly during the late fetal period, actively stores glycogen, and in mammals, this function has been shown to be under the control of adrenocortical steroid hormones in conjunction with an influence by the pituitary (Jost, 1962). The system of enzymes involved in the synthesis of urea from nitrogenous metabolites gradually assumes prominence in the fetal liver and attains its full functional capacity close to the time of birth. As we shall see later, the embryonic liver also serves as a transient site of blood-cell formation.

The great synthetic activity of the postnatal liver is served by an unusual adaptation of the vascular system, the *portal vein* (page 507). This is a vein which arises from a confluence of smaller veins and their capillaries along much of the intestinal tract and which breaks up into a capillarylike network within the parenchymal substance of the liver. By this arrangement, the postnatal liver is

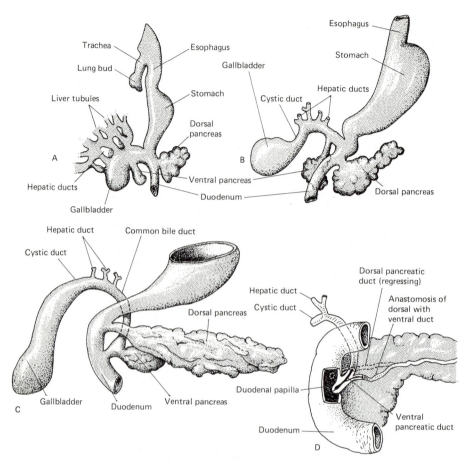

Fig. 14-6 Development of hepatic and pancreatic primordia. All diagrams are viewed from the ventral aspect. (A) Semischematic diagram based, in part, on Thyng's reconstructions of a 5.5-mm pig embryo. This is comparable to a human embryo early in the fifth week. (B) Reconstruction from a 9.4-mm pig embryo. This is comparable to a human embryo of 7 weeks. (C) Schematized drawing equivalent to a 20-mm pig embryo or a human embryo of 7 weeks. (D) Manner in which common bile duct and pancreatic duct become confluent in the ampulla of Vater and discharge through the duodenal papilla.

the first organ to receive blood rich in protein, carbohydrate, and fat metabolites absorbed through the intestinal walls. The mammalian embryo possesses a different anatomical adaptation which fulfills a similar function. In embryonic life food materials are not carried through the portal vein because the digestive tract is nonfunctional with respect to nutrition. Rather, the umbilical vein, arising from the placenta, carries foodstuffs extracted from the maternal blood to the embryo. The umbilical vein goes directly to the embryonic liver, and much of its blood empties into a hepatic capillary network. By this adaptation in its circulation, the embryonic liver is thus afforded first access to metabolites essential for its synthetic functions.

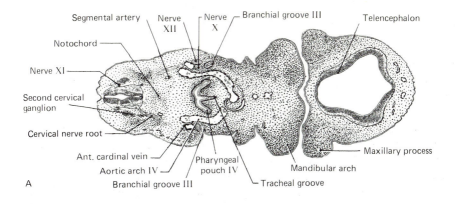

A

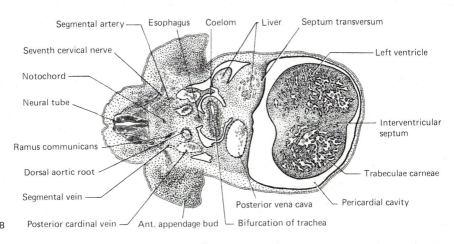

B

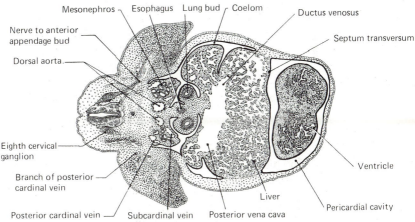

C

Fig. 14-7 Three transverse sections from the series of the 9.4-mm pig embryo used in making the reconstruction illustrated in Figure 17-13. By laying a straightedge across this reconstruction at the level of the marginal line bearing the serial number of a cross section, the relations of that section within the body as a whole are precisely indicated. (A) Section 180 at the level of the laryngotracheal groove; (B) section 293 at the level of the bifurcation of the trachea; (C) section 309 through the cephalic part of the liver and the lung buds.

Pancreas   The pancreas makes its appearance in the same region and at about the same time as the liver. It is derived from two separate primordia which later become fused. One primordium arises dorsally, directly from the duodenal endoderm; the other arises ventrally, from the endoderm of the hepatic diverticulum (Fig. 14-6). As they increase in size, these two buds approach each other and eventually fuse (Fig. 14-6C).

Like the salivary glands and the other organs which bud off from the gut, the appearance of the pancreas is the result of a coordinated interplay between localized endodermal cells and their surrounding mesenchyme. The earliest stages in pancreatic development are not well understood, but in the very early gut, before obvious differentiation has occurred, a small population of approximately 300 cells (Wessels, 1977) has become specified to form pancreas. The glandular epithelium of the pancreas is formed by the budding and rebudding of cords of cells derived from this population of primordial cells (Fig. 14-8). Although surrounding mesenchyme is required for epithelial outgrowth and branching, the developing pancreas, in contrast to the liver, seems content with mesoderm from a variety of sources, as shown by in vitro recombination experiments.

The mature pancreas is a dual organ, consisting of an exocrine and an endocrine portion, the latter being embodied in the million or so small clusters of secretory cells, the *islets of Langerhans*, which are scattered among the acini of the exocrine portion. The acinar cells, the principal secretory cells of the exocrine part of the pancreas, produce a variety of digestive enzymes which are carried into the small intestine via a system of ducts. The islets consist of several varieties of cells, the most prominent of which are the glucagon-secreting α cells and the insulin-secreting β cells. These hormones are secreted directly into the capillaries, which provide a rich blood supply to the islets. Insulin lowers and glucagon raises the levels of glucose in the blood.

In the developing pancreas there is a close correlation between morphogenesis and the synthesis of its specific secretory proteins. Rutter has described several phases of maturation (Fig. 14-8). The first phase precedes outgrowth and consists principally of establishing the primordial population of pancreatic cells. A transition to the second level of maturation occurs with the appearance of the original pancreatic diverticulum. This phase is characterized by the synthesis of low levels of many of the hydrolytic enzymes of the exocrine cells as well as low levels of insulin and relatively high levels of glucagon. During this phase mesenchymal cells form a close association with the budding epithelial cells. The third phase of differentiation involves the production of an elaborate protein-synthesizing and secreting mechanism by the acinar cells and a marked increase in the synthesis of digestive enzymes. Meanwhile, the islets of Langerhans are forming, probably by budding off of the developing acini. Both the α and β cells elaborate large numbers of secretory granules containing glucagon and insulin, and some quantity of the newly synthesized hormones enters into the fetal circulation.

In the fetus, the duct systems of the exocrine part of the pancreas drain into

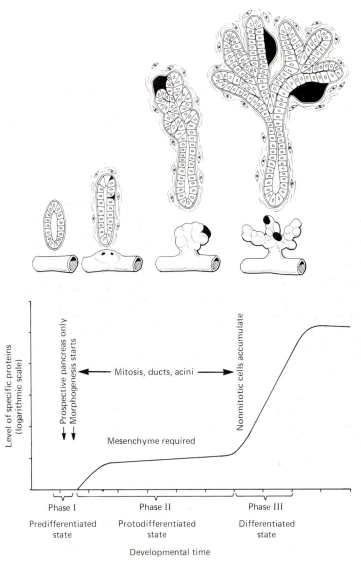

Fig. 14-8 Stages of structural and functional differentiation of the pancreas *(After Rutter)*. The graph and drawings stress the exocrine component of the pancreas. The dark areas in the drawings represent primitive islets. *(Drawings modified from Rutter, 1972, in* Handbook of Physiology. Endocrinology I, *p. 32.)*

a main duct running most of the length of the dorsal pancreas. In humans, the dorsal pancreatic duct becomes anastomosed with the main duct of the original ventral component of the pancreas and through it (the duct of Wirsung) the pancreatic enzymes are emptied into the duodenum. The terminal part of the dorsal pancreatic duct regresses (Fig. 14-6D). In some species (e.g., dog and horse) there are two ducts—the ventral pancreatic duct and the nonregressed dorsal pancreatic duct (duct of Santorini), which also empties into the duodenum.

THE RESPIRATORY SYSTEM

Trachea The first indication of the differentiation of the respiratory system is the formation of the *laryngotracheal groove*. This midventral furrow in the primitive gut tract appears at the posterior limit of the pharyngeal region (Figs. 8-14D and 15-2). As it becomes deepened, it is constricted off from the gut except for a narrowed communication cephalically. Once established as a separate diverticulum, it grows caudad as the trachea, ventral to, and roughly parallel with, the esophagus (Figs. 8-14E and 11-11).

The anatomical relations of the trachea in the embryo, even in early stages, are quite similar to those in the adult. We can recognize its communication with the posterior part of the pharynx as the future glottis (Fig. 13-5C and D) and the slightly dilated portion of the embryonic trachea just caudal to the glottis as foreshadowing the larynx.

Only the epithelial lining of the adult trachea is derived from foregut endoderm. The cartilage, connective tissue, and muscle of its wall are formed by mesenchymal cells which become massed about the growing endodermal tube (Figs. 15-8D and 17-14C).

Bronchi and Lungs As the tracheal outgrowth lengthens, it bifurcates at its caudal end to form the two lung buds (Fig. 14-9A and B). These in turn continue to grow and rebranch, giving rise to the bronchial trees of the lungs (Fig. 14-9C to 14-9F). The characteristic budding pattern of the endodermal bronchial tree is the result of a continuous inductive effect by the surrounding mesoderm. In the absence of its mesodermal investment, budding of the bronchial tree does not occur (Rudnick, 1933). More recent work (Wessels, 1970) has shown that there are two forms of mesoderm involved in the formation of the respiratory system. Tracheal mesoderm, possessing highly ordered sheaths of cells and collagen fibers, supports outgrowth of the trachea but inhibits branching, whereas the more loosely organized bronchial mesoderm promotes budding. In exchange experiments, bronchial mesoderm induces budding from the trachea and, conversely, tracheal mesoderm inhibits bronchial budding.

The terminal portions of the branches where cell proliferation is exceedingly active tend to remain somewhat bulbous (Fig. 14-10B and C). Later in development these terminal portions of the bronchial buds become still more dilated, their epithelium thins markedly, and they give rise to the characteristic air sacs of the lungs. As was the case with the trachea, the connective-tissue framework of the lung is derived from mesenchyme, which collects about the endodermal buds during their growth. The endoderm gives rise only to the lining epithelium of the bronchi, their glands, and the air sacs. The pleural covering of the lungs is derived from splanchnic mesoderm pushed ahead of the lung buds in their growth (Fig. 14-10C and D).

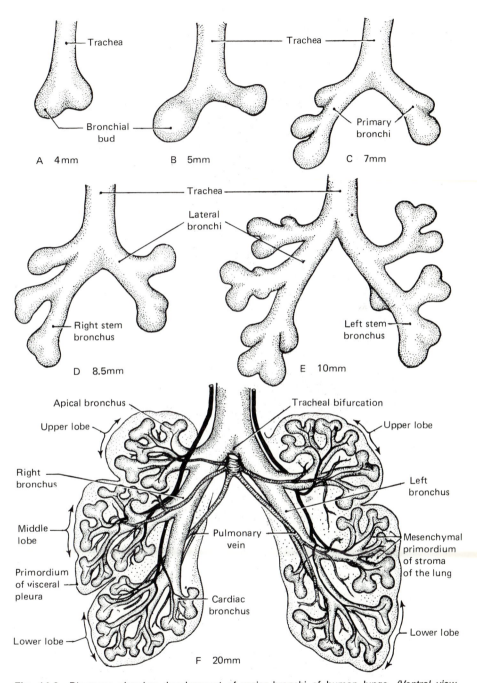

Trachea

Bronchial bud

A   4mm

Trachea

B   5mm

Trachea

Primary bronchi

C   7mm

Trachea

Lateral bronchi

Right stem bronchus

D   8.5mm

Left stem bronchus

E   10mm

Apical bronchus

Upper lobe

Right bronchus

Middle lobe

Primordium of visceral pleura

Lower lobe

Tracheal bifurcation

Upper lobe

Left bronchus

Pulmonary vein

Cardiac bronchus

Mesenchymal primordium of stroma of the lung

Lower lobe

F   20mm

Fig. 14-9 Diagrams showing development of major bronchi of human lungs. *(Ventral view, adapted from various sources, chiefly Corning and Arey.)*

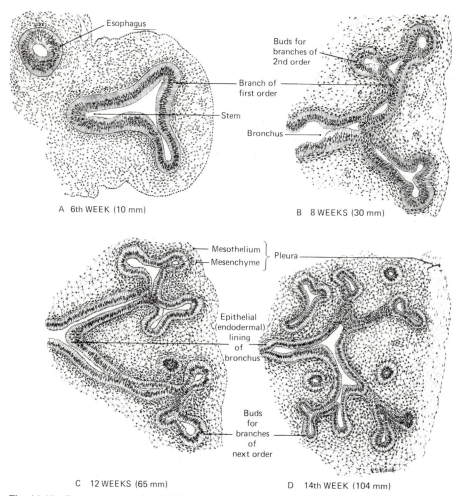

Fig. 14-10 Projection drawings (X100) showing four early stages in the histogenesis of the human lung. (A) Embryo of sixth week (EH 56; C–R 10 mm). (B) Embryo of 8 weeks (EH 15; C–R 30 mm). (C) Embryo of 11 weeks (EH 173, B; C–R 65 mm). (D) Embryo of fourteenth week (EH 145, C; C–R 104 mm). All embryos from University of Michigan Collection.

The lungs do not at first occupy the position characteristic of adult anatomy. In very young embryos they lie dorsal to the heart (Figs. 8-13 and 8-19). A little later when they have extended caudad, they are situated dorsal to the heart and liver (Figs. 14-13 and 14-14). The changes by which they eventually come to occupy their definitive position in the thorax will be taken up in connection with the partitioning of the primitive coelom to form the body cavities of the adult.

As is also true of the embryonic heart and circulatory system, the lungs must prepare themselves well before birth for the important physiological role which they must assume with the newborn infant's first breath. Within the first few minutes after birth they must become transformed from organs resembling waterlogged sponges to air-filled sacs which are capable of retaining an adequate air space without collapsing at each exhalation.

The fetal human lung is filled with a fluid resembling blood plasma, instead of amniotic fluid, as was often formerly supposed. Secreted by the lungs, this fluid undergoes some distinct changes, particularly in its lipid composition, during the latter weeks of pregnancy. The lipids, primarily lecithin and sphingomyelin, comprise what is known as *pulmonary surfactant*. Surfactant acts to reduce the surface tension of the fluid layer lining the interior of the air sacs (alveoli) within the lungs. This enables the air sacs to remain expanded with minimal effort in breathing. Although surfactant in the human can be detected as early as 26 weeks, it is not accumulated in sufficient quantities in the lungs until shortly before the normal time of birth. It is for this reason that prematurely born infants have a tendency to develop respiratory difficulties.

Another adaptation of the fetal respiratory tract for its function at birth is the relatively great size of the trachea and other respiratory passages in relation to the total size of the fetus. Although a normal newborn infant is approximately one twenty-fifth the size of an adult, the respiratory passages are considerably larger, being between one-fourth and one-third the adult diameter. Such precocious growth is a physical necessity because if the respiratory passages were any narrower, the physical resistance to breathing would be too great to be overcome (Avery, Wang, and Taeusch, 1973). By the time of the baby's first breath, the fluid that was present in the lungs has been eliminated by a combination of physical expulsion during delivery and by resorption into the lymphatic and blood vessels of the lungs. If the newborn infant is able to overcome the great physical resistance involved in taking the first breath or two, the blood flow in the lungs is quickly increased to postnatal levels with the closing of the ductus arteriosus (see Chap. 17), and the normal breathing pattern becomes quickly stabilized.

THE BODY CAVITIES AND MESENTERIES

The body cavities of adult mammals are the *pericardial cavity*, containing the heart, the paired *pleural cavities*, containing the lungs, and the *peritoneal cavity*, containing the viscera lying caudal to the diaphragm. All three of these regional divisions of the body cavity are derived from the coelom of the embryo.

Primitive Coelom   The coelom arises by the splitting of the lateral mesoderm on either side of the body into splanchnic and somatic layers (Figs. A-51A and B and 14-11A). It is, therefore, primarily a paired cavity bounded on one side by splanchnic mesoderm and on the other by somatic mesoderm. In forms such as birds and mammals, which have highly developed extraembryonic membranes, the coelom extends between the mesodermal layers of the extraembryonic membranes beyond the confines of the developing body. In

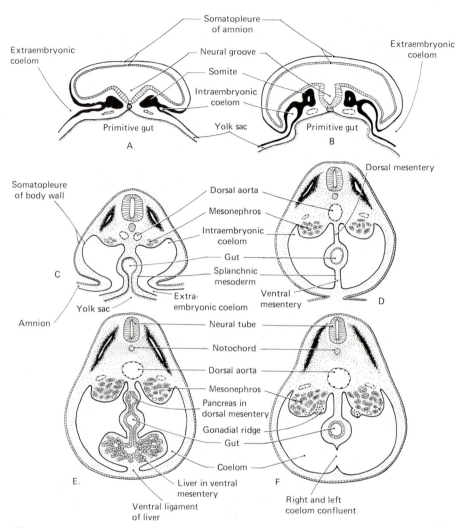

Fig. 14-11   Diagrams illustrating early stages in the development of the coelom and mesenteries.

Mammalia, where the nutrition of the embryo depends on the uterine relations established by the extraembryonic membranes, they develop exceedingly precociously. It is not surprising, in view of this fact, that the splitting of the mesoderm in mammals occurs first extraembryonically and progresses thence toward the embryo (Fig. 6-16A and B). When the body of the embryo is folded off from the extraembryonic membranes, the extra- and intraembryonic portions of the coelom are thereby separated from each other, the last place of confluence to be closed off being in the region of the belly stalk (Figs. 14-11C and 14-12). It is the intraembryonic portion of the primitive coelom, thus delimited, which gives rise to the body cavities of the adult.

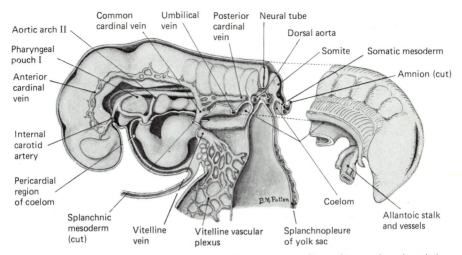

Common cardinal vein
Umbilical vein
Posterior cardinal vein
Neural tube
Aortic arch II
Pharyngeal pouch I
Dorsal aorta
Anterior cardinal vein
Somite
Somatic mesoderm
Amnion (cut)
Internal carotid artery
Pericardial region of coelom
Coelom
B.M.Patten
Allantoic stalk and vessels
Splanchnic mesoderm (cut)
Vitelline vein
Vitelline vascular plexus
Splanchnopleure of yolk sac

Fig. 14-12 Schematic plan of lateral dissection of young mammalian embryo to show the relations of the pericardial region of the coelom to the primary paired coelomic chambers caudal to the level of the heart. The proportions of the illustration were based in part on Heuser's study of human embryos about 3 weeks old, but all the essential relationships shown are equally applicable to pig embryos of 3 to 4 mm.

Mesenteries   The same folding process that separates the embryo from the extraembryonic membranes completes the floor of the gut (Figs. 8-10 and 14-11). Coincidently the splanchnic mesoderm of either side is swept toward the midline, enveloping the now tubular digestive tract. The two layers of splanchnic mesoderm which thus become apposed to the gut and support it in the body cavity are known as the *primary* or *common mesentery*. The part of the mesentery ventral to the gut, attaching it to the ventral body wall, is the *ventral mesentery* (Fig. 14-11D). The primary mesentery, while intact, keeps the original right and left halves of the coelom separate. But the part of the mesentery ventral to the gut breaks through very early, bringing the right and left coelom into confluence and establishing the unpaired condition of the body cavity characteristic of the adult (Fig. 14-11F).

In the hepatic region the ventral mesentery does not disappear. The liver arises, as we have seen, from an outgrowth of the gut and in its development pushes into the ventral mesentery (Fig. 14-11E). The portion of the ventral mesentery between the liver and the stomach persists as the *gastrohepatic omentum* (*ventral mesogastrium*), and the portion between the liver and the ventral body wall, although reduced, remains in part as the *falciform ligament* of the liver (Fig. 14-14).

Whereas the ventral mesentery, except in the region of the liver, eventually disappears, almost the entire original dorsal mesentery persists. It serves at once as a membrane supporting the gut in the body cavity and as a path over which nerves and vessels reach the gut from main trunks situated in the dorsal body wall. Its different regions are named according to the part of the digestive tube with which they are associated; for example, *mesogastrium*, that part of the dorsal mesentery supporting the stomach; *mesocolon*, that part of the dorsal mesentery supporting the colon (Fig. 14-14).

Partitioning of the Coelom   The structure which initiates the division of the coelom into separate chambers is the *septum transversum*. The septum transversum appears very early in development (Figs. 8-13 and 14-13B) and is already a conspicuous structure in embryos of 9 to 12 mm (Fig. 14-14). Extending from the ventral body wall dorsad, it forms a sort of semicircular shelf. Fused to the caudal face of the shelf is the liver, and on its cephalic face rests the ventricular part of the heart. The septum transversum is the beginning

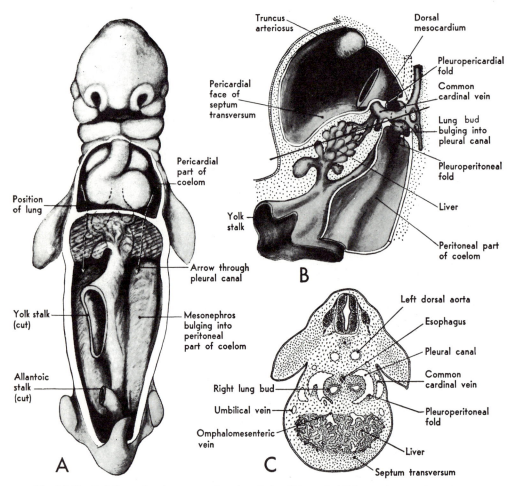

Fig. 14-13 Relations of various parts of coelom during fifth week. (A) Semischematic frontal plan. Arrows indicate location of pleural canals, on either side, dorsal to liver. (B) Lateral dissection to show left pleural canal opened with lung bud bulging into it medially. *(Modified from Kollmann.)* (C) Section through body of an 8-mm embryo, schematized to show relations at level of pleural canal. Level of section is indicated by heavy line across B.

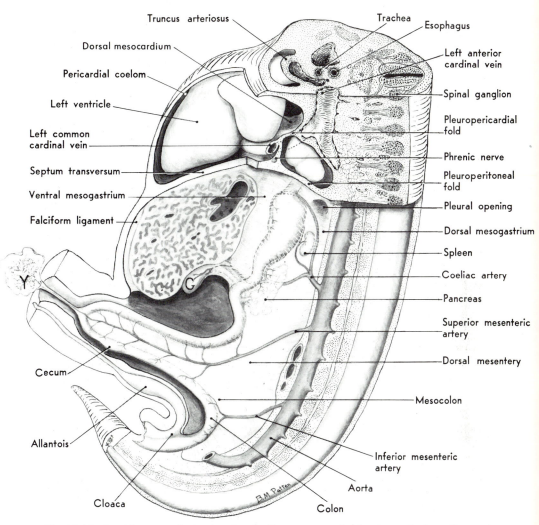

Fig. 14-14 Semidiagrammatic drawing showing the arrangement of the viscera, body cavities, and mesenteries in a human embryo early in the seventh week of development. In all essentials the conditions here represented will be found in pig embryos of 15 mm. In the region of the developing lungs the body is cut parasagittally, well to the left of the midline, in order to show the relations of pleuropericardial and pleuroperitoneal folds. Below the developing diaphragm, dissection has been carried to the midline. Abbreviations: *G*, gallbladder; *Y*, yolk sac.

of the diaphragm; however, it must be kept in mind that the diaphragm is a composite structure of which only the ventral portion is derived from the septum transversum.

As it expands dorsad from the ventral body wall, the septum transversum acts as a partial partition between the pericardial and peritoneal portions of the coelom. When it has grown to the extent that it makes contact with the floor of

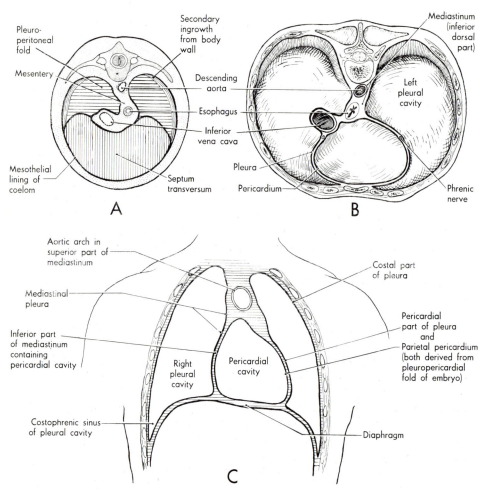

Fig. 14-15 Diaphragm in embryo and in adult. (A) Diagram indicating embryological origin of various regions of diaphragm. *(Modified from Broman.)* (B) Thoracic face of adult diaphragm. *(Modified from Rauber-Kopsch.)* (C) Relations of diaphragm as seen in a frontal section of adult body. *(Modified from Rauber-Kopsch.)*

the foregut, the septum transversum has almost cut the coelomic cavity into two parts. The separation is not complete, however, for on either side of the gut dorsal to the septum transversum small portions of the coelomic cavity remain (Fig. 14-13C). These remnants of the coelom are actually short channels called the *pleural canals.* They connect the thoracic portion of the coelom, which is the single pericardial cavity, with the peritoneal part of the coelom, also a single cavity (Fig. 14-13A and B). At first the pleural canals are relatively small channels, but as the developing lung buds grow into them, they increase greatly in size and become the third major component of the coelom, the pleural cavities.

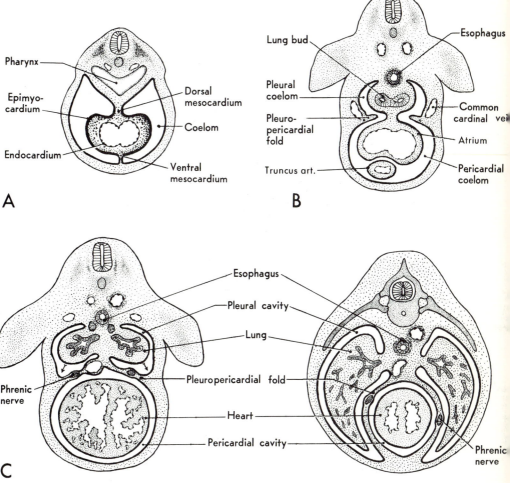

Fig. 14-16 Schematic diagrams showing the manner in which the pleural and pericardial regions of the coelom become separated.

The pleural canals are, in effect, partially bounded by two paired folds of tissue, the pleuropericardial and pleuroperitoneal folds. The *pleuropericardial folds* are ridges of tissue arising from the dorsolateral body walls where the common cardinal veins bulge into the coelom as they swing toward the midline to enter the sinus venosus of the heart (Figs. 14-13B, 14-14, and 14-16B and C). In time these folds effectively partition the anterior ends of the pleural canals from the pericardial cavities. Posteriorly, the pleural canals are delimited by the *pleuroperitoneal folds* (Fig. 14-13B), which project from the body wall to demarcate the posterior pleural canals from the peritoneal coelom.

The formation of the definitive pleural cavities and their isolation from the pericardial and peritoneal cavities involves the expansion of the lungs into the pleural cavities as well as the sealing off of both ends of the pleural canals by the pleuropericardial and pleuroperitoneal folds. The early lung buds expand caudally and laterally in the mesenchyme ventral to the esophagus. This expansion results in a bulging of the lung buds into the pleural canals (Fig. 14-13C). As the developing lungs continue to grow, they expand dorsally, laterally, and ventrally, thus displacing the mesenchyme of the body wall (Fig. 14-16B to 14-16D). This expansion of the lungs greatly increases the size of the original pleural canals, so that they now partially surround the pericardial cavity and also extend somewhat dorsal to the esophagus (Fig. 14-16D).

As the lungs are expanding, the anterior portion of the pleural canals is becoming separated from the pericardial cavity by the pleuropericardial folds. It will be recalled that the pleuropericardial folds are actually ridges of mesenchymal tissue which accompany the common cardinal veins as they converge toward the heart (Fig. 14-16B and C). As the lungs expand into the lateral body walls and the common cardinal veins shift toward the midline with the descent of the heart into its final position, the pleuropericardial folds converge and fuse with one another, thus effecting a dorsal separation between the lungs and the heart.

Meanwhile, the pleuroperitoneal folds are converging from the body walls toward the septum transversum to complete the separation of the posterior ends of the pleural canals from the peritoneal cavity. The dorsal mesentery is caught between the septum transversum and the pleuroperitoneal folds. Fusion along the lines of contact of these structures essentially completes the diaphragm (Fig. 14-15A). Later in development the edges of the diaphragm, especially dorsolaterally, are invaded by body-wall mesenchyme, which contributes the marginal part of the diaphragmatic musculature (Fig. 14-15A).

Chapter 15

Ductless Glands, Pharyngeal Derivatives, and the Lymphoid Organs

Some functionally integrated systems in the body defy neat anatomical packaging because the individual organs are widely scattered in location and quite diverse with respect to their mode of origin. Because of their functional ties, however, it is preferable to present them as groups. Two such systems, the endocrine system and the lymphoid system, will be presented in this chapter. Because of the intimate association between the origin of many endocrine and lymphoid organs and the development of the pharynx, the embryology of the entire pharyngeal region has been left to this chapter.

Development of the Hypophysis The hypophysis is formed from two separate primordial ectodermal parts, which unite secondarily. One of these primordia, known as *Rathke's pocket*, is an outgrowth from the stomodeal depression. This ectodermally lined space, located externally to the oral membrane, is shaped somewhat like a glove finger and extends in the midline toward the diencephalic floor (Fig. 15-1A).

The other primordial part of the hypophysis is the infundibular process, which is formed in the floor of the diencephalon. This tissue is also of ectodermal origin, but its histology, in keeping with its derivation by way of the neural tube, differs from that of the epithelial portions derived from the lining of Rathke's

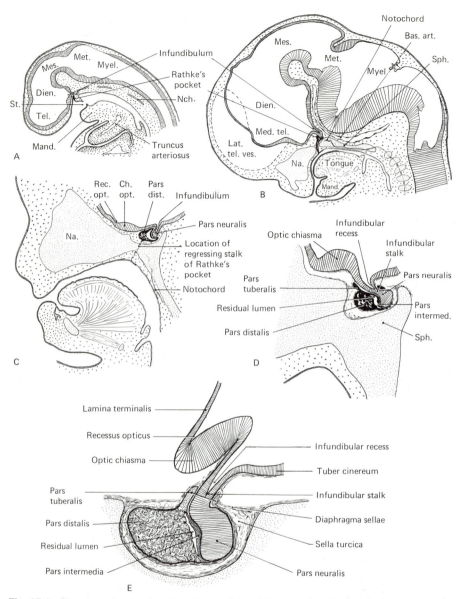

Fig. 15-1   Diagrams showing the changing relations of the neural and stomodeal portions of the hypophysis during its development. (A) Semischematic sagittal section of head of 4-week (4 to 5 mm) human embryo. (B) Projection diagram of sagittal section of head of human embryo of about 6½ weeks. *(C-R 15 mm, University of Michigan Collection, EH 4.)* (C) Projection diagrams of sagittal section of head of human embryo of eighth week. *(C-R 25 mm, University of Michigan Collection, EH 33.)* (D) Sagittal section of hypophyseal region of embryo of eleventh week. *(C-R 60 mm, University of Michigan Collection, EH 23.)* (E) Schematic plan of adult hypophysis.

pocket. Eventually, with very little change in its relations, the infundibular process comes to constitute the so-called *pars neuralis* of the adult hypophysis (Fig. 15-1D and E).

The later changes in the stomodeal portion of the hypophysis are more striking and extensive. Rathke's pocket elongates and its blind end becomes closely applied to the infundibular process (Fig. 15-1B). During this period the original stalk by which Rathke's pocket arose from the stomodeum first becomes narrowed (Fig. 15-1B), and then loses its connection with the stomodeal epithelium (Fig. 15-1C and D). Meanwhile, the originally shallow stomodeal depression has been greatly deepened by the forward growth of the adjacent nasal and maxillary processes and the mandibular arch. A vivid picture of the extent of this deepening process can be obtained by comparing the original superficial location of Rathke's pocket (Fig. 15-1A) with the location of the line along which its stalk may be seen regressing in older embryos (Fig. 15-1C).

As the stomodeal portion of the hypophysis becomes molded into a double-layered cup about the pars neuralis, the rostral portion of its outer layer increases rapidly in thickness and takes on an obviously glandular appearance. In it we can now readily recognize the *pars distalis* (*anterior lobe*) of the adult hypophysis (Fig. 15-1C and D). The inner layer of Rathke's pocket makes contact with the pars neuralis and tends to fuse with it. The layer thus secondarily applied to the pars neuralis constitutes what is called the *pars intermedia* of the adult organ. Between the pars distalis and the pars intermedia there remains for a time a slitlike lumen, which is appropriately called the *residual lumen* because it is all that remains of the original lumen of Rathke's pocket (Fig. 15-1C and D).

Functional relationships between the hypophysis and other endocrine organs, such as the thyroid and the adrenal, are established during embryonic life. In the absence of the hypophysis these organs attain neither full morphological development nor normal hormonal function.

Branchial Region   In 1-month human embryos (4- to 6-mm pig embryos) the pharyngeal part of the foregut and the tissues surrounding it are organized in a manner reminiscent of the branchial region of the lower vertebrates, with a paired series of branchial arches alternating with branchial clefts and pharyngeal pouches along either side of the pharynx (Figs. 8-14B and 15-2). This stage of development of the branchial region is a recapitulation of conditions which had an obvious functional significance in water-living ancestral forms, for the branchial clefts and the pharyngeal pouches of the mammalian embryo are homologous with the outer and inner portions of the gill slits in lower forms. As so often happens in the embryo, the repetition of phylogenetic history is slurred over. Although in the mammalian embryo the tissue closing the gill clefts becomes reduced to a thin membrane consisting of a layer of endoderm and ectoderm with little or no intervening mesoderm (Fig. 11-16B), this membrane rarely disappears altogether. Occasionally in mammals, as happens quite frequently in the chick, the more cephalic of the pharyngeal pouches break

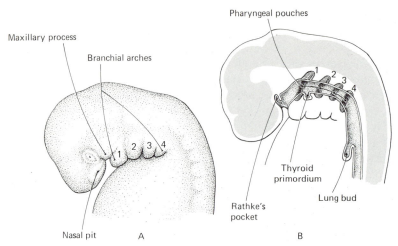

Fig. 15-2 Topography of the branchial region in a human embryo in the fifth week. *(In part after Langman, Medical Embryology, 1963, The Williams & Wilkins Company, Baltimore.)* (A) External configuration to show the branchial arches. (B) Head opened medially to show the pharyngeal pouches.

through to the outside, establishing open gill slits, but in such cases the opening is very short-lived and the clefts promptly close again.

Although aspects of the development of certain branchial structures have been dealt with in other chapters, this section will summarize development of the region as a whole. Figure 15-3 presents in diagrammatic form the main adult derivatives of key branchial structures.

Branchial Arches   There are four recognizable branchial arches, each of which contains a cranial nerve, a blood vessel (aortic arch; Fig. 15-4), a skeletal primordium of neural crest origin, and premuscle mesenchyme (Fig. 15-5).

The first arch (mandibular), which gives rise to both the mandibular and the maxillary portions of the face, is innervated by the trigeminal (V) nerve. Its cartilaginous rod (Meckel's cartilage) persists in the lower jaw as Meckel's cartilage, later to be surrounded by the intramembranously formed bone that forms the mandible. The dorsal part of the rod breaks up to form one of the middle ear ossicles (malleus) (see Chap. 12). The premuscular mesenchyme (mesodermal) of the first arch forms the muscles of mastication, among others (see Fig. 15-3), all innervated by the fifth cranial nerve.

The second arch (hyoid), innervated by the seventh cranial nerve, forms a string of skeletal structures, starting with the stapes of the middle ear and extending down to the body of the hyoid bone (Fig. 15-5). Much of the mesoderm migrates to the face to form the muscles of facial expression, and the other second-arch muscles are associated with the skeletal derivatives. All muscles of the second arch are innervated by the seventh cranial nerve.

The skeletal and muscular derivatives of the third arch are related to the hyoid bone and upper pharynx. The one muscle derived from this arch is

BRANCHIAL ARCH STRUCTURES

Adult derivative of BRANCHIAL GROOVE		Arch Number	Aortic Arch	Cranial Nerve	Examples of Branchiomeric Muscles	Skeletal Derivatives		Adult Derivative of PHARYNGEAL POUCH and Lining Structures
External ear	External auditory meatus	I (Mandibular)	I Maxillary a.	V Trigeminal	Muscles of mastication tensor tympani, mylohyoid, tensor veli palatini, ant. belly digastric	Malleus, incus, sphenomandibular ligament, Meckel's cartilage	I	Middle ear auditory tube
		II (Hyoid)	II Hyoid a. Stapedial a.	VII Facial	Muscles of facial expression, stapedius stylohyoid, post. belly digastric	Stapes, styloid process, stylohyoid ligament, lesser cornu of hyoid, part of body of hyoid	II	Supra-tonsillar fossa
		III	III Internal carotid a.	IX Glosso-pharyngeal	Stylo-pharyngeus	Greater cornu of hyoid, part of body of hyoid	III	Thymus, inferior parathyroid gland
		IV	IV R. subclavian a. and aorta	X Vagus	Pharyngeal and laryngeal musculature	Laryngeal cartilages	IV	Superior parathyroid gland, postbranchial body

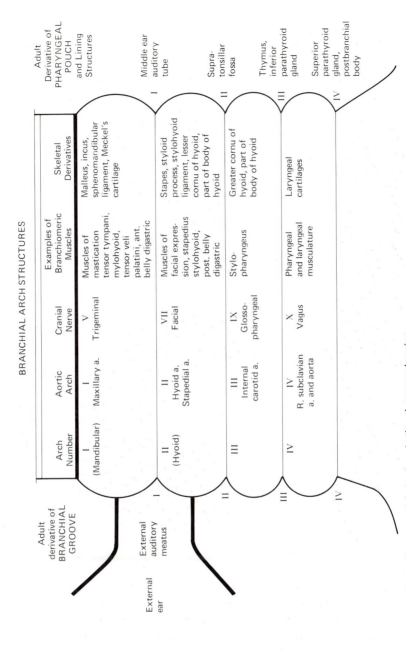

Fig. 15-3 Diagram of derivatives in the pharyngeal region.

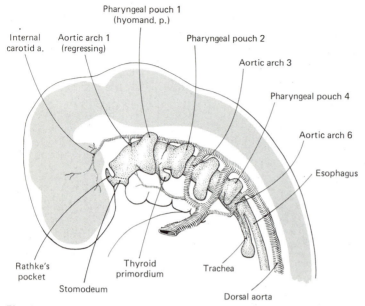

Fig. 15-4 Schematic diagram of the branchial region with the overlying ectoderm and mesenchyme removed to show the pharyngeal pouches and the relations of the aortic arches to them.

innervated by the ninth cranial nerve. The fourth branchial arch, innervated by the vagus (X) nerve, supplies some cartilaginous and muscular structures to the larynx and lower pharynx, but the vagus nerve grows into the thoracic and abdominal cavities, as well.

Branchial Clefts Only the first branchial cleft contributes to a recognizable structure, persisting as the external auditory meatus. Clefts II to IV temporarily become overshadowed by the hyoid arch (the homologue of the operculum, or gill cover, of fishes) and are collectively called the *cervical sinus* (Figs. 8-7 and 15-6A). Later in embryonic development, the external contours of the neck smooth out and the cervical sinus disappears without a trace.

Pharynx The main pharyngeal chamber of the embryo becomes converted directly into the pharynx of the adult. In this process the configuration of its lumen is simplified and relatively reduced in extent. An important factor in these changes is that of the separation of various pouches from the main part of the pharynx. The cell masses thus originating migrate into the surrounding tissues and there undergo divergent differentiation.

The first pair of pharyngeal pouches, extending between the mandibular and hyoid arches, comes into close relation at the distal ends with the auditory vesicles. They give rise, on either side, to the *tympanic cavity* of the middle ear and to the *auditory (eustachian) tube* (Fig. 12-14).

The second pair of pouches becomes progressively shallower and less conspicuous. Late in fetal life the faucial (palatine) tonsils are formed by the aggregation of lymphoid tissue in their walls, and vestiges of the pouches themselves persist as the *supratonsillar fossae*.

From the floor of the pharynx, at about the level of the constriction between the first and the second pair of pharyngeal pouches, there is formed a midline diverticulum destined to give rise to the thyroid gland (Figs. 15-4 and 15-6). The endodermal cells making up the walls of this evagination push into the

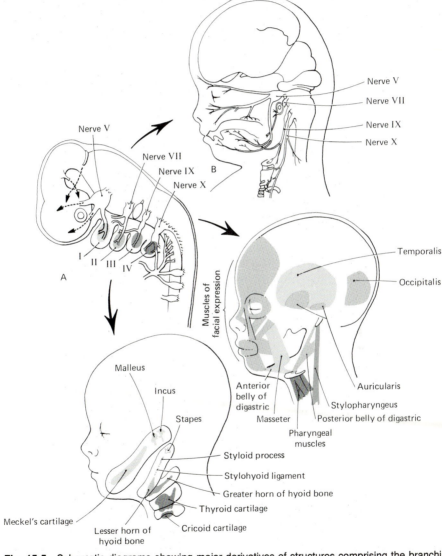

Fig. 15-5 Schematic diagrams showing major derivatives of structures comprising the branchial arches. (A) 5-week embryo. (B–D) 4- to 5-month fetuses. The gray tones of structures in C and D correspond to those of the branchial arches depicted in A.

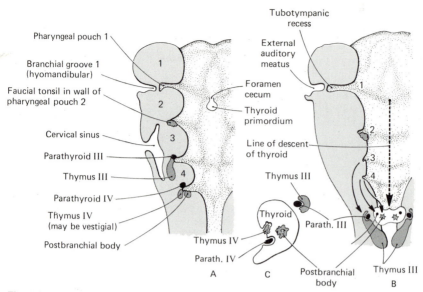

Fig. 15-6 Diagrams showing the origin of the pharyngeal derivatives. *(Adapted from several sources.)* (A) The primary relations of the several primordia to the pharyngeal pouches. (B) Course of migration of some of the primordia from their place of origin. (C) Definitive relations of parathyroids, postbranchial body, thymus, and thyroid as they appear in a transverse section of the right lobe of the thyroid taken above the level of the isthmus. Abbreviation: *Parath.,* parathyroid.

underlying mesenchyme, break away from the parent pharyngeal epithelium, and migrate down into the neck (Figs. 15-7, 15-8D, and 17-33). As with most developing glands, differentiation of the epithelial follicles of the thyroid is dependent upon a specific inductive interaction with the surrounding mesenchyme. Only after arriving in its definitive location, relatively late in development, does the thyroid primordium undergo its final characteristic histogenetic changes.

The third and fourth pairs of pharyngeal pouches give rise to outgrowths which are involved in the formation of the parathyroid glands, the thymus, and the postbranchial bodies. In most mammals there are two pairs of parathyroid glands, usually spoken of as parathyroids III and parathyroids IV because they arise from the third and the fourth pharyngeal pouches (Figs. 15-6A and 15-7). As was the case with the thyroid, the parathyroid primordia soon break away from their points of origin and migrate into the neck. Here, as their name implies, they are positionally more or less closely associated with the thyroid. Parathyroids IV are particularly likely to become adherent to the thyroid capsule or even to become partially embedded in the substance of the gland (Fig. 15-6B).

The thymus in the mammalian group is derived from outgrowths from the more ventral portions of the third and fourth pharyngeal pouches (Figs. 15-6A and 15-7). In different species there is considerable difference in the relative

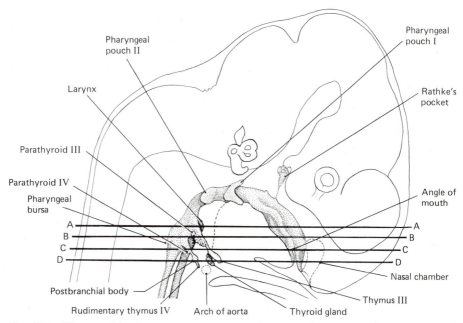

Fig. 15-7 Pharynx of 15-mm pig embryo schematically represented in relation to the outlines of other cephalic structures. *(Adapted from several sources.)* The heavy horizontal lines indicate the levels of the correspondingly lettered sections in the following figure.

conspicuousness of the two pairs of primordia. In most of the higher mammals the primordia arising from the third pouches are much more important as thymic contributors. This is the situation for the pig as well as for humans. In pig embryos of 15 to 17 mm, however, it is often possible to make out a rudimentary thymus IV (Fig. 15-7). The characteristic histogenetic changes in the thymus occur relatively late in development. The original epithelial character of the thymus becomes obscured because of an extensive ingrowth of mesenchyme. Soon thereafter lymphoid cells invade the organ and proliferate extensively, thus imparting to the thymus the histological characteristics of a true lymphoid organ.

For years the *postbranchial (ultimobranchial) bodies*, structures of probable neural crest origin, remained structures of problematical significance. In the early 1960s investigations by a number of laboratories established that cells derived from the postbranchial bodies produce a polypeptide hormone, *calcitonin*, which acts to reduce the concentration of calcium in the blood. The activity of this newly discovered hormone serves to counter the long-recognized function of *parathyroid hormone*, which causes an increase in calcium blood levels. In mammalian development, the calcitonin-producing cells of the postbranchial bodies become incorporated into the thyroid gland, whereas in birds and the other lower vertebrates the postbranchial bodies persist as distinct organs.

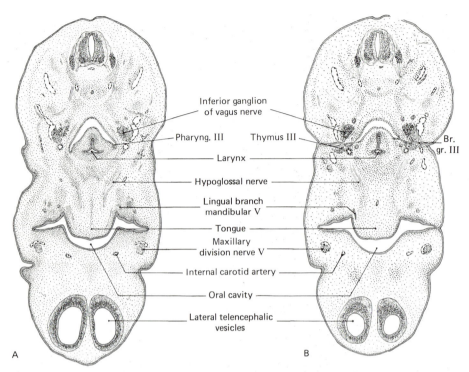

Inferior ganglion of vagus nerve

Pharyng. III   Thymus III   Br. gr. III

Larynx

Hypoglossal nerve

Lingual branch mandibular V

Tongue

Maxillary division nerve V

Internal carotid artery

Oral cavity

Lateral telencephalic vesicles

A   B

Fig. 15-8 Drawings (X11) of transverse section through the pharyngeal region of a 15-mm pig embryo. (A) Upper laryngeal level; (B) level of third pharyngeal pouch; (C) level of fourth pharyngeal pouch; (D) through neck, caudal to level of pharynx and larynx. The level of each of the sections represented in this figure is indicated by the correspondingly lettered line in Fig. 15-7. Abbreviations: *Br. gr. III,* third branchial groove; *N. X,* vagus nerve; *N. XII,* hypoglossal nerve; *Pharyng. III, Pharyng. IV,* third and fourth pharyngeal pouches; *Postbranch.,* postbranchial body; *Premusc.,* premuscular concentration of mesenchyme.

Adrenal Glands   Certain cells that migrate ventrally from the neural crest when the sympathetic ganglia are formed do not become nerve cells but rather gland cells which are active in the production of the specific hormones epinephrine and norepinephrine. They exhibit a characteristic reaction with chromic acid salts, which has led to their designation as *chromaffin cells.* Clusters of chromaffin cells become located close to each sympathetic ganglion, and other masses of chromaffin tissue from the same source appear in various places beneath the mesoderm lining the coelom. The largest mass of extrasympathetic chromaffin tissue appears just cephalic to the kidney and becomes converted into the *medulla of the adrenal.*

The cortical portion of the adrenal gland appears very early in development. Even in embryos of the sixth week, there is accelerated local proliferation of cells from the splanchnic mesoderm around the notch on either side of the base of the primary dorsal mesentery adjacent to the cephalic pole of the mesonephros. These cells push into the underlying mesenchyme and begin to

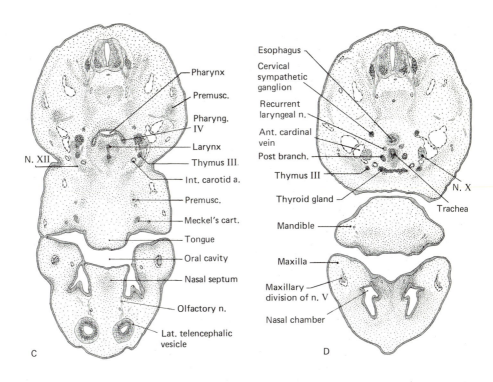

show a tendency to become arranged in cords. By the 15- to 17-mm stage the aggregation of cell cords which constitutes the primordium of the adrenal cortex is quite conspicuous. Later in development the migrating cells which give rise to the medulla of the adrenal invade the cortical primordium and become encapsulated within it (Fig. 15-9).

The cells first forming the adrenal cortex proliferate extensively to form a large fetal cortex, the function of which is still poorly understood. A few days after the fetal cortex first forms, a second wave of cells surrounds it. These cells, comprising the germinal layer of the definitive cortex, also proliferate and ultimately form the three cortical layers characteristic of the mature adrenal gland. The fetal cortex, on the other hand, remains quite prominent until birth, after which it rapidly involutes and ultimately disappears (Fig. 15-9).

Lymphoid System   A major functional characteristic of the vertebrate body is the ability to defend itself against foreign cells and organisms. In response to exposure to disease-producing bacteria or viruses, higher vertebrates produce *antibodies*, protein molecules which react very specifically against the microorganisms or their toxic products and thus prevent recurrence of the disease. Another form of defense is operating when a transplanted kidney in a human becomes infiltrated by lymphocytes and is rejected by the host. These defense reactions are examples of the functioning of the *lymphoid system*, or, as it is sometimes called, the *immune system*. The lymphoid system is not

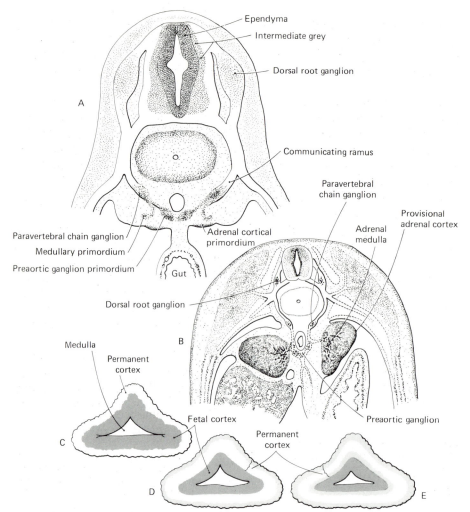

Fig. 15-9 Stages in the formation of the sympathetic ganglia and the adrenal gland. (A) Cortical and medullary components still separate; (B) Medullary cells migrating into the provisional adrenal cortex; (C–E) Later stages in development of the adrenal cortex; (C) Fifth month; (D) Eighth month; (E) Newborn infant (all zones of mature cortex present). *(C–E after Gray and Skandalakis, 1972, Embryology for Surgeons, W.B. Saunders Company, Philadelphia.)*

embodied in a single anatomical structure. Like the endocrine system, it is structurally diffuse but functions as an integrated unit. Very little was known about the ontogeny of the lymphoid system before 1960. Since that time, our understanding has advanced rapidly, and although a number of important questions remain unanswered, it is now possible to present a general scheme of the development of the lymphoid system.

Wherever it is found, lymphoid tissue presents a roughly similar microscopic appearance. It contains a stroma consisting of stellate mesenchymelike reticular cells which produce a loose meshwork of very fine reticular fibers. Embedded in this meshwork are lymphocytes and other cells involved in immunological defense reactions. Detailed anatomical features will not be considered here.

There are several organs, such as the thymus gland, lymph nodes, and tonsils, which are primarily lymphoid in nature, but the remainder of the lymphoid tissue in the body is closely associated with tissues fulfilling other functions. Thus in bone marrow it is intermingled with hematopoietic tissue, whereas in the spleen it is associated with tissue primarily concerned with the destruction of old blood cells. The epithelial linings of the gastrointestinal tract, and to a lesser extent those of the respiratory and female genital tracts, are underlain by varying amounts of lymphoid cells which produce a special type of antibody and effectively serve as a front line of immunological defense for these internal body surfaces.

Development of the Lymphoid System The ontogeny of the immune system has been studied intensively in both birds and mammals, and a general scheme of lymphoid development based upon findings in both of these groups is presented in Fig. 15-10.

In terms of function, the immune system can be divided into two parts, one of which is concerned with *humoral immunity*, that is, the production of antibodies, and the other with *cell-mediated immunological responses*. Cell-mediated responses are those in which immunological reactivity is intimately associated with the cells themselves and not with a secretion of these cells, as is the case in humoral immunity.

The origin of lymphoid cells is not definitely established, but it is thought that they may arise from the mesoderm of the liver or bone marrow. The *lymphoid stem cells* then migrate into what are called the *central lymphoid organs* (Fig. 15-10), where they are subjected to an inductionlike influence which determines the subsequent functional properties of the cells. There appear to be two central lymphoid organs. One is the thymus. Not long after its formation from the endoderm of the third and fourth pharyngeal pouches, the epithelial thymus is invaded by lymphoid stem cells. While in the thymus these cells become conditioned to take part in cell-mediated immune responses rather than to produce antibodies. In birds, lymphoid stem cells not entering the thymus migrate to the *bursa of Fabricius*, the other central lymphoid organ. There they are transformed into the line of cells which will produce humoral antibody. The mammalian equivalent to the bursa of Fabricius has not yet been positively identified, but many suspect that the lymphoid tissues associated with the gut fulfill the same function as the bursa.

From the central lymphoid organs the lymphoid cells, recognizable as lymphocytes, migrate to *peripheral lymphoid tissues*, for example, the lymph nodes, spleen, tonsils, and appendix (Fig. 15-10). Cells of the humoral part of the lymphoid system settle down into aggregations called *germinal centers*. When stimulated by foreign antigens, cells of the germinal centers undergo additional changes and then begin to produce antibodies. Lymphocytes emanat-

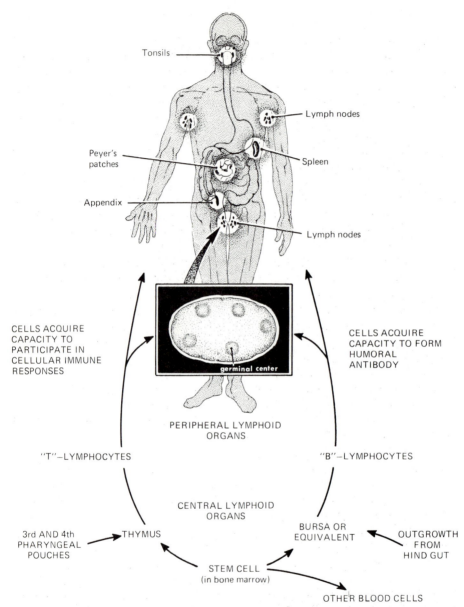

Fig. 15-10   Simplified summary of the functional development of the immune system. Hemopoietic stem cells pass from the bone marrow into either the thymus or the bursa. During passage through these central lymphoid organs they acquire the capacity to participate in either cell-mediated or humoral immune responses. The "T" (thymic) and "B" (bursal) lymphocytes then migrate to the various peripheral lymphoid organs (*top of figure*) where they await a specific antigenic stimulus. Further interactions between T and B lymphocytes are necessary for the development of full immunological function.

ing from the thymus go to the same peripheral lymphoid tissues as those derived from the bursa or its mammalian equivalent, but instead of settling down into discrete germinal centers, they form dense masses of small cells around the germinal centers or in other areas. These represent the cells that can be mobilized to face the challenge of a mass of cells, for example, a transplanted kidney, which differs immunologically from the host.

In early embryos, the protective function of the immune system is minimal, and foreign cells or large molecules are not met with an immunological rejection. Later in fetal development and shortly after birth, the embryo begins to distinguish "self" from "nonself" (Burnet, 1969), and at that time the immune defense system begins to become functional. The T-lymphocytes of humans are capable of generating cell-mediated immune responses at birth, but the antibody-secreting function of the B-lymphocytes does not mature until varying periods after birth, depending upon the specific class of antibody. Part of the deficiency in antibody production in newborn infants is made up by the presence of maternal antibody (IgG), which passed from the mother through the placenta in utero. The maternal antibody is gradually degraded over the first 2 to 3 months of life, after which it is replaced by antibody synthesized by the infant.

Chapter 16

The Development of the Urogenital System

In both their adult anatomy and their embryological development, the urinary and reproductive systems are so closely linked that they must inevitably be considered together. Both the urinary and genital tracts share a common origin from the intermediate mesoderm of the early embryo. The urinary system begins to take shape first, following a series of stages that reveal its phylogenetic history. Only later, after a certain degree of function already exists within the developing urinary system, does the genital system appear. Genital development is initiated in the gonad, which differentiates into an ovary or a testis. The formation of genital ducts and glands occurs later, their male or female configuration coming about as the result of specific secretions, or lack thereof, from the gonads.

THE URINARY SYSTEM

General Relationships of Pronephros, Mesonephros, and Metanephros As a preface to the account of the development of the urinary system in mammals, it is worthwhile to review certain facts about the structure and development of the excretory organs in the vertebrates generally. Without such information as a background, the story of the early stages of the formation of these organs in any specific group seems utterly without logical sequence. With it, the progress of events encountered in mammalian development seems natural, because it is so clearly an abbreviated recapitulation of conditions which existed in the adult stages of ancestral forms.

There occur in adult vertebrates three major types of urinary organs. Although they are often treated as distinct structures in elementary textbooks, one must recognize, when reviewing both the phlogenetic aand ontogenetic history of the vertebrate kidney, that the major types of kidneys actually represent dominant forms, with intergrades between them. Some authors feel that the amount of intergradation is so great that the vertebrate kidney should be looked upon as a true continuum of varying morphological and functional units. According to this concept, the aggregate of all forms of renal units is called the *holonephros*. For a more specialized treatment of this concept, the reader is referred to the works of Torrey (1965, 1971).

The most primitive of the major types of kidney is the *pronephros*, which exists as a functional excretory organ only in some of the lowest fishes. It is located far cephalically in the body. In all the higher fishes and in the amphibia the pronephros has degenerated, and its functional role has been assumed by the *mesonephros*, a new organ located farther caudally in the body. In birds and mammals a third urinary organ develops caudal to the mesonephros. This is the *metanephros*, or permanent kidney. All three of these organs are paired structures located retroperitoneally in the dorsolateral body wall.

In a recapitulation of phylogeny, three major types of kidneys, the pronephros, mesonephros, and metanephros, are formed in succession during the development of avian and mammalian embryos. The pronephros is the most anterior of the three and the first to be formed. It is wholly vestigial, appearing only as a slurred-over recapitulation of structural conditions which exist in the adults of the most primitive of the vertebrate stock. The mesonephros represents the adult excretory organs of the higher fishes and amphibia. It makes its appearance in the embryo somewhat later than the pronephros and is formed caudal to it. The mesonephros is the principal organ of excretion during early embryonic life, but like the pronephros, it also disappears in the adult, except for parts of its duct system, which become associated with the male reproductive organs. The metanephros is the most caudally located of the excretory organs and the last to appear. It becomes functional late in embryonic life when the mesonephros is regressing, and it persists permanently as the functional kidney of the adult.

As a general introduction, the main stages in the ontogenetic history of the nephric organs are schematically illustrated in Fig. 16-1. Kidney formation begins with the appearance of the pronephros, several pairs of tubules located in the cephalic portion of the intermediate mesoderm. These small tubules, forming in a cephalocaudal sequence, empty laterally on each side into a common duct called the *primary nephric* (*pronephric*) *duct* (Fig. 16-1B). The primary nephric ducts elongate toward the cloaca, and with their progression caudad, new pairs of tubules form and join the ducts.

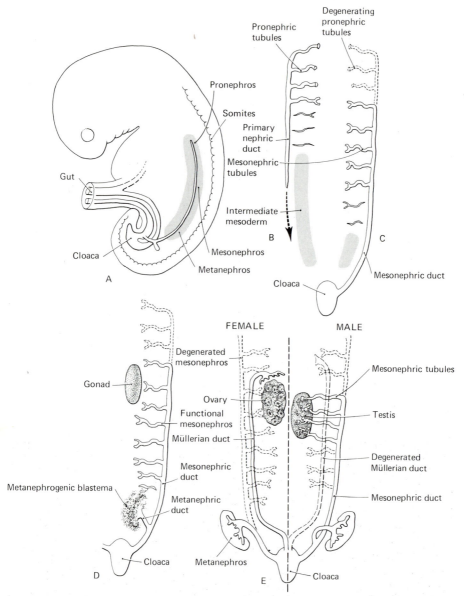

Fig. 16-1 Schematic diagrams to show the relations of pronephros, mesonephros, and metanephros at various stages of development. (A) Diagram showing the subdivision of the intermediate mesoderm into pronephric, mesonephric, and metanephric segments. (B) Early stage, showing the primary nephric duct extending toward the cloaca. (C) Establishment of the mesonephros. (D) Early appearance of the metanephros. (E) Urogenital system after sexual differentiation. The Müllerian ducts are structures which arise later in development and are not part of the urinary system proper.

After only a few segments, the character of the newly forming tubules changes, providing a gradual transition to the next major type of tubule, the mesonephric tubule (Fig. 16-1C). As the mesonephric tubules continue to form, the pronephric tubules, which are rudimentary in all higher vertebrates, degenerate. With the mesonephric tubules now constituting the dominant variety of nephric tubule, the primary nephric ducts are now called the *mesonephric (Wolffian) ducts* because of their new associations.

While the posterior parts of the mesonephros are still becoming established in their final form, a small outgrowth appears at the cloacal end of each mesonephric duct (Fig. 16-1D). These ducts, called the *metanephric diverticula* or *ureteric buds*, grow into the most posterior part of the intermediate mesoderm and stimulate the formation of the third major type of tubule, the *metanephric tubule*. Instead of being strung out along the dorsal wall of the body cavity like the pronephric and mesonephric tubules, the metanephric tubules become aggregated into a compact mass, which ultimately becomes the definitive kidney. With the establishment of the metanephroi, or permanent kidneys, the mesonephroi begin to degenerate. The only parts of the mesonephric system to persist, except in vestigial form, are some of the ducts and tubules which, in the male, are appropriated by the testis as a duct system.

Despite the variations in morphology, the different varieties of nephric tubules all function in a roughly similar manner. Fluid enters the proximal end of the tubule either as a filtrate from the blood, in the case of tubules supplied with glomeruli, or as coelomic fluid, in the case of pronephric and primitive mesonephric tubules which have openings (*nephrostomes*) into the body cavity (Fig. 16-2). All functional tubules are closely associated with networks of capillaries, and as the fluid flows down the tubules, it is processed by the actions of the epithelial cells lining the tubules and the associated capillaries. Ultimately the fluid is transformed into urine and flows down the appropriate duct systems to the exterior. The composition of the urine varies in accordance with both functional needs and the arrangement of the tubules which produce the urine.

Pronephros   The pronephros is well developed and functional in embryos and larvae of fishes and amphibians, but in embryos of birds and mammals it is represented only by several pairs of rudimentary tubules or cords of cells. In amphibians, the pronephros is made up of only three to five pairs of tubules, arising at the levels of the second through fourth somites.

A typical pronephric tubule arises from a *nephrotome*, a hollow mass of intermediate mesodermal cells which comes off of the somite like a stalk. A ventral opening of the nephrotome, called the *nephrostome*, is continuous with the coelom, which separates the visceral and parietal layers of the lateral plate mesoderm (Fig. 16-2B). From the nephrostome the pronephric tubule extends laterally. The lateral (distal) ends of several pronephric tubules come together to form the primordium of the primary nephric duct. In its early form it is a solid cord of cells, but later it hollows out into a duct.

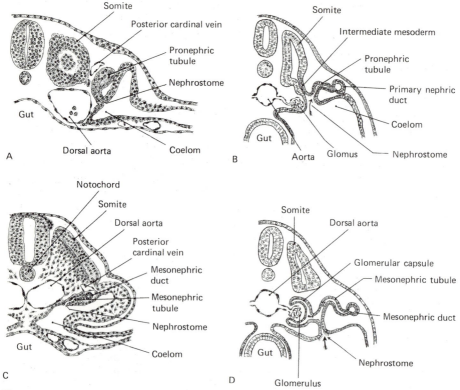

Fig. 16-2 Drawings to show nephric tubules. (A) Drawing from transverse section through twelfth somite of 16-somite chick to show pronephric tubule. *(After Lillie.)* (B) Schematic diagram of functional pronephric tubule. *(After Wiedersheim.)* (C) Drawing from transverse section through seventeenth somite of 30-somite chick to show primitive mesonephric tubule. (D) Schematic diagram of functional mesonephric tubule of primitive type. *(After Wiedersheim.)*

The primary nephric duct, growing from an apical proliferation center, extends backwards toward the cloaca, following environmental cues as it pushes caudally. If the normal pathway of extension is disturbed by partially transecting the embryo or even rotating the posterior end (Fig. 16-3), the primary nephric duct often makes its way back to the correct path and continues its course toward the cloaca (Holtfreter, 1943). In some species removal of the tip of the primary nephric duct prevents its further extension..

The pronephros in the chick is represented by tubules which first appear at about 36 hours of incubation. The pronephric tubules arise from the intermediate mesoderm lateral to the somites. They are paired, segmentally arranged structures, a tubule appearing on either side opposite each somite from the fifth to the sixteenth. Each tubule arises as a solid bud of cells organized from the intermediate mesoderm near its junction with the lateral mesoderm (Fig. 16-2A). At first the free ends of the buds grow dorsad, close to the posterior cardinal veins. Later the end of each tubule is bent caudad, coming in con-

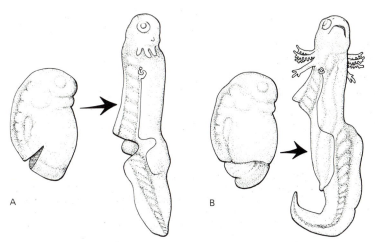

Fig. 16-3 Experiments involving interference with the pathway of migration of the pronephric duct (black lines) in amphibian larvae. (A) After partial transection of the embryo, the duct courses around the defect and then returns to its customary route of extension. (B) After rotation of the posterior part of the embryo, the duct assumes its normal course through the tissues of the rotated segment. *(After Holtfreter, 1943, Rev. Canad. Biol. 3:220.)*

tact with the tubule lying posterior to it. In this manner the distal ends of the tubules on each side give rise to a continuous cord of cells, the primordium of the primary nephric duct. In the manner already described for amphibians, each primary nephric duct extends caudad until it reaches the cloaca (Fig. 16-1B and C).

The significance of the rudimentary structures which represent pronephric tubules in the chick can most readily be understood by comparing them with fully developed and functional pronephric tubules. Figure 16-2B shows the scheme of organization of a functional pronephric tubule. The ciliated *nephrostome* draws in fluid from the coelom. As the coelomic fluid passes close to the capillaries of a vascular ridge known as the *glomus*, waste materials from the blood are transferred to it. The primary nephric duct serves to collect and discharge the fluid passing through the tubules. In the pronephric tubules of the chick embryo there are vestiges of a nephrostome opening into the coelom (Fig. 16-2A), but the tubules never become completely patent and never acquire the vascular relations characteristic of the functional pronephros in primitive vertebrates. Shortly after their initial appearance the pronephric tubules begin to undergo regressive changes, and by the end of the fourth day of incubation a few isolated epithelial vesicles are all that remain to chronicle the transitory appearance of the avian pronephros.

Mesonephros The mesonephric tubules develop from the intermediate mesoderm caudal to the pronephros. The early steps in their formation are well shown in transverse sections of chicks of 29 to 30 somites (about 55 hours). In the caudal region conditions are less advanced than they are more cephalically.

A possible explanation for the cephalocaudal gradient of development in the mesonephric tubules lies in the relationship between the primary nephric duct and the mesonephrogenic mesoderm. Waddington (1938) has found that if the caudal extension of the primary nephric duct is interrupted, mesonephric tubules either do not form or develop only poorly caudal to the level of interruption. It is likely, therefore, that as the primary nephric duct pushes back toward the cloaca, it stimulates or induces the intermediate mesoderm along the way to form mesonephric tubules.

A mesonephric tubule differs from a pronephric one chiefly in its relation to the blood vessels associated with it. It develops a cuplike outgrowth into which a knot of capillaries is pushed. The cup-shaped outgrowth from the tubule is called the *glomerular capsule* or *capsule of Bowman*, and the tuft of capillaries is termed a *glomerulus* (Fig. 16-2D). The diminutive form *glomerulus* is used to distinguish such small, localized capillary tufts from the continuous vascular ridge of the pronephros called a *glomus*.

Chick embryos show some cephalically located mesonephric tubules suggestive of the more primitive type such as that diagramed in Fig. 16-2D, with a ciliated nephrostome which draws in fluid from the coelom. This fluid mingles in the tubule with the fluid from the glomerulus. In the embryos of higher vertebrates only a few of the more cephalic of the mesonephric tubules ever show nephrostomes, and these are rudimentary (Fig. 16-2C). The great bulk of the functioning mesonephros is made up of tubules which form no nephrostomes (Fig. 16-4) but depend entirely on their glomerular apparatus for their fluid intake. Nevertheless, the primitive mesonephric tubule with a nephrostome is of considerable interest as a transition form between the pronephric tubules and the final type of mesonephric tubule. Figuratively speaking, it is as if nature conservatively held on to the old way of bringing in fluid through the nephrostome, as a safety factor, until the new glomerular source of fluid intake has proved itself in actual operation.

The early stages in the formation of the definitive type of mesonephric tubules without nephrostomes are much the same in mammalian material as those seen in 2- to 4-day chicks. Once having established their connection with the primary nephric duct (Fig. 16-4B), the primordial tubules elongate and take on a S-shaped configuration (Fig. 16-4C and D). Continued increase in length brings about a series of secondary bendings (Fig. 16-4E) which increase their surface exposure, thereby enhancing their capacity for interchanging materials with the blood in the adjacent capillaries.

Easy to overlook in the study of sections, but of vital functional significance, is the course of the blood after it has passed through the glomerulus. Leaving the glomerulus, the blood goes out over one or more vessels which promptly break up into a rich capillary plexus closely applied to the mesonephric tubules. The capillaries empty into collecting veins (Fig. 16-4E), which ultimately return the blood into the general circulation via the subcardinal and posterior cardinal veins.

The mesonephric tubules function much like the nephrons of the adult kidney. A filtrate of blood from the glomerulus enters the capsule and flows into

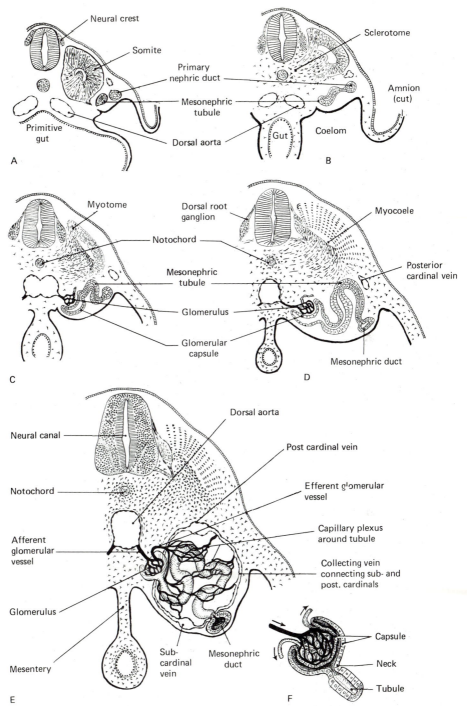

Fig. 16-4 Development of mesonephric tubules and their vascular relations. (A) Tubule primordium still independent of duct. (B) Union of tubule with primary nephric duct. (C) Early stage in development of glomerulus and capsule. (D) Further development of capsule and lengthening of tubule. (E) Relations of blood vessels to well-developed mesonephric tubule. (F) Glomerulus and capsule, enlarged.

the tubule, where selective resorption of ions and other substances occurs. A major difference between the mesonephric kidney and the permanent kidney of higher vertebrates is the relative inability of the mesonephros to concentrate urine. This is related to the elongated structure of the mesonephros and the absence of a well-developed renal medulla, a structural adaptation of land animals to preserve water by concentrating it through an elaborate countercurrent exchange mechanism. Such a fluid-conserving mechanism is not needed by the embryo, which lives in a bath of amniotic fluid, just as preservation of body water is not a problem for the mesonephric kidney of fishes and aquatic amphibians. Many details of the function of the mesonephros in the embryo are still poorly understood.

Although it is relatively more conspicuous early in development, the mesonephros does not attain its greatest actual bulk until later. In human embryos it reaches its maximum size toward the close of the second month. In pig embryos, which have relatively much larger mesonephros, its greatest size is attained in embryos of about 60 mm. When the metanephros becomes well developed, the mesonephros undergoes rapid involution and ceases to be of importance in its original capacity, but in the male, its ducts and some of its tubules still persist and give rise to structures of vital functional importance.

Metanephros The development of the metanephros begins with the appearance of a tiny budlike outgrowth from the mesonephric duct just cephalic to the point where the duct opens into the cloaca (Fig. 16-5A). The outgrowth, called the *metanephric diverticulum*, pushes into the posterior portion of the intermediate mesoderm, which condenses around the diverticulum to form the *metanephrogenic blastema*. Thus, from the very beginning the permanent kidney has a dual origin—the metanephric diverticulum, which gives rise to the ureter, the renal pelvis, and the collecting duct system, and the intermediate mesoderm, from which the tubular units of the kidney arise.

The essential feature of metanephric development is the elongation and dichotomous branching (up to 14 or 15 times in the human) of the metanephric diverticulum and the formation of tubular excretory units around the tips of the branches. This is accomplished through a series of reciprocal inductive interactions between the two components of the metanephros. The terminal portions of the metanephric duct induce the formation of metanephric tubules in the metanephrogenic mesoderm which surrounds it (Grobstein, 1955). In the absence of the metanephric duct, tubules do not appear. Conversely, the metanephrogenic mesoderm, acting in turn on the metanephric duct, induces the characteristic branching pattern of the duct system (Erickson, 1968). These reciprocal inductive interactions continue throughout the tubule-forming stage of kidney development. The morphological characteristics of the tubules formed from the metanephrogenic mesoderm are not rigidly fixed, however, for if this mesoderm is brought into contact with mesonephric ducts, tubules of the mesonephric variety are formed. This lability of developmental fate has been used by Torrey (1971, p. 340) as evidence in favor of the holonephric concept of the vertebrate kidney.

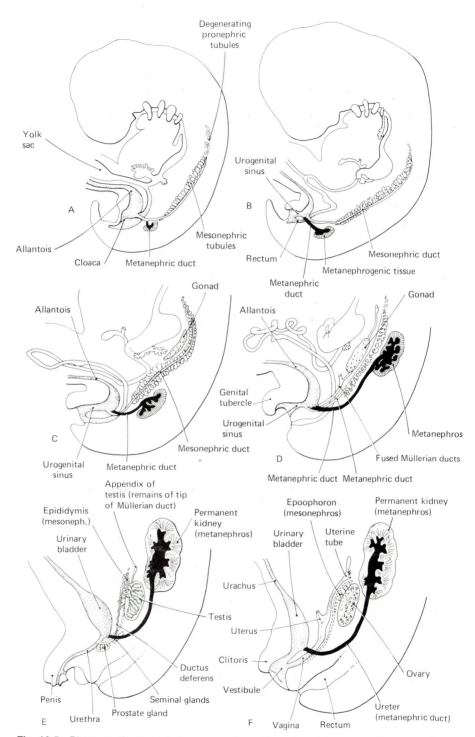

Fig. 16-5 Diagrams showing relative sizes and positions of nephric organs of human embryo at various stages of development. (A) Early in fifth week. *(Adapted from several sources covering 5- to 6-mm embryos.)* (B) Early in sixth week. *(Modified from Shikinami's 8-mm embryo.)* (C) Seventh week. *(Modified from Shikinami's 14.6-mm embryo.)* (D) Eighth week. *(Adapted from Shikinami's 23-mm embryo and the Kelly and Burnam 25-mm stage.)* (E) Male at about 3 months—schematized. (F) Female at about 3 months—schematized.

The developing metanephros has been an important model for experimental embryologists trying to understand the nature of inductive interactions. Analysis of this inductive process has been greatly facilitated through an experimental model by which metanephric induction can be demonstrated in vitro. Grobstein (1955) isolated tissue of the metanephric diverticulum and mesenchyme of the metanephrogenic blastema and placed them in culture, with a porous filter separating the two tissues. Despite the interposition of the filter, metanephric tubules formed in the mesenchyme, suggesting that the inductive reaction was mediated by diffusion of a chemical. More recent work with different varieties of filters has shown that cellular processes penetrate the small (< 1.0 µm) pores of the filter so that the cells on either side of the filter are actually in close contact with one another (Lehtonen, 1976). Nevertheless, the experimental evidence suggests that the inductive effect is mediated by the short-range transmission of active compounds from one component of the system to the other (Saxén and Lehtonen, 1978).

Almost from its first appearance, the blind end of the metanephric diverticulum is dilated, foreshadowing its subsequent enlargement to form the lining of the pelvis of the kidney. The portion of the diverticulum near the mesonephric duct remains slender, presaging its eventual fate as the ureter (Fig. 16-5F). The metanephric diverticulum, or duct, elongates rapidly. While this is occurring, the cells of the metanephrogenic blastema become concentrated around the distal end of the metanephric duct and lose their original relations with the intermediate mesoderm (Fig. 16-6B). As the pelvic end of the diverticulum expands within its investing mass of mesoderm, it takes on a shape suggestive of the pelvic cavity of the adult kidney (Figs. 16-7 and 16-8). From this early pelvic dilation arise the numerous outgrowths of the future system of collecting tubules (Fig. 16-8E). These push radially into the surrounding mass of nephrogenic mesoderm, and the histogenesis of the renal tubules begins.

Histogenesis of the Metanephros Individual uriniferous tubules arise near the distal ends of the terminal branches of the collecting duct system. Cells of the metanephric blastema become arranged into small vesicular masses which lie close to the blind end of a collecting duct (Fig. 16-9). Each of these masses becomes an elongated and highly convoluted tubule. One end of the tubule becomes attached to the collecting duct, the site of junction being the boundary between the distal convoluted tubule and the arched collecting duct in the differentiated kidney. The other end of the developing tubule becomes associated with an arteriolar branch of the renal artery. The arteriole breaks up into a small glomerulus and the tubule expands around it to form the glomerular capsule. The glomerulus reforms into an arteriole, which again breaks up into a network of capillaries which intertwine about the tortuous uriniferous tubule to form the structural basis for the formation of urine. One portion of the tubule becomes greatly elongated into a hairpin shape to form the *loop of Henle*, which extends toward or into the renal medulla.

As the kidney grows in mass, additional generations of tubules are formed

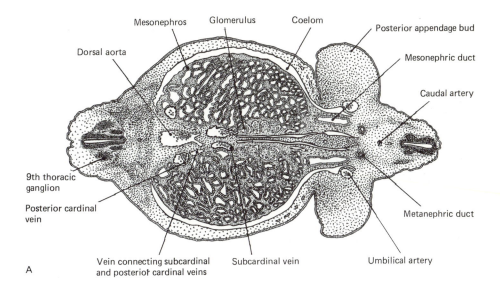

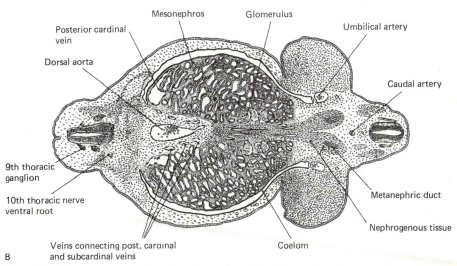

Fig. 16-6 Two transverse sections from the series of the 9.4-mm pig embryo used in making the reconstructions illustrated in Fig. 17-13. *(Projection drawings, X17.)* By laying a straightedge across either of these reconstructions at the level of the marginal line bearing the serial number of a cross section, the relations of that section within the body as a whole are precisely indicated. (A) Section 518 passing through meso- and metanephric ducts; (B) section 529 passing through the main concentration of metanephrogenous tissue.

in its peripheral zone. The development of the internal architecture of the kidney is structurally very complex. Basically, the collecting duct system expands outwards as its pattern of branching continues, and new tubules continue to form at the ends. As the result of a careful microdissection study, Osathanondh and

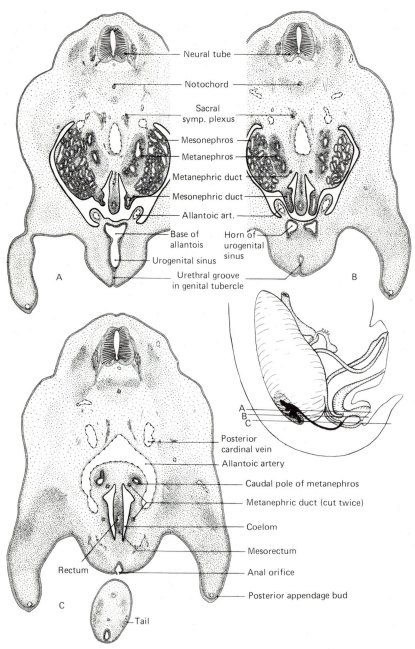

Neural tube

Notochord

Sacral
symp. plexus

Mesonephros

Metanephros

Metanephric duct

Mesonephric duct

Allantoic art.

Base of
allantois

Horn of
urogenital
sinus

Urogenital sinus

Urethral groove
in genital tubercle

A

B

Posterior
cardinal vein

Allantoic artery

Caudal pole of metanephros

Metanephric duct (cut twice)

Coelom

Mesorectum

Anal orifice

Posterior appendage bud

Rectum

C

Tail

Fig. 16-7 Projection drawings (X15) of transverse sections through the pelvic region of a 15-mm pig embryo. The level of each section is indicated on the inset.

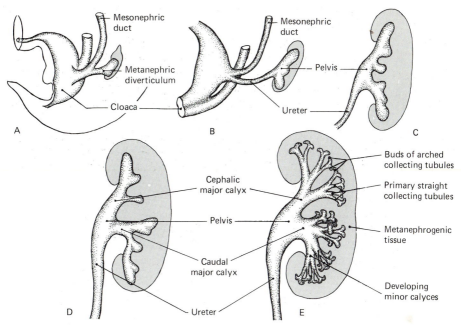

Fig. 16-8 Diagrams showing a series of stages in the growth and differentiation of the metanephric diverticulum.

Potter (1963, 1966) have divided the histogenesis of the human metanephros into four stages: period 1—a period of dichotomous branching of the metanephric duct and the formation of tubules (nephrons) at the end of each branch; period 2—the formation of arcades of nephrons, connected to the elongating ends of the collecting ducts by small canals; period 3—the continued formation of nephrons, which connect directly with the elongating collecting ducts rather than forming arcades; period 4—a period of interstitial growth, in which the expanded ampullary ends of the collecting ducts lose prominence and no new nephrons form. A detailed, well-illustrated summary of renal histogenesis can be found in Hamilton, Boyd and Mossman (1972, p. 384). Upon completion of its development, the metanephros has about 15 generations of nephrons, and Osathanondh and Potter (1966) reported that the kidneys at birth contain about 822,300 nephrons.

Later Positional Changes of the Kidney   When the metanephroi are first established they are located far caudally in the growing body (Fig. 16-5A and B), but as development progresses they come to lie relatively much farther cephalad (Fig. 16-5C to 16-5F). Their own actual movement headward is not quite so great as their change in relative position would indicate. Part of their apparent migration is due to the marked expansion of the portion of the body caudad to them. It is so much easier to describe the changing relations of the kidney as if

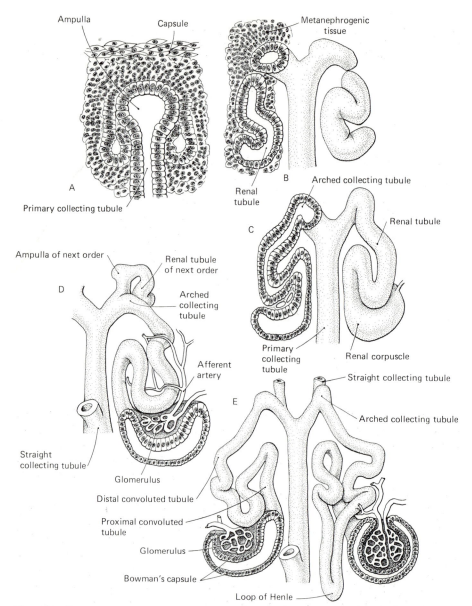

Fig. 16-9 Diagrams showing the development of the metanephric tubules of mammalian embryos. *(After Huber, from Kelly and Burnam,* Diseases of Kidneys, Ureters, and Bladder. *Courtesy of Appleton-Century-Crofts, Inc., New York.)*

they were entirely accounted for by its own migration that this factor of differential growth in the caudal region is frequently ignored. The metanephros in human embryos arises opposite the twenty-eighth somite (fourth lumbar segment). At term it has moved up to the level of the first lumbar vertebra, or even as high as the twelfth thoracic.

More striking than the change in segmental level is the shift of the kidneys out of the pelvic part of the coelom. In young embryos the kidneys lie retroperitoneally, bulging into the narrow pelvic cavity, caudal to the bifurcation of the aorta where it gives rise to the umbilical arteries. During the seventh week (Fig. 16-10A and B) the kidneys start to slide forward over the ridges formed by the umbilical arteries. By the ninth week they have cleared this narrowed portion of the coelomic chamber and come to lie against the dorsolateral body walls above the arterial fork (Fig. 16-10C and D). In this part of their ascent they are rotated about a quarter turn so that their convex borders, originally directed dorsally, become directed laterally. After this their changes in position are

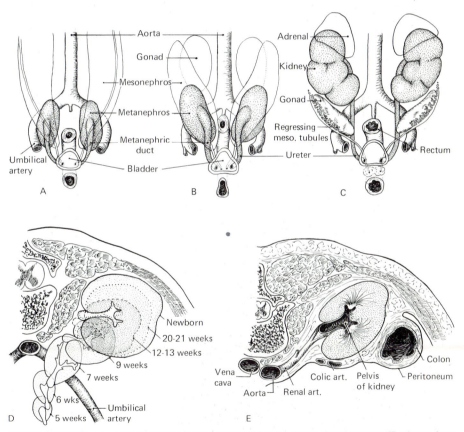

Fig. 16-10 Diagrams showing changes in position of kidney during development. *(Redrawn from Kelly and Burnam,* Diseases of Kidneys, Ureters, and Bladder. *Courtesy Appleton-Century-Crofts, Inc., New York.)* (A–C) Frontal views showing ascent of kidneys out of pelvis. Note their rotation, occurring chiefly as they rise above the common iliac arteries. (D) Schematic composite diagram showing rotation of kidney as seen in a cross-section of the body. (E) Location of kidney as seen in a cross section of the adult body at lumbar level.

slower. They move cephalad about two more segments and settle more deeply into the subperitoneal fat and connective tissue on the inner side of the dorsal body wall (Fig. 16-10D), gradually reaching their characteristic adult position (Fig. 16-10E).

Formation of the Bladder and Early Changes in the Cloacal Region   In dealing with the development of the extraembryonic membranes, we have already taken up the formation of the allantois as an evagination from the caudal end of the primitive gut (Figs. 8-10C and D, and 7-7). Shortly after this occurs, the gut caudal to the point of origin of the allantois becomes enlarged to form the cloaca (Fig. 16-11). When the cloacal dilation is first formed, the hindgut (postcloacal gut) still ends blindly. Under the root of the tail, the ectoderm sinks in toward the gut to form the *proctodeum*. The thin plate of tissue formed by the apposition of the proctodeal ectoderm and the endoderm of the hindgut is known as the *cloacal membrane* or *plate*, which eventually ruptures, establishing a caudal outlet for the gut in much the same manner that rupture of the oral plate has previously established communication between the stomodeum and the foregut.

Before this occurs important changes take place internally. The cloaca begins to be divided into two parts, a dorsal part, which forms the *rectum*, and a ventral part, the *urogenital sinus* (Fig. 16-11). This division is effected by the growth of the *urorectal fold*, a crescentic fold which cuts into the cephalic part of

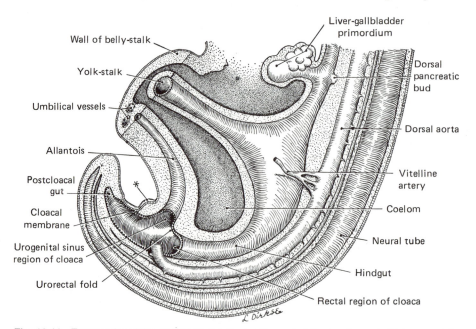

Fig. 16-11   Topography of the cloacal region toward the end of the fourth week. Note especially that the primordial genital tubercle is paired at this stage. The right member of the pair, indicated here by an asterisk, is shown in perspective just beyond the midline.

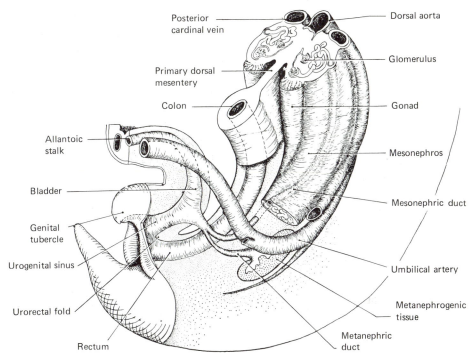

Fig. 16-12 Schematic ventro lateral view of the urogenital organs of a young mammalian embryo. *(Redrawn from Kelly and Burnam,* Diseases of Kidneys, Ureters, and Bladder. *Courtesy of Appleton-Century-Crofts, Inc., New York.)* The figure as originally drawn was based on human embryos of 12 to 14 mm. In all essentials the relations shown are equally applicable to 14- to 15-mm pig embryos.

the cloaca where the allantois and the gut meet (Fig. 16-11). The two limbs of the fold bulge into the lumen of the cloaca from either side, eventually meeting and merging with each other. The progress of this partitioning fold toward the proctodeal end of the cloaca (Fig. 16-12) makes it difficult to keep track of the original limits of the allantois, because as the urogenital sinus is lengthened, it is, in effect, added onto the allantois (Fig. 16-5A to 16-5D). The point of entrance of the mesonephric ducts, however, affords a landmark which is sufficiently accurate for all practical purposes.

At about the same time that the cloacal membrane ruptures, separation of the cloaca is complete and its two parts open independently (Fig. 16-13). The opening of the rectum is the *anus*, and that of the urogenital sinus is the *ostium urogenitale*.

Meanwhile the proximal part of the allantois has become greatly dilated and can now quite properly be called the *urinary bladder*. We should remember, however, that the neck of the bladder has been formed from tissue which was originally part of the cloaca.

In the growth of the bladder the caudal portion of the mesonephric duct is absorbed into the bladder wall. This absorption progresses until the part of the

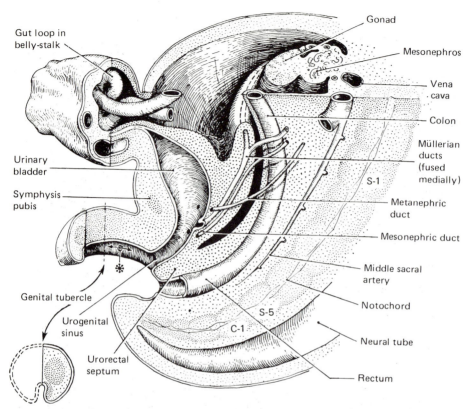

Fig. 16-13 ' Semischematic drawing of reconstruction of urogenital system of a human embryo of eighth week. The arrow, keyed by the asterisk, indicates the urethral groove. *(After Patten and Barry, 1952,* Am. I. Anat., *vol. 90.)*

mesonephric duct caudal to the point of origin of the metanephric diverticulum has disappeared. The end result of this process is that the mesonephric and metanephric ducts open independently into the urogenital sinus. The metanephric duct, possibly because of traction exerted by the kidney in its migration headward, acquires its definitive opening somewhat laterally and cephalically to that of the mesonephric duct. It then discharges into the part of the urogenital sinus which was incorporated into the bladder. The mesonephric ducts open into the part of the urogenital sinus which remains narrower and gives rise to the urethra (Fig. 16-5D to 16-5F). The urethra acquires quite different relations in the two sexes. It is, therefore, desirable to defer consideration of it and take it up in connection with the external genitalia.

THE DEVELOPMENT OF THE REPRODUCTIVE ORGANS

Factors Involved in Sexual Differentiation   Differentiation of the reproductive system is a complex process, involving a number of different mechanisms operating at several stages of development. Two important generalizations regarding mammalian sexual differentiation should be kept in mind. One is that several of the major genital structures (e.g., gonads, sexual ducts, external genitalia) first pass through a morphologically *indifferent stage* in which they cannot be identified as being either male or female (Figs. 16-14 and 16-15). Later, a course of development characteristic of one or the other sex occurs (Table 16-1). The other major generalization is the inherent tendency of genital structures to develop into the female type in the absence of specific masculinizing influences.

The first critical stage of sexual differentiation occurs at the moment of fertilization, when the genetic sex of the zygote is determined by the nature of the sex chromosome contributed by the sperm. Although an XY zygote is destined to become a male, no distinctive differences between the early development of male and female embryos have been noted. The principal function of the Y chromosome is to direct the differentiation of the indifferent gonad into a testis. This is accomplished after the migration of the primordial germ cells into the early gonad (see page 58).

The sex-determining function of the Y chromosome is intimately bound with the activity of the H-Y antigen, an antigen which for many years had been recognized as a minor Y-linked histocompatibility antigen that was known to result in the rejection of male skin grafted onto females of the same inbred strain of mice. This role of the H-Y antigen seems, in a sense, to be an immunological

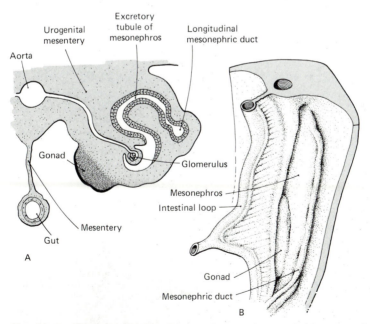

Fig. 16-14 Diagrams showing the organization of urogenital structures in a 5-week human embryo. (A) Cross-section through the thoracic region. (B) Dissection of an embryo. *(After Langman, 1975, Medical Embryology, 3rd ed., The Williams & Wilkins Co., Baltimore.)*

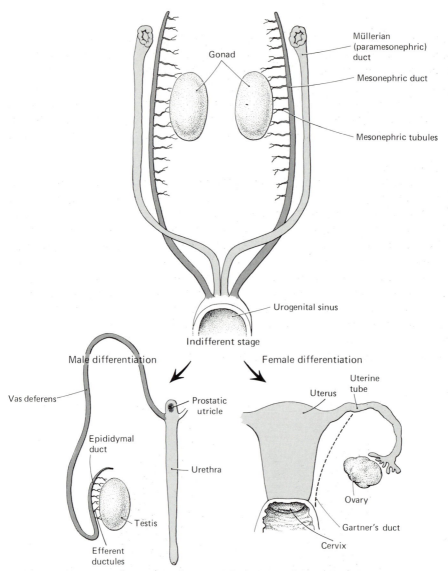

Fig. 16-15 Schematic diagram showing the differentiation of the male and female reproductive organs from an indifferent stage. The prostatic utricle in the male represents the caudal remains of the Müllerian duct system. Gartner's duct in the female is the remains of the mesonephric duct.

byproduct of its major function, which is to cause the organization of the primitive gonad into the testis in male animals (Wachtel et al., 1975; Ohno, 1978). In the absence of the H-Y antigen, the gonad later becomes transformed into the ovary. The transformation of the gonad into a testis or an ovary completes the second critical stage of sexual differentiation, namely, the establishment of gonadal sex.

The next, and most obvious, phase in sexual differentiation of the embryo

Table 16-1   Major Homologies in the Urogenital System

Male derivative	Indifferent structure	Female derivative
Testis	Gonad	Ovary
Spermatozoa	Primordial germ cells	Ova
Seminiferous tubules	Sex cords	Follicles
Efferent ductules, paradidymis	Mesonephric tubules	Epoöphoron; other rudiments
Epididymal duct, ductus deferens	Mesonephric (Wolffian) duct	Degenerates, canals of Gartner
Degenerates, appendix of testis	Paramesonephric (Müllerian) duct	Uterine tubes, uterus, part of vagina (?)
Bladder, prostatic urethra	Early urogenital sinus (upper)	Bladder, urethra
Lower urethra	Definitive urogenital sinus (lower)	Vestibule
Penis	Genital tubercle	Clitoris
Floor of penile urethra	Genital folds	Labia minora
Scrotum	Genital swellings	Labia majora

is the differentiation of somatic sex. The early embryo develops a dual set of potential genital ducts, one the original mesonephric duct, which persists after degeneration of the mesonephros as an excretory organ, and another, newly formed pair of ducts, called the *paramesonephric (Müllerian) ducts*. Under the influence of testosterone secreted by the testes, the mesonephric ducts develop into the duct system through which spermatozoa are conveyed from the testes to the urethra. Differentiation of the major glands associated with the ducts (prostate and seminal vesicle) also depends upon testosterone. The potentially female paramesonephric ducts regress under the influence of another secretion of the embryonic testes, the *Müllerian inhibitory factor*. In genetically female embryos, neither testosterone nor Müllerian inhibitory factor is secreted by the gonads. In the absence of testosterone the mesonephric ducts regress, and the lack of Müllerian inhibitory factor permits the paramesonephric ducts to develop into the oviducts, uterus, and part of the vagina (Jost, 1972). The external genitalia also first take form in a morphologically indifferent condition and then develop either in the male direction under the influence of testosterone or in the female direction if the influence of testosterone is lacking.

A good example of the natural tendency of the body to develop along female lines in the absence of other modifying influences is seen in *Turner's syndrome*, a rare human condition characterized by the deletion of one of the sex chromosomes (XO). Although individuals with Turner's syndrome are sterile and gonadally undifferentiated, the internal and external genitalia are easily recognizable as female in type. There is increasing evidence that testosterone acts upon the developing brain and may affect behavior in a sexually dimorphic manner.

The last stages of sexual differentiation occur after birth. The newborn baby is assigned a sex on the basis of its sexual phenotype, and under normal circumstances it develops psychologically as a member of that sex. The final major events in sexual differentiation occur at puberty, when the overall body configuration becomes transformed from what is essentially an indifferent form to a mature male or female type with the appearance of the secondary sexual characteristics. For either sex the development of secondary sexual characteristics requires specific input from gonadal hormones. In the male, this follows the pattern set early in embryonic life, but in the female this is the first stage at which specific ovarian hormonal influences come into play in the process of sexual differentiation.

It has long been recognized that sometimes the genital structures of an embryo differentiate in a way counter to that which would be predicted by the genetic sex. A classical example of this is the freemartin in cattle (Lillie, 1917). If a heterosexual pair of twin cattle develops in utero with fusion of the extraembryonic blood vessels and intermingling of the blood, the male develops normally. In contrast, there is a large scale reversal of many sexual structures in the female, resulting in their resembling those of the male. Such a modified female is sterile and is called a *freemartin*. Lillie attributed the sex reversal in freemartins to an overcoming of normal female sexual development by hormones produced by the male, and his theory provided the major impetus to subsequent studies of factors controlling sexual differentiation.

Another example of a disparity between genital and somatic sex is a condition called *testicular feminization*. Here the genetic sex of the individual is male and the gonads (internal testes) produce large amounts of testosterone, but the external somatic sex develops into that of a typical female. This condition is occasionally seen in humans and is present in a strain of mice. It is due to the lack of development of specific cellular receptors for testosterone, so that despite a plentiful supply of circulating testosterone, the hormone cannot be utilized by the testosterone-sensitive tissues which would normally become male genital structures. The testes, however, still produce Müllerian inhibitory factor. As a result, the uterine tubes and uterus fail to form.

Early Differentiation of the Gonads   From their earliest appearance the gonads are intimately associated with the nephric system. While the mesonephros is still the dominant excretory organ, the gonads arise as ridgelike thickenings (*gonadal ridges*) on its ventromesial face (Fig. 16-14). Histologically, the early gonad consists essentially of a mesenchymal core covered by a mesothelium, called the *germinal epithelium* because it was at one time presumed to give rise to the germ cells.

Differentiation of the indifferent gonads into ovaries or testes occurs after the arrival of the primordial germ cells. The mechanisms underlying gonadal differentiation have been the object of investigation and debate for many years, and even the morphology of the process is not agreed upon by all. Two somewhat different viewpoints have long been dominant. According to one concept (Fig. 16-16) *primary sex cords* grow from the germinal epithelium into the gonadal mesenchyme. In males these cords become prominent components of the central medullary part of the gonad and become the testis cords, or seminiferous tubules. The testis cords are then separated from an almost nonexistent cortex by a layer of connective tissue called the *tunica albuginea*. In ovarian differentiation the primary sex cords regress in the medulla and outside a rudimentary tunica albuginea *cortical*, or *secondary sex*, *cords* again grow in from the germinal epithelium to form the ovarian follicles. When this concept first arose it was thought that the sex cords originally gave rise to the germ cells,

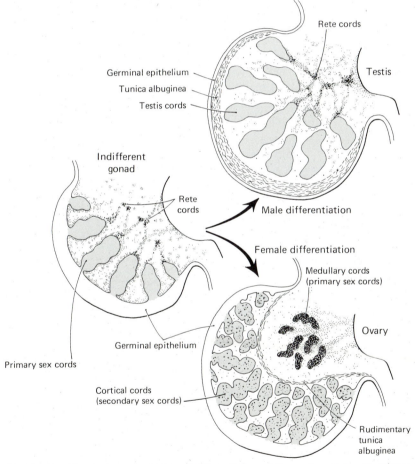

Fig. 16-16 Classical representation of differentiation of the male and female gonads from an indifferent gonad *(after Burns)*. The primary sex cords represent the male (medullary) component. The germinal epithelium represents the female (cortical) component. In the male medullary development dominates whereas in the female cortical development is dominant. See Fig. 16-17 for a more contemporary view of gonadal differentiation.

but this viewpoint was modified so that only the Sertoli cells in the male and the follicular epithelial cells in the female were assumed to originate from the sex cords.

Another concept, more concerned with mechanisms than with morphological detail, was proposed by Witschi (1939), who looked upon gonadal differentiation as the result of a natural antagonism between the cortical and medullary parts of the gonad. The genetic sex of the embryo was said to determine that one component of the gonad would prevail and that development of the other would be suppressed. In males, the medulla, and in females, the cortex, would be dominant.

According to a more recent viewpoint held by a number of investigators (reviewed by Jost, 1972), the mesenchyme of the indifferent gonad shows no obvious organization (Fig. 16-17). Testicular cells, under the influence of the H-Y antigen, aggregate very early into primitive seminiferous tubules, containing both germ cells and Sertoli cells (Ohno, 1978). The testes soon become hormonally functional, and by their secretions they influence the development of the genital ducts and the external genitalia to differentiate in the male direction. Not until late in pregnancy, after meiosis has begun, do the ovaries depart from their indifferent configuration and primordial follicles begin to form. By the time

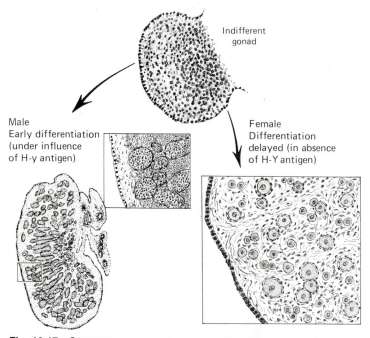

Indifferent gonad

Male
Early differentiation
(under influence
of H-y antigen)

Female
Differentiation
delayed (in absence
of H-Y antigen)

Fig. 16-17 Contemporary view of gonadal differentiation. According to Jost (1972) the internal structure of the indifferent gonad is not so highly organized as was once believed. The testis (*lower left*) is that of a 14-week human embryo. The box in the center shows details of the structure of the early seminiferous tubules. The section of ovary is from a newborn infant.

this has occurred, differentiation of the internal and external genitalia into the female type has already largely been accomplished.

Early in development a *rete* system appears in gonads of both sexes. The origin of the rete system is not completely clear. An extragonadal component may represent persisting portions of mesonephric tubules, and the intragonadal component may either be an extension of the extragonadal rete or arise in situ. In the testis, the *rete testis* persists as a network of channels which connect the seminiferous tubules to the efferent ductules, whereas in the ovary, the *rete ovarii* gradually loses prominence.

Recent anatomical and experimental evidence suggests that the embryonic rete may secrete a diffusible factor which acts as a trigger for meiosis (Byskov, 1978). According to this hypothesis the rete in both the ovaries and testes of the early gonad secrete a *meiosis-inducing factor*, but the earlier isolation of the male germ cells within the seminiferous tubules prevents them from being exposed to the effects of this factor (and therefore from entering meiosis), whereas the exposed female germ cells begin the first meiotic prophase in the embryo. Experiments involving the culture of fetal mouse ovaries and testes (Byskov and Saxén, 1976) have also provided some evidence in favor of a meiosis-preventing effect (or substance) by the cells of the seminiferous epithelium. By analogy, the first meiotic block in the female germ cells has been attributed to a similar blocking effect by the follicular epithelium which surrounds the ova later in the fetal period.

THE SEXUAL DUCT SYSTEM OF THE MALE

The ducts which convey the spermatozoa away from the testis are, with the exception of the urethra, appropriated from the mesonephros—a developmental opportunism facilitated by the proximity of the growing testes to the degenerating mesonephros (Fig. 16-18). The mesonephric structures which are taken over by the testes are shown schematically in Fig. 16-19.

Epididymis   Some of the mesonephric tubules which lie especially close to the testes are retained as the *efferent ductules* (Fig. 2-5). They, together with that part of the mesonephric duct into which they empty, become the epididymis. Cephalic to the tubules which are converted into efferent ductules, a few mesonephric tubules sometimes persist in vestigial form as the *appendix of the epididymis* (Fig. 16-19). Caudal to the efferent ductules a cluster of mesonephric tubules almost invariably persists in rudimentary form as the *paradidymis*.

Ductus Deferens, Seminal Vesicle, and Ejaculatory Duct   Distal to the epididymis the mesonephric duct receives a thick investment of smooth muscle and becomes the ductus (vas) deferens. A short distance before the ductus deferens enters the urethral part of the urogenital sinus, it develops local dilations which become elaborately sacculated and form the seminal vesicle (Fig.

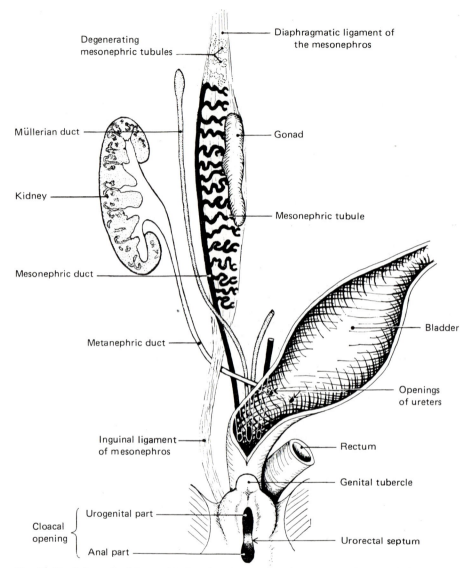

Fig. 16-18 Schematic diagram showing plan of urogenital system at an early stage when it is still sexually undifferentiated. *(Modified from Hertwig.)*

2-5). The short part of the mesonephric duct between the seminal vesicle and the urethra constitutes the ejaculatory duct. From this point on, the spermatozoa traverse the urethra, which serves as a common passageway to the exterior for both the sexual cells and the renal excretion.

Prostate and Cowper's Glands   Under the influence of fetal testosterone, the prostate and bulbourethral glands develop from the urethral epithelium

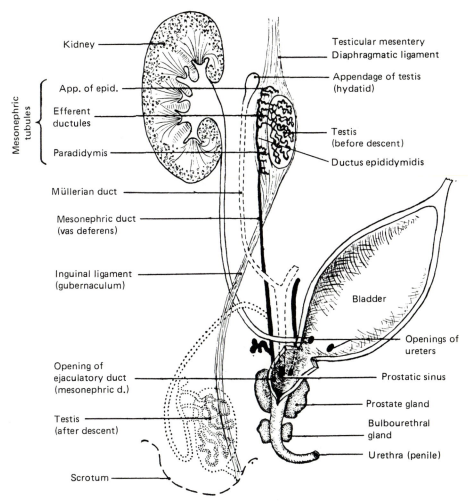

Fig. 16-19 Diagram of the male sexual-duct system in mammalian embryos. *(Modified from Hertwig.)* The dotted lines indicate the position of the testis and its ducts after descent into the scrotum.

(Price, 1926). The prostate surrounds the urethra near the neck of the bladder; the bulbourethral (Cowper's) glands lie adjacent (Figs. 2-5 and 16-19). Their secretions, discharged into the urethra together with that of the seminal vesicles, serve as a conveying and activating fluid for the spermatozoa.

THE SEXUAL DUCT SYSTEM OF THE FEMALE

The Müllerian (paramesonephric) ducts first appear close beside and parallel to the mesonephric ducts (Figs. 16-20 and 16-22A and B). They are the primordial structures from which the uterine tubes (oviducts) and uterus arise in the female

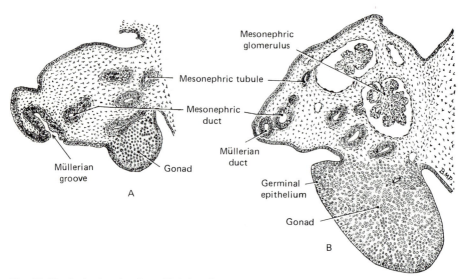

Fig. 16-20 Projection drawings (X125) at the upper mesonephric level of a 6½-week embryo to show the formation of the Müllerian duct. *(University of Michigan Collection, EH 707, CR 15 mm.)* (A) Open groove in the coelomic mesothelium of the mesonephros, near its cephalic pole. (B) Slightly caudal to the open groove, the Müllerian duct has become a completed tube, lying in the margin of the mesonephros just lateral to the mesonephric duct.

(Fig. 16-21). The Müllerian ducts come together caudally and approach the urogenital sinus at a point where the wall of the latter has thickened to form a *Müllerian tubercle*. Flanking the fused Müllerian ducts are the unfused ends of the mesonephric ducts, which enter the urogenital sinus and the Müllerian tubercle. In females, the mesonephric ducts degenerate.

Vagina   The origin of the vagina has not yet been satisfactorily determined. O'Rahilly (1977) has reviewed the varying opinions found in the literature. At present a widely accepted viewpoint is that the vagina originates almost entirely from a paired epithelial thickening (sinovaginal bulbs) arising from the dorsal wall of the urogenital sinus and connecting with the fused ends of the Müllerian ducts (Fig. 16-22). Other authors, however, still feel that the terminal portions of the Müllerian ducts make some contribution to the vagina.

Uterus   The extent of fusion of the Müllerian ducts varies from one group of mammal to another. Marsupials have paired uteri formed by enlargement of the Müllerian ducts cephalic to their entrance into the vagina (Fig. 16-23A). In all the higher mammals, fusion of the Müllerian ducts involves the caudal end of the uterus so that it opens into the vagina in the form of an unpaired neck or cervix. Toward the ovary from the cervix there is great variation in the degree of fusion encountered in the different groups (Fig. 16-23B to 16-23D). In the sow the fusion is carried only a short way beyond the cervix to form a typical *bicornate uterus* (Fig. 16-23C). In the human female, fusion of the paired Müllerian ducts is complete in the uterine region. As a result the uterus is

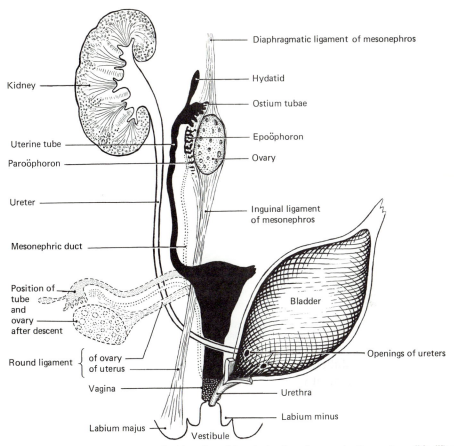

Kidney

Uterine tube

Paroöphoron

Ureter

Mesonephric duct

Position of
tube
and
ovary
after descent

Round ligament { of ovary
of uterus

Vagina

Labium majus

Vestibule

Diaphragmatic ligament of mesonephros

Hydatid

Ostium tubae

Epoöphoron

Ovary

Inguinal ligament
of mesonephros

Bladder

Openings of ureters

Urethra

Labium minus

Fig. 16-21 Schematic diagram showing plan of developing female reproductive system. *(Modified from Hertwig.)* The dotted lines indicate the position of the ovary and uterine tube after their descent into the pelvis.

pear-shaped with a single lumen (Fig. 16-23D). In distinction to a bipartite, or bicornate type, this is called a *simplex uterus.*

Uterine Tubes The part of the Müllerian duct between the uterus and the ovary remains slender and forms the uterine tube (oviduct). Near its cephalic end, but not usually at the extreme tip, a more or less funnel-shaped opening, or *ostium*, develops. In different forms the detailed configuration of the ostium and its relation to the ovary are quite variable. Conditions range all the way from a pouchlike dilation which almost completely invests the ovary (sow) and an elaborately fringed, funnel-shaped ostium which opens in the general direction of the ovary (humans). Whatever the morphological eccentricities of the ostium may be, they apparently make less difference in its efficiency in picking up the discharged ovum than one might suppose. Even in species possessing the least intimate relation of the ostium to the ovary, abdominal pregnancies resulting

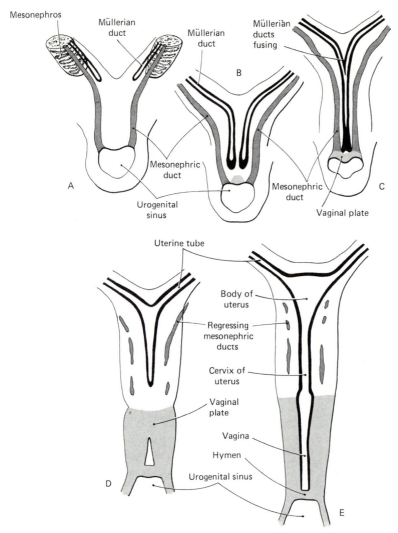

Fig. 16-22 Fusion of Müllerian ducts in human embryos and formation of the uterus and vagina. (A) Early eighth week; (B) mid-eighth week; (C) ninth week; (D) end of the third month; (E) mid-fetal period. *(A-C after Koff.)*

from the fertilization of an ovum which failed to enter the ostium are comparatively uncommon.

Vestigial Structures in the Genital Duct System   In the conversion of the primordial duct systems to their definitive conditions, some of the parts which are not utilized in the formation of functional structures persist in vestigial form even in the adult. Mention has already been made of the rudimentary mesonephric tubules which persist in the male as the paradidymis and the appendix of the epididymis. Traces of the old Müllerian duct system, also, can usually be found in the male. Attached to the connective tissue investing the

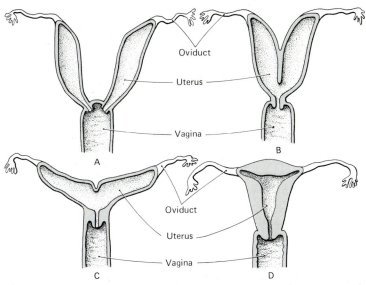

Fig. 16-23 Four types of uteri occurring in different groups of mammals. (A) Duplex, the type found in marsupials; (B) bipartite, the type found in certain rodents; (C) bicornate, the type found in most ungulates and carnivores; (D) simplex, the type characteristic of the primates. *(After Wiedersheim.)*

testis there is sometimes a well-marked vesicular structure called the *appendix of the testis* (*hydatid*), which represents the cephalic end of the Müllerian duct. These ducts also leave a vestige at their opposite ends in the form of a minute diverticulum embedded in the prostate (*prostatic utricle, vagina masculina*) which persists where the fused Müllerian ducts opened into the urogenital sinus (Fig. 16-19).

In the female the ostium of the oviduct does not ordinarily develop at the extreme cephalic end of the Müllerian duct. The tip of the duct is likely to persist in rudimentary form as a stalked vesicle (*hydatid*) attached to the oviduct (Fig. 16-21).

The mesonephric tubules and ducts may remain recognizable to a variable extent. Usually there is embedded in the mesovarium a cluster of blind tubules and traces of a duct, corresponding to the part of the mesonephric duct and tubules which in the male forms the epididymis. These vestiges are called the *epoöphoron* (Fig. 2-2). Less frequently the more distal portion of the mesonephric duct (the part which in the male forms the vas deferens) leaves traces known as the *canals of Gartner* in the broad ligament close to the uterus and vagina.

DESCENT OF THE GONADS

Descent of the Testes Neither the testes nor the ovaries remain located in the body at their place of origin. The excursion of the testes is particularly extensive. Many factors are involved in their descent from the mesonephric region, where they first appear, to their definitive position in the scrotal sac. We can only sketch very briefly the course of events.

When the mesonephros begins to grow rapidly in bulk, it bulges out into the coelom, pushing ahead of itself a covering of peritoneum. At either end of the mesonephros the peritoneum is, in this process, thrown into folds. One of them extends cephalad to the diaphragm and is known as the *diaphragmatic ligament of the mesonephros* (Fig. 16-19). The other, which extends to the extreme caudal end of the coelom, becomes fibrous and is then known as the *inguinal ligament of the mesonephros* (Fig. 16-19). The inguinal ligament is destined to play an important part in the descent of testes.

When the testis develops it causes a local expansion of the peritoneal covering of the mesonephros to accommodate its increasing mass. As the testis grows, the mesonephros decreases in size and the testis takes to itself more and more of the peritoneal coat of the mesonephros. In this process it becomes closely related to the inguinal ligament of the mesonephros. In effect, the inguinal ligament extends its attachment to include the growing testis as well as the shrinking mesonephros. With this change the ligament is spoken of as the *gubernaculum* (Figs. 16-19 and 16-24).

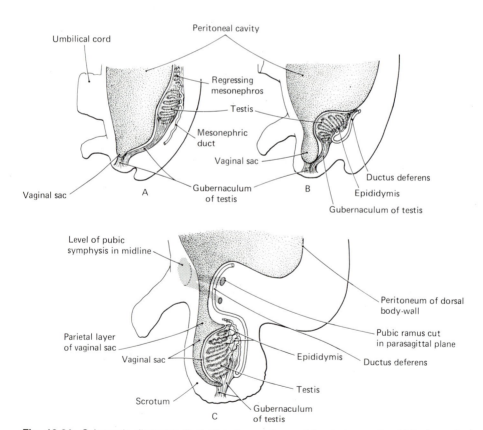

Fig. 16-24 Schematic diagrams illustrating the relations of the testes and epididymis to the peritoneum during the descent of the testes.

In the meantime bilateral coelomic evaginations are formed, one in the inguinal region of each side of the pelvis where the caudal end of the gubernaculum is attached. These are the *scrotal pouches*. Perhaps in part because of traction exerted by the gubernaculum, but primarily through differential growth, the testes and the mesonephric structures which give rise to the epididymis begin to shift their relative position progressively farther caudad (cf. Figs. 16-25 and 16-26). Eventually they come to lie in the scrotal pouches.

In its entire descent, the testis moves caudad beneath the peritoneum. It does not, therefore, enter the lumen of the scrotal pouch directly but slips down

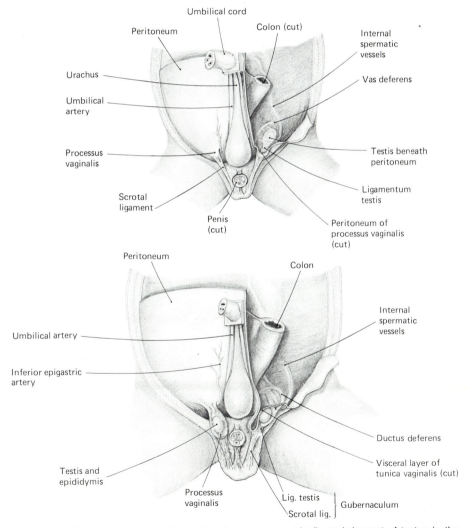

Fig. 16-25 Dissections to show formation of processus vaginalis and descent of testes in the human fetus. (A) At about 20 weeks (C-R 180 mm); (B) early in seventh month (C-R about 280 to 290 mm).

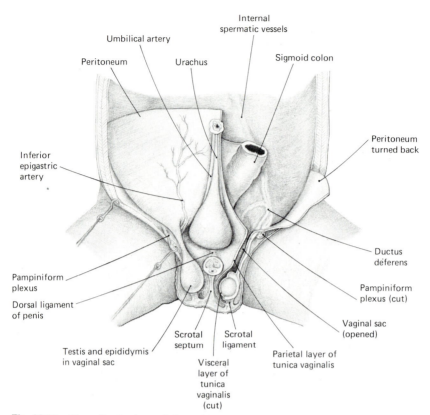

Fig. 16-26 Dissection to show relations of the testis in a human fetus of ninth month. The left testis has been rotated through 90° to expose the epididymis. *(Modified from Corning.)*

under the peritoneal lining and protrudes into the lumen, reflecting a peritoneal layer over itself (Figs. 16-25 and 16-26). This layer of reflected peritoneum is known anatomically as the *visceral tunica vaginalis*. In most mammals when the testis has come to rest in the scrotal sac, the canal connecting the sac with the abdominal cavity becomes closed. In some rodents, however, it remains patent and the testes descend into the scrotum only during the breeding season, to be retracted again into the abdominal cavity until the next period of sexual activity.

Formation of the Broad Ligament In the young female embryo, as in the male, the mesonephroi and the gonads arise retroperitoneally and bulge into the coelom, carrying a fold of peritoneum about themselves (Fig. 16-27A). The mesonephroi degenerate more completely in the female than in the male, and their decreasing bulk leaves the peritoneal folds quite thin. At this stage they more or less resemble a pair of mesenteries suspending the Müllerian ducts in their ventral margins and the ovaries on their mesial faces (Fig. 16-27B). With further degeneration of the mesonephros and its replacement by fibrous tissue these folds become the part of the broad ligament supporting the uterine tubes.

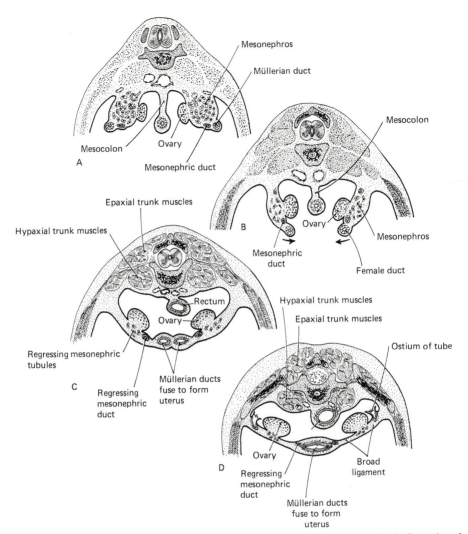

Fig. 16-27 Schematic cross-sectional diagrams to show some of the main steps in the formation of the broad ligament.

Farther caudally in the body, where the Müllerian ducts fuse with each other in the midline to form the uterus, the supporting peritoneal folds coalesce medially to form the part of the broad ligament supporting the uterus (Fig. 16-27C and D).

Descent of the Ovaries Although the ovaries do not move as far as the testes, their change in position is quite characteristic and definite. As they increase in size, both the gonads and the ducts sag farther into the body cavity. In so doing they pull with them the broad ligament, which, as it is stretched out, allows the ovaries, uterine tubes, and uterus to move caudally and somewhat ventrally (Fig. 16-21). The inguinal ligament of the mesonephros, which in the male forms the gubernaculum, in the female is embedded in the broad ligament. When the ovaries move caudad and laterad, the inguinal ligament is bent into angular form. Cephalic to the bend it becomes the *round ligament of the ovary*, and caudal to it, the *round ligament of the uterus* (Fig. 2-2 and 16-21). It should be noted that the caudal end of the round ligament of the uterus is embedded in the connective tissue of the labium majus in a positon homologous with the anchorage of the gubernaculum in the scrotal pouch of the male (cf. Figs. 16-19 and 16-21).

THE EXTERNAL GENITALIA

Indifferent Stages   In very young embryos, a vaguely outlined elevation known as the *genital eminence* (Fig. 16-28) can be seen in the midline, just cephalic to the proctodeal depression. This is soon differentiated into a central prominence (*genital tubercle*) closely flanked by a pair of folds (*genital folds*) extending toward the proctodeum. Somewhat farther to either side are rounded elevations known as the *genital swellings* (Figs. 16-29A and 16-30A). Between the genital folds is a depression which attains communication with the urogenital

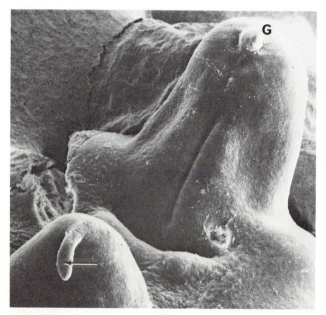

Fig. 16-28 Scanning electron micrograph of a human embryo (stage 18) showing the genital tubercle (G) and the regressing tail (arrow). X38. *(From Fallon and Sinandl, 1978, Am. J. Anat.* **152**:*111. Courtesy of the authors.)*

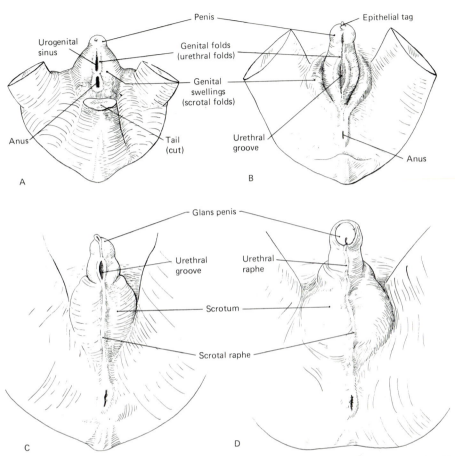

Fig. 16-29 Stages in the development of the external genitals in the male. (A) At 7 weeks, 17 to 20 mm; (B) in tenth week, 45 to 50 mm; (C) early in twelfth week, 58 to 68 mm; (D) toward close of gestation. *(Adapted from several sources, especially Spaulding, 1921, in* Carnegie Cont. to Emb., *vol. 13.)*

sinus to establish the urogenital orifice (*ostium urogenitale*). This opening is separated from the anal opening by the urorectal fold (Figs. 16-12 and 16-13). From this common starting point the external genitalia of either sex differentiate.

Male Genitalia If the individual develops into a male, the genital tubercle becomes greatly elongated to form the penis and the genital swellings become enlarged to form the scrotal pouches (Fig. 16-29B to 16-29D). During the growth of the penis there develops on its caudal face a groove extending throughout its entire length (Fig. 16-13). Proximally the groove is continuous with the slitlike opening of the urogenital sinus. This groove in the penis later becomes closed over by a ventral fusion of the genital folds, establishing the penile portion of the urethra. That portion of the urogenital sinus between the neck of the bladder and the original opening of the urogenital sinus becomes the prostatic urethra. With the closure of the urethral groove in the penis, the prostatic urethra and the penile urethra become continuous. The most distal portion of the male urethra arises as a solid cord of ectodermal cells growing in from the glans to meet the penile urethra. The cord then canalizes, completing the urogenital outlet in the male. The line of fusion in the urogenital sinus region and along the caudal surface of the penis is clearly marked by the persistence of a ridgelike thickening known as the *raphe* (Fig. 16-29C and D).

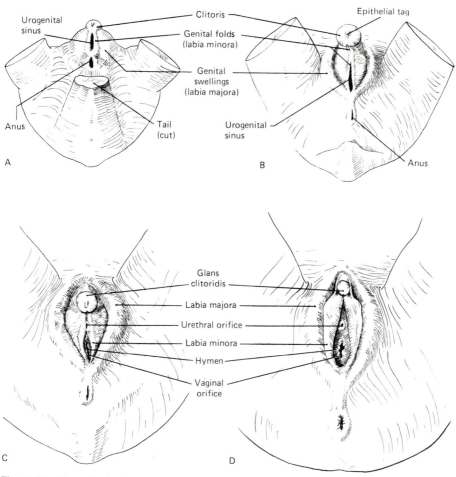

Fig. 16-30 Stages in the development of the external genitals in the female. (A) At 7 weeks, 17 to 20 mm; (B) in tenth week, 45 to 50 mm; (C) at 12 weeks, 75 to 80 mm; (D) toward close of gestation. *(Adapted from a number of sources, especially Spaulding, 1921, in* Carnegie Cont. to Emb., *vol. 13.)*

Female Genitalia   In the female the genital tubercle becomes the clitoris, the genital folds become the labia minora, and the genital swellings the labia majora (Fig. 16-30). The original opening of the urogenital sinus undergoes no such changes as occur in the male but persists nearly in its original position. Its orifice, enlarged and flanked by the labia, becomes the vestibule into which open the vagina and the urethra (Fig. 16-5F). The urethra in the female is derived from the urogenital sinus, being homologous with the prostatic portion of the male urethra.

Chapter 17

The Development of the Circulatory System

INTRODUCTION TO THE EMBRYONIC CIRCULATION

The plan of the embryonic heart and vascular system is bound by three major constraints. First it must meet the immediate needs of the embryo at its various stages of development by supplying it with oxygen, nutrients, and other essential materials for growth while at the same time removing CO_2 and other metabolic wastes. To meet these needs only a simple unidirectional pumping action is required of the heart to move the blood along channels supplying all the developing organs of the embryo, as well as through other channels designed to bring in and remove the necessary substrates and wastes of its metabolism.

To meet the requirements of respiration, nutrition, and excretion, the two extraembryonic circulatory arcs have developed (Fig. 8-17). The arc to the yolk sac, the vitelline arc, supplies foodstuffs to all large-yolked vertebrate embryos, but in mammals the yolk sac and vitelline circulation persist for what were originally probably subsidiary functions, the origin and transport of primordial germ cells and primitive blood cells. The large allantois of amniote eggs (with its circulatory arc) was originally the chief organ of respiration and deposition of excretory wastes, but as the placenta in mammals evolved into a more efficient organ of exchange, the allantois correspondingly became reduced in prominence. The allantoic circulation, however, has been incorporated into the placenta and continues to serve its original functions.

In addition to satisfying the relatively simple requirements of the embryo, the plan of the embryonic circulation in amniotes must also anticipate the immediate needs of the embryo once it hatches from the egg or is delivered from the mother's uterus (Fig. 17-1). The most immediate and critical adjustment to birth is the need of the newly born individual to breathe independently. Breathing requires that the lungs, which have developed tardily from the morphological standpoint and are untested functionally, begin to function at full capacity immediately after birth. Because of this, the embryonic heart cannot be content with remaining in its original condition, as a simple tube with the blood passing through it in an undivided stream. Early in embryonic life it must become converted into an elaborately valved, four-chambered organ, partitioned in the midline and pumping from its right side a pulmonary stream which is returned to the left side and pumped out again as the systemic bloodstream. And the heart cannot cease work while making its internal alterations; there can be no interruption in the current of blood which it pumps to the growing embryo. The systemic, as well as the pulmonary, part of the circulation must be prepared. Because of the delayed development and the restricted vascular bed of the embryonic lungs, the left side of the heart receives less blood from· the pulmonary veins than the right side of the heart receives from the venae cavae. Yet after birth the left ventricle is destined to do more work than the right ventricle. On the other side of the coin, the right ventricle receives more blood than can be accommodated by the pulmonary circulation. These and other problems of the embryonic heart are solved by the presence of shunts, which act like safety valves, allowing the various chambers of the heart to obtain the exercise that they need for their required development, but yet not overloading the pulmonary vasculature beyond its limited carrying capacity.

Besides meeting immediate and future physiological needs, the circulatory plan of the embryo cannot escape the influence of its phylogenetic history. This third constraint causes the pattern of the embryonic circulatory system to take on a form that would not be predicted by its physiological needs alone. This is particularly evident in the pharyngeal region where in the system of aortic arches there is an unmistakable phylogenetic impress in the arrangement and manner of development of the blood vessels in that region. The blood leaving the ventrally located heart must pass around the gut to reach the dorsally located aorta. In primitive fishes six pairs of aortic arches encircle the pharynx, breaking up in the gills into capillaries which carry out the indispensable function of oxygenating the blood. Once an animal has replaced its gills with lungs it makes little difference functionally whether or not the blood passes by each gill cleft on its way from heart to aorta. In adult birds and mammals we find this communication simplified to a single main aortic arch. But in the embryos of both birds and mammals a whole series of symmetrical aortic arches appears, for a time, encircling the pharynx and passing in close relation to vestigial gill clefts. This can be interpreted only as a recapitulation of ancestral conditions—

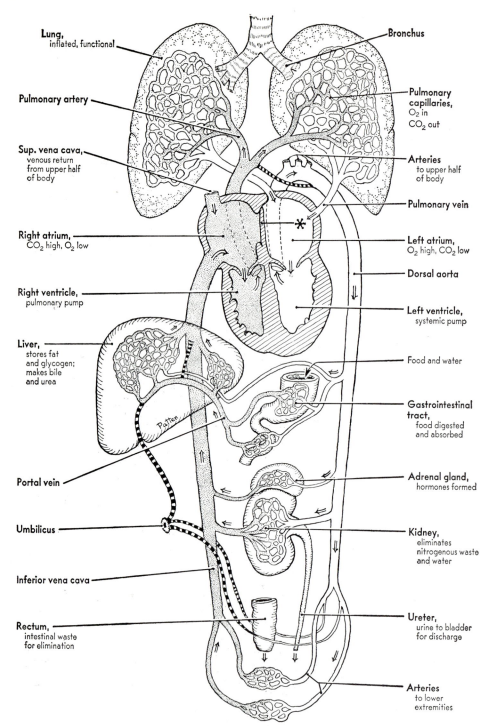

Lung,
inflated, functional

Bronchus

Pulmonary artery

**Pulmonary
capillaries,**
O_2 in
CO_2 out

Sup. vena cava,
venous return
from upper half
of body

Arteries
to upper half
of body

Pulmonary vein

Right atrium,
CO_2 high, O_2 low

Left atrium,
O_2 high, CO_2 low

Dorsal aorta

Right ventricle,
pulmonary pump

Left ventricle,
systemic pump

Liver,
stores fat
and glycogen;
makes bile
and urea

Food and water

**Gastrointestinal
tract,**
food digested
and absorbed

Portal vein

Adrenal gland,
hormones formed

Umbilicus

Kidney,
eliminates
nitrogenous waste
and water

Inferior vena cava

Rectum,
intestinal waste
for elimination

Ureter,
urine to bladder
for discharge

Arteries
to lower
extremities

Fig. 17-1 Plan of the postnatal circulation. *(After Patten, in Fishbein,* Birth Defects, *1963. Courtesy of the National Foundation and the J. B. Lippincott Company, Philadelphia.)* The heavily cross-banded structures were important fetal vessels (cf. Fig. 17-35) which after birth ceased to carry blood and gradually became reduced to fibrous cords. The asterisk indicates the valvula foraminis ovalis in the closed position characteristic for postnatal life. (See colored insert.)

conditions which, although they have ceased to be of functional importance, appear nevertheless as a developmental phase on the way to a more highly differentiated plan of adult structure. Our interpretation of the aortic arches, then, would take into consideration first the fundamental functional necessity of a channel connecting the ventrally located heart with the dorsally located aorta. Second, looking at the striking arrangement of the series of aortic arches in relation to the gill arches and clefts, one sees a repetition of the structural plan which existed in water-living ancestral forms. Thus in the end we are led back again to functional significance; for the relationship of the aortic arches to the gill arches writes into the story of individual development an unequivocal record of the evolutionary phase when the gills were a center of primary metabolic importance. Other examples of recapitulation in the circulatory system are the already mentioned persistence of the vitelline circulatory arc long after the yolk sac has abandoned its original nutritive function and the series of highly developed venous channels in the mesonephros, even when it is destined to degenerate later in development.

Whatever peculiarities may be impressed on the course of the circulation by the appearance of ancestral structures or by the development of special fetal organs such as the yolk sac and the placenta, the main blood currents will at any time be found concentrated at the centers of activity. Changes in these main currents as one center retrogresses and another becomes dominant must take place gradually. Large vessels become smaller, and what was formerly an irregular series of small vessels becomes excavated to form a new main channel; but the circulation of blood to all parts of the body never ceases. Even slight curtailment of the normal blood supply to any region would stop its growth. Any marked local decrease in the circulation would result in local atrophy or malformation. Complete interruption of any important circulatory channel, even for a short time, would inevitably mean the death of the embryo.

EMBRYONIC HEMATOPOIESIS

The first blood cells are produced in extraembryonic sites as small groups of mesodermal cells called *blood islands* (Fig. 17-2). The early blood islands are located next to the endodermal wall of the yolk sac (Fig. 17-3), and in the chick the differentiation of blood islands has been shown to depend upon an interaction between the splanchnic mesoderm and the underlying endoderm (Wilt, 1965). Cells in the outer zone of the primordial blood island become flattened as a young vascular endothelium, and enclose the more centrally located cells, which become *hematopoietic stem cells* (Fig. 17-3B). Within the endothelial vesicles, fluid accumulates and suspends the developing blood cells.

There is a temporal progression of major sites of hematopoiesis in the embryo. The first hematopoietic activity is seen in the yolk sac. Later in development dominant sites of hematopoiesis are found in the liver and spleen, body mesenchyme, and, finally, the bone marrow. Over the years there have been two principal viewpoints regarding the origin of the definitive blood cells of

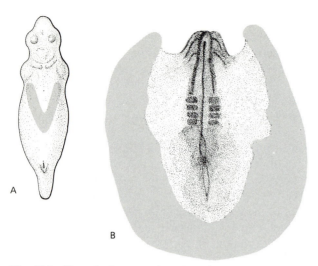

Fig. 17-2 Sites of primary erythropoiesis (shaded gray) in amphibian (A) and bird (B) embryos.

the embryo. According to one, blood islands from the yolk sac provide the cells that seed the lymphoid organs and the blood-forming organs of the embryonic body. According to the other viewpoint, most of the blood cells have an intraembryonic origin. Recent experiments on chimeric bird embryos (quail body grafted to a chick yolk sac) have shown that there are two separate populations of hematopoietic precursor cells in the embryo (Dieterlen-Lievre, 1978; Le Douarin, 1978). The first population (extraembryonic) arises in the yolk sac and seeds the liver and spleen and the central lymphoid organs (thymus and bursa of Fabricius) of the embryo. Later, a new population of hematopoietic stem cells, arising from an unknown origin (possibly the embryonic mesenchyme, in general), gradually replaces the early hepatic and splenic blood cells which come from the yolk sac. Intraembryonic blood cells populate the yolk sac, as well. The cells in the liver and spleen, and ultimately the bone marrow, are principally those which give rise to the erythroid series (cells forming red blood cells, or erythrocytes) and the granulocytic series (cells forming the granulated white blood cells, or neutrophils, eosinophils and basophils). The thymus and bursa of Fabricius remain, for the most part, populated by cells which are derived from the yolk sac and which later differentiate into lymphoid cells (lymphocytes and monocytes.) After further developmental processing in the central lymphoid organs, they secondarily seed the peripheral lymphoid tissues (see Chap. 15). In later life other pathways of seeding the central lymphoid organs, e.g., from the bone marrow, may also be set up.

Erythropoiesis and Hemoglobin Formation *Erythropoiesis* is the process by which a mature red blood cell loaded with hemoglobin molecules differentiates from primitive hemocytoblastic stem cells. Erythropoiesis can be viewed

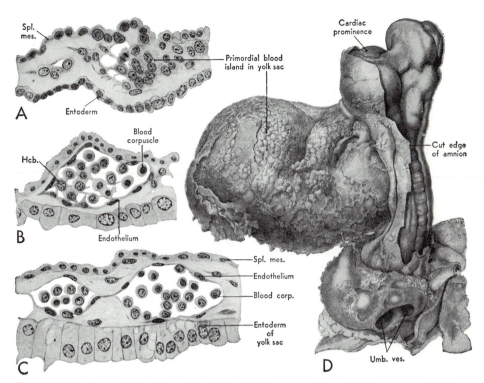

Fig. 17-3 Development of yolk sac blood islands in human embryos. A–C are camera lucida drawings, reproduced X355. (A) Early stage in aggregation of cells between endoderm and splanchnic mesoderm in yolk sac of an embryo early in fourth week (17 somites). (B) Beginning of differentiation of endothelium and primitive blood cells, from an embryo of about 4 weeks (4.5 mm). (C) A more advanced area from a 4-week embryo showing endothelium well differentiated and corpuscles suspended, free, in plasma. (D) The Corner 10-somite embryo showing location of young blood islands on yolk sac. (Carnegie Cont. to Emb., *1929, vol. 20.*) Abbreviation: *Hcb.,* primitive blood-mother-cell or hemocytoblast.

from several aspects and levels of organization. At the tissue and organ level, there are three major phases in erythropoiesis (Fig. 17-5, bottom). For a brief period the only blood-forming activity occurs extraembryonically, in the yolk sac. After the second month the major sites of erythropoiesis shift from the yolk sac to intraembryonic organs. The second major phase is the hepatic period (roughly the third through seventh month of human pregnancy), during which the liver and spleen are the dominant hematopoietic organs. Finally, late in pregnancy the bone marrow takes over as the definitive site for erythropoiesis in higher vertebrates.

The differentiation of individual erythrocytes must also be viewed in the context of their site of origin. The bulk of current evidence suggests that there are two main populations of erythrocyte-forming cells. The first consists of the cells that are derived from the original population of yolk-sac erythrocyte precursors. These cells differentiate relatively synchronously and they are released into the bloodstream at an early stage of differentiation. Maturation is completed in the bloodstream, and in mammalian embryos these yolk sac-derived erythrocytes are nucleated.

Differentiation of the other population of erythrocytes, derived from intraembryonic precursor cells, has been extensively studied with respect to both cellular morphology and production of hemoglobin. Both light and electron microscopic preparations show a series of developmental stages which are classical for a cell that is heavily engaged in the production of intracellular protein (Fig. 17-4). From the *hemocytoblast*, a primitive cell that is considered by many to be the stem cell for all families of blood cells, the erythrocyte line first appears as a highly basophilic cell called a *proerythroblast*. These cells have

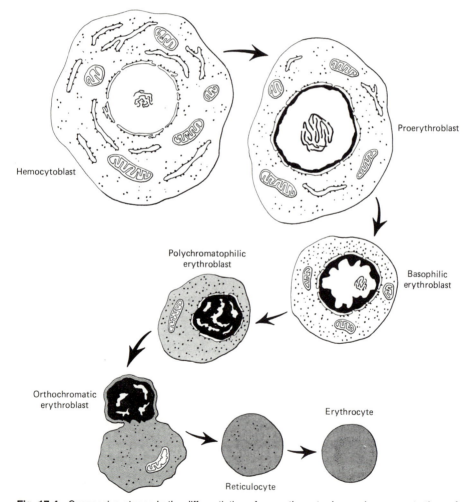

Hemocytoblast

Proerythroblast

Polychromatophilic erythroblast

Basophilic erythroblast

Orthochromatic erythroblast

Erythrocyte

Reticulocyte

Fig. 17-4 Successive stages in the differentiation of an erythrocyte. Increasing concentrations of hemoglobin are indicated by the intensity of the gray shading in the cytoplasm. *(Adapted from Rifkind, 1974, in Lash and Whittaker, eds.,* Concepts of Development, *Sinauer Assoc.)*

undergone sufficient restriction to be firmly committed to the production of red blood cells, but they have not yet begun to produce hemoglobin in amounts sufficient to be detected by cytochemical analysis. They have large nuclei with prominent nucleoli and largely uncondensed nuclei chromatin. RNA synthesis is high. The cytoplasm contains aggregates of mainly free ribosomes, which will be used in intracellular protein synthesis.

Subsequent stages of erythroid differentiation (basophilic, polychromatophilic, and orthochromatic erythroblasts) show a progressive change in the balance between the accumulation of newly synthesized hemoglobin molecules and the decline of first the RNA-producing machinery and, later, the protein-synthetic apparatus. During these stages the cytoplasm continuously stains less basophilically, as its eosinophilic staining (characteristic of accumulated protein) increases in intensity. Corresponding to these changes is a decrease in the concentration of ribosomes in the cytoplasm. The nucleus shows signs of its inexorable pathway toward inactivation and elimination by its steadily decreasing size, the progressive condensation of its chromatin and the elimination of the nucleolus. Finally, the late *orthochromatic erythroblast* extrudes its pycnotic nucleus. In contrast to yolk-sac erythropoiesis, the changes normally occur within the hematopoietic tissues, and only after it has lost its nucleus is the cell, now called a *reticulocyte,* released into the bloodstream. Reticulocytes still contain small numbers of ribosomes, and they continue to produce hemoglobin for the first day or two after their release into the bloodstream. The mature erythrocyte is a terminally differentiated cell, lacking both a nucleus and the intracellular apparatus for macromolecular synthesis. It has often been likened to a bag of hemoglobin, but such an appellation does not do justice to the extent to which the structure of the erythrocyte is adapted to fulfilling its function.

There are developmental changes in the hemoglobin molecule itself. The *hemoglobin* molecule (m.w. 64,500) contains four polypeptide chains complexed to a molecule of heme. At different stages of development and in the two different populations of erythrocytes the structures of some of the polypeptide chains vary, reflecting the activity of different genes during development. In the major normal varieties of hemoglobins, regardless of the type or period of synthesis, two of the polypeptide chains are the same type, the α chain. The other two chains vary and determine the type of hemoglobin molecule. The relative patterns of formation of the hemoglobin polypeptide chains during embryonic development are illustrated in Fig. 17-5. The earliest variety of hemoglobin, associated with yolk-sac erythropoiesis, is *embryonic hemoglobin,* in which two α chains and two ε polypeptide chains are attached to the heme molecule. Embryonic hemoglobin is present for only a couple of months, after which it is supplanted by *fetal hemoglobin,* a form which contains two α chains and two γ chains and is dominant throughout the rest of embryonic life. Lastly, the formation of adult hemoglobins gains ascendency shortly after birth. In addition to the constant α chains, the major varieties of adult hemoglobins contain either two β chains or two δ chains.

Embryonic and fetal hemoglobin have a higher affinity for oxygen than does

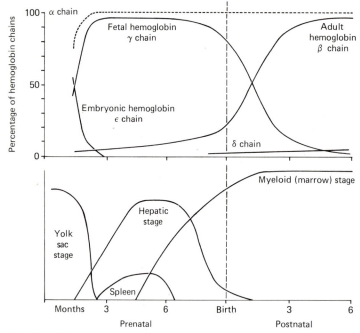

Fig. 17-5 Periods of erythropoiesis and hemoglobin synthesis in the human. The lower graph highlights dominant sites of erythropoiesis. The upper graph shows the percentages of hemoglobin polypeptide chains present in the blood at a given time. The α chain is placed in a separate category in this graph. *(Upper graph after Huehns et al., 1964,* Cold Spring Harbor Symp. Quant. Biol. **29***:327. Lower graph after Wintrobe et al., 1974,* Clinical Hematology, *7th ed., Lea and Febiger, Philadelphia.)*

the adult form. This represents a significant adaptation for intrauterine life because the fetal hemoglobin is able to extract oxygen diffusing across the placental barrier more efficiently than adult hemoglobin. During late fetal life increasing amounts of adult hemoglobin are manufactured. After birth the proportion of fetal hemoglobin rapidly decreases until by 4 to 6 months it is no longer present in the blood.

The production of new erythrocytes is regulated by a humoral substance called *erythropoietin*. In response to hypoxia, which could result from blood loss, a relative deficiency of erythrocyte production, or a move to a higher altitude, the levels of erythropoietin in the blood rise and stimulate the proliferation of erythroid stem cells. Erythropoiesis within the body of the embryo is responsive to the effects of erythropoietin, but yolk-sac erythropoiesis is not. In the chick embryo there is evidence in favor of a separate variety of erythropoietin to which the erythroid stem cells in the yolk sac respond (Knezevic et al., 1971).

THE ARTERIES

Derivatives of the Aortic Arches   In vertebrate embryos six pairs of aortic arches connect the ventral with the dorsal aorta (Fig. 17-6A). The portions of

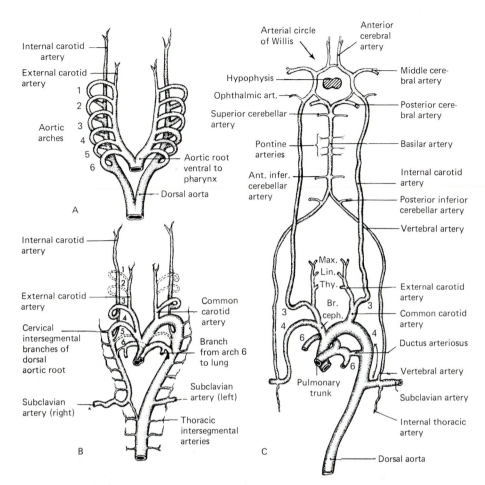

Fig. 17-6 Diagrams illustrating the major changes which occur in the aortic arches of mammalian embryos. *(Adapted from several sources.)* (A) Ground plan of complete set of aortic arches. (B) Early stage in modification of arches. (C) Adult derivatives of aortic arches. Abbreviations: *Br. ceph.,* brachiocephalic (innominate) artery; *Lin.,* lingual artery; *Max.,* maxillary artery; *Thy.,* thyroid artery. Figures 17-7 to 17-10 show, less schematically, some of the changes summarized in this illustration. Arrow in (C) indicates change in position of origin of left subclavian artery which occurs in the later stages of development.

the primitive paired aortae which bend around the anterior part of the pharynx constitute the first (i.e., the most cephalic) of these aortic arches. In its course around the pharynx the first aortic arch is embedded in the tissues of the mandibular arch (Figs. A-26 and 17-7A to 17-7C). The other aortic arches develop later, in sequence, one aortic arch in each branchial arch caudal to the mandibular (Fig. 17-7C to 17-7F). But in mammalian embryos we never find the entire series of aortic arches well developed at the same time. The two most cephalic arches degenerate as main channels before the most caudal arches have been established (Fig. 17-7E and F). It should be noted, however, that their disappearance is neither abrupt nor complete, as the schematic diagrams (Fig.

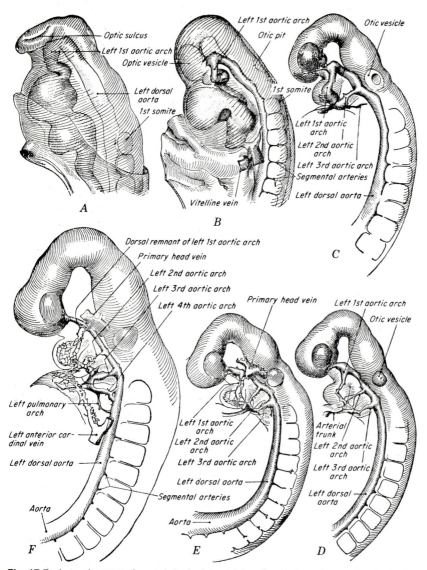

Fig. 17-7 Lateral aspect of vessels in the branchial region of pig embryos of various ages. (A) 10 somites; (B) 19 somites; (C) 26 somites; (D) 28 somites; (E) 30 somites; (F) 36 somites (6 mm). *(After Heuser,* Carnegie Cont. to Emb., *vol. 15, 1923.)* The drawings in this and the three following figures were made directly from injected specimens rendered transparent by treatment with oil of wintergreen. (See colored insert.)

17-6) summarizing the changes in the aortic arches might lead one to believe. They break down as main channels but leave behind small vessels appropriated by the local tissues as their source of nutrition (Fig. 17-8A).

The sequence of events in the development of the aortic arches has long challenged experimental embryologists. Rychter (1962) used minute silver clips to occlude single aortic arches and various combinations of aortic arches in 3-

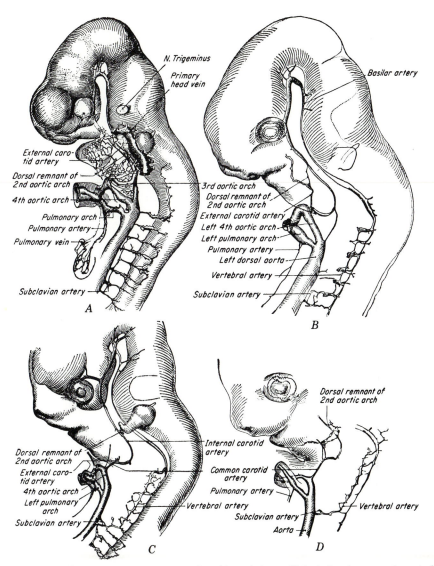

Fig. 17-8 Continuation of the same series of lateral views of injected embryos as shown in figures 17-7. (A) 14 mm; (B) 17 mm; (C) 19.3 mm; (D) 20.7 mm. (See colored insert.)

and 4-day chick embryos. The embryos were then maintained in culture until the results of the operations were clearly manifested. For details of the consequences of specific disturbances, Rychter's review should be consulted. Of more general interest were the findings on the manner in which alternative channels were retained and enlarged when the normal route was interrupted. Some cardiac anomalies occurred, as well, as the result of disturbances in major blood routes.

The early degeneration of the first two aortic arches and the fact that the fifth arch never appears in mammalian embryos, except transitorily as a vestigial vessel appended to one of the neighboring arches, leaves only the ventral and dorsal aortic roots and the third, fourth, and sixth arches to play an important role in the formation of adult vessels.

The portions of the ventral aortic roots which formerly acted as feeders to the first two arches are retained as the *external carotid arteries.* These vessels, in part through the small channels left by the disintegration of the aortic arches with which they were originally associated and in part through the formation of new branches to subsequently formed structures, nourish the oral and cervical regions (Figs. 17-6C, and 17-8A and B).

The *internal carotid arteries* arise as prolongations of the dorsal aortic roots and extend to the brain (Figs. 17-12 and 17-13). When the part of the dorsal aortic root which lies between arches 3 and 4 dwindles and drops out, the third arch is left, constituting the curved proximal part of the internal carotid artery (Figs. 17-6B and C, and 17-8A to 17-8C). The part of the ventral aortic root which, from the first, has fed the third aortic arch becomes somewhat elongated and persists as the *common carotid artery* (Figs. 17-6 and 17-8).

The fourth aortic arch has a different fate on opposite sides of the body. On the left it is greatly enlarged and persists as the arch of the adult aorta (Figs. 17-6, 17-7, and 17-10). On the right, the fourth arch forms the root of the *subclavian artery.* The short section of the right ventral aortic root proximal to the fourth arch persists as the *innominate (brachiocephalic) artery,* from which both the right subclavian and the right common carotid arteries arise (Fig. 17-6C).

The sixth aortic arch changes its original relationships somewhat more than the others. At any early stage of development branches extend from its right and left limbs toward the lungs (Fig. 17-9D and E). After these pulmonary vessels have been established[1] the right side of the sixth aortic arch loses communication with the dorsal aortic root and disappears (Fig. 17-10A and B). On the left, however, the sixth arch retains its communication with the dorsal aortic root. The portion located between the dorsal aorta and the point where the pulmonary artery is given off is called the *ductus arteriosus* (Figs. 17-6C and 17-33). During the fetal period, when the lungs are not inflated, the ductus arteriosus shunts the excess blood from the pulmonary circulation directly into the aorta. The functional importance of this channel will be more fully appreciated when we have given it further consideration in connection with the development of the heart and the changes which take place in the circulation at the time of birth.

---

[1]The details of the formation of the pulmonary arteries differ somewhat in different mammals. In most forms which have been carefully studied (human, cat, dog, sheep, cow, opossum) the pulmonary arteries maintain their original paired condition throughout their entire length. In these forms part of the right sixth arch is retained as the proximal portion of the adult right pulmonary artery (Fig. 17-6). The pig is unusual in having its pulmonary branches fuse with each other proximally, forming a median vessel ventral to the trachea (cf. Figs. 17-9E and 17-10A). Distal to this short median trunk the pulmonary vessels retain their original paired condition, each running to the lung on its own side of the body (Fig. 17-14B). Proximally the median trunk becomes associated with the left sixth arch and the right sixth arch drops out altogether.

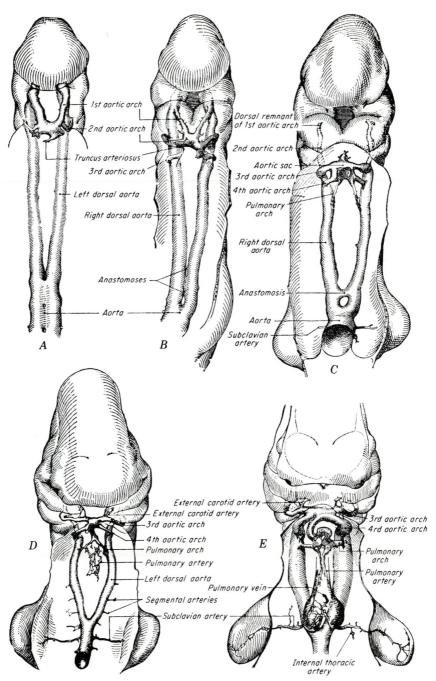

Fig. 17-9   Ventral aspect of vessels in the branchial region of pig embryos of various ages. (A) 24 somites; (B) 4.3 mm; (C) 6 mm; (D) 8 mm; (E) 12 mm. (See colored insert.) (*After Heuser,* Carnegie Cont. to Emb., *vol. 15, 1923.*)

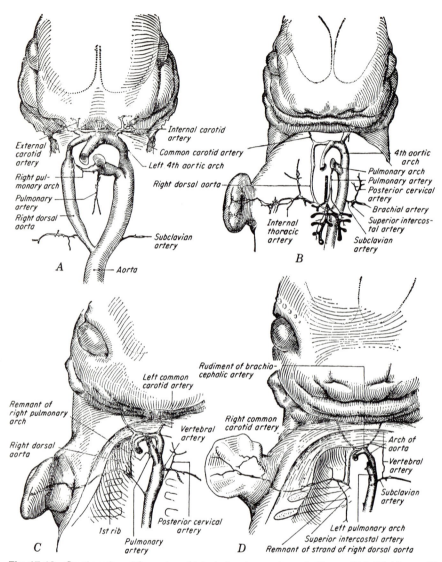

Fig. 17-10   Continuation of the series of injected embryos shown in Figure 17-9. (A) 14 mm; (B) 17 mm; (C) 19.3 mm; (D) 20.7 mm. (See colored insert.)

While these changes have been taking place in the more peripheral part of the vascular channels which lead to the lungs, a fundamental alteration has occurred in the truncus arteriosus. Formerly a single channel leading away from the undivided ventricle of the primitive tubular heart, the truncus now becomes divided lengthwise into two separate channels, one channel leading from the right ventricle to the lungs by way of the sixth aortic arches and the other leading from the left ventricle to the dorsal aorta by way of the left fourth aortic arch.

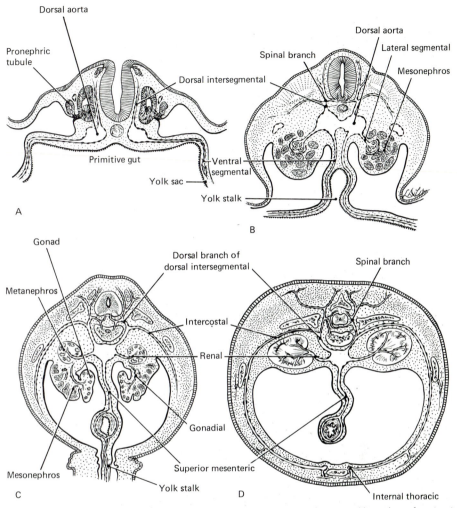

Fig. 17-11 Cross-sectional plans of the body showing relations of segmental branches of aorta at different stages of development. (See colored insert.)

Derivatives of the Intersegmental Branches of the Aorta A series of metamerically arranged small vessels coming off the dorsal aorta is a conspicuous feature of the vascular pattern. They are called *dorsal segmental vessels* or, if one wishes to emphasize the fact that they emerge between adjacent somites, *dorsal intersegmental vessels* (Fig. 8-17). When they are first established the dorsal intersegmental vessels are related primarily to the developing neural tube (Fig. 17-11A). As other adjacent structures are formed new branches grow from the original vessels. They vary in size and course at different levels in the body, according to the structures with which they become associated. In the thoracic region, for example, conspicuous branches are formed as the body wall is developed. These branches lie between the ribs and are appropriately known as the *intercostal arteries* (Figs. 17-11C and D, and 17-12).

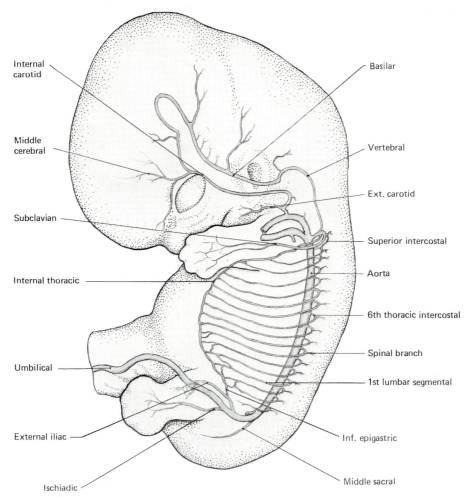

Internal carotid

Middle cerebral

Subclavian

Internal thoracic

Umbilical

External iliac

Ischiadic

Basilar

Vertebral

Ext. carotid

Superior intercostal

Aorta

6th thoracic intercostal

Spinal branch

1st lumbar segmental

Inf. epigastric

Middle sacral

Fig. 17-12   Arteries of the body wall in a human embryo of 7 weeks. *(Modified after Mall.)* (See colored insert.)

When the anterior appendage bud of a human embryo starts to take shape during the fifth week, its growing mass lies at the level of the seventh of the intersegmental arteries which arise in relation to the cervical somites. It is quite natural, therefore, that this vessel becomes enlarged to form the *subclavian artery*. As the left fourth aortic arch enlarges to form the main channel leading from the heart to the dorsal aorta, the dorsal aortic root on the right side becomes much reduced (Fig. 17-10A and B). An early stage in this reduction is shown by the unequal sizes of the right and left dorsal aortic roots as seen in transverse sections (Fig. 17-14B). Caudal to the level of the subclavian it drops out entirely (Fig. 17-10C and D). It will be recalled that the sixth aortic arch also drops out on this side. This leaves the right subclavian communicating with the dorsal aorta by way of a considerable section of the old dorsal aortic root and the fourth aortic arch.

Cephalic to the subclavian arteries, series of longitudinal anastomoses appear connecting the cervical intersegmentals to form the *vertebral arteries* (Fig. 17-8). When the vertebral arteries are thus established, all the intersegmental roots back to the subclavian drop out, leaving the vertebral arteries as a branch of the subclavian (Figs. 17-6C and 17-8).

Once established at cervical levels (Fig. 17-8), the vertebral arteries are rapidly extended cephalad. At the level of the auditory vesicles they bend sharply mesiad and fuse with each other ventral to the metencephalon. The new vessel thus formed extends rostrad in the midline as the *basilar artery* (Fig. 17-13). Just caudal to the hypophysis the basilar artery ends, where it makes connection on either side with recurrent branches from the internal carotid arteries (Fig. 17-6C).

Enteric Arteries The first of the three enteric arteries to appear is the *superior mesenteric.* We have already followed its origin as a pair of arteries, originally called the *vitelline arteries* or *omphalomesenterics,* which in young embryos extended to the surface of the yolk sac (Fig. 8-17). When the yolk sac degenerates and the ventral part of the body closes in, these paired channels fuse with each other to form a median vessel situated in the mesentery (Fig. 17-11C and D). The superior mesenteric artery forms the pivotal point around which the early rotation of the gut takes place.

The *coeliac artery* arises from the aorta in a manner basically similar to the origin of the superior mesenteric artery, but at a slightly later stage of development (Fig. 17-13). The embryonic body has, therefore, become more nearly closed ventrally, and the primary dorsal mesentery has been established. Later it becomes extensively branched, being the main artery which feeds the gastrohepatopancreatic region of the digestive system and also the spleen, which arises in its territory (Fig. 14-14).

The *inferior mesenteric artery* has an origin similar to that of the coeliac. It is established caudal to the superior mesenteric artery slightly later in development than the time at which the coeliac appears and is the main vessel to the posterior part of the intestinal tract (Fig. 14-14).

Renal Arteries The mesonephros is supplied by many small arteries which arise ventrolaterally from the aorta. While the metanephroi, or permanent kidneys, are still very small, they lie in close proximity to the mesonephroi and are fed by small arteries which arise from the aorta along with the mesonephric vessels (Fig. 17-11C). As the kidneys migrate cephalad from their early position deep within the pelvic cavity, they become supplied by more cephalic aortic branches. The vessels associated with the kidneys are progressively enlarged as the kidneys themselves grow in bulk and become the renal arteries of the adult (Fig. 17-11D).

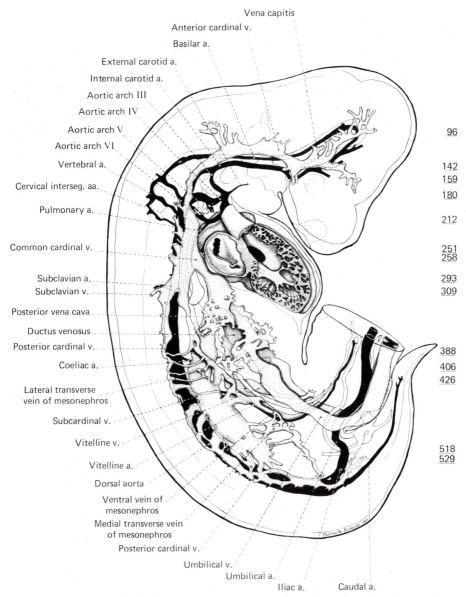

Vena capitis
Anterior cardinal v.
Basilar a.
External carotid a.
Internal carotid a.
Aortic arch III
Aortic arch IV
Aortic arch V
Aortic arch VI
Vertebral a.
Cervical interseg. aa.
Pulmonary a.
Common cardinal v.
Subclavian a.
Subclavian v.
Posterior vena cava
Ductus venosus
Posterior cardinal v.
Coeliac a.
Lateral transverse
vein of mesonephros
Subcardinal v.
Vitelline v.
Vitelline a.
Dorsal aorta
Ventral vein of
mesonephros
Medial transverse vein
of mesonephros
Posterior cardinal v.
Umbilical v.
Umbilical a.
Iliac a.   Caudal a.

96

142
159
180

212

251
258

293
309

388

406
426

518
529

Fig. 17-13 Reconstruction (X14) of the circulatory system of a 9.4-mm pig embryo. The numbered horizontal lines on the right indicate the levels of the cross sections drawn in Figures 11-16, 14-7, 16-6, 17-14 and 17-17. The use of a transparent straightedge placed at the level of the appropriate line will greatly help in the correlation of each of the cross sections with the lateral plan of the reconstruction.

Arteries Arising from the Caudal End of the Aorta   The main aortic trunk decreases abruptly in size where the large umbilical (allantoic) arteries turn off into the belly stalk. Beyond this point the aorta is continued toward the tail as a slender median vessel called the *caudal artery* (Fig. 17-13).

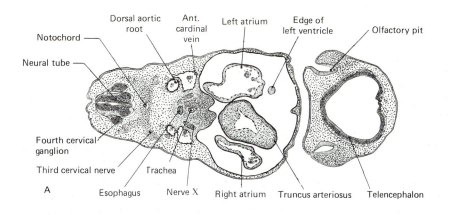

A

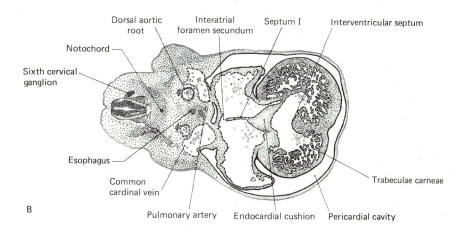

B

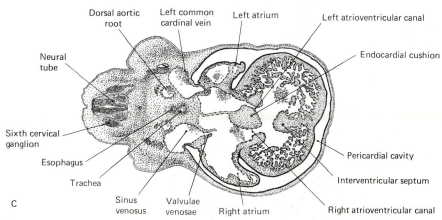

C

Fig. 17-14 Three transverse sections from the series of the 9.4-mm pig embryo used in making the reconstructions illustrated in figure 17-13. (Projection drawings, X14.) By laying a straightedge across either of these reconstructions at the level of the marginal line bearing the serial number of a cross section, its relations within the body as a whole are precisely indicated. (A) Section 212 passing through the olfactory pits and the truncus arteriosus; (B) section 251 passing through the heart at the level of interatrial foramen secundum; (C) section 258 passing through the heart at the level of the orifice of the sinus venosus and the atrioventricular canals.

The posterior appendage buds arise some time after the placental circulation has been established. The umbilical arteries are consequently of considerable size, and the small vessels which branch off from them to feed the appendage buds are by comparison quite insignificant. As the appendage buds grow, these small vessels grow with them to become the *external iliac arteries* (Fig. 17-12). When, at birth, the placental circulation stops, the umbilical arteries are reduced to small vessels nourishing the local tissues between their point of origin and the umbilicus. We then know their proximal portions as the *internal iliac,* or *hypogastric, arteries* and the fibrous cords which still mark their course along the wall of the bladder (the old allantoic stalk) as the *lateral,* or *medial, umbilical ligaments.*

THE VEINS

There is a natural grouping of the veins according to their relationships which is convenient to follow in discussing their development. Under the term *systemic veins* we can include all the vessels which collect the blood distributed to various parts of the body in the routine of local metabolism. In young embryos these would be the cardinal veins and their tributaries, that is, the return channels of the primitive intraembryonic circulatory arc. In older embryos and adults the systemic veins would include the anterior (superior) caval system, which is evolved from the anterior cardinals, and the posterior (inferior) caval system, which takes the place of the postcardinals and their tributaries.

We can set apart from the general systemic circulation three special venous arcs—the *umbilical,* returning the blood from the placenta; the *pulmonary,* returning the blood from the lungs; and the *hepatic portal,* carrying blood from the intestinal tract to the liver. The specialized nature of the placental and pulmonary circulations is obvious. The peculiarities of the *hepatic portal system* call, perhaps, for a word of explanation. Ordinarily veins collect blood from local capillaries and pass it on directly to the heart. Their bloodstream is away from the organ with which they are associated; once collected within a vein the blood is not redistributed in capillaries until it has again passed through the heart. The portal vein arises in typical fashion by collecting the blood from capillaries in the digestive tube. But then, contrary to the usual procedure, its blood flows, not directly to the heart, but to the liver, where it enters a second capillary bed and is returned by a second set of collecting vessels to the heart. With reference to the plexus of capillaries in the liver this vein is afferent; hence, its designation as a portal (translated, *carrying to*) vein, setting it apart from other veins, which carry blood only away from the organ with which they are associated.

Changes in the Systemic Veins Resulting in the Establishing of the Anterior (Superior) Vena Cava   The main tributaries draining the cephalic parts of the adult body are the external and internal jugulars and the subclavians. The *internal jugular vein* (Fig. 17-15) is merely the original anterior cardinal under a new name. The *external jugular vein* develops from the small vessel draining the

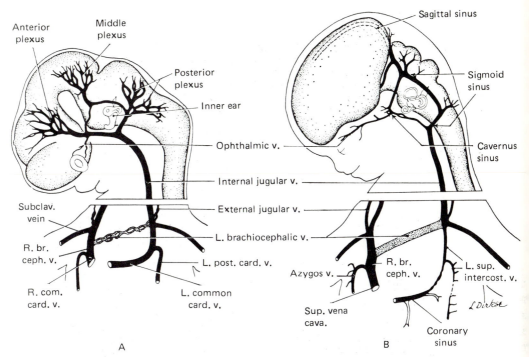

Fig. 17-15 Schematic diagrams to show some of the major events in the formation of the superior vena cava. (A) Conditions late in the seventh week when the left brachiocephalic vein is just beginning to establish a cross anastomosis between the right and left anterior cardinal (internal jugular) veins. (B) Conditions at about 10 weeks when the left brachiocephalic anastomosis has radically reduced the blood flow to the left common cardinal. *(The scheme of rotating the head in reference to the trunk was borrowed from Hamilton, Boyd, and Mossman. The arrangement of vessels in the cephalic region was based on the work of Streeter and of Padget.)* Abbreviations: *br. ceph.*, brachiocephalic; *com. card.*, common cardinal; *L.*, left; *R.*, right; *sup.*, superior; *v.*, vein.

mandibular region, and the *subclavian vein* arises as an enlargement of one of the segmental tributaries at the level of the anterior appendage bud (Fig. 17-13).

When the appendage buds first appear, the heart lies far forward in the body. As development progresses it is carried caudad. With this change in the position of the heart, the common cardinal veins change their relative position in the body and come to lie caudal to the anterior appendage buds. As a result of this altered relation, the subclavian veins from the anterior appendages, which early in development drain into the posterior cardinals (Fig. 17-16B), eventually empty into the anterior cardinals (Fig. 17-16C and D).

The outstanding characteristic of the systemic venous plan of a young embryo is its bilateral symmetry. Paired vessels from the anterior and from the posterior parts of the body become confluent to enter a mesially located sinus venosus (Figs. 17-16B and 17-22A to 17-22C). The still unpartitioned tubular heart then pumps the blood out by way of the aortic arches to be distributed to the body. This is in essence the circulatory plan of gill-breathing forms in that the heart is partitioned into right and left sides and the blood must pass through the heart twice in each cycle (Fig. 17-1). The right atrium is the collecting

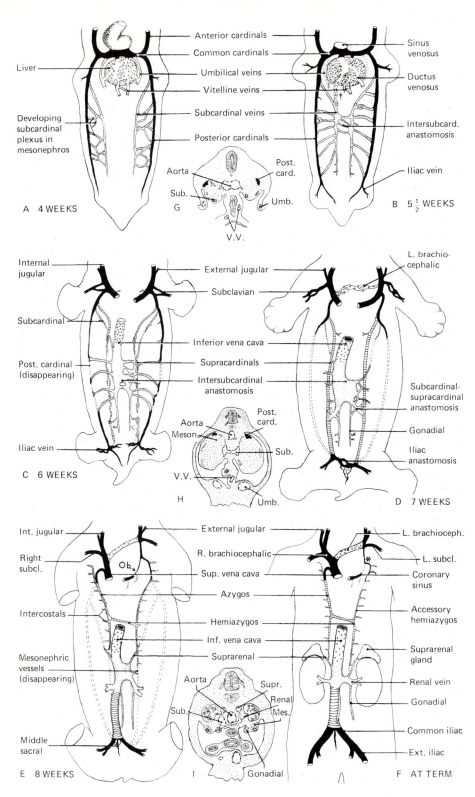

Anterior cardinals
Common cardinals
Umbilical veins
Vitelline veins
Subcardinal veins
Posterior cardinals

Liver

Developing
subcardinal
plexus in
mesonephros

A 4 WEEKS

Aorta
Sub.
G
V.V.

Post.
card.
Umb.

Sinus
venosus

Ductus
venosus

Intersubcard.
anastomosis

Iliac vein

B 5½ WEEKS

Internal
jugular

External jugular
Subclavian

Subcardinal

Inferior vena cava
Supracardinals
Intersubcardinal
anastomosis

Post. cardinal
(disappearing)

Iliac vein

C 6 WEEKS

Aorta
Meson.

Post.
card.

Sub.

V.V.

H

Umb.

L. brachio-
cephalic

Subcardinal-
supracardinal
anastomosis

Gonadial

Iliac
anastomosis

D 7 WEEKS

Int. jugular

External jugular
R. brachiocephalic
Sup. vena cava
Azygos
Hemiazygos
Inf. vena cava
Suprarenal

Right
subcl.

Ob.

Intercostals

Mesonephric
vessels
(disappearing)

Aorta
Sub.

Supr.
Renal
Mes.

Middle
sacral

E 8 WEEKS

I

Gonadial

L. brachioceph.
L. subcl.
Coronary
sinus

Accessory
hemiazygos

Suprarenal
gland

Renal vein

Gonadial

Common iliac

Ext. iliac

F AT TERM

502

chamber of the blood returning from the body for its first passage through the heart. The blood flows thence to the right ventricle, which pumps it to the lungs. After being oxygenated in the capillaries of the lungs the blood is returned to the left atrium and then to the left ventricle, to be pumped by way of the aorta and its branches to all parts of the body. This is the circulatory path toward which the developing vascular plan of a mammalian embryo must be shaped. The shift of the venous return to the right side of the heart early in prenatal development is a vital step toward establishing a vascular pattern adaptable to lung breathing and recognition of this trend is vital in dealing with any of the local changes in the venous system.

In the anterior systemic channels the shift to the right is accomplished very simply and directly. A new vessel forms between the right and left anterior cardinals (jugular veins) and shunts the left anterior cardinal bloodstream across to the right (Figs. 17-15 and 17-16D). With the establishment of this new channel the proximal part of the left anterior cardinal becomes progressively smaller (Fig. 17-16E) and finally loses its connection with the left common cardinal (Fig. 17-16F).

Meanwhile, shunting of the blood from left to right has also occurred in the drainage territory of the posterior cardinal vein. Deprived thus of its blood flow from both anterior and posterior tributaries, the distal part of the left common cardinal is reduced to a mere fibrous cord (see dotted line in Fig. 17-16F). Its proximal part, however, where it lies along the atrioventricular sulcus, persists as the coronary sinus, receiving blood from new tributary veins which drain the heart wall (Fig. 17-22D to 17-22F). If these changes have been followed, one has the basic pattern of the adult systemic veins of the upper part of the body. It is only necessary to apply the nomenclature of adult anatomy to vessels already familiar from following their development. The new connecting vessel is the *left brachiocephalic (left innominate)*. The old right anterior cardinal between the union of the subclavian with the jugulars and the transverse connection is the *right brachiocephalic (right innominate)*. From the confluence of the brachiocephalics to the heart is the *superior (anterior) vena cava* (Fig. 17-15). The superior vena cava is thus composed of the most proximal part of the right anterior cardinal and the right common cardinal vein. The small *azygos (cervicothoracic) vein,* the most anterior part of which is the reduced posterior cardinal, indicates the old point of transition from anterior cardinal to common cardinal. On the left, the small persisting portion of the anterior cardinal below the left brachiocephalic is called the *left superior intercostal vein* (Fig. 17-15B, and see asterisk in Fig. 17-16F).

Changes in the Systemic Veins Resulting in the Formation of the Posterior (Inferior) Vena Cava   The changes in the systemic veins of the posterior part of

Fig. 17-16 Schematic diagrams showing some of the steps in the development of the inferior vena cava. Cardinal veins are shown in black; subcardinals are stippled; supracardinals are horizontally hatched. Vessels arising independently of these three systems are indicated by small crosses. *(Based on the work of McClure and Butler.)* Abbreviations: *Ob.,* oblique vein of left atrium; *,* left superior intercostal; †, mesenteric portion of inferior vena cava; *subcl.,* subclavian vein.

the body are much more radical than they are anteriorly. The posterior cardinal veins, which are the primitive systemic drainage channels, are associated primarily with the mesonephroi. When the mesonephroi degenerate, it is but natural that the posterior cardinals should degenerate with them. The posterior vena cava replacing the cardinals is a composite vessel which gradually takes shape by the enlargement and straightening of small local channels which are pressed into service as the posterior cardinals degenerate.

The subcardinal veins initiate the diversion of the postcardinal bloodstream. The subcardinals are established as vessels lying along the ventromesial border of the mesonephros parallel with, and ventral to, the postcardinals. Taking origin from an irregular plexus of small vessels emptying from the mesonephroi into the posterior cardinals, the subcardinals from their first appearance have many channels connecting them with the postcardinals (Fig. 17-16A to 17-16C). These pursue tortuous courses among the mesonephric tubules, a relationship indicative of the fact that the subcardinal venous system of mammalian embryos is a fragmentary vestige of what was in remote phylogeny a functional renal portal system.

As the mesonephroi increase in size and bulge toward the midline, the subcardinals are brought very close together. In the midmesonephric region they anastomose with each other to form a large median vessel, the *subcardinal sinus,* or *intersubcardinal anastomosis* (Figs. 17-16C and D, and 17-17C). When this sinus is established the small vessels connecting sub- and postcardinals drain into the capacious sinus rather than toward the posterior cardinals. The result of this change soon becomes apparent in the disappearance of the posterior cardinals at the level of this sinus. The blood from the posterior part of the body is still collected by the distal ends of the postcardinals, but it returns to the heart by way of the subcardinal sinus. Consequently, the anterior portions of the posterior cardinals, although they persist, are much reduced in size (Fig. 17-16D).

Meanwhile the increased volume of blood entering the subcardinal sinus is finding a new and more direct route to the heart. The cephalic pole of the right mesonephros lies close to the liver. A fold of dorsal body wall tissue, just to the right of the primary dorsal mesentery, early makes a sort of bridge between these two organs. This fold is known as the *caval plica (caval mesentery).* In it, as everywhere in the growing body, are numerous small vessels. Connection of

Fig. 17-17 Three transverse sections from the series of the 9.4-mm pig embryo used in making the reconstructions illustrated in Figure 17-13. By laying a straightedge across either reconstruction at the level of the marginal line bearing the serial number of a cross section, its relations within the body as a whole are precisely indicated. The sections in this figure are especially helpful in working out the relations of the major veins of the abdominal region. (A) Section 388 passing through the body just cephalic to the attachment of the belly stalk. The belly stalk and the tip of the tail have not been included in the drawing. (B) Section 406 passing through the level of the pancreatic primordia. Tail and part of belly stalk not drawn. (C) **Section 426 at the level of the intersubcardinal** anastomosis. By cross-referring these sectional diagrams to the schematic plans of Figure 17-16, it will be evident that the part of the posterior vena cava labeled in (A) and (B) is its *mesenteric portion* and that the part labeled in (C) is an early stage in the formation of the *interrenal portion* from the intersubcardinal anastomosis.

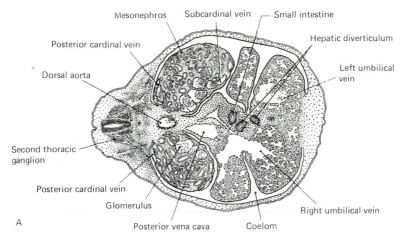

Mesonephros — Subcardinal vein — Small intestine

Posterior cardinal vein

Hepatic diverticulum

Dorsal aorta

Left umbilical vein

Second thoracic ganglion

Posterior cardinal vein

Glomerulus

Posterior vena cava

Coelom

Right umbilical vein

A

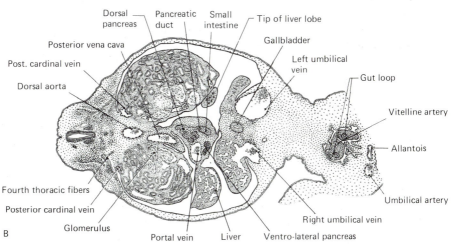

Dorsal pancreas — Pancreatic duct — Small intestine — Tip of liver lobe

Gallbladder

Posterior vena cava

Left umbilical vein

Post. cardinal vein

Dorsal aorta

Gut loop

Vitelline artery

Allantois

Fourth thoracic fibers

Posterior cardinal vein

Umbilical artery

Glomerulus

Portal vein

Liver

Ventro-lateral pancreas

Right umbilical vein

B

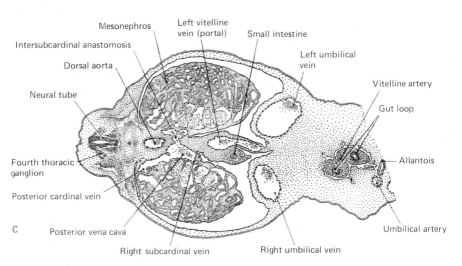

Mesonephros — Left vitelline vein (portal) — Small intestine

Intersubcardinal anastomosis

Left umbilical vein

Dorsal aorta

Vitelline artery

Neural tube

Gut loop

Allantois

Fourth thoracic ganglion

Posterior cardinal vein

Umbilical artery

Posterior vena cava

C

Right subcardinal vein

Right umbilical vein

505

these small vessels with the plexus of channels in the liver cephalically and in the mesonephros caudally provides the entering wedge. Once a current of blood finds its way from the mesonephros to the liver through these small vessels, enlargement of the channel proceeds with rapidity. This new channel becomes the *mesenteric part of the inferior vena cava.* (See relations of inferior vena cava in Fig. 17-17A and B, and portion of cava indicated with small crosses in Fig. 17-16B and C).

Within the liver this new bloodstream at first finds its way by devious small channels, eventually entering the sinus venosus along with the vitelline circulation. As its volume of blood increases, it excavates a main channel through the liver; a channel which gradually becomes walled in. As this new vessel becomes more and more definitely organized, it crowds toward the surface and eventually appears as a great vein lying in a notch along the dorsal side of the liver. This is the *hepatic part of the inferior vena cava* (Fig. 17-18C).

While the mesenteric and hepatic regions of the inferior vena cava are taking shape, the region where the anastomosis between the subcardinals resulted in the formation of the subcardinal sinus is changing its configuration. Originally a highly irregular venous space (Fig. 17-16C and D), it is gradually molded into a clearly outlined main channel. With the regression of the mesonephros and the ascent of the metanephros this part of the developing channel comes to lie between the kidneys. It is, therefore, known as the *interrenal portion of the inferior vena cava* (stippled region in Fig. 17-16E and F).

Caudal to its developing interrenal portion, still another set of veins enters into the formation of the inferior vena cava. These are the *supracardinal veins,* which appear relatively late in development as paired channels draining the dorsal body wall (Fig. 17-16C, D, and I). At the midmesonephric level, the supracardinals are diverted into the subcardinal sinus just as happened with the postcardinals earlier in development. Cephalic to the sinus, parts of the supracardinals persist as the azygos group draining in a somewhat variable manner into the reduced proximal part of the postcardinals. Caudal to the anastomosis with the subcardinal sinus, the right supracardinal becomes the principal drainage channel of the region. Its appropriation of the tributary vessels from the posterior appendages establishes it as the *postrenal portion of the inferior vena cava* and is the last step in the formation of that composite vessel (Fig. 17-16D to 17-16F).

Coronary Sinus The ultimate fate of the left common cardinal vein is a result of the shift in the course of the systemic blood so that it all enters the right side of the heart. Formerly returning a full half of the systemic bloodstream to the heart, the left common cardinal vein is finally left almost without a tributary from the body. Nevertheless, the proximal part of the old left cardinal channel is utilized. Pulled around the heart in the course of the migration of the sinus venosus toward the right, the left common cardinal vein lies close against the heart wall for a considerable distance (Fig. 17-22D). As the heart muscle grows in bulk it demands a greater blood supply for its metabolism. The returning veins of this circulation find their way into this conveniently located main vessel (Fig.

17-22E). Thus even when its peripheral circulation is cut off, the proximal part of the left common cardinal vein still persists as the coronary sinus into which the vessels of the cardiac wall drain (Fig. 17-22F).

Pulmonary Veins Phylogenetically the lungs are relatively new structures. It is not surprising, therefore, that we find the pulmonary veins arising independently and not by the conversion of old vascular channels. They originate as vessels which drain the various branches of the lung buds and converge into a common trunk (Fig. 14-9F), entering the left atrium dorsally (Fig. 17-22D). In the growth of the heart this trunk vessel is gradually absorbed into the atrial wall and usually four of its original branches open directly into the left atrium as the main pulmonary veins of the adult (Fig. 17-22D to 17-22F).

Portal Vein The blood supply to the intestines is first established through the vitelline arteries, which later become modified to form the superior (anterior) mesenteric artery. Likewise, the drainage of the intestinal tract is provided for by vessels which were originally the return channels of the primitive vitelline circulatory arc (Fig. 8-17). Very early in development the growing cords of hepatic tissue break up the proximal portion of the vitelline veins into a maze of small channels ramifying through the substance of the liver (Fig. 17-18A and B). Cephalic to the liver the stubs of both of the original vitelline veins for a time recollect the blood from the small new hepatic channels and empty it into the sinus venosus (Fig. 17-18C). Eventually the stub of the left vitelline vein drops out (Fig. 17-18D), leaving the stub of the right vitelline vein persisting as the right hepatic vein. Distal to the liver, the original veins are for a time retained, bringing blood from the yolk sac and intestines to the liver. With the regression of the yolk sac, the omphalic (yolk sac) portions of these veins necessarily disappear, but the mesenteric branches persist and become more extensive concomitantly with the increased length and complexity of the intestinal tract.

The original vitelline trunks into which these tributaries converge become the unpaired portal vein by forming transverse anastomoses and then abandoning one of the original channels. The curious spiral course of the portal vein is due to the dropping out of the original left channel cephalic to the middle anastomosis and the original right channel caudal to the anastomosis (Fig. 17-18D).

Umbilical Veins When they are first established the umbilical (allantoic) veins are embedded in the lateral body walls throughout their course from the belly stalk to the sinus venosus (Fig. 8-17). As the liver grows in bulk, it fuses with the lateral body wall. Where this fusion occurs, vessels develop, connecting the umbilical veins with the plexus of vessels in the liver (Fig. 17-18B). Once these connections are established, the umbilical stream tends more and more to pass by way of them to the liver, and the old channels to the sinus venosus gradually degenerate (Fig. 17-18C).

Meanwhile the umbilical veins distal to their entrance into the body become

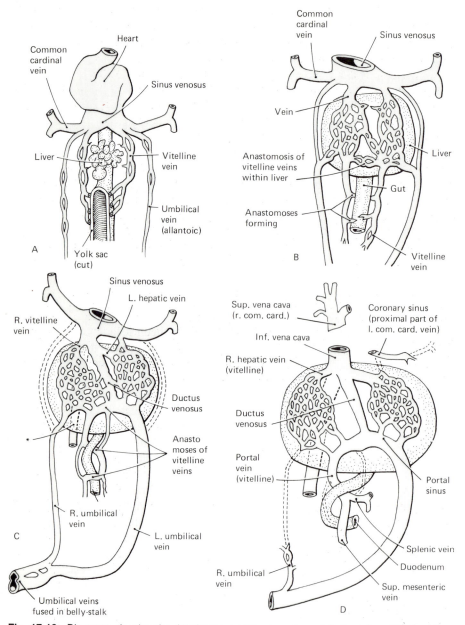

Fig. 17-18 Diagrams showing the development of the hepatic portal circulation. (A) Based on conditions in pig embryos of 3 to 4 mm; (B) about 6 mm; (C) 8 to 9 mm; (D) 20 mm and older. The asterisk in (C) indicates the location of the hepatic portion of the inferior vena cava, which is formed by the enlarging and straightening of originally small and irregular hepatic sinusoids. When fully formed this portion of the cava lies in a notch on the dorsal side of the liver. *(Collaboration of Dr. Alexander Barry is gratefully acknowledged.)* Abbreviation: *umb.,* umbilical.

fused with each other so that there comes to be but a single vein in the umbilical cord (Fig. 17-18C). Following this fusion in the cord, the intraembryonic part of the umbilical channel also loses its original paired condition. The right umbilical vein is abandoned as a route to the liver, and all the placental blood is returned over the left umbilical vein.

When first diverted into the liver, the umbilical bloodstream passes through by way of a meshwork of small anastomosing sinusoids. As its volume increases it excavates a main channel through the substance of the liver which is known as the *ductus venosus* (Fig. 17-18C and D). Leaving the liver, the ductus venosus becomes confluent with the hepatic veins which drain the maze of small sinusoids in the liver. The vena cava also joins the others at this point. Thus the bloodstreams from the posterior systemic circulation, from the portal circulation, and from the placental circulation all enter the heart together. Embryologically this great trunk vessel represents the proximal part of the old right vitelline vein enlarged by the placental blood from the ductus venosus and by the systemic blood from the vena cava (Fig. 17-18).

THE HEART

Like blood vessels, the heart arises from splanchnic mesoderm, but in both amphibians and birds prospective heart cells have been localized in the early gastrula stage (Fig. 5-20). The evidence for an inductive effect by the endoderm on the precardiac mesoderm in the initiation of cardiac mesoderm in higher vertebrates is equivocal, but evidence of an inductive interaction between these two tissues exists in amphibians (Jacobson, 1960). In a recently investigated "cardiac lethal" mutant in the axolotl an endodermal induction has been shown to be responsible for the formation of an adequate amount of myofibrillar material for beating of the heart (Lemanski et al., 1979).

The early establishment of the heart as two simple endocardial tubes on either side of the body and their subsequent fusion into a single cardiac tube was discussed in Chap. 8 (Figs. 8-15 and 8-16). The bilateral origin of the cardiac primordia is strikingly emphasized in a condition known as *cardia bifida* (Fig. 17-19). If the cardiac primordia are prevented from fusing by chemical or surgical means, separate independently beating hearts form on each side of the body (Gräper, 1907).

Formation of the Cardiac Loop and Establishing of the Regional Divisions of the Heart   The primary factor which brings about the regional differentiation of the heart is the rapid elongation of the primitive cardiac tube, resulting in its bending into an S-shape. The factors that cause the heart to bend are still incompletely understood. It is now known that much of the impetus to looping of the heart is intrinsic to the cardiac tube, for when the early heart is explanted and allowed to develop in vitro or as a free graft, the characteristic curving configuration appears (Orts Llorca and Gil, 1965). More recent studies

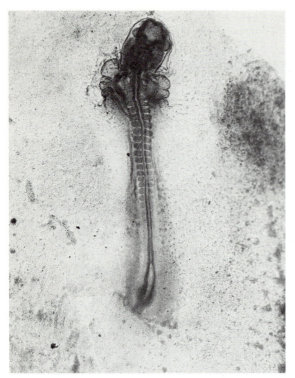

Fig. 17-19 Chick embryo with surgically produced cardia bifida. This condition was produced by cutting the floor of the foregut at an early stage. The cardiac primordia are prevented from fusing. *(Courtesy of R. L. DeHaan.)*

(Manasek et al., 1972) have shown that regional changes in cell shape along the early cardiac tube are to a great extent responsible for the looping.

During the period in which the cardiac loop is being formed, the primary regional divisions of the heart become clearly differentiated. The *sinus venosus* is the thin-walled chamber in which the great veins become confluent, entering the heart at its primary caudal end (Fig. 17-22C). The *atrial region* is established by transverse dilation of the heart tube just cephalic to the sinus venosus (Figs. 17-21C to 17-21E and 17-22C to 17-22E).

The ventricle is formed by the bent midportion of the original cardiac tube. As this *ventricular loop* becomes progressively more extensive, it at first projects ventrally beneath the attached aortic and sinus ends of the heart (Fig. A-48F and G). Later it is bent caudally so that the ventricle, formerly situated cephalic to the atrium, is brought into its characteristic adult position caudal to the atrium (Fig. A-48I to A-48L). Between the atrium and ventricle the heart remains relatively undilated. This narrow connecting portion is the *atrioventricular canal* (Fig. 17-24A). The most cephalic part of the cardiac tube undergoes the least change in appearance, persisting as the *truncus arteriosus*, which connects the ventricle with the ventral aortic roots (Fig. 17-21). The transitional region where

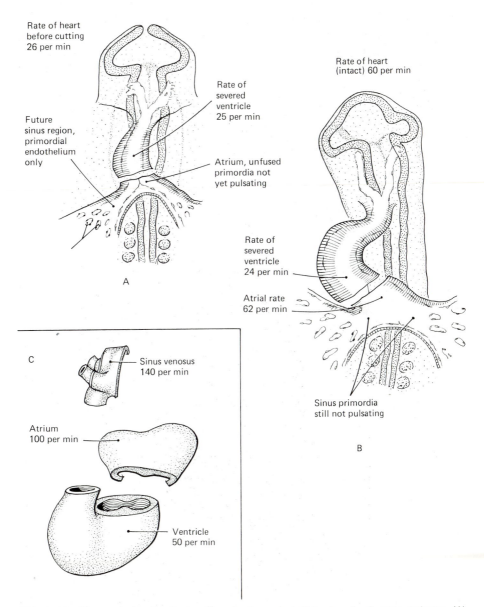

Rate of heart
before cutting
26 per min

Rate of
severed
ventricle
25 per min

Future
sinus region,
primordial
endothelium
only

Atrium, unfused
primordia not
yet pulsating

A

Rate of heart
(intact) 60 per min

Rate of
severed
ventricle
24 per min

Atrial rate
62 per min

Sinus primordia
still not pulsating

B

C

Sinus venosus
140 per min

Atrium
100 per min

Ventricle
50 per min

Fig. 17-20 Diagrams showing the location of cuts made in living hearts of young embryos. (A) Embryos of 10-12 somites. At this stage only the ventricle is pulsating, and cutting it away from the nonactive atrium has no appreciable effect on its rate. (B) Embryos of 13 to 15 somites. By this stage the atrium has begun to pulsate and, in the intact heart, acts as a pacemaker. Transection between atrium and ventricle shows the atrium maintaining essentially the rate of the intact heart, whereas the ventricle drops back to approximately the same slow rate it exhibited before the atrium became active and drove the ventricle at its own faster rate. (C) When the heart of a 4-day embryo is cut into three parts, each part beats at its characteristic rate. *(From Patten, 1956,* Univ. of Mich. Med. Bull. **22:***1.)*

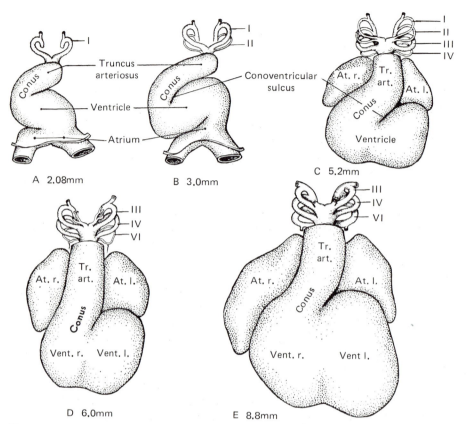

Fig. 17-21 Ventral views of human embryonic hearts to show bending of the cardiac tube and the establishing of its regional divisions. *(After Kramer, 1942, Am. J. Anat., vol. 71.)* Abbreviations: *At. l., At. r.,* atrium left and right; *Roman numerals I to VI,* aortic arches of corresponding numbers; *Tr. art.,* truncus arteriosus; *Vent. l., Vent. r.,* ventricle left and right.

the ventricle narrows to join the truncus arteriosus is known as the *conus* (Fig. 17-21).

Almost from their earliest appearance the atrium and the ventricle show external indications of the impending division of the heart into right and left sides. A distinct median furrow appears at the apex of the ventricular loop (Figs. 17-21D and E, and 17-22D and E). The atrium meanwhile has undergone rapid dilation and bulges out on either side of the midline (Fig. 17-22). Its bilobed configuration is emphasized by the manner in which the truncus arteriosus compresses it midventrally (Fig. 17-21C to 17-21E). These superficial features suggest the more important changes going on internally. Among these internal factors are localized cell death (Pexieder, 1972) and relationships between the extracellular matrix and the early myocardial cells (Manasek, 1975).

Beginning of Cardiac Function The metabolic requirements of the embryo require the heart to commence beating early in its development. In the chick,

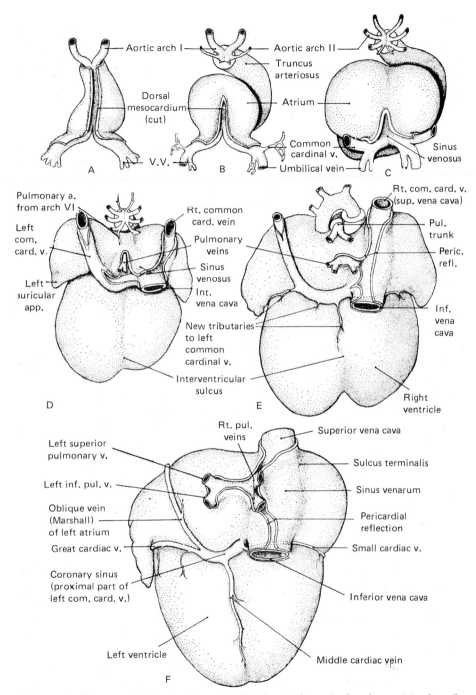

Fig. 17-22 Six stages in the development of the heart, drawn in dorsal aspect to show the changing relations of the sinus venosus and great veins entering the heart. Abbreviation: *Peric. refl.*, pericardial reflections; *V.V.*, vitelline vein.

the best-studied vertebrate embryo, the earliest tentative beats begin while the paired cardiac primordia are still fusing (about 29 hours). In the human embryo, the heart begins to beat at the end of the third week. The strength and regularity of the heartbeat increase dramatically as fusion of the cardiac primordia continues and the major divisions of the heart are laid down.

Pulsation of the myocardium appears in the major cardiac regions in the same sequence in which they are laid down. The first contractions appear in the ventricle before the atrium is fully established. The contraction rate of the ventricle is at first very slow (Fig. 17-20A). When the atrium begins to pulsate, the heart rate increases. Experiments involving transection of the early heart between the atrium and the ventricle have shown that the atrium has a faster inherent rate of pulsation (Fig. 17-20B) and takes control of cardiac rhythm as a pacemaker. In experimentally produced cardia bifida (DeHaan, 1959) the right heart begins to beat at a more rapid rate than the left at about the time when the sinus venosus would normally form. This suggests that the origin of the cardiac pacemaker is in the right cardiac primordium.

If transection experiments are carried out after the sinus venosus has been formed caudal to the atrium, one finds that its rate of contraction is higher than that of the atrium (Fig. 17-20C). This gradient in contraction rate within the myocardium is not merely a difference in rate from one chamber to another. In very small explants taken in cephalocaudal sequence from within a given region, there is a gradient *within* the cardiac chambers (Barry, 1942). In other words, the cephalocaudal gradient in the rate of myocardial contraction is continuous throughout the tubular heart. The importance of this situation is far-reaching. It means that the pacemaking center of the heart is, in all stages of development, at its intake end. Therefore, when a wave of contraction starts at this area and sweeps throughout the length of the cardiac tube, it picks up the incoming blood and forces it out at the other end of the heart into the vessels by which it will be distributed to the body. One would have great difficulty postulating a more efficient type of pumping action in a simple tubular heart without valves.

There is an interesting relationship between the morphological development of the heart and its electrical activity. This can be demonstrated by making electrocardiographic recordings. In the chick the first traces of electrical activity are noted in embryos with 15 somites (somewhat more advanced than Fig. A-48C). A pattern of electrical activity resembling the ventricular pattern in the adult heart appears in embryos of 16 to 17 somites (Fig. A-48D), at which time the embryonic heart is practically all ventricle. As the atrium appears and shifts toward its normal position, the pattern of the embryonic electrocardiogram resembles more closely that of the adult; by the fourth day of development the electrocardiogram has practically assumed its adult configuration (Hoff, Kramer, DuBois, and Patten, 1939). The blood pressure of an early (3-day) chick embryo is extremely low (0.61/0.43 mm Hg), and the heart rate is about 140 beats per minute (Girard, 1973). The blood pressure at first rises exponentially and later the rate of increase tapers off until just before hatching when, with a pulse rate of 220 beats per minute, the blood pressure is 30/19 mm Hg.

Partitioning of the Heart   As the wall of the ventricle increases in thickness it develops on its internal face a meshwork of interlacing muscular bands, the *trabeculae carneae* (Fig. 17-23). Opposite the external furrow in the ventricle these muscular bands become consolidated as a partition which appears to grow from the apex of the ventricle toward the atrium. This is the *interventricular septum* (Fig. 17-23).

Meanwhile, two conspicuous masses of loosely organized mesenchyme, called *endocardial cushion tissue,* develop in the walls of the narrowed portion of the heart between the atrium and the ventricle. The endocardial cushions consist of relatively large masses of an extracellular matrix, rich in glycosaminoglycans, which has long been called *cardiac jelly.* During their outgrowth, the cardiac jelly becomes seeded by mesenchymal cells originating from the endocardium of the cushions (Markwald et al., 1977, 1978). Ultimately the cushions become transformed into dense connective tissue. One of these endocardial cushions is formed in the dorsal wall of the atrioventricular canal (Fig. 17-24B), and the

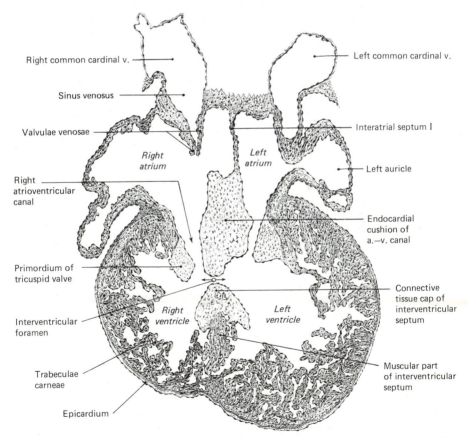

Fig. 17-23 Section (X45) through the heart of a 9.4-mm pig embryo, at the level of the atrioventricular canals. Its relations within the body can be worked out by laying a straightedge across Figure 17-13, 2 mm below the marginal line numbered 258.

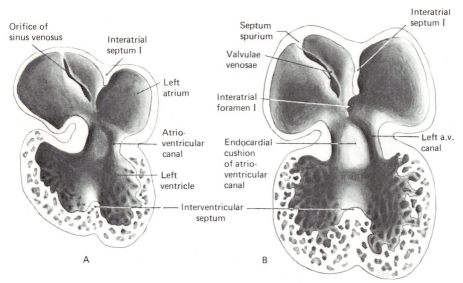

Fig. 17-24 Semischematic drawings of interior of heart to show initial steps in its partitioning. (A) Cardiac septa are represented at stage reached in human embryos early in fifth week of development. Note especially the primary relations of interatrial septum primum. Based on original reconstructions of the heart of a 3.7-mm pig embryo and on Tandler's reconstructions of corresponding stages of the human heart. (B) Cardiac septa as they appear in human embryos of sixth week. Note restriction of interatrial foramen primum by growth of interatrial septum primum. Based on original reconstructions of the heart of 6-mm pig embryo, on Born's reconstructions of rabbit heart, and Tandler's reconstructions of corresponding stages of the human heart.

other is formed on the ventral wall. When they meet, these two masses occlude the central part of the canal and separate it into right and left channels (Figs. 17-14C and 17-25).

At the same time a median partition appears in the cephalic wall of the atrium. Because another, closely related partition is destined to form here later, this one is called the *first interatrial septum* or the *septum primum*. It is composed of a thin layer of young cardiac muscle covered by endothelium. It is crescentic in shape, its concavity directed toward the ventricle and the apices of the crescent extending, one along the dorsal wall and one along the ventral wall of the atrium, all the way to the atrioventricular canal, where they merge, respectively, with the dorsal and ventral endocardial cushions (Fig. 17-24). This leaves the atria separated from each other except for an opening called the *interatrial foramen primum*.

While these changes have been occurring, the sinus venosus has been shifted out of the midline so that it opens into the atrium to the right of the interatrial septum (Figs. 17-22D and 17-24). The heart is now in a critical stage of development. Its simple tubular form has been altered such that the four chambers characteristic of the adult heart are clearly recognizable. Partitioning of the heart into right and left sides is well under way. But there is as yet incomplete division of the bloodstream because there are still open communica-

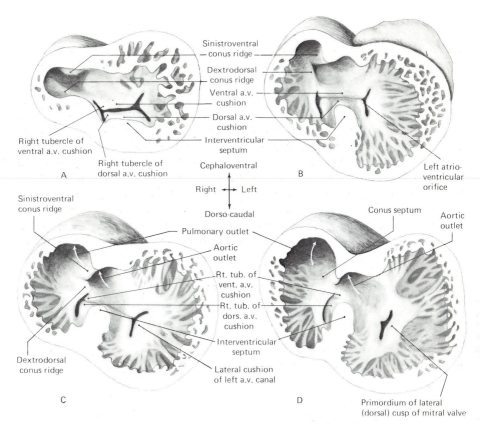

Fig. 17-25 Four stages in partitioning of atrioventricular canal. Apex of ventricle cut off, and heart viewed from below. In this aspect, the relations of conus ridges to atrioventricular canal cushions and to the upper part of interventricular septum are especially instructive. *(After Kramer, 1942, Am. J. Anat., vol. 71.)*

tions from the right to the left side in both atrium and ventricle. Further progress in the growth of the partitions, however, would leave the two sides of the heart completely separated. Were this to occur now, the left side of the heart would become almost literally dry, for the sinus venosus, into which systemic, portal, and placental currents all enter, opens on the right of the interatrial septum, and not until much later do the lungs and their vessels develop sufficiently to return any considerable volume of blood to the left atrium. The partitions in the ventricle and in the atrioventricular canal do progress rapidly to completion (Figs. 17-26 to 17-30), but an interesting series of events takes place at the interatrial partition which assures that an adequate supply of blood reaches the left atrium and thence the left ventricle.

Just when the septum primum is about to merge with the endocardial cushions of the atrioventricular canals, closing the interatrial foramen primum and isolating the left atrium, a new opening is established. This opening appears in the more cephalic part of the septum primum, first in the form of multiple

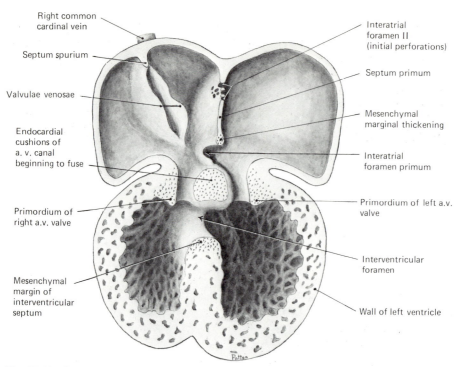

Fig. 17-26 Semischematic drawing showing the interior of the developing heart in the sixth week, at the stage when interatrial foramen primum is almost closed and interatrial foramen secundum is just being established by the appearance of multiple small perforations. *(From Patten, 1960, Am. J. Anat., vol. 107.)*

small perforations resulting from the presence of dying cells in the area (Fig. 17-26). These rapidly expand and unite to form a single opening of considerable size, known as the *interatrial foramen secundum* (Fig. 17-27). By the time the interatrial foramen primum has been closed, the interatrial foramen secundum offers an effective alternative passageway. By this route enough of the blood entering the right atrium can make its way across into the left atrium to equalize approximately the intake of the two sides of the heart.

Shortly after the formation of this secondary opening in the septum primum, a second interatrial partition makes its appearance just to the right of the first. Like the septum primum, the septum secundum is crescentic in form. However, the open end of its crescent is in a different place. Whereas the open part of the septum primum is directed toward the atrioventricular canal (Fig. 17-24B and 17-26), the open part of the septum secundum is directed toward the lower part of the sinus entrance (Fig. 17-27), which later becomes the opening of the inferior vena cava into the right atrium (Fig. 17-30). This is of vital functional significance, for it means that as the septum secundum grows, the opening remaining in it is carried out of line with the interatrial foramen secundum in the

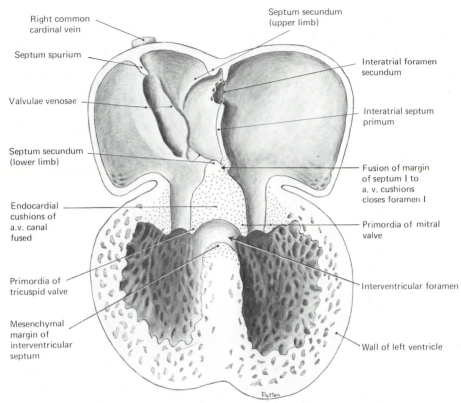

Right common
cardinal vein

Septum secundum
(upper limb)

Septum spurium

Interatrial foramen
secundum

Valvulae venosae

Interatrial septum
primum

Septum secundum
(lower limb)

Fusion of margin
of septum I to
a. v. cushions
closes foramen I

Endocardial
cushions of
a.v. canal
fused

Primordia of mitral
valve

Primordia of
tricuspid valve

Interventricular foramen

Mesenchymal
margin of
interventricular
septum

Wall of left ventricle

Patten

Fig. 17-27 Semischematic drawing showing the interior of the developing heart, at the stage when interatrial foramen primum has been closed and interatrial foramen secundum well opened. Septum secundum is just being established to the right of septum primum. *(From Patten, 1960, Am. J. Anat., vol. 107.)*

septum primum (Fig. 17-29). The opening in the septum secundum, although it becomes smaller as development progresses, is not destined to be completely closed but will remain open as the *foramen ovale* (Fig. 17-30).

The flaplike remains of the septum primum overlying the persistent oval opening in the septum secundum constitute an efficient valvular mechanism between the two atria. When the atria are filling, some of the blood returning by way of the great veins can pass freely through the foramen ovale by merely pushing aside the flap of the septum primum. The inferior caval entrance lies adjacent to, and is directed diagonally into, the orifice of the foramen ovale (Figs. 17-29 and 17-30). Consequently, it is largely blood from the inferior vena cava which passes through the foramen ovale into the left atrium. When the atria start to contract, pressure of the blood within the left atrium forces the flap of the septum primum against the foramen ovale, effectively closing it against return flow into the right atrium. Without some such mechanism affording a supply of blood to its left side, the developing heart could not be partitioned in

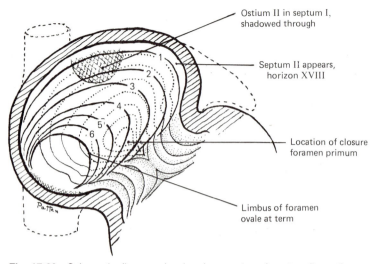

Ostium II in septum I,
shadowed through

Septum II appears,
horizon XVIII

Location of closure
foramen primum

Limbus of foramen
ovale at term

Fig. 17-28 Schematic diagram showing, by a series of contour lines, the progress of growth of septum secundum. By dotting in the contour lines for septum primum, the difference in the direction of growth of the two septa is emphasized. *(From Patten, 1960, Am. J. Anat., vol. 107.)*

the midline, ready to assume its adult function of pumping two separate bloodstreams.

While these changes are taking place in the main part of the heart, the truncus arteriosus is being divided into two separate channels. This process starts in the ventral aortic root between the fourth and sixth arches (Fig. 17-6). Continuing toward the ventricle, the division is effected by the formation of longitudinal ridges of readily molded young connective tissue of the same type as that making up the endocardial cushions of the atrioventricular canal. These ridges, called *truncus ridges,* bulge progressively farther into the lumen of the truncus arteriosus and finally meet to separate it into aortic and pulmonary channels (Figs. 17-31B and 17-32). The semilunar valves of the aorta and the main pulmonary trunk develop as local specializations of these truncus ridges. Toward the ventricles from the site of formation of the semilunar valves the same ridges are continued into the conus of the ventricles. The fact that proximal to the valves these ridges are, for descriptive purposes, called *conus ridges* should not be allowed to obscure their developmental and functional continuity with the truncus ridges. The truncoconal ridges follow a spiral course, and where they extend down into the ventricles they meet and become continuous with the margins of the interventricular septum. This reduces the size of the interventricular foramen but does not close it completely. Its final closure is brought about by a mass of endocardial cushion tissue from three sources. Bordering the interventricular foramen below is the interventricular septum, the crest of which, above its main muscular portion, is made up of endocardial cushion tissue (Fig. 17-23). Toward the atrioventricular canal lie the masses of

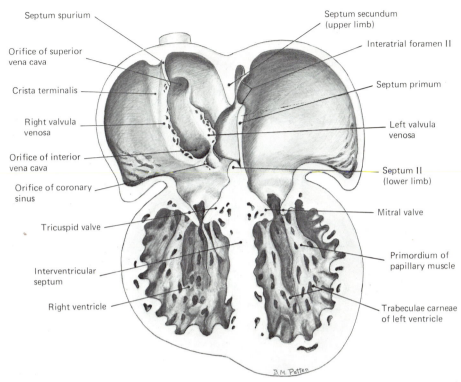

Septum spurium

Orifice of superior
vena cava

Crista terminalis

Right valvula
venosa

Orifice of interior
vena cava

Orifice of coronary
sinus

Tricuspid valve

Interventricular
septum

Right ventricle

Septum secundum
(upper limb)

Interatrial foramen II

Septum primum

Left valvula
venosa

Septum II
(lower limb)

Mitral valve

Primordium of
papillary muscle

Trabeculae carneae
of left ventricle

B.M. Patten

Fig. 17-29 Schematic drawing to show the internal configuration of the heart in human embryos of the third month. At this stage resorption has begun to involve the valvulae venosae and septum spurium, as indicated by the many small perforations in their margins. Note the way the left venous valve is coming to lie against septum secundum, with which it is already beginning to fuse. It usually leaves no recognizable traces in the adult, but occasionally delicate lacelike remains of it can be seen adherent to septum secundum and, more rarely, extending a way onto the valvula foraminis ovalis (cf. Fig. 17-30).

endocardial tissue which were responsible for its partitioning. In the conus outlet are the ridges we have just been considering. From all these adjacent areas endocardial cushion tissue encroaches on the interventricular foramen and finally completely plugs the opening with a loose mass of this young, readily molded connective tissue (Fig. 17-34E). Later this mass differentiates into the membranous portion of the interventricular septum and the septal cusps of the atrioventricular valves. When the interventricular septum has been completed, the right ventricle leads into the pulmonary channel and the left into the aorta. With this condition established, the heart is completely divided into right and left sides except for the interatrial valve at the foramen ovale, which must remain open throughout fetal life until, after birth, the lungs attain their full functional capacity and the full volume of the pulmonary stream passes through them to be returned to the left atrium.

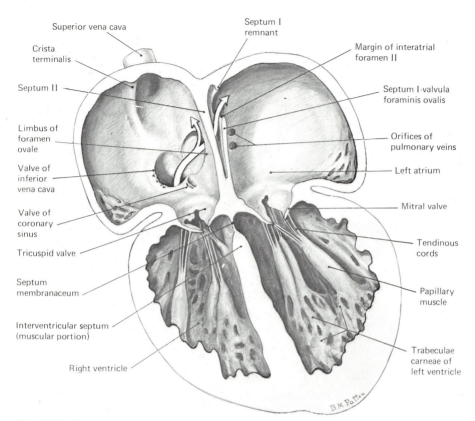

Superior vena cava

Crista terminalis

Septum II

Limbus of foramen ovale

Valve of inferior vena cava

Valve of coronary sinus

Tricuspid valve

Septum membranaceum

Interventricular septum (muscular portion)

Right ventricle

Septum I remnant

Margin of interatrial foramen II

Septum I-valvula foraminis ovalis

Orifices of pulmonary veins

Left atrium

Mitral valve

Tendinous cords

Papillary muscle

Trabeculae carneae of left ventricle

B M Patten

Fig. 17-30 Schematic drawing to show the interrelations of septum primum and septum secundum during the latter part of fetal life. Note especially the way in which the lower part of septum primum is situated so that it acts as a one-way valve at the oval foramen in septum secundum. The split arrow indicates the way a considerable part of the blood from the inferior vena cava passes through the foramen ovale to the left atrium while the remainder eddies back into the right atrium to mingle with the blood being returned by way of the superior vena cava.

This leaves but one of the exigencies of heart development still to be accounted for. If, during early fetal life, before the lungs were well developed, the pulmonary channel were the only exit from the right side of the heart, the right ventricle would have an outlet inadequate to develop its pumping power. It is only late in fetal life that the lungs and their vessels develop to a degree which prepares them for assuming their postnatal activity, and the power of the heart muscle must be built up gradually by continued functional activity. This situation is met by the ductus arteriosus leading from the pulmonary trunk to the aorta (Fig. 17-33). The right ventricle is not unprepared for its adult function because it pumps its full share of the blood throughout fetal life. Instead of all going to the lungs, however, part of the blood pumped by the right ventricle passes into the aorta by way of that portion of the left sixth aortic arch which persists as the ductus arteriosus (Fig. 17-33).

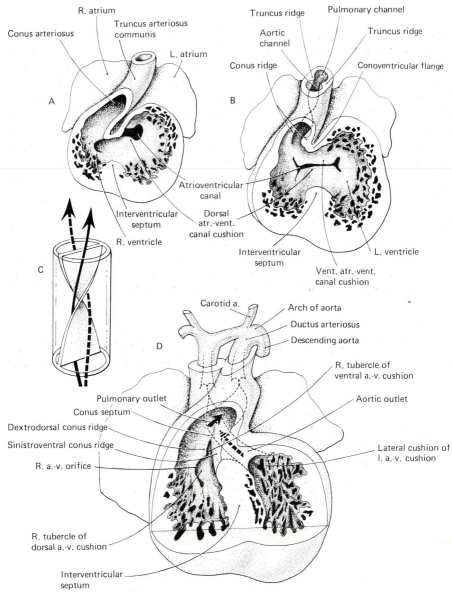

Fig. 17-31 Semischematic dissections of the developing heart viewed in frontal aspect to show relations of importance in the subdivision of the truncus arteriosus and in establishing the aortic and pulmonary outlets. (C) Diagram illustrating the 180° spiralling of the septum that subdivides the truncus. (*A,B,D after Kramer, 1942, Am. J. Anat. 71:343.*)
Abbreviations: A.-V. or Atr.-Vent., atrioventricular.

As the lungs increase in size, relatively more blood goes to them and relatively less goes through the ductus arteriosus. Obviously, by the end of gestation the lungs must be adequately developed to take over from the placenta

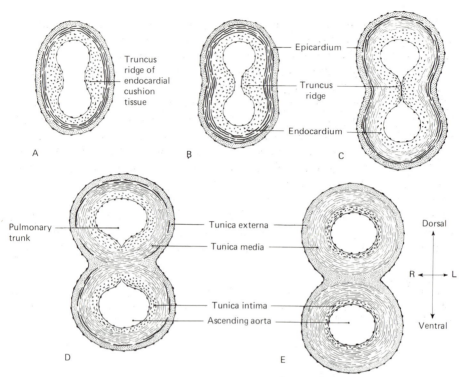

Fig. 17-32 Schematic diagrams showing the partitioning of the truncus arteriosus to form the ascending aorta and the pulmonary trunk. *(From Patten, 1960, in Gould,* The Pathology of the Heart, *2d ed. Courtesy of Charles C Thomas, Publisher, Springfield, Ill.)*

the function of oxygenating the blood. This means that the pulmonary vessels must be large enough to handle the requisite blood volume. Injection preparations clearly show that well before birth the vascular channels in the lungs are large enough to perform this function. What such postmortem material does not show is the fact that these large vessels, before birth, are not actually carrying anything like the blood volume their size would suggest. Lind and Wegelius (1949), utilizing angiocardiographic methods on living fetuses, have shown that the flow of blood through the pulmonary circuit is restricted to a volume much below its potential capacity. How much of this is due to mechanical factors such as the unexpanded condition of the lungs and how much of it depends on differential vasoconstriction remains to be determined. It is clear, however, that pulmonary vessels of the requisite size have been formed and are ready to increase their blood flow radically and promptly with the beginning of respiration. Within a short time after birth, under the stimulus of functional activity, the lungs are able to take all the blood from the right side of the heart, and the ductus arteriosus is gradually obliterated. The ductus arteriosus, therefore, serves during intrauterine life as a "compensatory exercising channel" for the right ventricle.

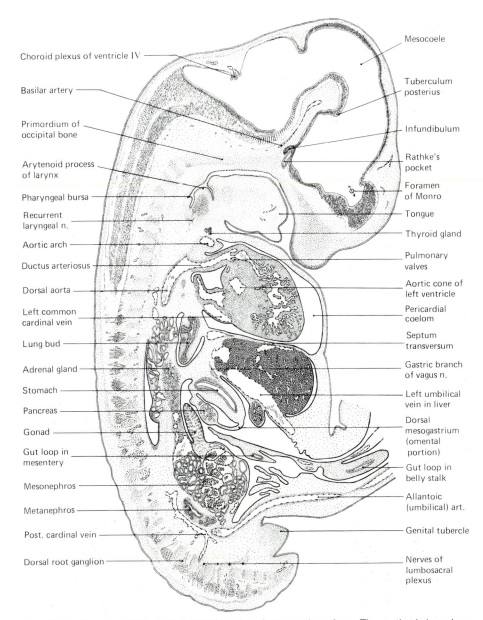

Choroid plexus of ventricle IV

Basilar artery

Primordium of
occipital bone

Arytenoid process
of larynx

Pharyngeal bursa

Recurrent
laryngeal n.

Aortic arch

Ductus arteriosus

Dorsal aorta

Left common
cardinal vein

Lung bud

Adrenal gland

Stomach

Pancreas

Gonad

Gut loop in
mesentery

Mesonephros

Metanephros

Post. cardinal vein

Dorsal root ganglion

Mesocoele

Tuberculum
posterius

Infundibulum

Rathke's
pocket

Foramen
of Monro

Tongue

Thyroid gland

Pulmonary
valves

Aortic cone of
left ventricle

Pericardial
coelom

Septum
transversum

Gastric branch
of vagus n.

Left umbilical
vein in liver

Dorsal
mesogastrium
(omental
portion)

Gut loop in
belly stalk

Allantoic
(umbilical) art.

Genital tubercle

Nerves of
lumbosacral
plexus

Fig. 17-33 Drawing (X10) of parasagittal section of 15-mm pig embryo. The section is in a plane
slightly to the left of the midline and passes through the ductus arteriosus, lung bud, stomach,
gonad, and metanephros.

Changes in the Sinus Region In following the story of the partitioning of
the heart, with emphasis on the functional standpoint, many things of less
striking significance have been passed over. Cephalic to the sinus orifice the
valvulae venosae which guard it against return flow fuse and are prolonged onto

the dorsal wall of the atrium in the form of a ridge called the *septum spurium* (false septum). This septum is, for a time, too conspicuous to ignore, but it is of little importance in the partitioning of the heart and soon undergoes retrogression (Fig. 17-34C to 17-34E). As the heart grows it absorbs the sinus venosus into its walls so that eventually the superior and inferior venae cavae and the coronary sinus all open separately into the right atrium (cf. Figs. 17-29 and 17-30). In this process the venous valves are rapidly reduced in size by resorption.

Atrioventricular Valves and Papillary Muscles   The point where the atrioventricular canals open into the ventricles is marked by early indications of the establishment of valves. From the partition in the atrioventricular canal and from the outer walls on either side, masses of tissue in the shape of thick, blunt flaps project toward the ventricle (Figs. 17-23, 17-26, and 17-34C and D). These masses of primitive connective tissue, similar histologically to the endocardial cushions of the canal, later become differentiated into the flaps of the adult valves (Figs. 17-29, 17-30, and 17-34F). The papillary muscles and tendinous cords which, in the adult, act as stays to these valves arise by modification of some of the related trabeculae carneae (cf. Figs. 17-23 and 17-34D to 17-34F).

Aortic and Pulmonary Valves   The valves which guard the orifices of the aorta and of the pulmonary artery arise from mesenchymal pads (Fig. 17-33) already mentioned as developing in connection with the truncus septum. In early histological appearance and in the manner in which this loose tissue gradually becomes organized into the exceedingly dense fibrous tissue characteristic of adult valves, they are similar to the atrioventricular valves. They do not, however, develop any supporting strands comparable to the papillary muscles and tendinous cords.

Course and Balance of Blood Flow in the Fetal Heart   All the steps in the partitioning of the embryonic heart lead gradually toward the final adult condition in which the heart is completely divided into right and left sides. Yet from the nature of its living conditions it is not possible for the fetus in utero to fully attain the adult type of circulation. The lungs, although fully formed and ready to function in the last part of fetal life, cannot actually begin their work until after birth. Yet in the first minutes of its postnatal life a fetus must change from an existence submerged in the amniotic fluid to air-breathing.

The rapid readjustment of the circulation at the time of birth is a dramatic and fascinating biological process. To understand how these changes are so promptly yet smoothly accomplished, it is necessary to have clearly in mind the manner in which the way for them has been prepared during intrauterine life. The very mechanisms which maintain cardiac balance during intrauterine life are perfectly adapted to keep the balance of the circulatory load on the new, postnatal basis. In the foregoing account of the development of the interatrial septal complex, emphasis was laid upon the fact that at no time were the atria

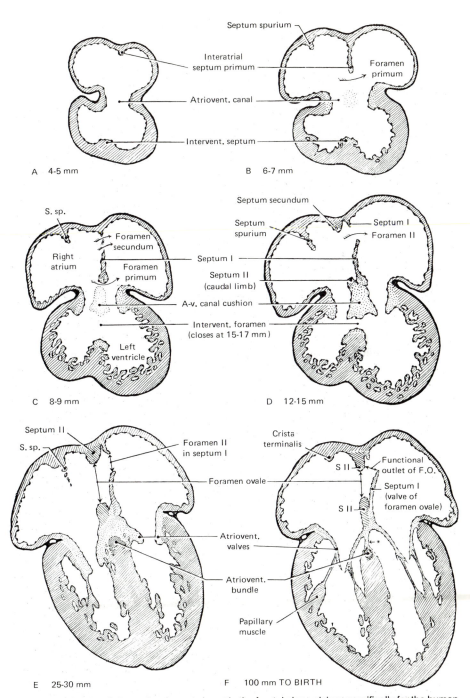

Fig. 17-34 Sectional plans of embryonic heart in the frontal plane giving, specifically for the human embryo, a more precise picture of the rate of progress of partitioning than do the preceding schematic drawings. Stippled areas indicate endocardial cushion tissue, muscle is shown in diagonal hatching, and epicardium in solid black. The lightly stippled areas in the atrioventricular canal in (B) and (C) indicate location of endocardial cushions of atrioventricular canal before they have grown sufficiently to fuse in the plane of the diagram. Abbreviations: *A-v.*, atrioventricular; *S. sp.*, septum spurium; asterisk and arrow in (F) indicate the membranous part of the interventricular septum.

completely separated from each other. This permits the left atrium, throughout prenatal life, to receive a contribution of blood from the inferior cava and the right atrium by a transseptal flow, which compensates for the relatively small amount of blood entering the left atrium of the embryo directly by way of the pulmonary circuit and maintains an approximate balance of intake into the two sides of the heart.

The manner in which transseptal flow occurs and where and to what extent the various bloodstreams of the fetal circulation are mixed was put into perspective by the studies of Barcroft and Barron and their coworkers. Using sheep fetuses, they performed quantitative analyses of blood samples drawn from various critical parts of the fetal circulation. The oxygen content of such samples gives important evidence as to what mixing of the currents is actually taking place in the living fetus. Later work involving the collaboration of Barclay and Franklin utilized angiocardiography, which involves injecting a radiologically opaque material such as Thorotrast or Diodrast into the bloodstream and following its course, with the aid of either a fluoroscopic screen or serially taken x-ray pictures. Later, Wegelius and Lind successfully applied angiocardiographic techniques to the study of living human fetuses. By using electronic intensification they were able to get images on the fluoroscopic screen brilliant enough to permit their recording on motion-picture film. Such records have now furnished us with dramatic evidence not only of the relative volume of circulatory arcs in the fetus but also of the changes that occur at the time of birth. These angiographic records are now being supplemented by catheterization studies. The 1956 Nobel Prize in medicine was awarded to the developers of this technique for following the course of the circulation in living subjects. It involves the insertion of a slender flexible tube through peripheral vessels into the heart. Its exact location is controlled by fluoroscopy, and through it blood samples may be withdrawn from known locations. By means of the catheter it is possible, also, to record precisely the pressure changes during the cardiac cycle.

The evidence of both anatomical and functional studies allows one to summarize the course followed by the blood in passing through the fetal heart as follows: The inferior caval entrance is so directed with reference to the foramen ovale that a considerable portion of its stream passes directly into the left atrium (Figs. 17-30 and 17-35). Under fluctuating pressure conditions—for example, following uterine contractions which send a surge of placental blood through the umbilical vein—the placental flow may temporarily hold back any blood from entering the circuit by way of either the portal vein or the inferior caval tributaries (Fig. 17-35). Under these conditions, the left atrium would be charged almost entirely with fully oxygenated blood. Such conditions, however, would be only temporary and would be counterbalanced by periods when the portal and systemic veins poured enough blood into the common channels to load the heart for a time with mixed or depleted blood. The important thing physiologically is not the fluctuations but the maintenance of the average oxygen content of the blood at adequate levels. Careful measurements have shown that the interatrial communication in the heart of the fetus at term is considerably

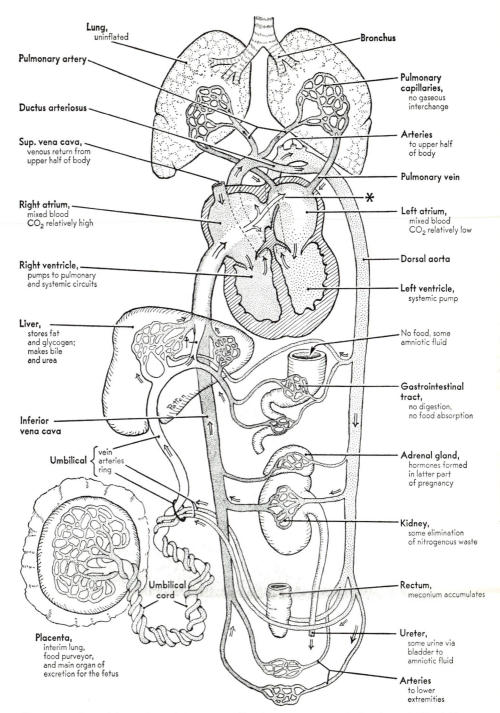

Lung, uninflated

Pulmonary artery

Ductus arteriosus

Sup. vena cava, venous return from upper half of body

Right atrium, mixed blood CO_2 relatively high

Right ventricle, pumps to pulmonary and systemic circuits

Liver, stores fat and glycogen; makes bile and urea

Inferior vena cava

Umbilical { vein arteries ring

Placenta, interim lung, food purveyor, and main organ of excretion for the fetus

Umbilical cord

Bronchus

Pulmonary capillaries, no gaseous interchange

Arteries to upper half of body

Pulmonary vein

Left atrium, mixed blood CO_2 relatively low

Dorsal aorta

Left ventricle, systemic pump

No food, some amniotic fluid

Gastrointestinal tract, no digestion, no food absorption

Adrenal gland, hormones formed in latter part of pregnancy

Kidney, some elimination of nitrogenous waste

Rectum, meconium accumulates

Ureter, some urine via bladder to amniotic fluid

Arteries to lower extremities

Fig. 17-35 Plan of the fetal circulation at term. The ductus venosus in the liver is marked by †. The arrow in the heart marked by * indicates the passage of blood from the right atrium to the left as it occurs during atrial diastole. This flow pushes the valvula into the open positions here represented. When the atria contract, the valvula moves back against the septum, closing the foramen ovale against the return flow and thus forcing all the blood in the left atrium to enter the left ventricle. (See colored insert.) *(After Patten, in Fishbein, 1963, Birth Defects. Courtesy of the National Foundation and the J. B. Lippincott Company, Philadelphia.)*

smaller than the inferior caval inlet. This would suggest that the portion of the inferior caval stream which could not pass through this opening into the left atrium must eddy back and mix with the rest of the blood in the right atrium. Angiocardiographic studies confirm this inference.

Compared with conditions in adult mammals, the mixing of oxygenated blood freshly returned from the placenta with depleted blood returning from a circuit of the body may seem inefficient. But this is a one-sided comparison. The fetus is an organism in transition. Starting with a simple ancestral plan of structure and living an aquatic life, it attains its full heritage slowly. It must be viewed in the light of both the primitive conditions from which it is emerging and the definitive conditions toward which it is progressing. Below the bird-mammal level circulatory mechanisms with partially divided and undivided hearts and correspondingly unseparated bloodstreams meet all the needs of metabolism and growth. Maintenance of food, oxygen, and waste products at an average level which successfully supports life does not depend on "pure currents," although such separated currents undoubtedly make for higher efficiency in the rate of interchange of materials. From a comparative viewpoint, the fact that the mammalian fetus is supported by a mixed circulation seems but natural.

From the standpoint of smooth postnatal circulatory readjustments, the larger the fetal pulmonary return becomes, the less will be the balancing transatrial flow and the less will be the change entailed by the assumption of lung breathing. Very early in development, before the lungs have grown to any great extent, the pulmonary return is negligible and the flow from the right atrium through the interatrial ostium primum constitutes practically the entire intake of the left atrium. After the ostium primum is closed and while the lungs are only slightly developed, flow through the interatrial ostium secundum must still be the major part of the blood entering the left atrium. During the latter part of fetal life the foramen ovale in the septum secundum becomes the gateway of the transseptal route. As the lungs grow and the pulmonary circulation increases in volume, a progressively smaller proportion of the left atrial intake comes by way of the foramen ovale and a progressively larger amount comes from the vessels of the growing lungs. But this circulation to the lungs, although increased as compared with earlier stages, has been shown by angiocardiographic studies to be much less in volume than the caliber of the pulmonary vessels would lead one to expect. Until the lungs are inflated there are evidently factors—vasomotor or mechanical or both—which restrict flow through the smaller pulmonary vessels. Therefore, even in the terminal months of pregnancy, a considerable right-left flow must still be maintained through the foramen ovale in order to keep the left atrial intake on a par with that of the right.

The balanced atrial intake thus maintained implies a balanced ventricular output. Although not in the heart itself, there is a mechanism in the closely associated great vessels which affords an adequate outlet from the right ventricle during the period when the pulmonary circuit is developing. When the pulmonary arteries are formed from the sixth pair of aortic arches, the right sixth arch soon loses its original connection with the dorsal aorta. On the left, however, a portion of the sixth arch persists as a large vessel connecting the pulmonary artery with the dorsal aorta (Figs. 17-33 and 17-35). This vessel, already familiar to us as the ductus arteriosus, remains open throughout fetal life and acts as a shunt, carrying over to the aorta whatever excess blood the pulmonary vessels at any particular phase of their development are not prepared to receive from the right ventricle. As has already been pointed out, the ductus arteriosus is the "exercising channel" of the right ventricle because it makes it possible for the right ventricle to carry its full share of work throughout development and thus to be prepared for pumping all the blood through the lungs at the time of birth.

CHANGES IN THE CIRCULATION FOLLOWING BIRTH

The two most obvious changes which occur in the circulation at the time of birth are the abrupt cutting off of the placental bloodstream and the immediate assumption by the pulmonary circulation of the function of oxygenating the blood. One of the most impressive things in embryology is the perfect preparedness for this event, which has been built into the very architecture of the circulatory system during its development. The shunt at the ductus arteriosus, which has been one of the factors in balancing ventricular loads throughout intrauterine development, and the valvular mechanism at the foramen ovale, which has at the same time been balancing atrial intakes, are perfectly adapted to effect the postnatal rebalancing of the circulation. The lessening of peripheral resistance in the small vessels of the lungs and the closure of the ductus arteriosus are the primary events, and the closure of the foramen ovale follows as a logical sequence (Fig. 17-1).

It has long been known that the lumen of the ductus arteriosus is gradually occluded postnatally by an overgrowth of its intimal tissue. This process in the wall of the ductus is as characteristic and regular a feature of the development of the circulatory system as the formation of the cardiac septa. Its earliest phases begin to be recognizable in the fetus as the time of birth approaches, and postnatally the process continues at an accelerated rate to terminate in complete anatomical occlusion of the lumen of the ductus about 6 to 8 weeks after birth.

Barcroft, Barclay, and Barron have conducted an extensive series of experiments on sheep delivered by cesarean section which indicate that the ductus arteriosus closes functionally almost immediately. Following birth there appears to be a contraction of the circularly disposed smooth muscle in the wall of the ductus which promptly reduces the flow of blood through it. This reduction in the shunt from the pulmonary circuit to the aorta, acting together with the lowering of resistance in the vascular bed of the lungs which accompanies their newly assumed respiratory activity, aids in raising the pulmonary circulation promptly to full functional level. At the same time the functional closure of the ductus paves the way for the ultimate anatomical obliteration of its lumen by overgrowth of intimal connective tissue. This concept of the immediate functional closure of the ductus by muscular contraction is so appealing on theoretical grounds that a little extra caution in evaluating the evidence is indicated. It should be borne in mind that an initial tendency on the part of the circular smooth muscle of the ductus to contract does not necessarily imply a contraction maintained strongly and steadily enough to shut off all blood flow during the 6 to 8 weeks occupied by morphological closure. The dramatic quality of an immediate muscular response should not cause us to forget the importance of the slower but more positive structural closure. Nor should we lose sight of the possibility that freer blood flow through the lungs may well take enough pressure off the shunt at the ductus arteriosus so that sustained muscular contraction is not a crucial factor in the functional abandonment which precedes structural closure.

The effects of increased pulmonary circulation with the concomitant increase in the direct intake of the left atrium are manifested secondarily at the foramen ovale. Following birth, as the pulmonary return increases, compensatory blood flow from the right atrium to the left decreases correspondingly and shortly ceases altogether. In other words, when equalization of atrial intakes has occurred, the compensating one-way valve at the foramen ovale falls into disuse and the foramen may be regarded as functionally closed. The abandonment of the shunt at the foramen ovale is indicated anatomically by a progressive reduction in the looseness of the valvula foraminis ovalis and the consequent diminution of the interatrial communication to a progressively narrower slit between the valvula and the septum.

Anatomical obliteration of the foramen ovale follows leisurely in the wake of its functional abandonment. There is a considerable interval following birth before the septum primum fuses with the septum secundum to seal the foramen ovale. This delay is, however, of no import because as long as the pulmonary circuit is normal and pressure in the left atrium equals or exceeds that in the right, the orifice between them is functionally inoperative. It is not uncommon to find the fusion of these two septa incomplete in the hearts of individuals who have, as far as circulatory disturbances are concerned, lived uneventfully to maturity.

With birth and the interruption of the placental circuit there follows the gradual fibrous involution of the umbilical vein and the umbilical arteries. The flow of blood in these vessels, of course, ceases immediately with the severing of the umbilical cord, but obliteration of the lumen is likely to take from 3 to 5 weeks, and isolated portions of these vessels may retain a vestigial lumen for much longer. Ultimately these vessels are reduced to fibrous cords. The old course of the umbilical vein is represented in the adult by the round ligament of the liver extending from the umbilicus through the falciform ligament, and by the ligamentum venosum within the substance of the liver. The proximal portions of the umbilical arteries are retained in reduced relative size as the hypogastric or internal iliac arteries. The fibrous cords extending from these arteries on either side of the urachus toward the umbilicus represent the remains of the more distal portions of the old umbilical arteries. They are known in the adult as the *lateral (medial) umbilical ligaments*.

Much yet remains to be learned regarding the more precise physiology of the fetal circulation and the interaction of various factors during the transition from intrauterine to postnatal conditions. Nevertheless, with our present knowledge it is quite apparent that the changes in the circulation which occur following birth involve no revolutionary disturbances of the load carried by different parts of the heart. The fact that the pulmonary vessels are already so well developed before birth means that the changes which must occur following birth have been thoroughly prepared for; and the compensatory mechanisms at the foramen ovale and the ductus arteriosus which have been functioning all during fetal life are entirely competent to effect the final postnatal rebalancing of the circulation with a minimum of functional disturbance. It is still true that as individuals we crowd into a few crucial moments the change from water-living to air-living that in phylogeny must have been spread over eons of transitional amphibious existence. But as we learn more about this change in manner of living, it becomes apparent that we should marvel more at the completeness and the perfection of the preparations for its smooth accomplishment and dwell less on the old theme of the revolutionary character of the changes involved.

Development of Chick Embryos from 18 Hours to 4 Days of Incubation

The Appendix has been designed as an aid for the students who use this book as the basis for laboratory study of chick or pig development. It begins with a short section introducing the student to the study of serial sections. The bulk of Appendix 1 is a section describing the development of chick embryos from 18 hours to 4 days of incubation. Appendix 2 is a summary of the figures that are of greatest use for the study of both chick and pig material.

Study of Serial Sections In certain respects the laboratory study of embryological material involves methods of work for which courses in general zoology do not entirely prepare the student. Some general suggestions as to methods of procedure are therefore not out of place.

In dissecting gross material it is not unduly difficult to appreciate the complete relationships of a structure. The nature of embryological material, however, introduces new problems. Embryos of the age when the establishment of the various organ systems and processes of body formation are being initiated are too small to admit of successful dissection, but they are not sufficiently small to permit the satisfactory microscopic study of an intact embryo, except for its more general organization. To study embryos of this stage with any degree of thoroughness one must cut them into sections which are sufficiently thin to allow effective use of the microscope to ascertain cellular organization and detailed structural relationships. In preparing such material the entire embryo is cut into sections which are mounted on slides in the order in which they were cut. A sectional view of any region of the embryo is then available for study.

Although sections readily yield accurate information about local regions, it is extremely difficult to construct a mental picture of any whole organism from a study of serial sections alone. For this reason it is necessary to work first on entire embryos which have been prepared by staining and clearing so they may be studied as transparent objects. From such preparations it is possible to map out the configuration of the body and the location and extent of the more conspicuous internal organs. The fact that embryos have three dimensions must be kept constantly in mind in this work and, by careful focusing, the depth at which a structure lies must be determined as well as its apparent position in surface view.

In studying a section from a series it is first necessary to determine the location in which it was cut through the embryo. The plane of the section under consideration, as well as the region of the embryo through which it passes, should be ascertained by comparing it with an entire embryo of the same age as that from which the section was cut. Only when the location of a section is known precisely can the structures appearing in it be correlated with the organization of the embryo as a whole. Probably nothing in the study of embryology causes students more difficulties than neglecting to locate sections accurately, with the consequent failure to appreciate the relationships of the structures seen in them. The vital importance of fitting the structures shown by sections properly into the general scheme of organization as it appears in whole-mounts (Figs. A-1 and A-2) cannot be too greatly emphasized.

Terms of Location   In embryology it is necessary to designate the location of structures and the direction of growth processes by terms which are referable to the body of the embryo regardless of the position it occupies. Our ordinary terms of location refer to the direction of the action of gravity, such as *above, over,* or *under.* These terms are not sufficiently accurate in this type of work because the embryo itself may lie in a great

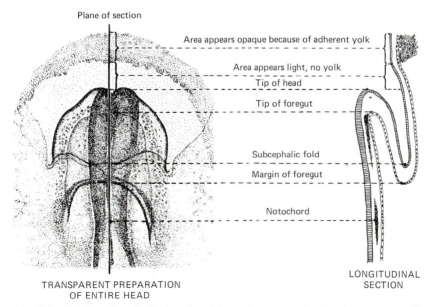

Plane of section

Area appears opaque because of adherent yolk

Area appears light, no yolk

Tip of head

Tip of foregut

Subcephalic fold

Margin of foregut

Notochord

TRANSPARENT PREPARATION
OF ENTIRE HEAD

LONGITUDINAL
SECTION

Fig. A-1   Relation of longitudinal section of the embryonic head to the picture presented by a head of the same age mounted entire as a transparent preparation.

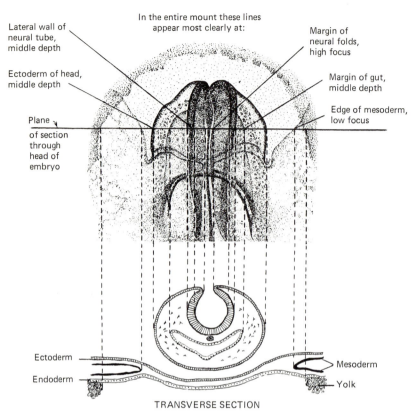

Fig. A-2 Relation of transverse section of the embryonic head to the picture presented by an entire head of the same age viewed as a transparent preparation. The two drawings may be brought into closer relation by looking at them with the top of the page tilted downward. This and the preceding figure show the aid that sections can be in interpreting entire embryos and, at the same time, the way in which the greater perspective lent by study of entire embryos increases the significance of sectional pictures. Note especially that neither the transverse nor the longitudinal section alone adequately interprets all the features of the entire mount. An embryo is a three-dimensional object and must be studied with that fact in mind. These figures show, also, how largely the appearance of lines or bands seen in studying transparent mounts of entire embryos is due to looking edgewise through a folded layer.

variety of positions.[1] The correct adjectives of position are *dorsal,* pertaining to the back; *ventral,* the belly; *cephalic,* the head; *caudal,* the tail; *mesial,* the middle part; and *lateral,* the side of the embryonic body. In dealing with relations within the head the adjective *cephalic* is inadequate and the term *rostral* (Latin, *beak of a bird; bow of a ship*) is used to designate the extreme anterior portion of the head or the relative location of intracephalic

[1]In gross human anatomy there still persist many terms that are referred to gravity. Such terms, because of the erect posture of man, are not applicable to comparative anatomy or to embryology. The most confusing of these are *anterior* and *posterior* as used to mean, respectively, *pertaining to the belly* and *to the back.* In comparative anatomy and in embryology, anterior has reference to the head region and posterior to the tail region. Because of this possible source of confusion, the terms anterior and posterior should be replaced by their more precise synonyms, *cephalic* and *caudal,* wherever the context is such that their significance might otherwise be doubtful.

structures such as the various parts of the brain. Adverbs of fixed position are made in the usual way by adding -*ly* to the root of the adjective.

In addition to adverbs of position, corresponding adverbs of motion or direction are formed by adding the suffix -*ad* to the root of the adjective, as *dorsad*, meaning *toward the back, cephalad*, meaning *toward the head,* etc. These adverbs should be applied only to the progress of processes or to the extension of structures toward the part indicated by their root. Thus, for example, we should say that the developing eye of an embryo was located in the late*ral* wall of the forebrain or that the forebrain was in the cepha*lic* part of the embryo; but if we wished to express the idea that as the eye increased in size it moved farther to the side, we should say it grew late*rad*. Cultivation of the use of correct and definite terms of position and direction in dealing with embryological processes will greatly aid you in accurate thinking and clear understanding.

FROM THE PRIMITIVE-STREAK STAGE TO THE APPEARANCE OF SOMITES

Because the earliest stages of embryonic development of the chick occur within the reproductive tract of the hen, the laboratory study of chick embryos commonly omits the early stages of cleavage and blastulation. Gastrulation and the formation of the primitive streak were described in Chap. 5 (Figs. 5-10 to 5-14) and will not be repeated here. By 18 hours' incubation the primitive streak is well established, and cells which have been invaginated by way of Hensen's node push cephalad, initiating the formation of the notochord (Fig. 5-16B). In chick embryos which have been incubated about 18 hours the notochord has become markedly elongated to form a well-defined midline structure (Fig. A-3). In older embryological treatises the young notochord was frequently called the "head process." Chick embryos of about 18 hours' incubation, when the notochord is one of the few prominent structural features, have often been spoken of as being in the "head-process stage."

Sections of embryos in the head-process stage (Fig. A-4) give much more

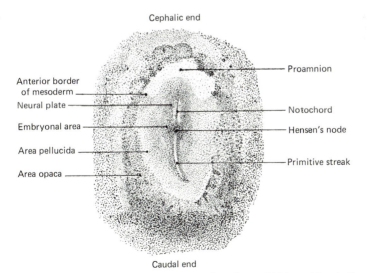

Fig. A-3 Dorsal view (X14) of entire chick embryo of 18 hours' incubation.

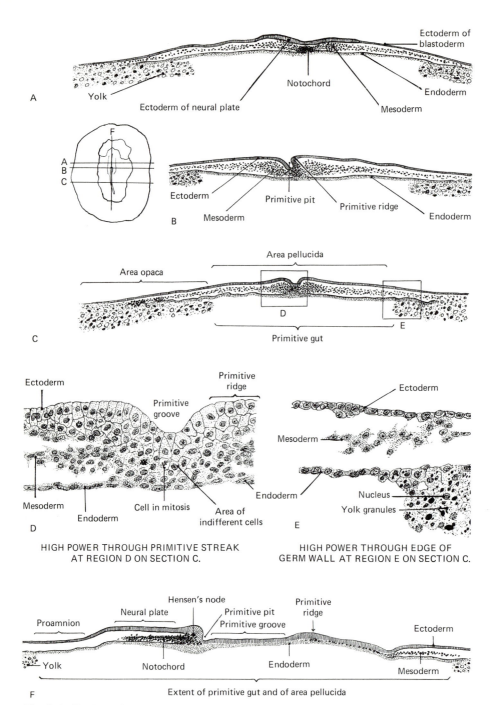

Fig. A-4 Sections of 18-hour chick. The location of each section is indicated by a line drawn on a small outline sketch of an entire embryo of corresponding age. The letters affixed to the lines on the sketch indicating the location of the sections correspond with the letters designating the section diagrams. Each germ layer is represented by a different conventional scheme: ectoderm by vertical hatching; endoderm by fine stippling backed by a single line; the cells of the mesoderm, which at this stage do not form a coherent layer, by heavy angular dots. (A) Diagram of transverse section through neural plate and notochord. (B) Diagram of transverse section through primitive pit. (C) Diagram of transverse section through primitive streak. (D) Drawing showing cellular structure in primitive-streak region. (E) Drawing showing cellular structure at inner margin of germ wall. (F) Diagram of median longitudinal section passing through notochord and primitive streak.

information on the structure and relations of the notochord than can be gleaned from the study of whole mounts. Near its origin at Hensen's node, the notochord is made up of diffusely arranged cells that merge caudally with the recently turned-under cells of the lateral portions of the mesoderm and the endoderm. Its cephalic tip gradually develops a more definitely circumscribed, rodlike configuration.

In the sections diagramed in Fig. A-4 a conventional scheme of representation has been employed to indicate each of the germ layers. The ectoderm is vertically hatched, the cells of the mesoderm are represented by heavy angular dots when they are isolated or by solid black lines when they lie arranged in the form of compact layers, and the endoderm is represented by stippling backed by a single line. A similar conventional representation of the different germ layers is observed in all diagrams of chick sections to facilitate following the way in which the organ systems of the embryo are constructed from the germ layers. The plane in which each of the sections diagramed passes through the embryo is indicated by a line drawn on a small outline sketch of an embryo at a corresponding stage.

The Primitive Streak   The primitive streak serves as a useful point of reference in assessing the extent and direction of early growth processes. Cross sections passing through the primitive streak a little caudal to Hensen's node are most suitable for studying the relations of the rest of the mesoderm to ectoderm and endoderm (Fig. A-4C). In the floor of the primitive groove the ectoderm dips down toward the endoderm, and in this region there is no line of demarcation between the two layers, since both merge in a mass of cells that can best be characterized as indifferent (Fig. A-4D). Emerging from either side of this indifferent cell area is the mesoderm, growing laterad into the space between ectoderm and endoderm.

These migrating mesodermal cells pass through the indifferent area on their way to take their appointed places. During all these cell movements there is concomitant cell division.

In Fig. A-4D, a small region at the primitive streak has been drawn at higher magnification to show the characteristic cellular structure of the undifferentiated region in the floor of the primitive groove and of the various layers continuous with one another at this place. The cells of the ectoderm are much more closely packed together and more sharply delimited than those of the other germ layers. Where the ectoderm is thickened in the primitive-ridge region (Fig. A-4D), it has become several cell layers in depth (stratified). In regions lateral to the primitive ridge it gradually becomes thinner until it consists of but a single cell layer (Fig. A-4E). The rapid extension that the mesoderm is undergoing at this time is indicated by the loose arrangement and irregular cytoplasmic processes of its cells. Between the germ layers of an embryo, what looks like empty space in a section was occupied in the living embryo by a noncellular ground substance, composed mainly of mucopolysaccharides. This serves as a substratum in which the young mesodermal cells can move by their own activity. When a cell thus breaks loose from the coherent mesodermal layer and starts to move to some other place where it will settle down and become differentiated, we speak of it as a *mesenchymal cell.* A loose aggregation of mesenchymal cells is called *mesenchyme.* The cells of the endoderm are neither so closely packed nor so clearly defined as the ectodermal cells. Nevertheless, in contrast to the condition of the mesoderm at this stage, the endodermal cells form a definite, unbroken layer.

Regression of the Primitive Streak   Following the phases of its greatest conspicuousness in young embryos, the primitive streak rapidly becomes both relatively and actually smaller. There is still proliferative activity in and around it, but the cells move into other areas and the streak itself is shortened and shifted relatively farther caudally in the embryo. The carbon marking experiments of Spratt have shown that even after the streak has begun to shorten there is still some convergence of cells toward it, although this activity is far less extensive than it was in earlier stages. Emigration of cells from the caudal half of the streak contributes to the lateral and caudal expansion of both the intra- and extraembryonic tissues and also accounts for much of the shortening of the streak. At the same time the notochord is greatly increasing in length. The major points in the changing relations between the primitive streak and notochord are summarized quantitatively in Fig. A-5.

As the primitive streak thus undergoes a sort of dismemberment from its more caudal portion, Hensen's node moves farther and farther back toward the area opaca. By the end of the second day of incubation little more than the node and a very short portion of the streak persist. At this stage what remains of the once prominent node and the primitive streak is usually given a new name, the *end bud* or *tail bud* (Fig. A-32). Cells of the tail bud ultimately contribute to the neural tube, somites, mesenchyme, and caudal arteries of the tail, but do not contribute to the notochord, ectoderm, or gut (Schoenwolf, 1977).

Growth of the Endoderm and Establishing of the Primitive Gut Sections of embryos which had been incubated from 18 to 20 hours show how the endoderm has spread out and become organized into a coherent layer of cells merging peripherally with the yolk and overlapping it to a certain extent (Fig. A-4). This marginal area where the expanding germ layers merge with the underlying yolk is known as the *germ wall* (Fig. A-4E). The cavity between the yolk and the endoderm is now termed the *primitive gut*. The yolk floor of the primitive gut does not show in sections prepared by the usual methods. The reasons for this are found in the relations of the embryo to the yolk before

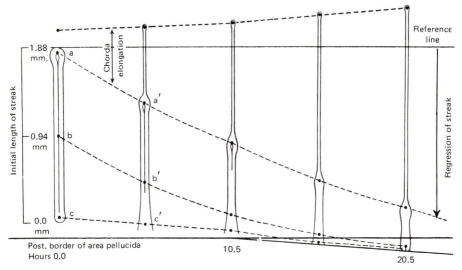

Fig. A-5 Graphic summary of the regression of the primitive streak and the growth of the notochord. The scale is indicated by the length of the primitive streak given in millimeters at the left. The time is given in hours beyond the start of the experiments. (*From Spratt, 1947, J. Exp. Zool.* **104:***69.*)

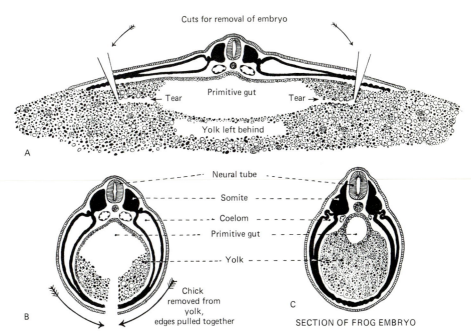

Cuts for removal of embryo

Primitive gut

Tear

Tear

Yolk left behind

A

Neural tube

Somite

Coelom

Primitive gut

Yolk

Chick
removed from
yolk,
edges pulled together

B

C

SECTION OF FROG EMBRYO

Fig. A-6 (A) is a diagram showing how the usual method of removing chick embryos from the yolk in order to prepare them for microscopic study makes the sections appear as if the primitive gut had no ventral boundary. (B) and (C) show how removing a chick from the yolk and pulling its edges together ventrally facilitates comparisons with forms which do not develop with their growing bodies spread out on a large mass of yolk.

it is removed for sectioning. Throughout the entire central region of the blastoderm the yolk is separated from the endoderm by the cavity of the primitive gut. When the embryo is removed from the yolk sphere, the yolk floor of the primitive gut, not being adherent to the blastoderm, is left behind (Fig. A-6A). In contrast, the periphery of the blastoderm lies closely applied to the yolk. Some yolk adheres to this part of the blastoderm when it is removed. This adherent yolk is shown in the section diagram of Fig. A-4C. Its presence clearly indicates why this region (area opaca) appears less translucent in surface views of entire embryos (Fig. A-3). This spread-out arrangement of the developing walls of the gut is so unlike conditions seen in adults or in embryos of forms with less yolk that it sometimes bothers students encountering it for the first time. To see these relations in proper perspective it may be helpful to picture a chick lifted off the yolk and the lateral margins pulled together ventrally (Fig. A-6B). If one chooses, the method of comparison may readily be reversed by imagining an amphibian embryo split open along its midventral line and spread out on the surface of a sphere as a chick lies on the yolk.

In embryos of 18 hours the primitive gut is a cavity with a flat roof of endoderm and a floor of yolk. Peripherally it is bounded on all sides by the germ wall (Fig. A-4C and F). The merging of the cells of the endoderm with the yolk mass is shown in the small area of the germ wall drawn to a high magnification in Fig. A-4E. In the germ wall the cell boundaries are incomplete and very difficult to distinguish, but nuclei surrounded by more or less definite areas of cytoplasm can be made out. This cytoplasm contains numerous yolk granules in various stages of absorption. It will be recalled that the nuclei of the germ wall arise by division from the nuclei of cells lying at the margins of the expanding blastoderm. These ill-defined cells appear to be concerned in breaking up the yolk in advance of the more definitely organized layers of the endoderm.

By about the twentieth hour of incubation we can see indications of a local differentiation of that region of the primitive gut which underlies the cephalic part of the embryo. By focusing through the ectoderm in the anterior region of a whole mount of this age a pocket of endoderm can be observed (Fig. A-7). This endodermal pocket is the first part of the gut to acquire a floor, other than the yolk floor, and is called, because of its cephalic position, the *foregut*.

Growth and Early Differentiation of the Mesoderm The mesoderm which arises from either side of the primitive streak spreads rapidly laterad, and at the same time each lateral wing of the mesoderm swings cephalad. Figure A-8A to A-8C shows schematically the extension of the mesoderm during the latter part of the first day of incubation. The diagonal hatching represents the mesoderm seen through the ectoderm, which is supposed to have been rendered transparent. The principal landmarks of the embryos are shadowed in lightly.

The manner in which the mesoderm spreads out leaves a mesoderm-free area in the cephalic portion of the blastoderm. This region, known as the *proamnion*, is clearly recognizable in entire embryos by reason of its lesser density (Figs. A-7 and A-9). Despite the implications of its name, this region is not the precursor of the amnion. It is merely the area in the cephalic end of the embryo where the ever-expanding mesodermal wings have not filled in the space between the ectoderm and the endoderm (Fig. A-8A to A-8C).

The laterally spreading mesoderm tends to pull away from the notochord, leaving the notochord sharply outlined in the midline in a territory that for a time is relatively free of other mesodermal elements (Figs. A-8C and A-13C). Sections passing through the

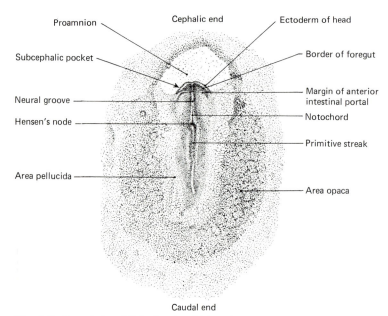

Fig. A-7 Dorsal view (X14) of entire chick embryo of about 20 hours' incubation.

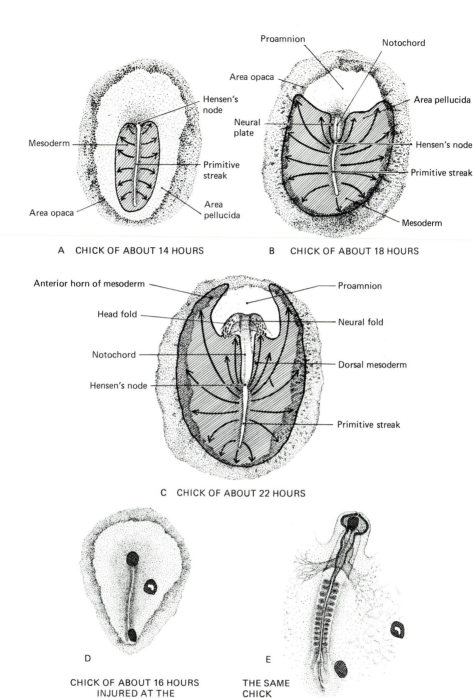

A CHICK OF ABOUT 14 HOURS

B CHICK OF ABOUT 18 HOURS

C CHICK OF ABOUT 22 HOURS

D

CHICK OF ABOUT 16 HOURS
INJURED AT THE
THREE POINTS MARKED

E

THE SAME
CHICK
AFTER FURTHER INCUBATION

Fig. A-8 Diagrams outlining direction of growth from the primitive streak (A–C) show the progress of the mesoderm during the latter part of the first day of incubation. Some of the more prominent structural features of the embryos are drawn in lightly for orientation, but the ectoderm is supposed to be nearly transparent, allowing the mesoderm to show through. The areas into which the mesoderm has grown are indicated by diagonal hatching. (D) and (E) show the direction of growth as demonstrated by experimental methods. *(After Kopsch.)* (D) shows the location at which three injuries were made close to the primitive streak of a 16-hour embryo. (E) shows the position to which the injured areas were carried by growth of the same embryo subsequent to the operation.

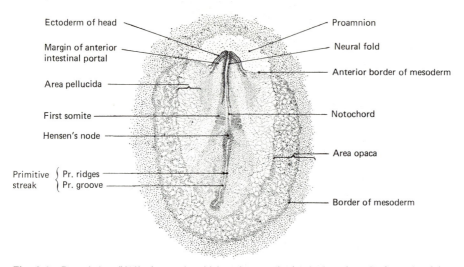

Ectoderm of head

Margin of anterior
intestinal portal

Area pellucida

First somite

Hensen's node

Primitive { Pr. ridges
streak { Pr. groove

Proamnion

Neural fold

Anterior border of mesoderm

Notochord

Area opaca

Border of mesoderm

Fig. A-9   Dorsal view (X16) of an entire chick embryo at the beginning of somite formation (about 22 to 23 hours of incubation).

primitive streak of embryos of about 18 hours' incubation age show loosely aggregated masses of mesoderm extending to either side between the ectoderm and endoderm (Fig. A-4C to A-4E). As would be expected from the method of its origin, little mesoderm in the form of a coherent layer appears in the midline except caudal to the primitive streak (Fig. A-4F). Immediately adjacent to the midline the mesoderm is markedly thicker than it is farther laterad (Fig. A-4A to A-4C). In whole mounts the positions of the regional thickenings of the mesoderm are evidenced by the greater opacity they impart locally to the embryo (Fig. A-7). These thickened zones of the mesoderm are the primordia of the dorsal mesodermal plates. Because of the way in which they are later divided into metamerically arranged cell masses or somites, they are frequently designated as the *segmental zones of the mesoderm*. The segmental zones are, in early stages, most clearly marked somewhat cephalic to Hensen's node, where the first somites will appear. In the embryo represented in Fig. A-9 there is just a suggestion of the lines of division separating the first somites. As the segmental zones of the mesoderm are followed from this region caudad on either side of the primitive streak, they gradually become less and less definite. Traced laterally the mesoderm rapidly becomes less thick. This thinner region along the sides of the embryo contains the primordial tissue from which the *intermediate zones* and *lateral plates of the mesoderm* are later differentiated.

The sheetlike layers of mesoderm that are characteristic of the midbody region do not extend to the cephalic part of the embryo. The mesoderm of the head is largely derived from cells which become detached from the more definitely organized layers of mesoderm lying farther caudally in the body and then migrate into the cephalic region. There is also a supplementary contribution of cells growing in from the ectoderm of the neural plate and the neural crest. For this reason the cephalic mesoderm consists of loosely aggregated mesenchymal cells and shows neither the regional differentiations nor the organization into definite layers which appear in the mesoderm of the midbody region.

Formation of the Neural Plate   In surface views of entire chicks incubated for about 18 hours (Fig. A-3) areas of greater density may be made out on either side of the notochord. These areas extend somewhat rostral to the cephalic end of the notochord, where they appear to blend with each other in the midline. Sections of this region (Fig. A-4A) show that the greater density seen in whole mounts is due to thickening of the ectoderm. This thickened area of pseudostratified epithelium[2] is known as the *neural* (*medullary*) *plate*. Laterally the thickened ectoderm of the neural plate blends without abrupt transition into the thinner ectoderm of the general blastodermal surface. Rostrally the neural plate is more clearly marked than it is caudally. At the level of Hensen's node the neural plate diverges into two elongated areas of thickening extending caudad, one on either side of the primitive streak. In embryos of about 22 hours (Fig. A-9) the neural plate becomes longitudinally folded to establish a trough known as the *neural groove* (Figs. A-13 and A-14). The formation of the neural plate and its subsequent folding to form the neural groove are the first indications of the differentiation of the central nervous system.

Differentiation of the Embryonal Area   Because of the thickening of the ectoderm to form the neural plate and also because of the thickening of the dorsal zones of the mesoderm, the part of the blastoderm immediately surrounding the primitive streak and notochord has become noticeably more dense than that in the peripheral portion of the area pellucida. Since it is the region in which the embryo itself is developed, this denser region is known as the *embryonal area*. Although the embryonal area is at this early stage directly continuous with the peripheral part of the blastoderm and has no definite line of demarcation, the two later become folded off from each other. The peripheral portion of the blastoderm is then spoken of as extraembryonic because it gives rise to structures which are not built into the body of the embryo, although they play a vital part in its nutrition and protection during development.

The cephalic region of the embryonal area is thickened and protrudes above the general surface of the surrounding blastoderm as a rounded elevation. This prominence marks the region in which the head of the embryo will develop (Figs. A-7 and A-9). The crescentic fold which bounds it is termed the *head fold* and is the first definite boundary of the body of the embryo. Throughout the course of development we shall find the cephalic region farther advanced in differentiation than other parts of the body.

STRUCTURE OF 24-HOUR CHICKS

Formation of the Head   In embryos of about 22 hours the cephalic part of the embryonal area is thickened and is elevated above the level of the surrounding blastoderm, with a well-defined crescentic fold marking its boundary (Fig. A-9). During the next 3 or 4 hours the cephalic region undergoes rapid growth. Its elevation above the blastoderm becomes much more marked, and it extends anteriorly, overhanging the proamnion region (Figs. A-10 and A-13E). The cephalic region which projects free from the blastoderm may now properly be termed the *head* of the embryo. The space formed between the head and the blastoderm is called the *subcephalic space*, or *pocket* (Fig. A-13).

---

[2]A pseudostratified epithelium looks as if it contains several discrete layers of cells, but in reality, all cells touch the underlying basement membrane.

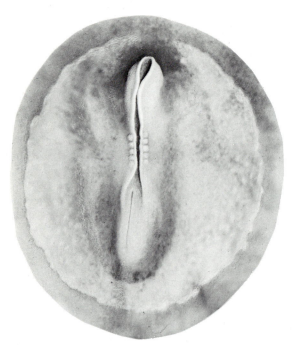

Fig. A-10   Chick embryo of 25 to 26 hours photographed by reflected light to show its external configuration. Compare with Fig. A-11 and A-12.

In the midline the notochord can be seen through the overlying ectoderm. It is larger caudally near its point of origin than it is cephalically. Nevertheless, it can readily be traced into the cephalic region where it will terminate somewhat short of the rostral end of the head (Fig. A-11).

Formation of the Neural Groove   At 24 hours of incubation the folding of the neural plate is much more clearly marked than it is at 22 hours of incubation. In transparent preparations of the entire embryo (Figs. A-11 and A-12) the neural folds appear as a pair of dark bands. (As an aid in the interpretation of such bands, review Fig. A-2.) The folding which establishes the neural groove takes place first in the cephalic region of the embryo. At its cephalic end the neural groove is, therefore, deeper and the neural folds are correspondingly more prominent than they are caudally. The folding has not, at this stage, been carried much beyond the cephalic half of the embryo. Consequently, as the neural folds are followed caudad, they diverge slightly from each other and become less and less distinct.

Study of transverse sections of an embryo at this stage affords a clearer interpretation of the conditions in neural-groove formation than the study of entire embryos. A section passing through the head region (Fig. A-13A) shows the neural plate folded, so that it forms a nearly complete tube. Dorsally the margins of the neural folds of either side have approached each other and lie almost in contact. The formation of the neural folds takes place first in about the center of the head region, and then progresses both rostrad and caudad. In following the sections of a transverse series caudad, the margins of the neural folds will be seen less and less closely approximated to each other.

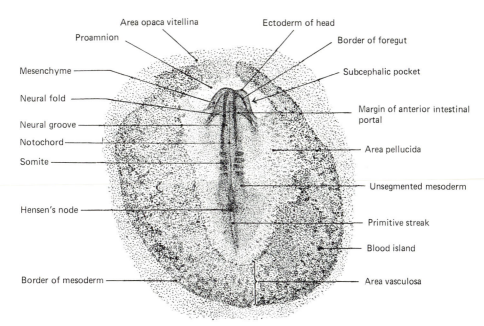

Fig. A-11 Dorsal view (X16) of entire chick embryo having four pairs of somites (about 24 hours' incubation). Compare this figure of an embryo, which has been stained and cleared and then drawn by transmitted light, with the preceding figure which shows an embryo of about the same age photographed by reflected light.

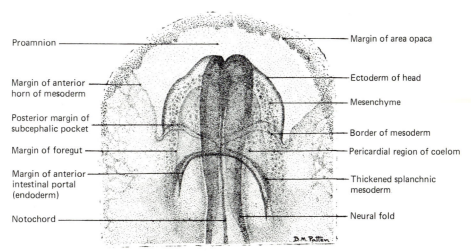

Fig. A-12 Ventral view (X40) of cephalic region of chick embryo having five pairs of somites (about 25 to 26 hours of incubation).

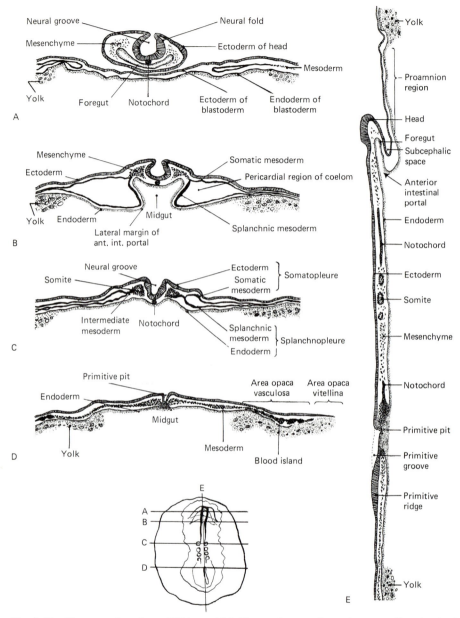

Fig. A-13   Diagrams of sections of 24-hour chick. The sections are located on an outline sketch of the entire embryo. The conventional representation of the germ layers is the same as that employed in Figure A-4 except that here, where its cells have become aggregated to form definite layers, the mesoderm is represented by solid black lines.

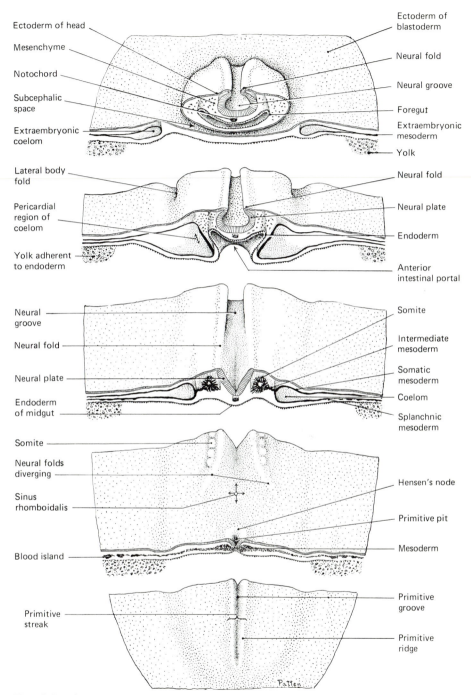

Ectoderm of head

Mesenchyme

Notochord

Subcephalic space

Extraembryonic coelom

Ectoderm of blastoderm

Neural fold

Neural groove

Foregut

Extraembryonic mesoderm

Yolk

Lateral body fold

Pericardial region of coelom

Yolk adherent to endoderm

Neural fold

Neural plate

Endoderm

Anterior intestinal portal

Neural groove

Neural fold

Neural plate

Endoderm of midgut

Somite

Intermediate mesoderm

Somatic mesoderm

Coelom

Splanchnic mesoderm

Somite

Neural folds diverging

Sinus rhomboidalis

Blood island

Hensen's node

Primitive pit

Mesoderm

Primitive streak

Primitive groove

Primitive ridge

Patten

Fig. A-14 Stereogram of 24-hour chick. Compare with Fig. A-13. (*This three-dimensional approach was suggested by Huettner's drawings in his* Fundamentals of Comparative Embryology of the Vertebrates, *The Macmillan Company, New York.*)

Establishing of the Foregut   In the outgrowth of the head, the endoderm as well as the ectoderm has been involved. As a result the endoderm forms a pocket within the ectoderm, much like a small glove finger within a larger one. This endodermal pocket, or foregut, is the first part of the digestive tract to acquire a definite cellular floor. That part of the gut caudal to the foregut, where the yolk still constitutes the only floor, is termed the *midgut.* The opening from the midgut into the foregut is called the *anterior intestinal portal* (Figs. A-13E and A-14).

The topography of the foregut region at this stage can be made out very well by studying the ventral aspect of entire embryos. The margin of the anterior intestinal portal appears as a well-defined crescentic line (Fig. A-12). The lateral boundaries of the foregut can be seen to join the caudally directed tips of the crescentic margin of the portal. Considerably cephalic to the intestinal portal an irregularly recurved line can be made out. On either side it appears to merge with the ectoderm of the head. This line marks the extent to which the head is free from the blastoderm. It is due to the fold at the bottom of the subcephalic pocket where the ectoderm of the undersurface of the head is continuous with the ectoderm of the blastoderm. Comparison of Fig. A-12 with a sagittal section as diagramed in Fig. A-13E, will aid in making clear the relationships of the foregut to the head. From the sagittal section it will also be apparent why the margins of the intestinal portal and of the subcephalic pocket appear as dark lines in the whole mount. Such correlative study of whole mounts and sections is essential for acquring a three-dimensional concept of embryonic structure.

Regional Divisions of the Mesoderm   The first conspicuous metamerically arranged structures to appear in the chick are the *mesodermal somites.* The somites arise by division of the mesoderm of the dorsal or segmental zone to form blocklike cell masses. In the embryo shown in Fig. A-11 three pairs of somites are completely delimited and one can make out a fourth pair, which is not yet completely cut off from the dorsal mesoderm caudal to it.

Cross-sections passing through the midbody region show the formation of the somites and the beginning of other changes in the mesoderm (Fig. A-13C; cf. also Fig. A-25E). If one follows the mesoderm from the midline toward either side, one can make out (1) the *dorsal paraxial mesoderm,* which at this level has been organized into somites, (2) the *intermediate mesoderm,* a thin and narrow plate of cells connecting the dorsal and lateral mesoderm, and (3) the *lateral mesoderm,* which is distinguished from the intermediate by being split into two sheetlike layers with a space between them.

The somites are compact cell masses lying immediately lateral to the neural folds. The cells composing them have a fairly definite radial arrangement about a central cavity which is very minute or wanting altogether when the somites are first formed but which later becomes enlarged (Fig. 6-17). Cephalic and caudal to the region in which somites have been formed, the dorsal mesoderm is differentiated from the rest of the mesoderm simply by its greater thickness and compactness.

In 24-hour embryos the intermediate mesoderm shows very little differentiation. In the chick it never becomes segmentally divided as does the dorsal mesoderm. The fact that it is potentially segmental in character is indicated, however, by the way in which it later gives rise to segmentally arranged nephric tubules. Because of the part it plays in the establishment of the excretory system the intermediate mesoderm is frequently called the *nephrotomic plate.*

In the chick the lateral mesoderm, like the intermediate mesoderm, shows no segmental division. In 24-hour embryos (Fig. A-13C) it is clearly differentiated from the intermediate mesoderm by being split horizontally into two layers with a space between them. The layer of lateral mesoderm lying next to the ectoderm is termed the *somatic mesoderm,* the layer next to the endoderm is termed the *splanchnic mesoderm,* and the cavity between somatic and splanchnic mesoderms is the *coelom.* Because in development the somatic mesoderm and ectoderm are closely associated and undergo many foldings in common, it is convenient to designate the two layers together by the single term *somatopleure.* Similarly the splanchnic mesoderm and the endoderm together are called the *splanchnopleure.*

The Coelom   The coelom, like the cell layers of the blastoderm, extends over the yolk peripherally beyond the embryonal area (Fig. A-13B and C). Later in development, foldings mark off the embryonic from the extraembryonic portion of the germ layers. This same folding process divides the coelom into intraembryonic and extraembryonic regions. In the 24-hour chick, however, embryonic and extraembryonic coelom have not been separated.

It is evident from the manner in which the coelomic chambers arise in the lateral mesoderm that the coelom of the embryo consists, at first, of a pair of bilaterally symmetrical chambers. It is not until later in development that the right and left coelomic chambers become confluent ventrally to form an unpaired abdominal cavity such as is found in adult vertebrates.

The Pericardial Region   In the region of the anterior intestinal portal the coelomic chambers on either side show very marked local enlargement. Later in development these dilated regions are extended mesiad and break through into each other ventral to the foregut to form the pericardial cavity. In their early condition these enlarged regions of the coelomic chambers are sometimes called *amniocardiac vesicles.* With their later fate in mind one may avoid multiplication of terms and speak of them from their first appearance as constituting the *pericardial region of the coelom.*

The relationships of the pericardial region of the coelom in embryos of 24 hours can most readily be grasped from a study of transverse sections. Figure A-13B shows the great dilation of the coelom on either side of the anterior intestinal portal as compared with its condition farther caudad (Fig. A-13C). Where the splanchnic mesoderm lies closely applied to the endoderm at the lateral margins of the portal it is noticeably thickened. It is from these areas of thickened splanchnic mesoderm that the paired primordia of the heart will later arise.

In entire embryos of this age the thickened splanchnic mesoderm can be made out as a dark band lying close against the crescentic endodermal border of the anterior intestinal portal (Fig. A-12). If the preparation is favorably stained, the boundaries of the pericardial regions of the coelom can be traced. Following mesiad from the easily located thickened areas, the mesodermal borders can be seen to extend from either side, parallel to the endodermal margins of the portal, nearly to the midline. They then turn cephalad for a short distance. When they encounter the ectodermal fold which constitutes the posterior boundary of the subcephalic pocket, they swing laterad parallel with it and can be traced outside the embryonic region where they constitute the mesial margins of the anterior horns of the mesoderm (Fig. A-12).

The portion of the coelom with its borders located between the subcephalic pocket and the anterior intestinal portal is an important landmark from a standpoint other than the part it is destined to play in the formation of the pericardial region. This is the most cephalic part of the coelom, for there is no coelom in the head. Attention has already been called to the fact that in the cephalic region the mesoderm is not aggregated into definite masses or coherent cell layers but is rather mesenchymal in nature. The cephalic mesenchymal cells have migrated there from more caudally located mesoderm and also from the neural crest. By careful focusing on the whole mount, the mesenchyme of the head can be seen as a vaguely defined mass lying between the superficial ectoderm and the endoderm of the foregut. The distribution of the mesenchymal cells and the characteristic irregularity of shape correlated with their active amoeboid movement may readily be made out from sections (Fig. A-13).

The Area Vasculosa   In a 24-hour chick the boundary between the area opaca and the area pellucida has the same appearance and significance as in chicks of 18 to 20 hours. There is, however, a very marked difference between the proximal portion of the area opaca adjacent to the area pellucida and the more distal portions of the area opaca. The proximal region is much darker and has a somewhat mottled appearance (Fig. A-11). The greater density of this region is due to its invasion by mesoderm, which makes it thicker and therefore more opaque in transmitted light (Fig. A-13D). The boundary between the inner and outer zones of the area opaca is established by the extent to which the mesoderm has grown peripherally. The distal zone is called the *area opaca vitellina,* because the yolk alone underlies it. The proximal zone into which mesoderm has grown is known as the *area opaca vasculosa,* because it is from the mesoderm in this region that the yolk-sac blood vessels arise. The mottled appearance of this region is due to the aggregation of mesoderm into cell clusters, or *blood islands* (Figs. A-11 and A-13D). These mark the initial step in the formation of blood vessels and blood corpuscles. Later in development the formation of blood islands and vessels extends in toward the body of the embryo from its place of earliest appearance in the area opaca and involves the mesoderm of the area pellucida.

CHANGES BETWEEN 24 AND 33 HOURS OF INCUBATION

Closure of the Neural Groove   In comparison with 24-hour chicks, entire embryos of 27 to 28 hours of incubation (Fig. A-15) show marked advances in the development of the cephalic region. The head has elongated rapidly and now projects free from the blastoderm for a considerable distance, with a corresponding increase in the depth of the subcephalic pocket and in the length of the foregut.

By 27 hours the neural folds in the cephalic region meet in the middorsal line and their edges become fused. This fusion is really a double one. Careful following of Fig. A-22A to A-22E will aid greatly in understanding the process. Each neural fold consists of a mesial component, which is thickened neural-plate ectoderm, and a lateral component, which is unmodified superficial ectoderm (Fig. A-22A). When the neural folds meet in the middorsal line (Fig. A-22B and C), the neural-plate components of the two folds fuse with each other and the outer layers of unmodified ectoderm also become fused (Fig. A-22D). Thus in the same process the neural groove becomes closed to form the neural tube and the superficial ectoderm closes over the place formerly occupied by the open neural groove. Shortly after this double fusion the neural tube and the superficial ectoderm become somewhat separated from each other, leaving no hint of their former continuity (Fig. A-22E).

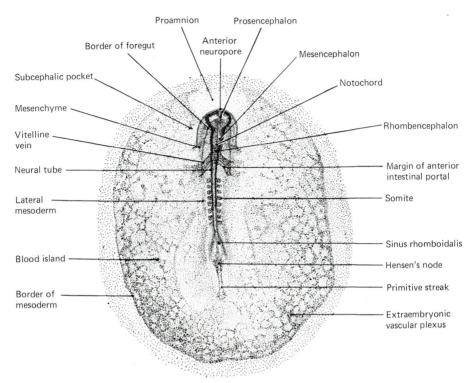

Proamnion
Prosencephalon
Border of foregut
Anterior neuropore
Mesencephalon
Subcephalic pocket
Notochord
Mesenchyme
Vitelline vein
Rhombencephalon
Neural tube
Margin of anterior intestinal portal
Lateral mesoderm
Somite
Sinus rhomboidalis
Blood island
Hensen's node
Border of mesoderm
Primitive streak
Extraembryonic vascular plexus

Fig. A-15 Dorsal view (X14) of entire chick embryo having 8 pairs of somites (about 27 to 28 hours' incubation).

Differentiation of the Brain Region By 27 hours of incubation the cephalic part of the neural tube is markedly enlarged, compared with the more caudal parts. Its thickened walls and dilated lumen mark the region that will develop into the brain. The undilated caudal part of the neural tube gives rise to the spinal cord. Three divisions, the three *primary brain vesicles,* can be distinguished in the enlarged cephalic region of the neural tube (Fig. A-15). Occupying most of the rostral part of the head is a conspicuous dilation known from its position as the *forebrain* or *prosencephalon.* Posterior to the prosencephalon and marked off from it by a constriction is the *midbrain* or *mesencephalon.* Posterior to the mesencephalon with only a very slight constriction marking the boundary is the *hindbrain* or *rhombencephalon.* The rhombencephalon is continuous posteriorly with the cord region of the neural tube without any definite point of transition.

In somewhat older embryos (Fig. A-16) the lateral walls of the prosencephalon become outpocketed to form a pair of rounded dilations known as the *primary optic vesicles.* When the optic vesicles are first formed there is no constriction between them and the lateral walls of the prosencephalon, and the lumen of each optic vesicle communicates mesially with the lumen of the prosencephalon without any definite line of demarcation.

Ventral to the brain the notochord extends rostrally as far as a depression in the floor of the prosencephalon, known as the *infundibulum* (Fig. A-16). As the result of an interaction with the ectoderm of the future oral cavity, the infundibulum develops into the neural hypophysis.

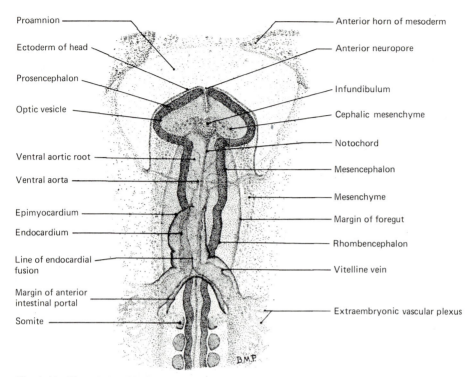

Proamnion — Anterior horn of mesoderm
Ectoderm of head — Anterior neuropore
Prosencephalon — Infundibulum
Optic vesicle — Cephalic mesenchyme
— Notochord
Ventral aortic root — Mesencephalon
Ventral aorta — Mesenchyme
Epimyocardium — Margin of foregut
Endocardium — Rhombencephalon
Line of endocardial fusion — Vitelline vein
Margin of anterior intestinal portal — Extraembryonic vascular plexus
Somite —

B.M.P.

Fig. A-16 Ventral view (X47) of cephalic and cardiac region of chick embryo of 9 somites (about 29 to 30 hours' incubation).

The Anterior Neuropore The closure of the neural folds takes place first near the rostral end of the neural groove and then progresses both cephalad and caudad. At the extreme rostral end of the brain, closure is delayed. As a result, the prosencephalon remains for some time in communication with the outside through an opening called the *anterior neuropore.* The anterior neuropore is still open in chicks of 27 hours (Fig. A-15). In embryos of 30 hours the neuropore appears much narrowed (Fig. A-16), and by 33 hours it is almost closed (Fig. A-18). A little later it becomes entirely closed. The anterior neuropore does not give rise to any definite brain structure, but for some time it leaves a scarlike fissure in the anterior wall of the prosencephalon, which serves to mark the point originally most rostral in the developing brain.

The Sinus Rhomboidalis The brain merges caudally without any definite line of demarcation into the region of the neural tube destined to become the spinal cord. After 27 hours of incubation the neural tube, as far caudally as somite formation has progressed, is completely closed and of nearly uniform diameter. Caudal to the most posterior somites, the neural groove is still open and the neural folds diverge to either side of Hensen's node (Fig. A-15). In their later growth caudad, the neural folds converge toward the midline and form the lateral boundaries of an unclosed region at the caudal extremity of the spinal cord, known because of its shape as the *sinus rhomboidalis* (Fig. A-18). Hensen's node and the primitive pit lie in the floor of this still unclosed region of the neural groove; subsequently they are enclosed within it when the neural folds here finally fuse to complete the neural tube.

This process in the chick is homologous with the enclosure of the blastopore by the neural folds in lower vertebrates (Fig. 5-2E). In forms where the blastopore does not become closed until after it is surrounded by the neural folds, it for a time constitutes an opening from the neural canal into the primitive gut, known as the *neurenteric canal.* In the chick the fact that there is never an open blastopore precludes the establishment of an open neurenteric canal, but the primitive pit represents its homologue.

Formation of Additional Somites   The division of the dorsal mesoderm to form somites begins to be apparent in embryos of about 22 hours (Fig. A-9). By the end of the first day three or four pairs of somites have been cut off (Fig. A-11). As development progresses new somites are added caudal to those first formed. In embryos which have been incubated about 27 hours, eight pairs of somites have been established (Fig. A-15).

Lengthening of the Foregut   Comparison of the relations of the crescentic margin of the anterior intestinal portal in embryos of between 24 and 30 hours shows it occupying progressively more caudal positions. (Note its location in relation to the first somites in Fig. A-23.) This change in the position of the anterior intestinal portal is the result of two distinct growth processes. The margins of either side of the portal are constantly converging toward the midline, where they become merged with each other. Their merging lengthens the foregut by adding to its floor and thereby displacing the crescentic margin of the portal caudad. At the same time the structures cephalic to the anterior intestinal portal are elongating rapidly, so that the portal becomes more and more remote from the rostral end of the embryo. This results in further lengthening of the foregut.

As a result of these two processes the space between the subcephalic pocket and the margin of the anterior intestinal portal is also elongated (Fig. A-23). This is of importance in connection with the formation of the heart, for it is into this enlarging space that the pericardial portions of the coelom extend, and within it that the heart develops.

Appearance of the Heart and Veins   In dorsal views of entire embryos the heart is largely concealed by the overlying rhombencephalon (Fig. A-15), but it may readily be made out by viewing the embryo from the ventral surface (Fig. A-16). At this stage the heart is a nearly straight tubular structure lying in the midline ventral to the foregut. Its dilated midregion has noticeably thickened walls. Cephalically the heart is continuous with a large median vessel, the *ventral aorta;* caudally it is continuous with the paired *vitelline veins.* The fork formed by the union of the vitelline veins in the caudal part of the heart lies immediately cephalic to the crescentic margin of the anterior intestinal portal, the veins lying within the fold of endoderm which constitutes the margin of the portal.

Organization in the Area Vasculosa   The extraembryonic vascular area at this stage is undergoing rapid enlargement and presents a netted appearance instead of a mottled one, as in the earlier embryos. The peripheral boundary of the area vasculosa is definitely marked by a dark band, the precursor of the *sinus terminalis (marginal sinus).* Its netted appearance is due to the extension and anastomosing of blood islands. The formation of this network is the first step in the organization of a maze of blood vessels on the yolk surface *(vitelline vascular plexus)* which will later be the means of absorbing and transferring food materials to the embryo. When this plexus of vessels developing in the yolk sac comes into communication with the vitelline (omphalomesenteric) veins—already developing within the embryo and extending laterad—the return vascular channels from the yolk sac are established. The main vessels to the yolk sac are established somewhat later when the vitelline arteries, arising from the aorta, extend peripherally and become connected with the yolk-sac plexus (cf. Figs. A-27, A-28, and A-29).

STRUCTURE OF CHICKS BETWEEN 33 AND 38 HOURS OF INCUBATION

Chicks which have been incubated from 33 to 38 hours are in a favorable stage to show some of the fundamental steps in the formation of the central nervous system and of the circulatory system. In this section, therefore, attention has been concentrated on these two systems.

Divisions of the Brain and Their Neuromeric Structure   The metameric arrangement of structures which is so striking a feature in the body organization of all vertebrates is masked in the head region of the adult by superimposed specializations. In the brain of young vertebrate embryos, however, the metamerism is still indicated. Dissections of the neural plate of chicks at the end of the first day of incubation show a series of eleven enlargements marked off from each other by constrictions (Fig. A-17). Complete agreement has not been reached regarding the precise homologies of individual enlargements with specific neuromeres in other forms. The controversies center about the question of neuromeric fusions in the more rostral parts of the brain.

With the closure of the neural tube and the establishment of the three primary brain vesicles, we can begin to trace the fate of the various neuromeric enlargements in the formation of the brain regions. The three most rostral neuromeres form the *prosencephalon;* neuromeres IV and V are incorporated into the *mesencephalon;* and neuromeres VI to XI become part of the *rhombencephalon* (Fig. A-17B). Rostrally, all but two of the interneuromeric constrictions soon disappear, namely, the one between the prosencephalon and the mesencephalon, and the one between the mesencephalon and the rhombencephalon. The rhombencephalic neuromeres, however, remain clearly marked for a considerable period.

By about 33 hours of incubation the *optic vesicles* are established as paired lateral outgrowths of the prosencephalon. They soon extend to occupy the full width of the head (Figs. A-17C and A-18). The distal portion of each of the vesicles thus comes to lie closely approximated to the superficial ectoderm, a relationship of importance in the later development of the lens. At first the cavities of the optic vesicles are broadly confluent with the cavity of the prosencephalon. Somewhat later, constrictions mark more definitely the boundaries between the optic vesicles and the prosencephalon (Figs. A-17D and A-19). The infundibulum remains as a depression in the floor of the prosencephalon (Figs. A-19 and A-20).

In chicks of about 38 hours, indications of the impending division of the three primary vesicles to form the five regions characteristic of the adult brain are already beginning to appear. In the establishing of the five-vesicle condition of the brain, the prosencephalon is subdivided to form the *telencephalon* and *diencephalon,* the mesencephalon remains undivided, and the rhombencephalon divides to form the *metencephalon* and *myelencephalon.*

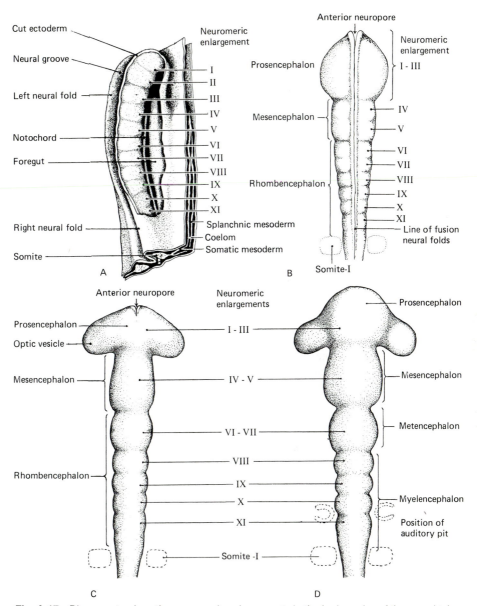

Fig. A-17 Diagrams to show the neuromeric enlargements in the brain region of the neural tube. (A) Lateral view of neural plate from dissection of chick of 4 somites (24 hours). (B) Dorsal view of brain dissected out of 7-somite (26- to 27-hour) embryo. (C) Dorsal view of brain from 10-somite (30-hour) embryo. (D) Dorsal view of brain from 14-somite (36-hour) embryo. *(Based on figures by Hill.)*

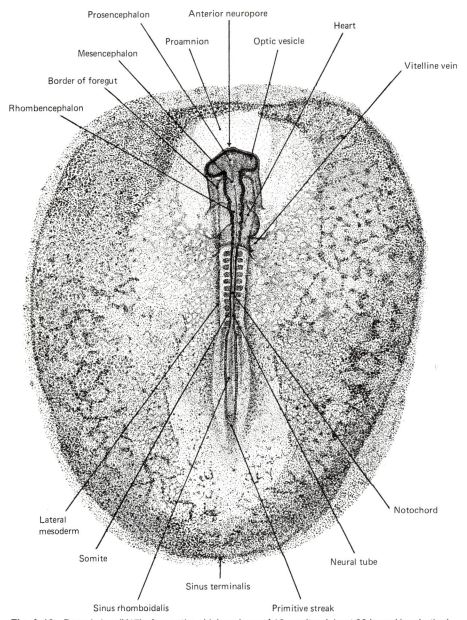

Prosencephalon
Anterior neuropore
Heart
Proamnion
Optic vesicle
Mesencephalon
Vitelline vein
Border of foregut
Rhombencephalon
Lateral mesoderm
Somite
Notochord
Sinus rhomboidalis
Sinus terminalis
Neural tube
Primitive streak

Fig. A-18 Dorsal view (X17) of an entire chick embryo of 12 somites (about 33 hours' incubation).

The division of the prosencephalon into telencephalon and diencephalon is not well marked until a much later stage of development, but the median enlargement at this stage extending rostrad beyond the level of the optic vesicles (Fig. A-17D) indicates where the telencephalon will be established. The optic vesicles and that part of the prosencephalon lying between them go into the diencephalon.

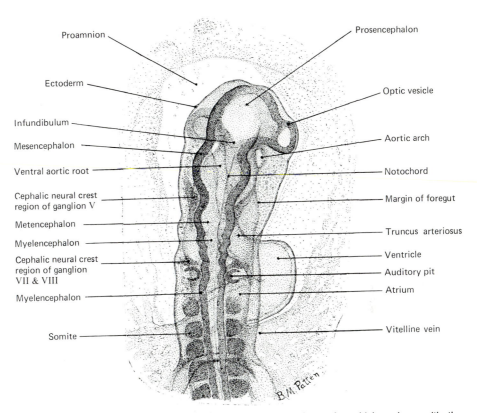

Fig. A-19 Dorsal view (X50) of cephalic and cardiac regions of a chick embryo with the seventeenth somite just forming (about 38 hours' incubation).

The mesencephalon, as stated above, undergoes no subdivision, so the original mesencephalic region of the three-vesicle brain gives rise to the mesencephalon of the adult. This region of the brain does not undergo any marked differentiation until relatively late in development.

At this stage the beginning of the subdivision of the rhombencephalon is clearly indicated (Figs. A-17D and A-19). The two most anterior neuromeres of the original rhombencephalon form the metencephalon, and the posterior four neuromeres are incorporated in the myelencephalon.

The Auditory Pits   As is the case with the central nervous system, the eyes and ears arise early in development. The first indication of the formation of the sensory part of the ear becomes evident at about 35 hours of incubation. At this age a pair of thickenings termed the *auditory placodes* arise in the superficial ectoderm of the head. They are situated on the dorsolateral surface opposite the most caudal interneuromeric constriction of the myelencephalon. By 38 hours of incubation (Fig. A-19) the auditory placodes have become depressed below the general level of the ectoderm and form the walls of a pair of cavities, the *auditory pits*. When first formed, the walls of the auditory pits are directly continuous with the superficial ectoderm and their cavities are wide open to the outside. In later stages the openings into the pits become narrowed and finally closed, so

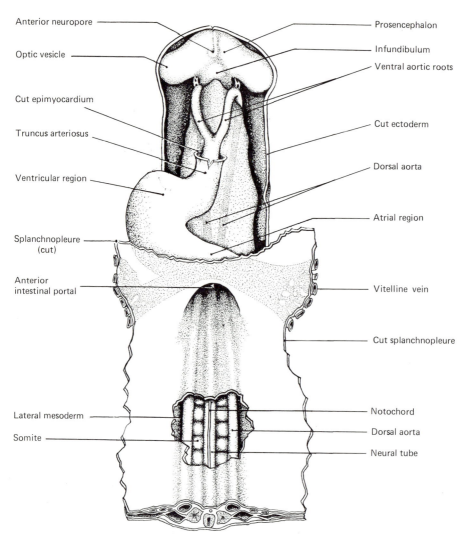

Anterior neuropore

Optic vesicle

Cut epimyocardium

Truncus arteriosus

Ventricular region

Splanchnopleure (cut)

Anterior intestinal portal

Lateral mesoderm

Somite

Prosencephalon

Infundibulum

Ventral aortic roots

Cut ectoderm

Dorsal aorta

Atrial region

Vitelline vein

Cut splanchnopleure

Notochord

Dorsal aorta

Neural tube

Fig. A-20 Diagrammatic ventral view of dissection of a 35-hour chick embryo. The splanchnopleure of the yolk sac cephalic to the anterior intestinal portal, the ectoderm of the ventral surface of the head, and the mesoderm of the pericardial region have been removed to show the underlying structures. Figure A-26 should be referred to for the relations of pericardial mesoderm. *(Modified from Prentiss.)*

the pits become vesicles lying between the superficial ectoderm and the myelencephalon. At this point they have no connection with the central nervous system.

Formation of Extraembryonic Blood Vessels In dealing with the circulation of young chicks we must recognize that there are two distinct circulatory arcs of which the heart is the common center. One complete circulatory arc is established entirely within the body of the embryo. A second arc, which has a rich plexus of terminal vessels, is established in the extraembryonic membranes enveloping the yolk. These are the *vitelline vessels*. The vitelline vessels communicate with the heart over main vessels which traverse the embryonic body. The chief distribution of the vitelline circulation is, however, extraembryonic. Later in development there arises a third circulatory arc involving another set of extraembryonic vessels in the allantois.

The formation of extraembryonic blood vessels is presaged by the appearance of blood islands in the vascular area of chicks toward the end of the first day of incubation (review Figs. A-11 and A-13D). At the time of their first appearance the blood islands are irregular clusters of mesodermal cells lying in intimate contact with the yolk-sac endoderm (Fig. A-21). Extension and anastomosis of neighboring blood islands results in the establishment of a network of communicating vessels.

Blood islands first appear in the peripheral part of the area vasculosa and from there

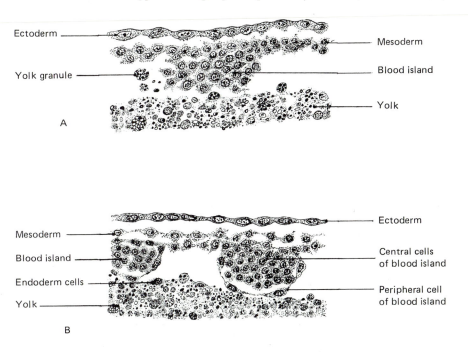

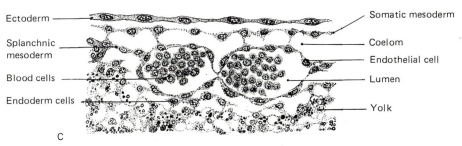

Fig. A-21 Drawings to show the cellular organization of blood islands at three stages in their differentiation. By referring to Fig. A-13D, the small areas here represented can be located in relation to the structure of the embryo as a whole. (A) From blastoderm of 18-hour chick. (B) From blastoderm of 24-hour chick. (C) From blastoderm of 33-hour chick.

extend toward the body of the embryo. By 33 to 35 hours of incubation the extraembryonic vascular plexus has extended inward and made connection with the vitelline (omphalomesenteric) veins which, originating within the body of the embryo, have grown outward (Fig. A-18). Thus are established the vascular channels leading from the vitelline plexus toward the heart.

The main arterial channels leading toward the yolk sac have not yet appeared, and there is no circulation of the blood corpuscles which are being formed in the area vasculosa. The developing intraembryonic blood vessels contain fluid but no corpuscles until the extraembryonic circuit is completed. The embryo meanwhile draws its nutrition from the yolk by direct absorption.

Formation of the Heart   The heart arises from paired primordia, which at first lie widely separated on either side of the midline. The paired condition of the heart at the time of its origin is due to the fact that the early embryo lies open ventrally, spread out on the yolk surface. The primordia of all ventral structures which appear at an early age are thus at first separated and lie on either side of the midline. As the embryo develops, a series of foldings undercut it and separate it from the yolk. At the same time, this folding-off process establishes the ventral wall of the gut and the ventral body wall of the embryo by bringing together in the midline the structures formerly spread out to right and left. As the embryo is completed ventrally the paired primordia of the heart are brought together in the midline and become fused (Figs. A-22 and A-23).

Although prospective cardiac tissue can be recognized much earlier, the first definite structural indications of heart formation appear in chicks of 25 to 26 hours in the region of the anterior intestinal portal. Where the splanchnopleure of either side bends toward the midline along the lateral margin of the intestinal portal, there is a marked regional thickening in the splanchnic mesoderm (Figs. A-12A, and A-22A). This pair of thickenings indicates where there has been rapid cell proliferation preliminary to the differentiation of the heart. Loosely associated cells can already be seen somewhat detached from the mesial face of the mesodermal layer. These cells soon become organized to form the *endocardial primordia.*

In a chick of about 26 hours, sections through a corresponding region show distinct differentiation of the endocardial and epimyocardial primordia (Fig. A-22B). The endocardial primordia are a pair of delicate tubular structures, with walls only a single cell in thickness, lying between the endoderm and the mesoderm. They arise from the cells seen separating from the adjacent thickened mesoderm in the 25-hour chick. As their name indicates they are destined to form the internal lining of the heart *(endocardium).* The greater part of each of the original mesodermal thickenings becomes applied to the lateral aspects of the endocardial tubes as the *epimyocardial primordium,* which will later differentiate into the external coat of the heart *(epicardium)* and the heavy muscular layers of the heart *(myocardium).* Under high magnification, careful study of well-stained sections of the developing heart at this stage reveals the presence of *cardiac jelly* between the epimyocardium and the endocardium.

In chicks of 27 hours the lateral margins of the anterior intestinal portal have been undergoing concrescence, thus lengthening the foregut caudally and at the same time elongating the pericardial region. In this process the lateral margins of the portal swing in to meet each other and merge in the midline, and the endocardial tubes of the right and left side are brought toward each other beneath the newly completed floor of the foregut (Figs. A-22C and A-23B). In the 28-hour chick embryo the endocardial primordia are approximated to each other (Figs. A-22D, and A-23C), and by 29 hours they fuse in their midregion to form a single tube (Figs. A-22E and A-23D).

At the same time the epimyocardial areas of the mesoderm are brought together, first ventrally (Fig. A-22D) and then dorsally to the endocardium (Fig. A-22E). Where the splanchnic mesodermal layers of the opposite sides of the body become apposed to each other dorsal and ventral to the heart, they form double-layered supporting membranes called, respectively, the *dorsal mesocardium* and the *ventral mesocardium.* The ventral mesocardium is a transitory structure, disappearing almost as soon as it is formed (Fig. A-22E). The dorsal mesocardium, although the greater part of it disappears in the next few hours of incubation, persists in embryos of the early part of the age range under consideration, suspending the heart in the pericardial region of the coelom.

The gross shape of the heart and its positional relations to other structures can be readily seen in entire embryos. The fusion of the paired cardiac primordia establishes the heart as a nearly straight tubular structure. It lies at the level of the rhombencephalon approximately in the midline, ventral to the foregut (Fig. A-16). By 33 hours of incubation the midregion of the heart is considerably dilated and bent to the right (Fig. A-18). At 38 hours the heart is bent so far to the right that it extends beyond the lateral body margin of the embryo (Fig. A-19). This bending process is correlated with the rupture of the dorsal mesocardium at the midregion of the heart.

Although there are not yet any sharply bounded subdivisions of the heart, its fundamental regions are beginning to take shape. From the heart's future intake end to its future discharging end, the regions are sinus venosus, atrium, ventricle, and truncus arteriosus. At the 36- to 38-hour stage the *sinus venosus* is only suggested. It is represented by the still paired primordia where the common cardinals enter the vitelline veins, and they in turn are becoming confluent with each other to enter the atrial portion of the tubular heart (Fig. A-26). The *atrium* is held close beneath the caudal part of the foregut by a persisting portion of the dorsal mesocardium. The *ventricle* is the part of the cardiac tube that makes a U-shaped bend to the right, bringing it into clear view at the side of the body in whole mounts (Fig. A-19). Swinging back to the midline the ventricle is narrowed to form the discharging part of the heart known as the *truncus arteriosus* (Figs. A-20 and A-26). The part of the ventricle that narrows abruptly to give rise to the truncus arteriosus is spoken of as the *ventricular cone* or, more briefly, as the *conus.* This is really just a zone of transition, rather than a basic cardiac region, for which it is convenient to have a name.

From the way the paired cardiac primordia are at first located on either side of the anterior intestinal portal, it is evident that they can fuse with each other in a sequential process only as the "flooring-in" of the foregut progresses (cf. Fig. A-22B to A-22D). The truncoventricular part of the heart is formed first (Fig. A-24A). Then the atrium is added caudal to the ventricle (Fig. A-24B and C). Finally, in stages older than those under discussion here, the sinus venosus is added caudal to the atrium (Fig. A-24D).

Formation of Intraembryonic Blood Vessels   Concurrently with the establishment of the heart, blood vessels have been arising within the body of the embryo. The large vessels connecting with the heart are the first of the intraembryonic channels established. At the cephalic end of the pericardial cavity the epimyocardial covering of the truncus arteriosus is reflected to become continuous with the general mesodermal lining of the pericardial coelom (Fig. A-26). The endothelial lining of the truncus is continued cephalad beneath the foregut as the *ventral aorta.* Almost immediately the ventral aorta bifurcates to form the paired *ventral aortic roots.* At the cephalic end of the foregut the ventral aortic roots turn dorsad, curve around the gut, and then extend caudad as the paired *dorsal aortae* (Figs. A-20, A-25B and C, and A-26). Few conspicuous branches

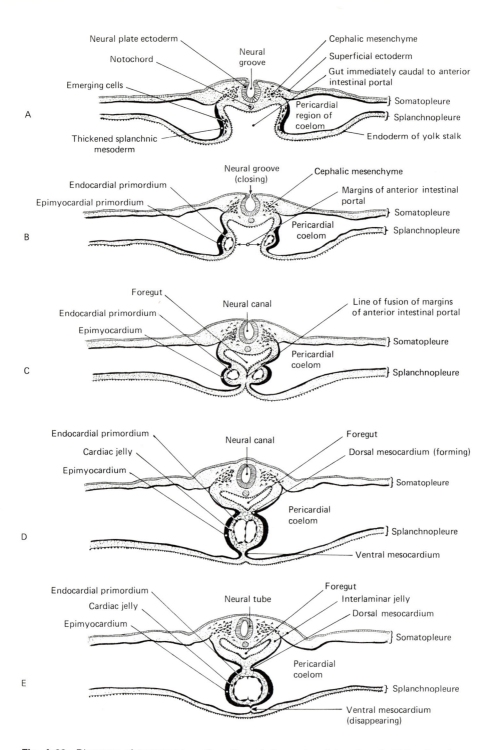

Fig. A-22 Diagrams of transverse sections through the pericardial region of chicks at various stages to show the formation of the heart. For location of the sections consult Figure A-23. (A) At 25 hours; (B) at 26 hours; (C) at 27 hours; (D) at 28 hours; (E) at 29 hours.

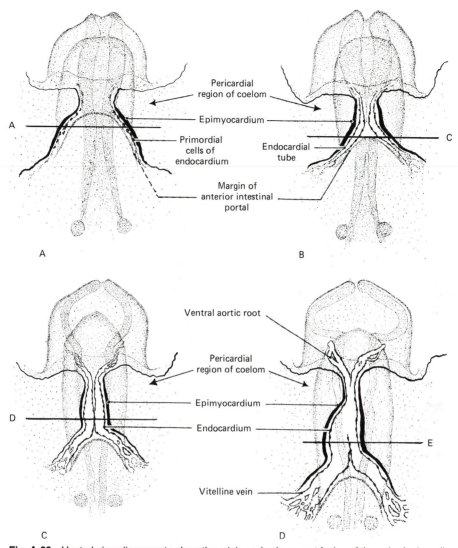

Pericardial region of coelom

Epimyocardium

Primordial cells of endocardium

Endocardial tube

Margin of anterior intestinal portal

A

B

Ventral aortic root

Pericardial region of coelom

Epimyocardium

Endocardium

Vitelline vein

C

D

Fig. A-23 Ventral-view diagrams to show the origin and subsequent fusion of the paired primordia of the heart. The lines *A, C, D,* and *E* indicate the planes of the sections diagramed in Figure A-22, A, C, D, and E, respectively. (A) Chick of 25 hours; (B) chick of 27 hours; (C) chick of 28 hours; (D) chick of 29 hours.

arise from the aortae at this stage, but as development progresses, branches extend to the various parts of the embryo and the aortae become the main efferent conducting vessels of the embryonic circulation. The curved vessels connecting the ventral aortic roots with the dorsal aortic roots are the *aortic arches.* At this age there is just a single pair, the first of the series of six which will appear as development progresses. Both the ventral aortic roots at the outlet end and the vitelline veins at the intake end are direct continuations of the paired endocardial primordia of the heart—which at this early stage consist of

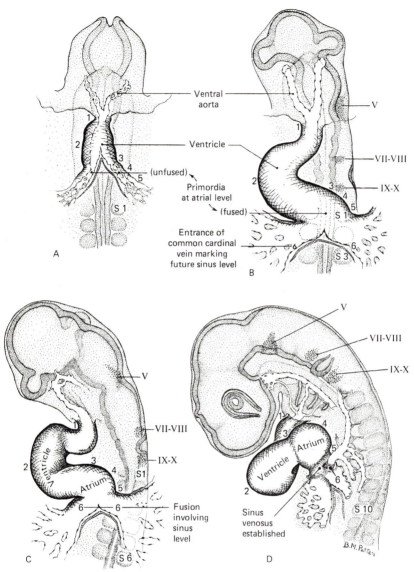

Fig. A-24   The formation of the fundamental regions of the chick heart by progressive fusion of its paired primordia. (A) Ventral view at the 9-somite stage (about 28-29 hours), when the first contractions appear. The ventricular part of the heart is the only region where the fusion of the paired primordia has occurred and the myocardial investment has been formed. (B) Ventral view at the 16-somite stage (about 36-37 hours), when the blood first begins to circulate. The atrium and ventricle have been established, but the sinus venosus exists only as undifferentiated primordial channels, still paired and still lacking myocardial investment. (C) Ventro-sinistral view at the 19-somite stage (about 43 hours). Fusion of the paired primordia is just beginning to involve the sinus region. (D) Sinistral view at the 26-somite stage (about 51-53 hours). The sinus venosus is definitely established and its investment with myocardium well advanced. To facilitate following the progress of fusion, Arabic numerals have been placed against approximately corresponding locations. The 6 is located at the point of entrance of the common cardinal vein as determined from injected specimens. *(From Patten and Kramer, 1933, Am. J. Anat. 53:349.)*

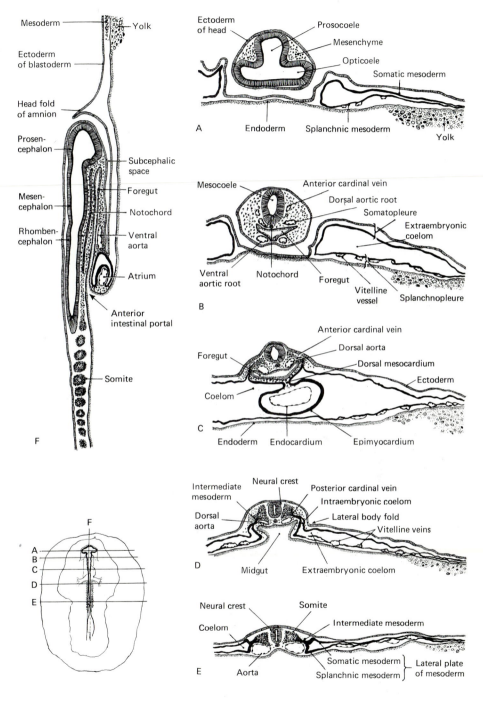

Fig. A-25 Diagrams of sections of 33-hour chick. The location of each section is indicated on a small outline sketch of the entire embryo.

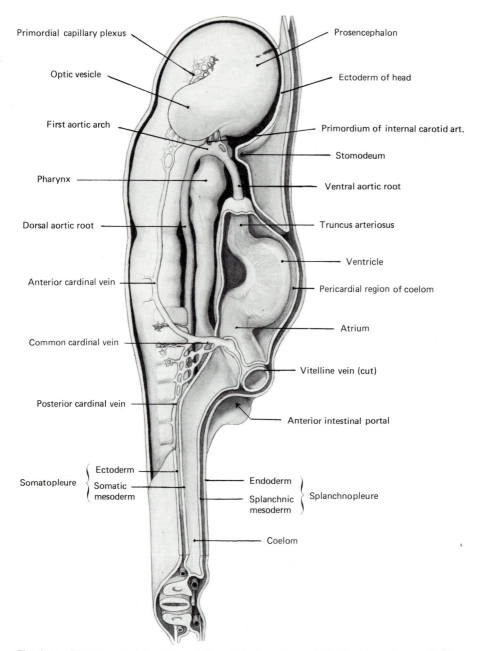

Primordial capillary plexus

Optic vesicle

First aortic arch

Pharynx

Dorsal aortic root

Anterior cardinal vein

Common cardinal vein

Posterior cardinal vein

Somatopleure { Ectoderm

Somatic mesoderm

Prosencephalon

Ectoderm of head

Primordium of internal carotid art.

Stomodeum

Ventral aortic root

Truncus arteriosus

Ventricle

Pericardial region of coelom

Atrium

Vitelline vein (cut)

Anterior intestinal portal

Endoderm }
Splanchnic mesoderm } Splanchnopleure

Coelom

Fig. A-26 Diagrammatic lateral view of dissection of a 38-hour chick. The lateral body wall of the right side has been removed to show the internal structures. Note especially the relations of the pericardial region to that part of the coelom which lies farther caudally and the small anastomosing channels of the developing posterior cardinal vein from which a single main vessel is later derived. (See colored insert.)

endothelium only. The epimyocardial coat is formed about the original endothelial tubes only where they are fused in the region destined to become the heart.

During early embryonic life the cardinal veins are the main afferent vessels of the intraembryonic circulation. The cardinal trunks are paired vessels, symmetrically placed on either side of the midline. There are two pairs, the *anterior cardinal veins,* which return the blood to the heart from the cephalic region of the embryo, and the *posterior cardinal veins,* which return the blood from the caudal region. The anterior and posterior cardinal veins of the same side of the body become confluent dorsal to the level of the heart, the vessels formed by their junction being the *common cardinal veins* (ducts of Cuvier, in the older literature). The right and left common cardinal veins turn ventrad, one on either side of the foregut, and enter the sinoatrial end of the heart along with the right and left vitelline veins (Fig. A-28).

In chicks of 33 hours of incubation the anterior cardinal veins can usually be made out in sections (Fig. A-25B and C). By 38 hours the anterior cardinals and the common cardinals are readily recognized. The posterior cardinals appear somewhat later than the anterior cardinals but are ordinarily discernible in the region of the common cardinals by 33 to 35 hours and are well established by 38 hours. For the sake of simplicity and clearness the cardinal veins have been represented in Fig. A-26 somewhat larger and more regularly formed than they are except in the oldest of the specimens in this age range. Like all the other blood vessels of the embryo they arise as irregular anastomosing endothelial tubes, only gradually taking on the regularity of shape characteristic of fully formed vessels.

CHANGES BETWEEN 38 AND 50 HOURS OF INCUBATION

Flexion and Torsion   Until 36 or 37 hours of incubation the longitudinal axis of the chick is straight except for slight fortuitous variations. Beginning at about 38 hours, positional changes of two distinct types, flexion and torsion, eventually change the entire configuration of the embryo and its relations to the yolk. As applied to an embryo, *flexion* means the bending of the body about a transverse axis, as one might bend the head forward at the neck or the trunk forward at the hips. *Torsion* means the twisting of the body, as one might turn the head and shoulders in looking backward without changing the position of the feet.

In chick embryos the first flexion of the originally straight body axis takes place in the head region. Because of its location it is known as the *cranial flexure.* The axis of bending in the development of the cranial flexure is a transverse axis passing through the midbrain at the level of the cephalic end of the notochord. The direction of the flexion is such that the forebrain becomes bent ventrally toward the yolk. Until the cranial flexure is well established it is inconspicuous in dorsal views of whole mounts, but even in its initial stages it appears plainly in lateral views (Fig. A-26).

To appreciate the correlation between the processes of flexion and torsion it is only necessary to bear in mind the relation of a chick of this age to the yolk. As long as the chick lies with its ventral surface closely applied to the yolk, the yolk constitutes a bar to flexion. Before extensive flexion can be carried out the chick must twist around on its side, i.e., undergo torsion, as a man lying face down turns on his side in order to flex his body.

Torsion begins in the cephalic region of the embryo and progresses caudad. The first indications of torsion appear almost as soon as the cranial flexure begins, and the two processes then progress synchronously. In the chick, torsion is normally carried out toward a definite side. The cephalic region of the embryo is twisted in such a manner that the left side comes to lie next to the yolk and the right side away from the yolk. The progress of torsion caudad is gradual, the caudal part of the embryo remains prone on the yolk for a considerable time after torsion has been completed in the head. Figure A-19 shows the head of an embryo of about 38 hours in which the cranial flexure and torsion are just becoming evident. In chicks of about 43 hours (Fig. A-27) the further progress of both flexion and torsion is well marked.

The processes of flexion and torsion thus initiated continue until the original orientation of the chick on the yolk is completely changed. As the body of the embryo becomes turned on its side the yolk no longer impedes the progress of flexion. Following the accomplishment of torsion in the cephalic region, the cranial flexure rapidly becomes greater until the head is practically doubled on itself (Fig. A-29). As development proceeds, torsion progresses caudad, involving more and more of the body of the embryo. Finally the entire embryo comes to lie with its left side on the yolk.

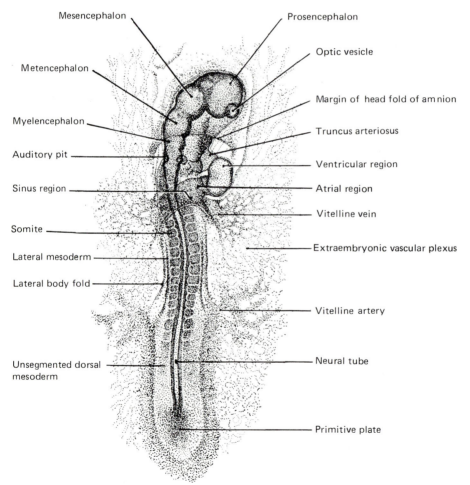

Fig. A-27 Dorsal view (X20) of entire chick embryo having 19 pairs of somites (about 43 hours' incubation). Because of torsion the cephalic region appears in dextrodorsal view.

Concomitantly with the progress of torsion, flexion also appears farther caudally, affecting in turn the cervical, dorsal, and caudal regions. The series of flexions which accompany torsion bend the head and tail of the embryo ventrally so that its spinal axis becomes C-shaped (Fig. A-38). The flexions which bend the embryo on itself so that the head and tail lie close together are characteristic not of the chick alone, but of the embryos of reptiles, birds, and mammals generally. Flexion would seem to be correlated with the spatial limitations of the egg or the uterine cavity within which these embryos undergo their development. Certainly there is no appreciable flexion in embryos of fishes and amphibia which develop free in water with no imposed limitations of space. The torsion which accompanies flexion in the chick is characteristic only of embryos developing on the surface of a very large yolk which would act as an impediment to flexion unless the embryo had first turned on its side.

Completion of the Vitelline Circulatory Channels   In chicks of 33 to 36 hours the main vitelline veins have been established as posterolateral extensions of the same endocardial tubes which are involved in the formation of the heart. As the vitelline veins extend laterad, the vessels developing in the vitelline plexus converge toward the embryo. Eventually these vessels attain communication, establishing the return route from the vitelline vascular plexus to the heart.

The vitelline arteries which are destined to carry blood from the dorsal aortae to the vitelline plexus develop in embryos of about 40 hours (Fig. A-27). Like the vitelline venous channels, the vitelline arterial channels have a dual origin. The proximal portions of the channels to the vitelline plexus arise within the embryo as branches of the dorsal aortae and extend peripherally. The distal portions of the channels arise in the extraembryonic vascular area and extend toward the embryo. Obviously the vitelline circulation cannot begin until these two sets of channels become confluent. At first their connection with each other is by way of a network of small channels rather than by large vessels. These preliminary channels are formed as exceedingly small, freely anastomosing vessels extending from the aorta to communicate laterally with the extraembryonic plexus. Later some of these channels become confluent, others disappear; and gradually, definite main vessels, the vitelline arteries, are established. For some time after their formation, the vitelline arteries are likely to retain traces of their origin from a plexus of small channels and arise from the aorta by several roots (Fig. A-30).

Beginning of the Circulation of Blood   As one might suspect, a mechanism as elaborate as the circulatory system does not go into full-scale operations all at once. Long before the actual circulation of blood commences, the heart has begun to pulsate tentatively and feebly. Its first contractions can usually be seen at the 9-somite stage (about 29 hours of incubation). At this age, the fusion of the paired cardiac primordia has not progressed beyond the ventricular region (Fig. A-24A). These first beats are, therefore, in the ventricular myocardium. Their rhythmic recurrence is at first very slow, and the amplitude of the contractions is small. The pulsations, watched in living embryos, are obviously far short of the power needed to set the blood in motion.

Within 3 or 4 hours of the time of the appearance of the first beats the rate and the amplitude of the pulsations are both strikingly increased. Examination of the structure of the heart at this stage shows that the fusion of the paired primordia has now established the atrium behind the ventricle (Fig. A-24B).

While the heart has been building up an efficient beat, blood corpuscles have been forming in the blood islands of the yolk sac (Fig. A-21). At the same time, the vessels from the area vasculosa to the heart have been formed so that there is an open path for the corpuscles to enter the heart. Just before the actual circulation of blood begins, some of these corpuscles can be seen in the afferent vessels floating in fluid and shuttling back and forth with each heart beat. The last links in the chain to be forged are the arterial channels from the dorsal aortae out to the yolk sac. At about the 16- to 17-somite stage (38 to 40 hours of incubation), these vessels open all the way out to the meshwork of small channels on the yolk sac, which is dotted with blood islands. If one is watching a living embryo when the final channels open, one sees the shuttling of corpuscles in the afferent vessels near the heart give way to a jerky progression. This dramatic series of events in the beginning of the circulation of blood can be described in detail because it has been watched step by step in living embryos, and micro-motion-picture records have been made of all the critical phases of the process (Patten and Kramer, 1933).

The routes followed by the blood that has begun circulating should make apparent the functional significance of the entire system. The heart is the logical point to begin tracing the course of either the embryonic or the vitelline circulation. From the heart the blood of the extraembryonic vitelline circulation passes through the ventral aortic roots, then through the aortic arches and along the dorsal aortae, and out through the vitelline arteries to the plexus of vessels on the yolk (Fig. A-28). In the small vessels which ramify

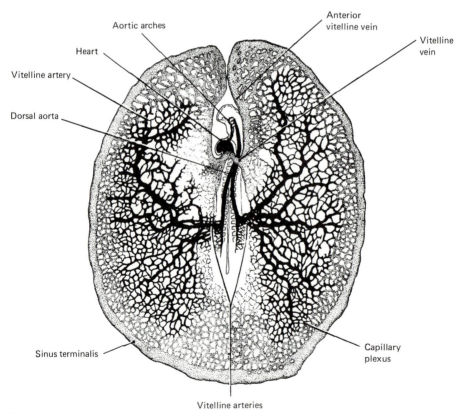

Fig. A-28   The vitelline circulation in a chick of about 44 hours' incubation. Diagrammatic ventral view based on Popoff's figures of injected embryos. The arteries are shown in solid black; the veins are stippled. Note the rich plexus of small, freely anastomosing vessels in the splanchnopleure of the yolk sac.

in the membranes enveloping the yolk, the blood absorbs food material which has been made soluble by the digestive action of the endodermal cells lining the yolk sac. In young embryos, before the allantoic circulation has appeared, the vitelline circulation is also involved in the oxygenation of the blood. The great surface exposure presented by the multitude of small vessels on the yolk makes it possible for the blood to take up oxygen which penetrates the porous shell and the albumen.

After acquiring food material and oxygen the blood is collected by the sinus terminalis and the vitelline veins, which converge toward the embryo from all parts of the vascular area and empty into the main vitelline veins, which return the blood to the heart (Fig. A-28).

The blood of the intraembryonic circulation, leaving the heart, enters the ventral aortic roots, passes by way of the aortic arches into the dorsal aortae, and is distributed through branches from the dorsal aortae to the body of the embryo. It is returned from the cephalic part of the body by the anterior cardinal veins, and from the caudal part of the body by the posterior cardinals. The anterior and posterior cardinals discharge together through the common cardinal veins into the sinoatrial region of the heart (Fig. A-26).

In the heart the blood of the extraembryonic circulation and of the intraembryonic circulation is mixed. The mixed blood in the heart is not so rich in oxygen and food material as the blood that comes to the heart from the vitelline circulation, nor is it as low in food and oxygen content as the blood returned to the heart from the intraembryonic circulation where these materials are drawn upon by the growing tissues of the embryo. Nevertheless, it carries a sufficient proportion of food and oxygen to supply the growing tissues as it is distributed to the body of the embryo.

STRUCTURE OF CHICKS FROM 50 TO 60 HOURS OF INCUBATION

External Features

In chicks which have been incubated about 55 hours (Fig. A-29) the entire head has been freed from the yolk by the caudal progression of the subcephalic fold. Torsion has involved the whole anterior half of the embryo and is completed in the cephalic region, so that the head now lies left side down on the yolk. The posterior half of the embryo is still in its original position, ventral surface prone on the yolk. At the extreme posterior end, the beginning of the caudal fold marks off the tail region of the embryo from the extraembryonic membranes. The head fold of the amnion, together with the lateral amniotic folds, has progressed caudad, inpocketing the embryo nearly to the level of the vitelline arteries.

The *cranial flexure,* which was seen beginning in chicks of about 38 hours, has increased rapidly until at this stage the brain is bent nearly double on itself. Since the axis of the bending is in the midbrain region, the mesencephalon comes to be the most anteriorly located part of the head and the prosencephalon and myelencephalon lie opposite each other, ventral surface to ventral surface (Fig. A-29). The original rostral end of the prosencephalon is thus brought in close proximity to the heart, and the optic vesicles and the auditory vesicles are brought opposite each other at nearly the same anteroposterior level.

At this stage flexion has involved the body farther caudally as well as in the brain region. It is especially marked at about the level of the heart in the region of transition from myelencephalon to spinal cord. Since this is the future neck region of the embryo, the flexure at this level is known as the *cervical flexure.*

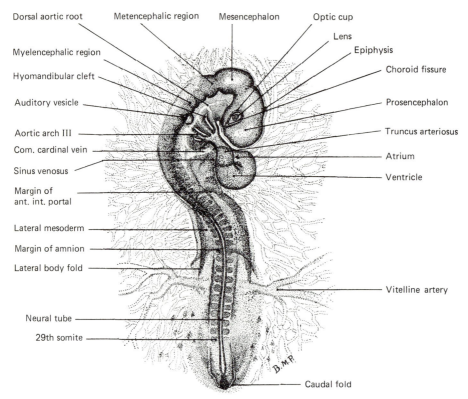

Dorsal aortic root Metencephalic region Mesencephalon Optic cup

Lens

Myelencephalic region Epiphysis

Hyomandibular cleft Choroid fissure

Auditory vesicle Prosencephalon

Aortic arch III Truncus arteriosus

Com. cardinal vein Atrium

Sinus venosus Ventricle

Margin of
ant. int. portal

Lateral mesoderm

Margin of amnion

Lateral body fold

Vitelline artery

Neural tube

29th somite

Caudal fold

Fig. A-29 Dextrodorsal view (X17) of entire embryo of 29 somites (about 55 hours' incubation).

The Nervous System

Growth of the Prosencephalic Region The completion of torsion in the cephalic region causes rapid changes in the configuration of the brain. The rostral part of the brain has undergone rapid enlargement. A slight constriction in the dorsal wall (Fig. A-30) indicates the impending division of the prosencephalon into *telencephalon* and *diencephalon* (Fig. A-33). A small invagination, called the *epiphysis*, has appeared in the middorsal wall of the diencephalic region (Figs. A-29 and A-30). The epiphysis forms the pineal body of the adult. In the floor of the diencephalon the infundibular depression has become deepened and lies close to a newly formed ectodermal invagination known as *Rathke's pocket* (Fig. A-30). The epithelium of Rathke's pocket is destined to be separated from the superficial ectoderm and to become permanently associated with the infundibular portion of the diencephalon to form the *hypophysis* or *pituitary body*.

The Optic Vesicles The optic vesicles have undergone changes which completely alter their appearance. In 33-hour chicks they were spheroidal vesicles connected by broad stalks with the lateral walls of the prosencephalon (Fig. A-18). At that stage the lumen of each optic vesicle *(opticoele)* was widely continuous with the lumen of the prosencephalon *(prosocoele)* (Fig. A-25A). The constriction of the optic stalk which begins to be apparent in 38-hour embryos (Fig. A-19) is much more marked in 55-hour chicks.

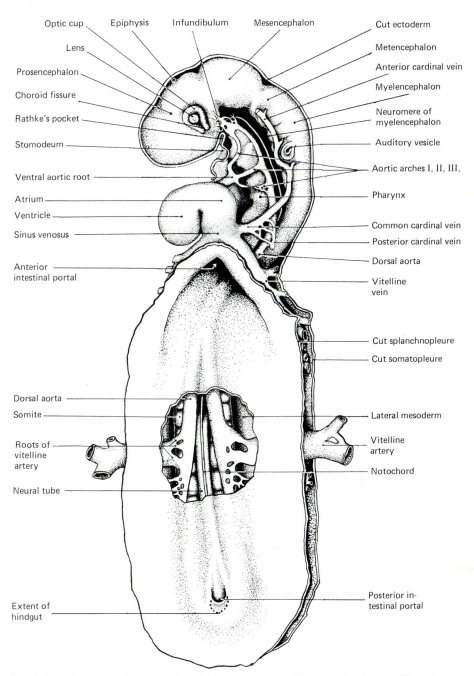

Fig. A-30 Diagram of dissection of chick of about 50 hours. The splanchnopleure of the yolk sac cephalic to the anterior intestinal portal, the ectoderm of the left side of the head, and the mesoderm in the pericardial region have been dissected away. A window has been cut in the splanchnopleure of the dorsal wall of the midgut to show the origin of the vitelline arteries. *(Modified from Prentiss.)*

The most striking and important advance in their development is the invagination of the distal ends of the single-walled optic vesicles to form double-walled *optic cups* (Fig. A-31B): The concavities of the cups are directed laterally. Mesially the cups are continuous, over the narrowed *optic stalks,* with the ventrolateral walls of the diencephalic region of the original prosencephalon. The invaginated layer of the optic cup is termed the *sensory layer* because it is destined to give rise to the sensory layer of the retina. The layer against which the sensory layer comes to lie after its invagination is termed the *pigment layer* because it gives rise to the pigmented layer of the retina. The double-walled cups formed by invagination are sometimes termed *secondary optic vesicles* to distinguish them from *primary optic vesicles,* as they are called before the invagination. The formerly capacious lumen of the primary optic vesicle is practically obliterated in the formation of the optic cup. What remains of the primary opticoele is now but a narrow space between the sensory and the pigment layers of the retina (Fig. A-31B). Later, when these two layers fuse, this space is entirely obliterated.

Although the secondary optic vesicles are usually spoken of as the optic cups, they are not complete cups. The invagination which gives rise to the secondary optic vesicles, instead of beginning at the most lateral point in the primary optic vesicles, begins at a point somewhat toward their ventral surface and is directed mesiodorsally. As a result the optic cups are formed with an incomplete lip on their ventral aspect. They may be likened to cups with a segment broken out of one side. This gap in the optic cup is the *choroid fissure* (Fig. A-30). Figure A-31B shows a section which passes through the head of the embryo on a slight slant so that the right optic cup, being cut to one side of the choroid fissure, appears complete; the left optic cup, being cut in the region of the fissure, shows no ventral lip.

The infolding process by which the optic cups are formed from the primary optic vesicles is continued to the region of the optic stalks. As a result the optic stalks are infolded so that their ventral surfaces become grooved. Later in development the optic nerves and blood vessels come to lie in the grooves thus formed in the optic stalks.

The Lens   The lens of the eye arises from the superficial ectoderm of the head overlying the optic vesicles. The first indications of lens formation appear in chicks of about 40 hours as local thickenings of the ectoderm immediately distal to the optic vesicles. These placodes of thickened ectoderm sink below the general level of the surface of the head to form small vesicles which extend into the secondary optic vesicles. Their openings to the surface are rapidly constricted, and eventually they are disconnected altogether from the superficial ectoderm. At this stage, however, the opening to the outside still persists, although it is very small (Fig. A-31B, right eye). In sections which do not pass directly through the opening, the lens vesicle appears as if it were completely separated from the overlying ectoderm (Fig. A-31B, left eye).

The Posterior Part of the Brain and the Cord Region of the Neural Tube   Caudal to the diencephalon the brain shows no great change as compared with the last stages considered. The mesencephalon is somewhat enlarged, and the constrictions separating it from the diencephalon rostrally and from the metencephalon caudally are more sharply marked. The metencephalon is more clearly delimited from the myelencephalon, and its roof is beginning to show thickening. In the myelencephalon the interneuromeric constrictions are still evident in the ventral and lateral walls (Figs. A-29 and A-30). The dorsal wall has become much thinner than the ventral and lateral walls (Fig. A-31A and B) and shows no trace of division between the neuromeres. As the result of continued invagination, the auditory pits are now called *auditory vesicles* (Figs. A-30 and A-31A).

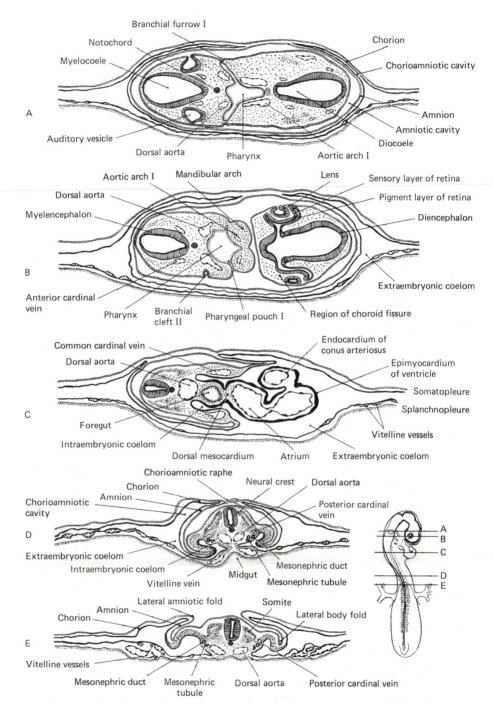

Fig. A-31 Diagrams of transverse sections of 55-hour (30-somite) chick. The location of the sections is indicated on an outline sketch of the entire embryo.

In the spinal cord region of the neural tube the lateral walls have become thickened at the expense of the lumen, so that the neural canal appears slitlike in sections of embryos of this age (Fig. A-31E). At this stage the neural tube is complete, with the closure of both the anterior neuropore and the sinus rhomboidalis.

At the extreme caudal end of the embryo, the future spinal cord in the tail is being formed in a different manner from the rest of the cord. At the area of the tail bud (Fig. A-32), the formation of the neural tube by closure of a neural groove ceases. To complete the remainder of the neural tube, the tail bud undergoes a process of cavitation, sometimes called *secondary neurulation* (Criley, 1969).

The Neural Crest   In the closure of the neural tube and its separation from the overlying ectoderm, groups of ectodermal cells, originally lying just lateral to the neural folds, are not involved in the final fusion of either the neural tube or the superficial ectoderm. These cells form a pair of longitudinal aggregations, extending on either side of the middorsal line in the angles between the superficial ectoderm and the neural tube (Fig. 6-12A). With the fusion of the edges of the neural folds to complete the neural tube and the fusion of the superficial ectoderm dorsal to the neural tube, these two longitudinal cell masses become for a time confluent in the midline (Fig. 6-12B). But because this aggregation of cells arises from paired components and soon again separates into right and left parts, it is considered potentially paired even in this phase of temporary fusion. Because of its position dorsal to the neural tube it is known as the *neural crest*. When first established, the neural crest is continuous cephalocaudally. As development progresses, the cells of the neural crest migrate ventrolaterally on either side of the neural tube (Fig. 6-12C).

The Digestive Tract

The Foregut   The manner in which the three primary regions of the gut tract are established has already been considered in a general way (Fig. 7-2). In 50- to 55-hour chicks the foregut has acquired considerable length. It extends from the anterior intestinal portal cephalad almost to the infundibulum (Fig. A-30).

As the first region of the digestive tract to be established, the foregut is naturally the most advanced in differentiation. One can already recognize a pharyngeal and an esophageal portion. The pharyngeal region lies ventral to the myelencephalon and is encircled by the aortic arches (Fig. A-33). The pharynx is somewhat flattened dorsoventrally and has a considerably larger lumen than the esophageal part of the foregut (cf. Fig. A-31B and C).

The Stomodeum   In the younger embryos of this age range, there is no mouth opening into the pharynx. However, the location where the opening will be formed is indicated by the approximation of a ventral outpocketing near the anterior end of the pharynx to a depression formed in the adjacent ectoderm of the ventral surface of the head (Fig. A-30). The ectodermal depression, known as the *stomodeum,* deepens until its floor lies in contact with the endoderm of the pharyngeal outpocketing (Fig. A-30). The thin layer of tissue formed by the apposition of stomodeal ectoderm to pharyngeal endoderm is known as the *oral plate*. A little later in development the oral plate breaks through, bringing the stomodeum and the pharynx into open communication.

The Preoral Gut   It will be noted by referring to Fig. A-30 that the oral opening is not established at the extreme cephalic end of the pharynx. The part of the pharynx which extends cephalic to the mouth opening is known as the *preoral gut*. After the rupture of the oral plate, the preoral gut eventually disappears, but an indication of it

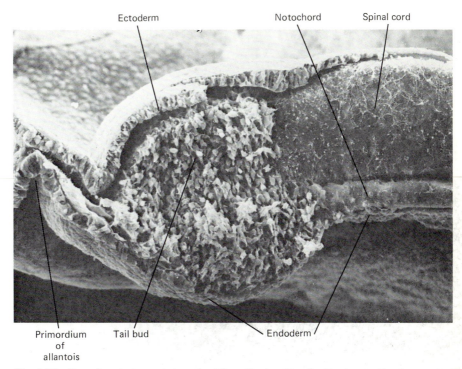

Fig. A-32 Scanning electron micrograph of the tailbud region of a Hamburger-Hamilton stage 15 (50-55 Hours) chick embryo. The spinal cord and notochord merge with the tail bud. (*From Schoenwolf, G. C., 1978 in Scanning Electron Microscopy, Vol. II. SEM Inc., AMF O'Hare, Ill. Courtesy of the author.*)

persists for a time as a small diverticulum with no known function termed *Seessel's pocket* (Fig. A-39).

The Midgut The midgut is still the most extensive of the three primary divisions of the digestive tract. It is nothing more than a region where the gut still lies open to the yolk and does not even have a fixed identity. As fast as any part of the midgut acquires a ventral wall by the closing-in process involved in the progress of the subcephalic and subcaudal folds it ceases to be midgut and becomes foregut or hindgut. Differentiation and local specialization appear in the digestive tract only in regions which have ceased to be midgut.

The Hindgut The hindgut first appears in embryos of about 50 hours (Fig. A-30). The method of its formation is similar to that by which the foregut was established. The subcaudal fold undercuts the tail region and walls off a gut pocket. The hindgut is lengthened at the expense of the midgut as the subcaudal fold progresses cephalad and is also lengthened by its own growth caudad. It shows no local specialization until later in development.

The Branchial Clefts and Branchial Arches

At this stage the chick embryo has unmistakable branchial (gill) arches and branchial (gill) clefts. Although only transitory, they are morphologically of great importance because of their significance as structures exemplifying recapitulation and also because

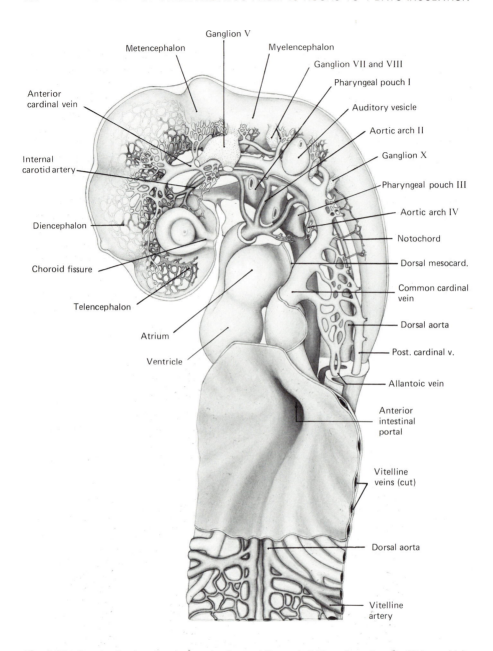

Fig. A-33 Drawing to show the deeper structures of the cephalothoracic region of a 60-hour chick, exposed from the left. The basis of the illustration was a wax-plate reconstruction made from serial sections; the smaller vessels and the primordial capillary plexuses were added from injected specimens. Note especially the relations of the aortic arches to the pharyngeal pouches and the way in which the large veins enter the sinus venosus. Although the vitelline veins are not fully exposed by this dissection, the bulges they cause in the endoderm on either side of the anterior intestinal portal clearly suggest the way the main right and left veins become confluent with each other to enter the sinus venosus as a short median trunk. (See colored insert.)

of their participation in the formation of some of the ductless glands, the eustachian tubes, and the face and jaws. Moreover, the location of the aortic arches within the correspondingly numbered branchial arches and their relations to the pharyngeal pouches are among the most characteristic and important structural features of young embryos.

The *branchial clefts* are formed by the meeting of ectodermal depressions, the *branchial furrows,* with diverticula from the lateral walls of the pharynx, the *pharyngeal pouches.* In sections the branchial clefts may be seen to be closed by a thin layer of tissue composed of the ectoderm of the floor of the branchial furrow and the endoderm at the distal extremity of the pharyngeal pouch (Fig. A-31A). The breaking through of this thin double layer of tissue brings the pharyngeal pouches into communication with the branchial furrows, thereby establishing open branchial (gill) clefts. In birds an open condition of any of the clefts is transitory. In the chick the most posterior of the series of clefts never becomes open. Although some of the clefts never become open and others open only for a short time, the term *cleft* is usually used to designate these structures which are potentially clefts, whether open or not.

The position of the branchial clefts is best seen in entire embryos. They are commonly designated by number, beginning with the first cleft posterior to the mouth and proceeding caudad. The first postoral cleft appears earliest in development and is discernible at about 46 hours of incubation. Branchial cleft II appears soon after, and by 50 to 55 hours three clefts have been formed (Fig. A-29).

Between adjacent branchial clefts, the lateral body walls about the pharynx are thickened. Each of these lateral thickenings meets and merges in the midventral line with the corresponding thickening of the opposite side of the body. Thus the pharynx is encompassed laterally and ventrally by a series of archlike thickenings, the branchial arches. The branchial arches, like the branchial clefts, are designated by number, beginning at the anterior end of the series. Branchial arch I lies cephalic to the first postoral cleft, between it and the mouth. Because of the part it plays in the formation of the mandible it is also designated as the *mandibular arch.* Branchial arch II is frequently termed the *hyoid arch,* and branchial cleft I, because of its position between the mandibular and hyoid arches, is known as the *hyomandibular cleft.* Posterior to the hyoid arch the branchial arches and clefts are ordinarily designated by numbers only. According to Witschi (1956) young chick embryos may show in their branchial arches transitory cell clusters that are suggestive of early stages in the formation of the external gill buds in amphibian embryos. These abortive gill bud primordia are most likely to be recognizable on the second and third arches of these young chick embryos.

The Circulatory System

The Heart During the period between 30 and 55 hours of incubation the heart is growing more rapidly than the body of the embryo in the region where it lies. The unattached midregion of the heart becomes at first U-shaped and then twisted on itself to form a loop. The atrial region of the heart is forced somewhat to the left, and the truncus is thrown across the atrium by being twisted to the right dorsally. The ventricular region constitutes the loop proper (Fig. A-24). This twisting process reverses the original cephalocaudal relations of the atrial and ventricular regions. Before the twisting, the atrial region of the heart was caudal to the ventricular region, as it is in the adult fish heart. In the twisting of the heart, the atrial region, by reason of its association with the fixed sinus region of the heart, undergoes relatively little change in position. The ventricular region is carried over the dextral side of the atrium and comes to lie caudal to

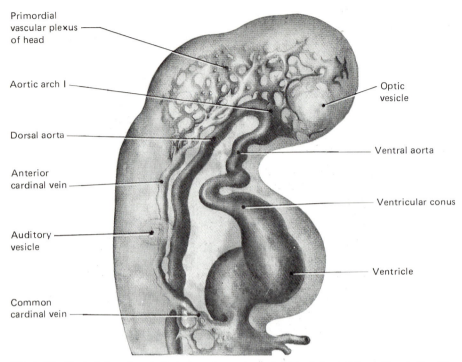

Primordial
vascular plexus
of head

Aortic arch I

Dorsal aorta

Anterior
cardinal vein

Auditory
vesicle

Common
cardinal vein

Optic
vesicle

Ventral aorta

Ventricular conus

Ventricle

Fig. A-34 Dextral view of cephalic and cardiac regions of chick of about 45 hours' incubation. The blood vessels have been injected to show the capillary plexus of the cephalic region. In the drawing the heart and arteries are differentiated from the veins and capillaries by darker shading. This figure, together with Fig. A-35, shows the manner in which main vessels develop in the embryo from a primary capillary plexus. It will be noted in this figure that the part of the capillary plexus which lies more superficially already shows an enlargement in diameter and a directness of path which is indicative of the fact that it is to become the main trunk vein of this region. *(From Minot after Evans.)*

it, thus arriving in the relative position it occupies in the adult heart of birds and mammals.

The bending and subsequent twisting of the heart lead toward its division into separate chambers. However, no indication of the actual partitioning of the heart into right and left sides is yet apparent. It is still essentially a tubular organ through which the blood passes directly, without any division into separate channels.

During the period from 50 to 80 hours of incubation, interesting changes become increasingly apparent in the ventricular portions of the cardiac wall. When the primordial tubular heart was first established, its endothelially lined lumen was smooth and fairly regular in diameter. Outside the endothelium, between it and the developing muscle, was a relatively thick layer of cardiac jelly (Fig. A-22E). Beginning in embryos of 50 to 55 hours and rapidly becoming more marked, the myocardium shows irregular projections extending into the cardiac jelly. These projections are the start of the *trabeculae carneae,* which are such conspicuous features of the interior of the adult ventricles. As the trabeculae grow, the endothelial lining tends to extend between them and follow closely the configuration of the muscular strands, being separated from them only by a thin layer of cardiac jelly.

Once established, the trabeculation shows an increasingly richly branching pattern,

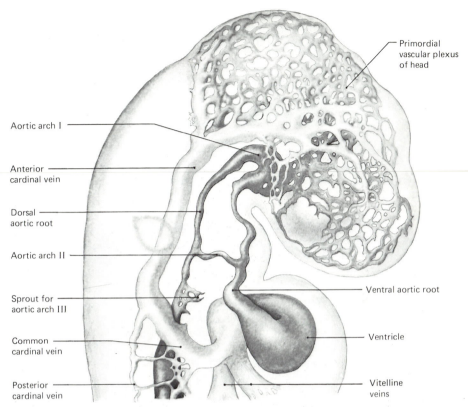

Fig. A-35   Dextral view of cephalic and cardiac regions of injected chick of about 50 hours' incubation. This figure shows a later stage in the development of the anterior cardinal vein from the primary capillary plexus of the cephalic region. The main channel, but vaguely suggested in the previous figure, is here quite definite. Here also a second aortic arch has been completed, and a plexiform outgrowth of vascular endothelium from the dorsal aortic root toward the ventral aorta indicates the impending formation of the third aortic arch. *(From Minot after Evans, redrawn with modifications in the cardiac region.)*

so that the ventricular wall becomes honeycombed by tortuous intertrabecular spaces, all of which are endothelially lined and all of which directly or indirectly communicate with the main lumen of the ventricle (Fig. 17-23). This structural pattern is of great functional significance, for it brings the blood in close relationship to the growing cardiac muscle during the period before the coronary circulation to the myocardium has been formed.

   The Aortic Arches   In 33- to 38-hour chicks the ventral aortae communicate with the dorsal aortae over a single pair of aortic arches which bend around the anterior end of the pharynx (Figs. A-20 and A-26). Even embryos as old as 43 to 45 hours may still show only their first pair of aortic arches (Fig. A-34), but soon thereafter, as the branchial arches caudal to the mandibular arch differentiate, new aortic arches appear within them. These aortic arches are designated by the postoral number of the branchial arch in which they lie. Thus the original aortic arches lying in the first (mandibular) arch are spoken of as the first aortic arches, those developing in the second (hyoid) arch are the second aortic arches, and so on. In chicks of about 50 hours' incubation, the second pair of aortic arches has usually made its appearance (Fig. A-35). The third arch at this stage is usually represented by some capillary sprouts (Fig. A-35). By 60 hours the third aortic arch has become quite sizable and sprouts indicating the start of the fourth arch are usually identifiable (Fig. A-33).

The Fusion of the Dorsal Aortae   The dorsal aortae arise as vessels paired throughout their entire length (Fig. A-20). As development progresses they fuse in the midline to form the unpaired dorsal aorta familiar in adult anatomy. This fusion takes place first at the cardiac level (Fig. A-31C), and progresses cephalad and caudad. Cephalically it never extends to the pharyngeal region. Caudally the whole length of the aorta is eventually involved. By 50 hours the fusion has progressed caudad well beyond the anterior intestinal portal (Fig. A-30).

The Cardinal and Vitelline Vessels   The basic relationships of the cardinal veins and the vitelline vessels are not radically changed from those seen at the time of their first appearance (Fig. A-26). The posterior cardinals have elongated, keeping pace with the caudal progress of differentiation in the mesoderm. They lie just dorsal to the nephric tubules arising from the intermediate mesoderm, in the angle formed between the somatic mesoderm and the somites (Figs. A-31D and 6-17C and D). Where the vitelline veins enter the sinoatrial part of the heart they lie progressively closer to the midline. Eventually they fuse with each other to form a median trunk. It is at this region of the convergence of the vitelline veins that the common cardinal veins enter them dorsolaterally (Figs. A-33 and A-48J). This region of confluence of the great veins, as we shall see in considering older stages, is being involved in the molding of the sinus venosus. The vitelline arteries, meanwhile, are tending to lose the multiple roots by which they emerged from the dorsal aortae in younger stages, but otherwise they exhibit essentially the same relationships.

The Somites   By 55 to 60 hours, the embryo possesses about 30 pairs of somites. At this time the cells of the ventromesial part of the somite (the sclerotome) have become mesenchymal in character and are extending toward the notochord. Ultimately they aggregate about the notochord and participate in the formation of the body of the vertebra.

The Urinary System

From the intermediate mesoderm, a segmentally arranged system of urine-secreting units gradually differentiates. Because of its complexity, the structural detail of the early urinary system is dealt with in Chap. 16. Starting at around 36 hours, solid buds of mesodermal cells, representing pronephric tubules, appear as pairs in sequence from the fifth to the sixteenth somite. They are connected to a longitudinal duct (the primary nephric duct) that makes its way toward the cloaca. By 55 hours, the newly developing nephric tubules are of a different morphological type—mesonephric tubules. In 55-hour chick embryos, the *mesonephros,* as the complex of mesonephric tubules and the mesonephric duct is called, begins to bulge into the dorsal part of the coelomic cavity (Fig. A-31D and E). There are yet no recognizable gonads.

DEVELOPMENT OF THE CHICK DURING THE THIRD AND FOURTH DAYS OF INCUBATION

External Features

Torsion Chicks of 3 days' incubation (Fig. A-36) have been affected by torsion throughout their entire length. Torsion is complete well posterior to the level of the heart, but the caudal portion of the embryo is not yet completely turned on its side. By 4 days the entire body has been turned through and the embryo lies with its left side on the yolk (Fig. A-38).

Flexion The cranial and cervical flexures which appeared in embryos during the second day have increased so that in 3-day and 4-day chicks the long axis of the embryo shows nearly right-angled bends in the midbrain and in the neck region. The midbody region of 3-day chicks is slightly concave dorsally. This is due to the fact that the embryo is still broadly attached to the yolk in that region. By the end of the fourth day the body

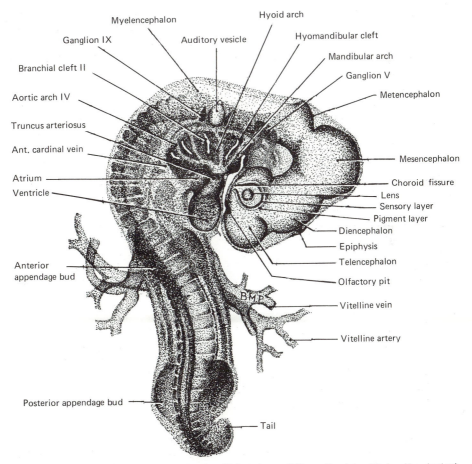

Fig. A-36 Dextrodorsal view (X16) of entire chick embryo of 36 somites (about 3 days' incubation).

folds have undercut the embryo so that it remains attached to the yolk only by a slender stalk. The yolk stalk soon becomes elongated, allowing the embryo to become first straight in the middorsal region and then convex dorsally. At the same time the caudal flexure is becoming more pronounced. The progressive increase in the cranial, cervical, dorsal, and caudal flexures results in the bending of the embryo on itself so that its originally straight long axis becomes C-shaped and its head and tail lie close together (Fig. A-38).

The Branchial Arches and Clefts A fourth branchial cleft has appeared caudal to the three that were already formed in 55-hour chicks. The branchial arches are thicker and more conspicuous than those in earlier embryos. In lightly stained whole mounts of a 3-day chick (Fig. A-36) it is still possible to make out the aortic arches running through the branchial arches. In a chick of 4 days (Fig. A-38) the branchial arches have become so much thickened that it is very difficult to see the vessels traversing them.

The Oral Region The cervical flexure presses the pharyngeal region and the ventral surface of the head so closely together that it is difficult to make out the topography of the oral region by studying entire embryos. If the head and pharyngeal region are cut from the trunk and viewed from the ventral aspect, the relations of the structures about the mouth are well shown (Fig. A-37). The mandibular arch forms the caudal boundary of the oral depression. Arising on either side in connection with the lateral part of the mandibular arch are paired elevations, the maxillary processes, which grow mesiad and form the cephalolateral boundaries of the mouth opening. The *nasal pits* appear as shallow depressions in the ectoderm of the rostral part of the head which overhangs the mouth region. Surrounding each nasal pit is a U-shaped elevation with its limbs directed

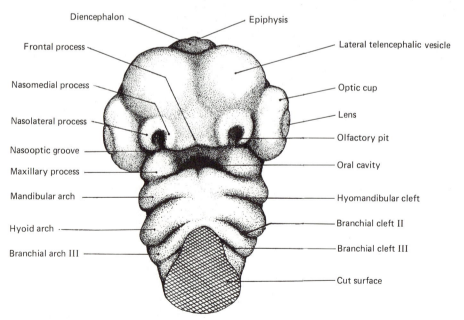

Fig. A-37 Drawing to show the external appearance of the structures in the oral region of a 4-day chick. Ventral aspect.

toward the oral cavity. The lateral limb of the elevation is the *nasolateral process,* and the median limb is the *nasomedial process.* As development proceeds, the two nasomedial processes grow toward the mouth and meet the maxillary processes which are growing in from either side. The merging of the two nasomedial processes with each other in the midline and the fusion of each of them laterally with the maxillary process of its own side give rise to the upper jaw (maxilla). The merging in the midline of the right and left components of the mandibular arch gives rise to the lower jaw (mandible). Formation of the beak does not occur until later in development.

The Allantois   The development of the extraembryonic membranes has already been considered (Chap. 7) and needs no further discussion here. In order to show the embryos more clearly, the extraembryonic membranes, except for the allantois, have been removed from the specimens drawn in Figs. A-36 and A-38. The cut edge of the amnion shows at its anterior attachment to the body, opposite the anterior appendage bud and just caudal to the tip of the ventricle (Fig. A-38). The allantois in the 3-day chick is still small and is concealed by the posterior appendage buds. In 4-day embryos it has undergone rapid enlargement and projects beyond the confines of the body as a stalked vesicle of considerable size (Fig. A-38).

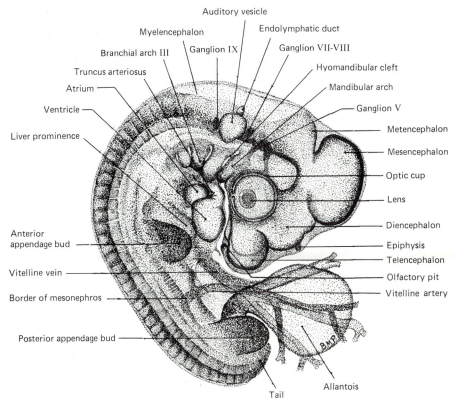

Fig. A-38   Dextral view (X16) of entire chick embryo of 41 somites (about 4 days' incubation). Stained and cleared preparation drawn by transmitted light.

The Appendage Buds   Both the anterior and the posterior appendage buds are grossly evident by the third day of development (Fig. A-36). By the end of the fourth day the appendage buds have increased in size to form conspicuous extensions from the sides of the body (Fig. A-38). The main mass of the buds is composed of closely packed mesenchymal cells; the outer covering is of ectoderm. In a section of an appendage properly prepared for microscopic study there appears a conspicuously thickened band of the ectodermal covering along the convex outer margin of the bud. This is the *apical ectodermal ridge* (Figs. 10-5B and A-41E).

The Nervous System

Formation of the Telencephalic Vesicles   By the end of the third day the anterolateral walls of the primary forebrain have been evaginated to form a pair of vesicles lying one on either side of the midline (Figs. A-36 to A-38). These lateral evaginations are known as the *telencephalic vesicles.* The openings through which their cavities are continuous with the lumen of the median portion of the brain are later known as the *foramina of Monro.* The telencephalic division of the brain includes not only these two lateral vesicles but also the median portion of the brain from which they arise. The lumen of the telencephalon has therefore three divisions, a *median telocoele,* broadly confluent posteriorly with the diocoele, and two lateral telencephalic vesicles, connecting with the median telecoele through the foramina of Monro (Fig. 11-14C).

The telencephalic vesicles become the cerebral hemispheres, and their cavities become the paired lateral ventricles of the adult brain. The hemispheres undergo enormous enlargement in their later development and extend dorsally and posteriorly as well as rostrally, eventually covering the entire diencephalon and mesencephalon under their posterior lobes.

The Diencephalon   The lateral walls of the diencephalon at this stage show little differentiation except ventrally where the optic stalks merge into the walls of the brain. The development of the *epiphysis* as a median evagination in the roof of the diencephalon has already been mentioned. Except for some elongation it does not differ from its condition when first formed in embryos of about 55 hours. The infundibular depression in the floor of the diencephalon has become appreciably deepened and lies in close proximity to Rathke's pocket (Fig. A-39) with which it is destined to fuse in the formation of the *hypophysis* (Fig. 15-1). Later in development the lateral walls of the diencephalon become greatly thickened to form the *thalami,* which, in the adult, function as relay stations for sensory pathways leading to the cerebral cortex. The enlarging thalami reduce the size and change the shape of the diocoele, which is known in adult anatomy as the third brain ventricle. The anterior part of the roof of the diencephalon remains thin and becomes richly vascular. Later these vessels, invaginating the roof with them, push into the third ventricle to form the *anterior choroid plexus* or *choroid plexus of third ventricle* (Fig. 11-18G). The choroid plexuses are highly vascular areas concerned with the formation of cerebrospinal fluid.

The Mesencephalon   The mesencephalon does not yet show any specializations beyond a thickening of its walls. The dorsal walls of the mesencephalon later increase rapidly in thickness and become the *corpora quadrigemina* of the adult brain. As the name implies, these are four symmetrically placed elevations. The anterior pair *(superior colliculi)* constitute the brain center for visual reflexes; the posterior pair *(inferior*

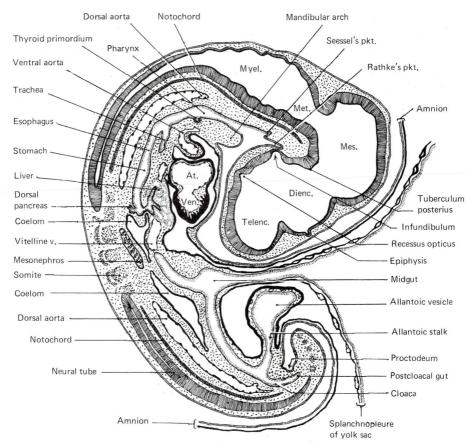

Fig. A-39 Diagram of median longitudinal section of 4-day chick. Because of a slight bend in the embryo the section is parasagittal in the middorsal region, but for the most part it passes through the embryo in the sagittal plane.

colliculi) are the center for auditory reflexes. The floor of the mesencephalon also becomes greatly thickened and is known in the adult as the *crura cerebri.* It serves as the main pathway of the fiber tracts which connect the cerebral hemispheres with the posterior part of the brain and the spinal cord. The originally capacious mesocoele is thus reduced by the thickening of the walls about it to a narrow canal, *the cerebral aqueduct,* or *aqueduct of Sylvius* (Fig. 11-18G).

The Metencephalon   The boundary between the mesencephalon and the metencephalon is indicated by the location of the original interneuromeric constriction which separated them at the time of their establishment (cf. Figs. A-17 and 11-14). Ventrally, the caudal boundary of the metencephalon is not positively defined. Dorsally, it is regarded as being located approximately at the point where the brain roof changes from the thickened condition characteristic of the metencephalon to the thin condition characteristic of the myelencephalon (Fig. 11-14A). The metencephalon shows practically no differentiation in 4-day chicks. Later there is ventrally and laterally an extensive ingrowth of fiber tracts giving rise to the pons and to the cerebellar peduncles. The roof of the metencephalon undergoes extensive enlargement and becomes the *cerebellum* of the adult brain, the coordinating center for complex muscular movements.

The Myelencephalon   The dorsal myelencephalic wall is reduced in thickness indicative of its final fate as the thin roof of the medulla. It later receives a rich supply of small blood vessels which, carrying the roof with them, grow into the myelocoele to form the *posterior choroid plexus,* or *choroid plexus of the fourth ventricle* (Figs. 11-11 and 11-18G). The ventral and lateral myelencephalic walls become the floor and side walls of the medulla. Functionally the medulla serves both as a conduction path between cord and brain and as a reflex center for involuntary activities such as breathing.

The Ganglia of the Cranial Nerves   In the brain region, cells derived from the cephalic portion of the neural crest have become aggregated to form the ganglia of the cranial nerves which have sensory components. The largest and the most clearly defined of the ganglia present in 4-day chicks is the semilunar (Gasserian) ganglion of the fifth (trigeminal) cranial nerve (Fig. A-40). It lies ventrolaterally, opposite the most anterior neuromere of the myelencephalon. From its cells sensory nerve fibers grow mesiad into the brain and distad to the face and mouth region. In 4-day chicks the beginning of the *ophthalmic division* of the fifth nerve extends from the ganglion toward the eye, and the beginning of the *mandibulomaxillary division* is growing toward the angle of the mouth (Fig. A-40). Immediately cephalic to the auditory vesicle is a mass of neural-crest cells which is the primordium of the ganglia of the seventh and eighth nerves. The separation of this double primordium to form the geniculate ganglion of the seventh nerve and the acoustic ganglion of the eighth nerve begins during the fourth day. Caudal to the auditory vesicle the superior ganglion of the ninth nerve can be clearly seen even in whole mounts (Fig. A-38). The ganglia of the tenth (vaugs) nerve can be recognized in sections of chicks during the fourth day or in reconstructions (Fig. A-40), but they are difficult to see in whole mounts.

The Spinal Cord   The spinal cord region of the neural tube, when first established, exhibits a lumen which is elliptical in cross section. As development progresses the lateral walls of the cord become greatly thickened in contrast with the dorsal and ventral walls which remain thin. In this process the lumen (central canal) becomes compressed laterally until it appears in cross section as little more than a vertical slit (Fig. A-38). The thin dorsal wall of the tube is known as the *roof plate,* the thin ventral wall as the *floor plate,* and the thickened side walls are known as the *lateral plates.*

The Spinal Nerves   During the fourth day the establishing of the spinal nerve roots has begun. The growth of nerve fibers from the neuroblasts can be traced only with the aid of special methods of staining. The more general steps in the development of the roots of the spinal nerves can, however, be followed in sections prepared by the ordinary methods.

In the adult each spinal nerve is connected with the cord by two roots, a *dorsal root* which is a pathway for sensory (afferent) nerve fibers and a *ventral root* which is a pathway for motor (efferent) nerve fibers. Lateral to the cord the dorsal and ventral roots unite. The *spinal ganglion (dorsal root ganglion)* is located on the dorsal root between the spinal cord and the point where dorsal and ventral roots unite. Distal to the union of

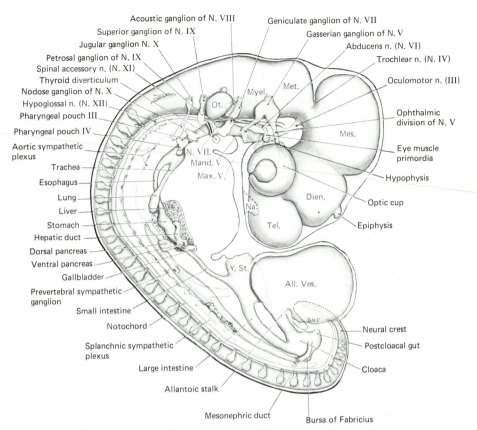

Fig. A-40 Reconstruction of nervous, digestive, and urinary systems of 4-day chick (original X51, reproduced X16). Abbreviations: *All. Ves.*, allantoic vesicle; *Dien.*, diencephalon; *Mand. V*, mandibular division of cranial nerve V (trigeminal); *Max. V*, maxillary division of cranial nerve V (trigeminal); *Mes.*, mesencephalon; *Met.*, metencephalon; *Myel.*, myelencephalon; *Na.*, location of nasal pit; *N. VII*, seventh cranial nerve (facial); *Ot.*, otic vesicle; *Tel.*, lateral telencephalic vesicle; *Y. St.*, yolk stalk.

dorsal and ventral roots is a branch, the *ramus communicans*, which extends ventrad to a ganglion of the sympathetic nerve cord (Fig. 11-1).

When first formed from the neural-crest cells, the spinal ganglion has no connection with the cord. The dorsal root is established by the growth of nerve fibers from cells of the spinal ganglion mesiad into the dorsal part of the lateral plate of the cord. At the same time fibers grow distad from these cells to form the peripheral part of the nerve. The fibers arising from cells of the dorsal root ganglion conduct sensory impulses toward the cord.

Coincident with the establishment of the dorsal root, the ventral root is formed by fibers which grow out from cells located in the ventral part of the lateral plate of the cord. Most of the fibers which thus arise from cells in the cord and pass out through the ventral root conduct motor impulses from the brain and cord to the muscles with which they are associated peripherally.

The Digestive and Respiratory Systems

Establishment of the Oral Opening As was seen in the study of younger chicks, when the gut is first established it ends as a blind pocket both cephalically and caudally. In embryos of 55 to 60 hours the processes leading toward the establishment of the oral opening were, however, clearly indicated. A midventral evagination of the pharynx had been established immediately cephalic to the mandibular arch. Opposite this outpocketing of the pharynx, and growing in to meet it, the stomodeal depression had been formed (Fig. A-30). The only bar to an actual opening was the *oral plate,* a thin membrane constituted by the apposition of the pharyngeal endoderm and the stomodeal ectoderm. The communication of the foregut with the outside is finally established, during the third day, by the breaking through of the oral plate. Following the rupture of the oral plate, growth of the surrounding structures (Fig. A-37) rapidly deepens the originally shallow stomodeal depression. The region where the oral plate was originally located in the embryo eventually becomes, in the adult, the region of transition from oral cavity to pharynx.

The formation of the oral opening in the manner described does not take place at the extreme anterior end of the foregut. A small gut pocket extends rostral to the mouth. This so-called preoral gut rapidly becomes less conspicuous after the breaking through of the oral plate. The small depression which in older embryos marks its location is known as *Seessel's pocket* (Fig. A-39). Even this small depression eventually disappears altogether. Its importance lies wholly in the fact that its margin indicates for some time the place at which ectoderm and endoderm originally became continuous in the formation of the oral opening.

The Pharynx Caudal to the oral opening the foregut has become flattened dorsoventrally and widened laterally to form the pharynx. On either side the pharyngeal lumen shows a series of extensions or bays known as the *pharyngeal pouches* (Fig. A-40). Each pharyngeal pouch lies opposite an external *branchial groove* (Fig. A-43B). This leaves in these areas only a thin layer of tissue (the *branchial plate*) separating the pharyngeal lumen from the outside. This layer is composed internally of endoderm and externally of the ectoderm of the bottom of the branchial groove (Fig. A-43B). Between adjacent branchial grooves or clefts the lateral walls are greatly thickened and filled with closely packed mesenchymal cells (Fig. A-41B). These thickened areas are known as the *branchial arches.* One of the most important relationships to grasp in the study of young embryos is the manner in which the *aortic arches* lie embedded in the tissue of the branchial arch of corresponding number. The stereogram appearing in Fig. A-42 was planned especially to emphasize these relationships.

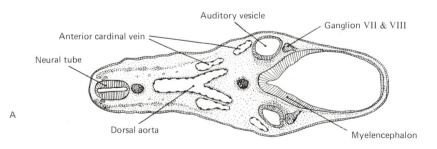

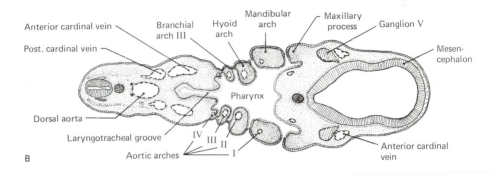

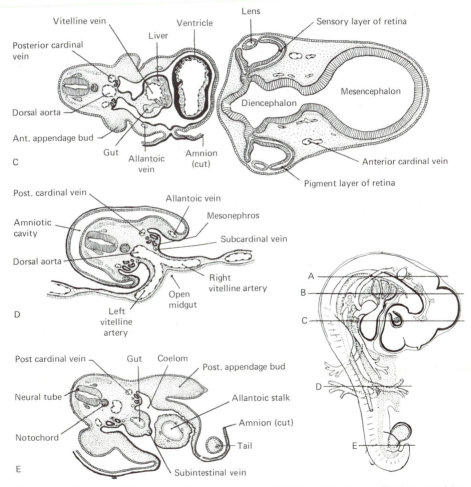

Fig. A-41 Diagrams of five representative transverse sections of a 3-day chick. The location of the sections is indicated on the small outline sketch of the entire embryo.

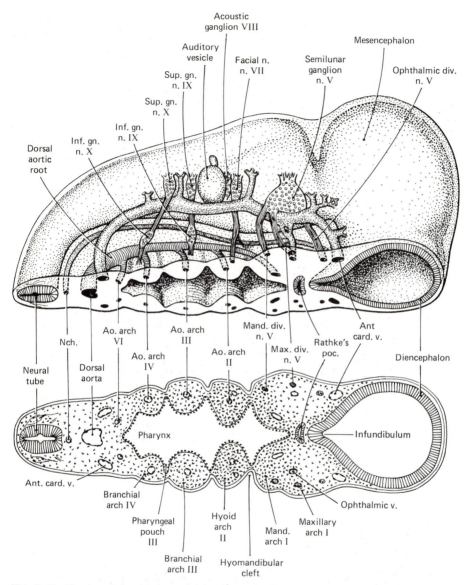

Fig. A-42 The lower drawing is a slightly schematized representation of a transverse section through the pharyngeal region of a 3½-day chick. This section is keyed to a stereogram showing cephalic structures above the level of the section. The illustration was planned primarily to show the relations of the aortic arches to the brancial arches. *(Method of presentation suggested by some of the drawings in Huettner's* Fundamentals of Comparative Embryology of the Vertebrates, The Macmillan Company, New York.) Abbreviations: *Mand. div. N. V,* mandibular division of fifth (trigeminal) cranial nerve; *Max. div. N. V,* maxillary division of fifth nerve; *Nch.,* notochord; *Rath. poc,* Rathke's pocket.

The Pharyngeal Derivatives   Several structures which arise in the pharyngeal region do not become parts of the digestive system. Nevertheless, the origin of their epithelial portions from foregut endoderm and their early association with this part of the gut tract make it convenient to consider them in connection with the digestive system.

The *thyroid gland* arises from the floor of the pharynx as a median diverticulum which makes its appearance at a cephalocaudal level between the first and second pair of pharyngeal pouches (Figs. A-39 and A-40). Toward the end of the fourth day the thyroid evagination has become saccular and retains its connection with the pharynx only by a narrow opening at the root of the tongue known as the *thyroglossal duct.*

The two pairs of *parathyroid glands* also arise as budlike outgrowths from the pharynx. One pair of parathyroids is budded off the caudal faces of the third pharyngeal pouches; the other pair arises in a similar manner from the fourth pouches. These parathyroid primordia are, however, not usually recognizable until stages of development later than those under consideration here.

The thymus of the chick is barely indicated, if present at all, on the fourth day of incubation. It takes its origin primarily from diverticula arising from the posterior faces of the third and fourth pharyngeal pouches.

The Trachea and Lungs   The first indication of the formation of the respiratory system appears in 3-day chicks as a midventral groove in the pharynx. Beginning just posterior to the level of the fourth pharyngeal pouches and extending caudad (Figs. A-39 and A-41B), this *laryngotracheal groove* deepens rapidly and by closure of its dorsal margins becomes separated from the pharynx except at its cephalic (laryngeal) end. The tube thus formed is the *trachea* (Fig. A-40), and the opening which persists between the laryngeal end of the trachea and the pharynx is the *glottis.* The original endodermal evagination gives rise only to the epithelial lining of the trachea, the supporting structures of the tracheal walls being derived from the surrounding mesenchyme.

The tracheal evagination grows caudad and bifurcates to form a pair of lung buds. In the caudolateral growth of the primordial lung buds the surrounding splanchnic mesoderm is pushed ahead of them and comes to constitute their outer investment (Fig. A-43D). The endodermal buds give rise only to the epithelial lining of the bronchi and the air passages and air chambers of the lungs. The connective-tissue stroma of the lungs is derived from mesenchyme immediately surrounding the endodermal buds, and their pleural covering is formed from their primary investment of splanchnic mesoderm (Figs. A-43D, 14-9, and 14-10).

The Esophagus and Stomach   Immediately caudal to the glottis is a narrowed region of the foregut which becomes the esophagus, and farther caudally a slightly dilated region which becomes the stomach (Fig. A-40). The concentration of mesenchymal cells about the endoderm of the esophageal and gastric regions foreshadows the formation of their muscular and connective tissue coats (Fig. A-43C and E).

The Liver   The liver arises as a diverticulum from the ventral wall of the gut immediately caudal to the stomach at about the end of the second day (22 somites). It appears just as the part of the gut from which it arises is acquiring a floor by the concrescence of the margins of the anterior intestinal portal. As a result the hepatic evagination is located for a short time on the lip of the intestinal portal and grows cephalad toward the fork where the vitelline veins enter the sinus venosus. As closure of

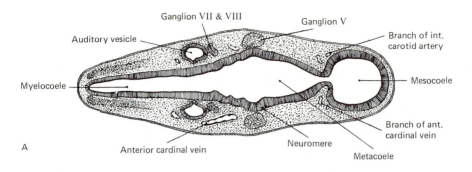

A

Ganglion VII & VIII

Auditory vesicle

Ganglion V

Branch of int. carotid artery

Myelocoele

Mesocoele

Branch of ant. cardinal vein

Anterior cardinal vein

Neuromere

Metacoele

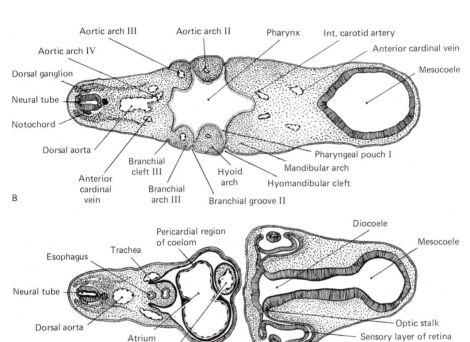

B

Aortic arch III

Aortic arch II

Pharynx

Int. carotid artery

Aortic arch IV

Anterior cardinal vein

Dorsal ganglion

Mesocoele

Neural tube

Notochord

Dorsal aorta

Branchial cleft III

Anterior cardinal vein

Branchial arch III

Hyoid arch

Branchial groove II

Hyomandibular cleft

Mandibular arch

Pharyngeal pouch I

C

Pericardial region of coelom

Diocoele

Trachea

Mesocoele

Esophagus

Neural tube

Dorsal aorta

Atrium

Truncus arteriosus

Ventral body wall

Lens

Optic stalk

Sensory layer of retina

Pigment layer of retina

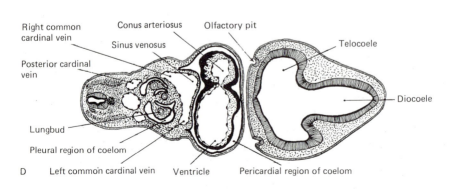

D

Right common cardinal vein

Conus arteriosus

Olfactory pit

Sinus venosus

Telocoele

Posterior cardinal vein

Lungbud

Pleural region of coelom

Left common cardinal vein

Ventricle

Pericardial region of coelom

Diocoele

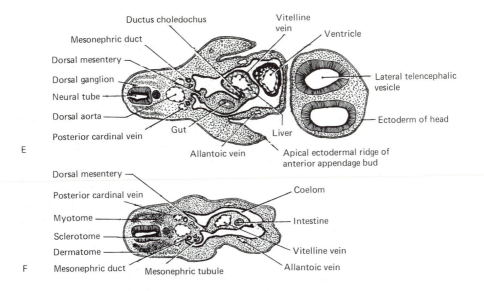

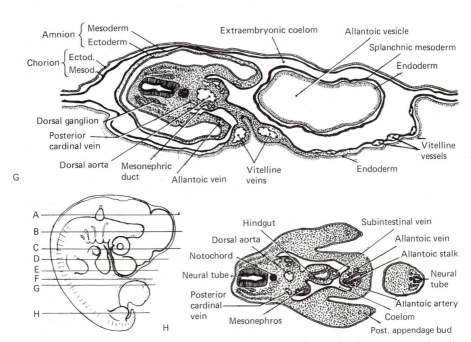

Fig. A-43 Diagrams of transverse sections of a 4-day chick. The location of the sections is indicated on a small outline sketch of the entire embryo.

the gut floor is completed, the hepatic diverticulum comes to lie in its characteristic position in the ventral wall of the gut. In embryos of 4 days the original evagination has grown out in the form of branching cords of cells and has become quite massive (Fig.

A-40). In its growth the liver pushes ahead of it the splanchnic mesoderm which surrounds the gut, with the result that the liver from its first appearance is invested by mesoderm (Figs. A-39, A-41C, and A-43E).

The proximal portion of the original evagination remains open to the intestine and serves as the duct of the liver. This primitive duct later undergoes regional differentiation and becomes in the adult the *common bile duct,* the *hepatic* and *cystic ducts,* and the *gallbladder* (Fig. 14-6). The cellular cords which bud off from the diverticulum become the secretory units of the liver *(hepatic tubules).*

The same process of concrescence that closes the floor of the foregut involves that proximal portion of the vitelline veins which, when they first appear, lie in the lateral folds of the anterior intestinal portal (Fig. A-30). As the intestinal portal moves caudad in the lengthening of the foregut, the proximal portions of the veins are brought together in the midline and become fused. The fusion extends caudad nearly to the level of the yolk stalk (Figs. A-46 and A-47). Beyond this point they retain their original paired condition. In its growth the liver surrounds the fused portion of the vitelline veins (Figs. A-41C and A-43E). This early association of the vitelline veins with the liver foreshadows the way in which the proximal part of the returning vitelline circulation is to be involved in the establishment of the hepatic portal circulation of the adult.

The Pancreas   The pancreas is derived from evaginations appearing in the walls of the intestine at the same level as the hepatic diverticulum. In birds there are three pancreatic buds, a single median dorsal and a pair of ventrolateral buds. The dorsal evagination appears at about 72 hours, the ventrolateral evaginations toward the end of the fourth day. The dorsal pancreatic bud arises directly opposite the diverticulum (Fig. A-40) and grows into the dorsal mesentery (Fig. A-51G). The ventrolateral buds arise close to the point where the duct of the liver connects with the intestine, so that the ducts of the liver and the ventral pancreatic ducts open into the intestine by a common duct *(ductus choledochus).* Later in development the masses of cellular cords derived from the three pancreatic primordia grow together and become fused into a single glandular mass, but in birds, usually two, and in rare cases all three, of the original ducts persist in the adult.

The Midgut Region   In chicks of 4 days the midgut has been practically replaced by the extension of the foregut and hindgut. It consists only of that restricted region where the yolk stalk leads from the gut tract to the yolk sac (Figs. A-39 and A-40).

The Cloaca   The beginning of the formation of the cloaca is indicated in chicks of 4 days' incubation by a dilation of the posterior portion of the hindgut (Fig. A-40). Although extensive differentiation in the cloacal region does not appear until later in development, certain of its fundamental relationships are established at this stage.

The cloaca of an adult bird is the common chamber into which the intestinal contents, the urine, and the products of the reproductive organs are received for discharge. The first appearance of the cloaca in the embryo as a dilated terminal portion of the gut establishes at the outset the relationship between cloaca and intestine that persists in the adult bird.

Arising from the dorsal wall of the cloaca in birds is an endodermal outgrowth known as the *bursa of Fabricius.* Later in development it becomes heavily infiltrated with lymphoid cells. Its function long remained obscure. It is now recognized that the bursa of Fabricius is one of the major components of the immune defense system of birds and that if it is extirpated, the ability of the bird to produce humoral antibodies is sharply reduced (Glick and Chang, 1957).

Although the urinary system is not at this stage developed to conditions which resemble those in the adult, the parts of it which have been established are already definitely associated with the cloaca. The proximal portion of the allantoic stalk, which is the homologue of the urinary bladder of mammals, opens directly into the cloaca (Fig. A-40).

The ducts which drain the developing excretory organs also open into the cloacal region on either side of the allantoic stalk. There is at this stage but little indication of the formation of the gonads. The relation of the sexual ducts to the cloaca can be discerned only by the study of older embryos.

The Circulatory System

The Vitelline Circulation The earliest indication of blood and blood-vessel formation is at the chick's source of food supply. As we have seen, blood islands appear in the extraembryonic splanchnopleure of the yolk sac toward the end of the first day of incubation and rapidly become differentiated to form vascular endothelium enclosing central clusters of primitive blood corpuscles (Fig. A-21). By extension and anastomosing of neighboring islands a plexus of blood channels is formed in the yolk sac. Further extension of the vitelline plexus brings it into communication with the main vitelline veins which have been developed in the embryo as caudal extensions of the endocardial primordia (Figs. A-18 and A-23).

Toward the end of the second day of development the vitelline arteries establish communication between the dorsal aortae and the vitelline plexus (Figs. A-27 and A-28). There is now a system of open channels leading from the embryo to the yolk sac and back again to the embryo. With the completion of these channels the heart, which has for some hours been building up the efficiency of its pulsation, is able to set the blood in circulation. At about the fortieth hour of incubation the blood cells formed in the yolk sac are for the first time carried into the body of the embryo.

The course of the vitelline circulation in chicks of 4 days is shown diagrammatically in Figs. A-44 and A-45. Circulating in the rich plexus of small vessels on the yolk, the blood finally makes its way either directly into one of the larger vitelline veins or into the sinus terminalis, which acts as a collecting channel, and then through the sinus terminalis to one of the vitelline veins. The vitelline veins converge toward the yolk stalk, where they empty into the main vitelline veins. These veins, at first paired throughout their entire length, have been brought together proximally by the closure of the ventral body wall and become fused to form a median vessel within the body of the embryo. Through this vessel the blood returning from the vitelline circuit eventually reaches the heart (Figs. A-46 and A-47). In the heart the blood of the vitelline, intraembryonic, and allantoic circulations is mingled. The mixed blood passes out by the ventral aorta and the aortic arches into the dorsal aorta. Leaving the dorsal aorta through the vitelline arteries, a certain portion of this blood is returned to the yolk sac for another circuit through it.

The Allantoic Circulation The allantoic arteries arise by the prolongation and enlargement of a pair of ventral segmental vessels arising from the aorta at the level of the allantoic stalk. Their size increases rapidly as the allantois increases in extent. From them the blood is distributed in a rich plexus of vessels which ramify in the mesoderm of the allantois (Figs. 7-6C and A-47).

The situation of the allantois directly beneath the porous shell is such that the blood

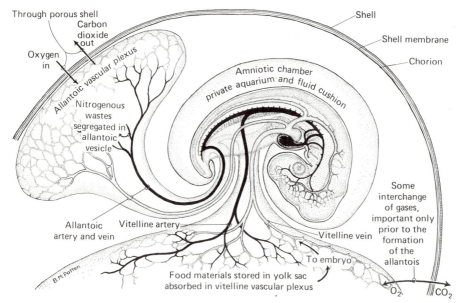

Fig. A-44 Schematic diagram showing arrangement of main circulatory channels in a young chick embryo. The sites of some of the extraembryonic interchanges important in its bioeconomics are indicated by the labeling. The vessels within the embryo carry food and oxygen to all its growing tissues and relieve them of the waste products incident to their metabolism. *(From Patten, 1951, The Am. Scientist, vol. 39.)*

can carry on interchange of gases with the outside air (Fig. A-44). It is in the rich plexus of small allantoic vessels where the surface exposure is very great that the blood gives off its carbon dioxide and takes up oxygen.

At a later stage of development the ducts of the embryonic excretory organs open into the allantoic stalk near its cloacal end. As the excretory organs become functional, the allantoic vesicle becomes the repository for the nitrogenous waste materials eliminated through them. The watery portion of the waste materials is passed off by evaporation. The remaining solids are deposited in the allantoic vesicle. They accumulate in the extraembryonic portion of the allantois and remain there until that portion of the allantois is discarded at the close of embryonic life.

The blood from the allantois is collected and returned to the heart by way of the allantoic veins. From the distal portion of the allantois the smaller veins converge and unite into two main vessels, right and left, which enter the body of the embryo with the allantoic stalk (Fig. A-44). After their entrance into the body the allantoic veins extend cephalad in the lateral body walls (Figs. A-43, E–G, and A-46). They enter the sinus venosus on either side of the entrance of the vitelline vein.

The Intraembryonic Circulation The earliest vessels of the intraembryonic circulation to appear are the large vessels communicating with the heart. In chicks of 33 hours the ventral aorta leads off from the heart cephalically and bifurcates ventral to the pharynx, giving rise to a single pair of aortic arches. The aortic arches pass dorsad around the anterolateral walls of the pharynx and are continued caudally along the dorsal wall of the gut as the paired dorsal aortae (Figs. A-20 and A-26).

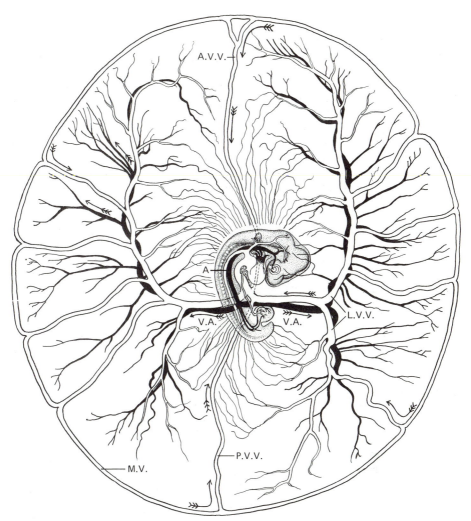

Fig. A-45   Diagram to show course of vitelline circulation in chick of about 4 days. *(After Lillie.)* For the intraembryonic vessels, see Figure A-46. Abbreviations: *A,* dorsal aorta; *A. V. V.,* anterior vitelline vein; *L. V. V.,* lateral vitelline vein; *M. V.,* marginal vein (sinus terminalis); *P. V. V.,* posterior vitelline vein; *V. A.,* vitelline artery. The direction of blood flow is indicated by arrows.

When, toward the end of the second day of incubation, branchial clefts and branchial arches appear, the original pair of aortic arches comes to lie in the mandibular arch. In each of the branchial arches posterior to the mandibular, new aortic arches are formed connecting the ventral aortae with the dorsal aortae. By 55 hours three pairs of aortic arches are present and a fourth is beginning to form (Figs. A-30 and A-33). By the fourth day of incubation two more pairs of aortic arches have appeared posterior to the four formed in 55- to 60-hour chicks (Fig. A-46). If they appear at all, the fifth aortic arches are very small. Frequently they are merely a tiny loop on the fourth or sixth arch, and they soon disappear altogether. The first and second pairs of aortic arches have by the

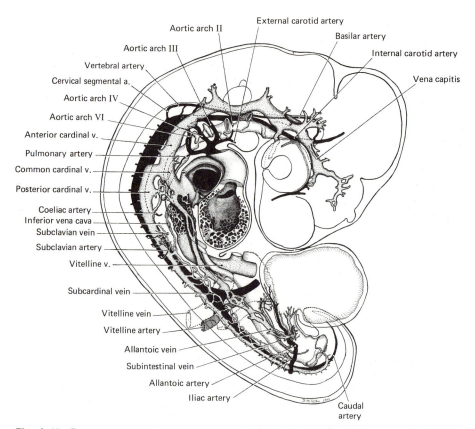

Fig. A-46 Reconstruction of circulatory system of 4-day chick (original X51, reproduced X18). From the same embryo as that represented in Figure A-40. These illustrations should be studied together.

fourth day suffered a marked diminution in size, which is indicative of their final disappearance. In many embryos of this age range the first arches, and in a few the second, also, have disappeared as main channels (Fig. A-46). This leaves only the third, fourth, and sixth pairs of aortic arches. These arches persist intact for some time, and parts of them remain permanently, being incorporated in the formation of the arch of the aorta and the main vessels arising from it and in the proximal parts of the pulmonary arteries.

In embryos late in the second day plexiform channels extend from the first aortic arch toward the forebrain (Figs. A-26 and A-34), foreshadowing the formation of the *internal carotid arteries*. By 60 hours the internal carotid artery is established as a definite trunk extending toward the brain from the point where the first aortic arch turns into the dorsal aortic root (Fig. A-33). With the regression of the first two aortic arches during the fourth day of development, the dorsal aortic root at this level is appropriated as a feeder to the original internal carotid artery, thereby making it appear to originate where the third aortic arch merges with the dorsal aortic root (Fig. A-46). The meshwork of small vessels fed by the internal carotid artery and developing in close relation to the brain is

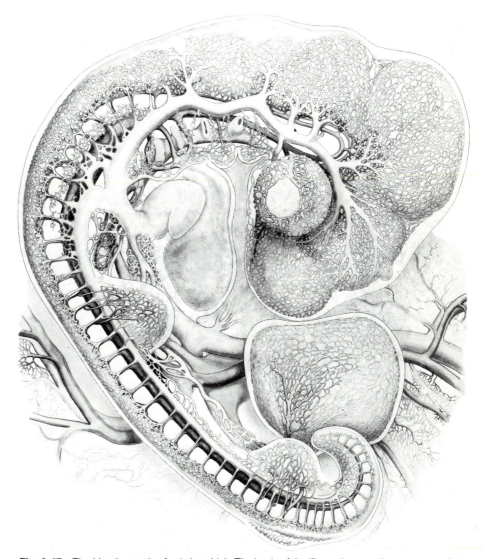

Fig. A-47 The blood vessels of a 4-day chick. The basis of the illustration was the same wax-plate reconstruction from which Figure A-46 was drawn. The smaller vessels and the primordial capillary plexuses were added from injected specimens. The labeled diagram of Figure A-46 will serve as a means of identifying the main vessels. (See colored insert.)

one of extraordinary richness (Fig. A-47). In embryos of this age it is possible to see suggestions of certain of the main named branches characteristic of the adult, but for the most part they are represented only by primordial capillary plexuses.

The regression of the first two aortic arches also plays a prominent role in the formation of the *external carotid arteries*. After these two arches lose connection with the dorsal aortic roots the parts of the ventral aortic roots which formerly fed them persist

(Fig. A-46). These vessels represent the main stems of the external carotid arteries which later develop many important branches supplying the region of the mandible and the front of the neck and the face.

In reptiles, birds, and mammals the main adult vessels which connect the heart with the dorsal aorta are derived from the fourth pair of aortic arches of the embryo. The paired condition of these arches persists as the adult condition in reptiles, but in birds and mammals one of the arches degenerates before the end of embryonic life. In birds the left fourth aortic arch degenerates, leaving the right one as the arch of the adult aorta; in mammals the right fourth aortic arch degenerates, leaving the left incorporated in the arch of the adult aorta.

The dorsal aortae, at first paired, later become fused to form a median vessel. The fusion begins at about the level of the tenth or eleventh somite and progresses cephalad and caudad. Fusion extends cephalad only a short distance, never involving the region of the dorsal aortic roots where they receive the aortic arches. Caudally the aortae eventually become fused throughout their entire length.

In the adult, three vessels arising from the dorsal aorta supply the abdominal viscera. They are the *coeliac,* the *superior mesenteric,* and the *inferior mesenteric arteries.* In 4-day chicks these vessels are usually represented only by the vitelline arteries. The vitelline arteries arise as paired vessels (Fig. A-30), but in the closure of the ventral body wall of the embryo they are brought together and fused to form a single vessel which runs in the mesentery from the aorta to the yolk stalk (Fig. A-41D). With the atrophy of the yolk sac the proximal part of the vitelline artery persists as the superior mesenteric of the adult. The coeliac and the inferior mesenteric arteries arise from the aorta independently, usually at a somewhat later stage. Occasionally, however, the coeliac artery can be identified in 4-day embryos (Fig. A-46).

The *anterior* and *posterior cardinal veins* are the principal afferent systemic vessels of the early embryo. They appear toward the end of the second day as paired vessels extending cephalically and caudally on either side of the midline. At the level of the heart the anterior and posterior cardinal veins of the same side of the body become confluent in the common cardinal veins, or ducts of Cuvier, and turn ventrad to enter the sinus venosus (Figs. A-26 and A-33). Chicks of 4 days show little change in the relationships of the cardinal veins (Figs. A-46 and A-47). Later in development the proximal ends of the anterior cardinals become connected by the formation of a new transverse vessel and empty together into the right atrium of the heart. Their distal portions remain in the adult as the principal afferent vessels *(internal jugular veins)* of the cephalic region.

The posterior cardinals lie at first in the angle between the somites and the lateral mesoderm (Fig. A-31D and E). When the mesonephroi develop from the intermediate mesoderm, the posterior cardinal veins lie just dorsal to them throughout their length (Figs. A-41D and E, and A-43E to A-43H). Situated ventrally in the mesonephroi are small irregular vessels roughly paralleling the posterior cardinals and anastomosing freely with them. These are the subcardinal veins (Fig. A-46). They are a relic of the renal portal circulation of more primitive ancestral forms and are of importance in the embryos of birds and mammals chiefly because of the way they are involved in the formation of the posterior vena cava. In 4-day chicks only the upper part of the posterior vena cava is indicated. It appears as a slender vessel extending from the right subcardinal vein cephalically to anastomose with veins within the liver. The formation of the vena cava, in part as a new vessel and in part by appropriation of already existing vessels, and the changes by which the posterior cardinals become reduced and broken up to form small vessels with new associations is considered in Chap. 17.

The Heart   As a review of early morphogenesis of the heart, its early changes in shape have been summarized in Fig. A-48. As the heart is lengthened, the unattached ventricular region becomes dilated and is bent out of the midline toward the embryo's right, while the fixed outlet of the truncus arteriosus and the firmly anchored sinoatrial end are held in their original median position. This bending of the heart to form a U-shaped tube begins to be apparent in embryos of 30 hours and becomes rapidly more conspicuous, until by 40 hours the ventricular region of the heart lies well to the right of the embryo's body (Fig. A-48A to A-48D).

The lateral bending of the heart attains its greatest extent at about 40 hours of incubation. At this stage torsion of the body of the embryo changes the mechanical limitations in the cardiac region. As the embryo comes to lie on its left side the heart is no longer pressed against the yolk (cf. Fig. A-48D and E). As a result the bent ventricular region begins to swing somewhat ventrad and lies less closely against the body of the embryo.

At about this stage of development the closed part of the U-shaped bend becomes twisted on itself to form a loop (Figs. A-48G to A-48I, A-49G to A-49I, and A-50G to A-50I). In the formation of the loop the atrial region is forced slightly to the left (i.e., toward the yolk) and the truncus arteriosus is thrown across the atrial region by being bent to the right (i.e., away from the yolk) and then dorsocaudad. The ventricular region constituting the closed end of the loop swings dorsad and toward the tail, possibly being crowded in this direction by the increasing flexion of the cephalic part of the embryo (Fig. A-48H and I). Thus the original cephalocaudal relations of the atrial and ventricular regions are reversed, and the atrial region which was at first caudal to the ventricle now lies cephalic to it as it does in the adult heart.

The atrial region and the ventricular region, which formerly were continuous without any line of demarcation, are by this time beginning to be marked off from each other by a constriction (Fig. A-49I). As both the atrium and the ventricle become enlarged, this constriction is accentuated (Fig. A-49L). The constricted region is now termed the *atrioventricular canal*.

During the fourth day the truncus arteriosus becomes closely applied to the ventral surface of the atrium. As the atrium grows it tends to expand on either side of the depression made in it by the pressure of the truncus (Figs. A-49J to A-49L and A-50J to A-50L). These lateral expansions are the first indication of the division of the atrium into right and left chambers, which are later completely separated from each other. At the same time a slight longitudinal groove appears in the surface of the ventricle (Fig. A-49L). This *interventricular groove* indicates the beginning of the separation of the ventricle into right and left chambers.

Concurrent with the formation of the atrioventricular constriction is the development of a groove between the atrial region and the sinus venosus into which the great veins empty the blood they are returning to the heart (Figs. A-48J to A-48L and A-50H to A-50L). With the progressive deepening of the sinoatrial groove the sinus is increasingly clearly delimited. This establishes the basic regional divisions of the heart—in order of blood flow, sinus venosus, atrium, and ventricle.

The Urinary System

It is possible to recognize pronephric tubules in chick embryos of 38 to 40 hours (Fig. 16-2A). In chicks of 50 to 55 hours the mesonephric ducts and primordial tubules are clearly identifiable just lateral to the somites (Fig. A-31D and E). By the fourth day mesonephric tubules are well under way in their development. The metanephric tubules,

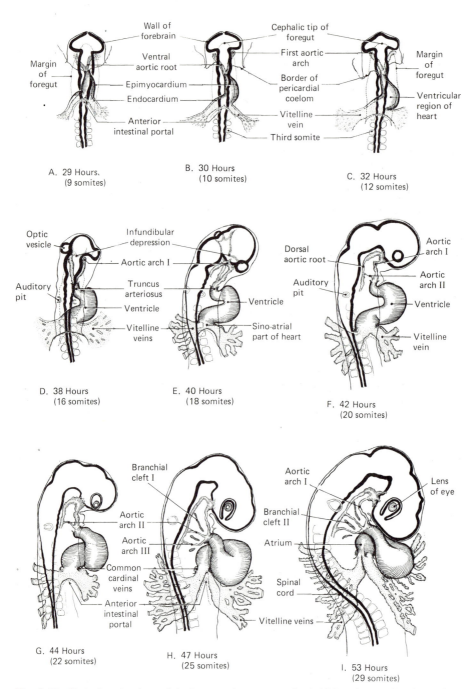

A. 29 Hours.
(9 somites)

B. 30 Hours
(10 somites)

C. 32 Hours
(12 somites)

D. 38 Hours
(16 somites)

E. 40 Hours
(18 somites)

F. 42 Hours
(20 somites)

G. 44 Hours
(22 somites)

H. 47 Hours
(25 somites)

I. 53 Hours
(29 somites)

Fig. A-48 Projection drawings of the heart and great vessels of chick embryos of various ages. Some of the major topographical features of the embryos have been outlined to emphasize the relations of the heart within the body. Abbreviations. *roman numerals I-VI,* aortic arches of the designated numbers; *Sin. ven.,* sinus venosus; *Vent.,* ventricle.

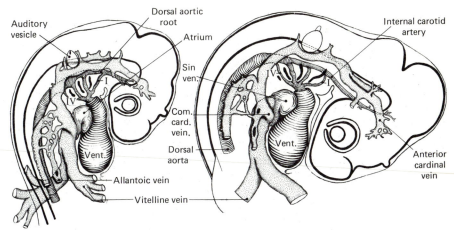

J. 65 HOURS (33 SOMITES)

K. 76 HOURS (38 SOMITES)

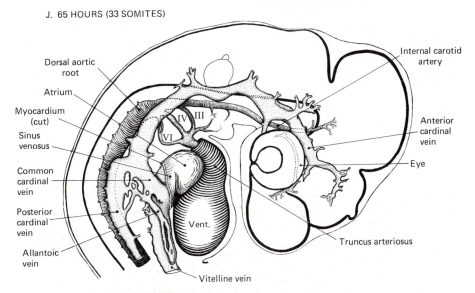

L. 100 HOURS (45 SOMITES)

which constitute the excretory units of the permanent kidneys, do not appear until considerably later in development, and the genital organs which become intimately interrelated with the urinary organs are also relatively late in making their appearance. The later stages in the development of the urinary system as well as the entire sequence of stages in the formation of the genital organs are covered in Chap. 16.

The Coelom and Mesenteries

In adult birds and mammals the body cavity consists of three regions, pericardial, pleural, and peritoneal. The *pleural cavities* are paired, each of the pleural chambers being a laterally situated sac containing one of the lungs. The *pericardial chamber* containing the

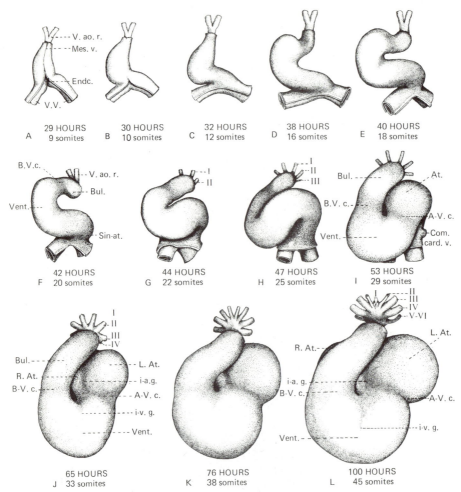

Fig. A-49 Ventral views of the heart at various stages to show its changes of shape and its regional differentiation. All the drawings were made from dissections with the aid of camera lucida outlines. In the stages represented in Figures E–H, torsion of the embryo's body is going on at the level of the heart. Since torsion involves the more cephalic regions first and progresses caudad, the transverse axis of the body of the embryo is at different inclinations to the yolk at the cephalic end and at the caudal end of the heart. In drawing these figures their orientation was taken from the body at the level of the conus region of the heart; the sinus region therefore appears inclined. *(From Patten, 1922,* Am. J. Anat., vol. 30.) Abbreviations: I-VI, aortic arches I-VI; *At.,* atrium; *L. At.,* left atrium; *R. At.,* right atrium; *A-V.c.,* atrioventricular constriction; *Bul.,* bulbus cordis; *B-V.c.,* bulboventricular constriction; *Com. card. v.,* common cardinal vein; *Mes. v.,* ventral mesentery; *Sin-at.,* sino-atrial region; *Vent.,* ventricle; *V. ao. r.,* ventral aortic root; *V. v., vitelline vein.*

heart and the *peritoneal chamber* containing the viscera, other than the lungs and heart, are unpaired. These regions of the adult body cavity are formed by the reshaping and partitioning of the primary body cavity, or coelom, of the embryo.

In the chick the coelom arises by a splitting of the lateral mesoderm of either side of the body (Fig. A-51A and B). It is, therefore, at first a paired cavity. Unlike the coelom of

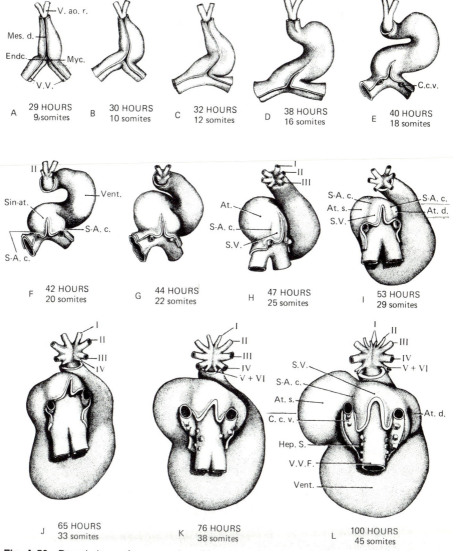

Fig. A-50 Dorsal views of same series of hearts as that shown in Figure A-49 in ventral view. Abbreviations: *I–VI,* aortic arches I to VI; *At. d.,* atrium, right; *At. s.,* atrium, left; *C. c. v.,* common cardinal vein; *Endc.,* endocardium; *Hep. S.,* stubs of some of the larger hepatic sinusoids; *Mes. d.,* dorsal mesocardium; *Myc.,* cut edge of epimyocardium; *S-A. c.,* sinoatrial constriction; *Sin-at.,* sinoatrial region (before its definite division); *S. V.,* sinus venosus; *V. ao. r.,* ventral aortic roots; *Vent.,* ventricle; *V. V.,* vitelline veins; *V. V. F.,* fused vitelline veins. *(From Patten, 1922, Am. J. Anat., vol. 30.)*

some of the more primitive vertebrates, the coelom of the chick never shows any indications of segmental pouches corresponding in arrangement with the somites. Instead, the right and left coelomic chambers extend cephalocaudally without interruption through the entire lateral plates of mesoderm.

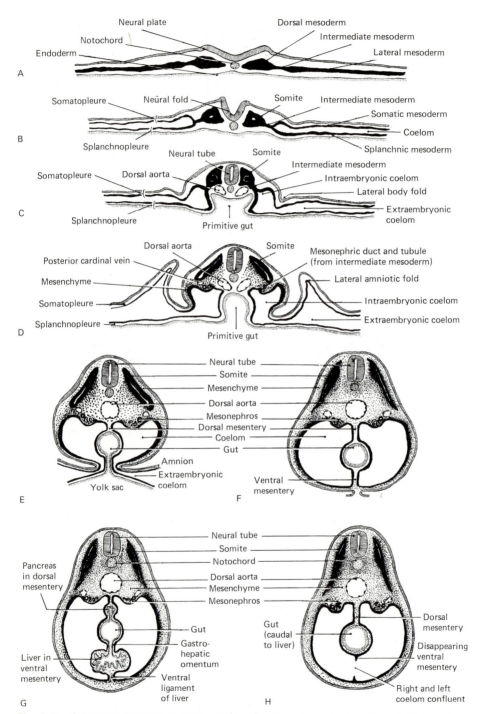

Fig. A-51 Schematic diagrams of cross sections at various stages to show the establishment of the coelom and mesenteries. For explanation see text.

The coelomic chambers are not limited to the region in which the body of the embryo is developing. They extend on either side into the mesoderm, which, in common with the other germ layers, spreads out over the yolk surface. Large parts of the primitive coelomic chambers thus come to be extraembryonic in their associations (Chap. 7 and Figs. 7-1 and 7-4). The portion of the coelom which gives rise to the embryonic body cavities is first marked off by the series of folds which separate the body of the embryo from the yolk (Fig. A-51C and D). As the closure of the ventral body wall progresses (Fig. A-51E and F) the embryonic coelom becomes completely separated from the extraembryonic. The delayed closure of the ventral body wall in the yolk-stalk region results in the embryonic and extraembryonic coelom retaining an open communication at this point for a considerable time after they have been completely separated elsewhere.

The same folding process which establishes the ventral body wall completes the gut ventrally (Fig. A-51C to A-51F). Meanwhile the right and left coelomic chambers are expanded mesiad. As a result the newly closed gut comes to lie suspended between the two layers of splanchnic mesoderm which constitute the mesial walls of the right and left coelomic chambers. The double layer of splanchnic mesoderm which thus becomes apposed to the gut and supports it in the body cavity is known as the *primary mesentery.* The part of the mesentery dorsal to the gut, suspending it from the dorsal body wall, is the *dorsal mesentery,* and the part ventral to the gut, attaching it to the ventral body wall, is the *ventral mesentery.*

When the dorsal and ventral mesenteries are intact they constitute a complete membranous partition dividing the body cavity into right and left halves. The primary dorsal mesentery persists in large part, but the ventral mesentery soon is extensively resorbed (Fig. A-51H), bringing the right and left coelomic chambers into confluence ventral to the gut and establishing the unpaired condition of the body cavity characteristic of the adult.

In their relation to the other mesenteries of the body, the mesocardia may be regarded as special regions of the ventral mesentery. In the most cephalic part of the body cavity, the gut lies embedded in the dorsal body wall instead of being suspended by a dorsal mesentery as it is farther caudally (cf. Figs. A-22E and A-51F). A ventral mesentery is, however, developed in the same manner anteriorly as it is posteriorly, and when the heart is formed it is suspended in the most anterior part of this ventral mesentery. The dorsal and ventral mesocardia may, therefore, be thought of as the parts of the primary ventral mesentery lying dorsal to the heart and ventral to the heart, respectively (Fig. A-22D). When the ventral mesocardium, and a little later the dorsal mesocardium, break through, the primary right and left coelomic chambers become confluent to form the pericardial region of the body cavity (Fig. A-26).

The entire dorsal mesentery persists and forms the membranes supporting the digestive tube. In the adult its different regions are named according to the parts of the digestive tube with which they are associated; for example, *mesogaster,* that part of the dorsal mesentery which suspends the stomach; *mesocolon,* that part of the dorsal mesentery supporting the colon, etc. The separation of the body cavity into pericardial, pleural, and peritoneal chambers is accomplished by the formation of septa growing in from the body wall at later stages of development.

Summary of Figures

Students using this text as the basis for laboratory study of chick or pig development have sometimes experienced difficulty in locating the appropriate illustrative material because of its incorporation into areas of the text concerned with a particular organ system. To increase the usefulness of this text in the laboratory, this Appendix summarizes the figures illustrating key stages in the development of both the chick and the pig. The chick material is subdivided into groups of both whole mounts and sectioned preparations, whereas the pig material is largely based upon serial sections as well as reconstructions derived from them.

Chick Whole Mounts

Figure Number	Incubation Age
5-10A	3–4 hours
5-10B	5–6 hours
5-10C	7–8 hours
5-10D	10–12 hours
5-12	16 hours
A-3	18 hours
A-7	20 hours
A-9	22–23 hours
A-11	24 hours

A-10	25–26 hours
A-12	25–26 hours—ventral view, cephalic region
A-15	27–28 hours
A-16	29–30 hours—ventral view, cephalic and cardiac region
A-18	33 hours
A-19	38 hours—dorsal view, cephalic and cardiac region
A-26	38 hours—lateral view, dissection
7-3	40 hours
A-27	43 hours
A-28	44 hours—vitelline circulation
A-34	45 hours—lateral view, cephalic and cardiac region
A-30	50 hours—ventral view, dissection
A-35	50 hours—lateral view, cephalic and cardiac region
A-29	55 hours
A-33	60 hours—ventrolateral view, dissection
A-36	72 hours
A-38	96 hours
A-40	96 hours—lateral view, reconstruction of nervous, digestive, and urinary systems
A-46, A-47	96 hours—lateral view, reconstruction of circulatory system

Chick Sections

18 Hours
Figure A-4—longitudinal and transverse sections
24 Hours
Figure A-13—longitudinal and transverse sections
Figure A-14—stereogram
33 Hours
Figure A-25—longitudinal and transverse sections
55 Hours
Figure A-31—transverse sections
72 Hours
Figure A-41—transverse sections
96 Hours
Figure A-39—longitudinal section
Figure A-43—transverse sections

Pig

Figure Number	*Length*
	5 mm
8-9	whole mount
17-7E and F	branchial arches, lateral aspect
17-9B and C	branchial arches, ventral aspect
8-13	sagittal section
8-18	parasagittal section
8-19	parasagittal section
8-14	transverse sections through 8 levels

References for Collateral Reading

TEXTS AND GENERAL TREATISES

Elementary Embryologies

Austin, C. R., and R. V. Short, 1972. *Reproduction in Mammals.* Cambridge University Press, Cambridge. Vols. 1–6.

Balinsky, B. I., 1975. *An Introduction to Embryology.* W. B. Saunders Company, Philadelphia. 4th ed., 648 pp.

Bellairs, R., 1971. *Developmental Processes in Higher Vertebrates.* Logos Press, London. 366 pp.

Berrill, N. J., and G. Karp, 1976. *Development.* McGraw-Hill Book Company, New York, 566 pp.

Bodemer, C. W., 1968. *Modern Embryology.* Holt, Rinehart and Winston, Inc., New York.

Davenport, R., 1979. *An Outline of Animal Development.* Addison-Wesley, Reading, Mass. 412 pp.

Deuchar, E. M., 1975. *Cellular Interactions in Animal Development.* Chapman & Hall, London. 298 pp.

Ebert, J. D., and I. M. Sussex, 1970. *Interacting Systems in Development.* Holt, Rinehart and Winston, Inc., New York. 2d ed., 338 pp.

Ede, D. A., 1978. *An Introduction to Developmental Biology.* Blackie, Glasgow. 246 pp.

Grant, P., 1978. *Biology of Developing Systems*. Holt, Rinehart and Winston, Inc., New York. 720 pp.

Ham, R. G., and M. J. Veomett, 1980. *Mechanisms of Development*. The C. V. Mosby Company, St. Louis. 843 pp.

Huettner, A. F., 1949. *Fundamentals of Comparative Embryology of the Vertebrates*. The Macmillan Company, New York. 2d ed., 309 pp.

Lash, J., and J. R. Whittaker, eds., 1974. *Concepts of Development*. Sinauer Associates, Stamford, Conn. 469 pp.

Marrable, A. W., 1971. *The Embryonic Pig*. Pitman, London. 130 pp.

Patten, B. M., 1948. *Embryology of the Pig*. McGraw-Hill Book Company, Blakiston Division, New York. 3d ed., 352 pp.

———, 1971. *Early Embryology of the Chick*. McGraw-Hill Book Company, New York. 5th ed., 284 pp.

Rugh, R., 1951. *The Frog—Its Reproduction and Development*. McGraw-Hill Book Company, Blakiston Division, New York. 336 pp.

———, 1964. *Vertebrate Embryology*. Harcourt, Brace & World, Inc., New York. 600 pp.

Saunders, J. W., 1970. *Patterns and Principles of Animal Development*. The Macmillan Company, New York. 296 pp.

Tokin, B. P., 1977. *General Embryology* (Russian). Vysshaya Shkola, Moscow. 509 pp.

Torrey, T. W., 1971. *Morphogenesis of the Vertebrates*. John Wiley & Sons, Inc., New York. 3d ed., 529 pp.

Waddington, C. H., 1956. *Principles of Embryology*. The Macmillan Company, New York. 3d printing, 1960, 510 pp.

Wessells, N. K., 1977. *Tissue Interactions and Development*. W. A. Benjamin, Inc., Menlo Park. 276 pp.

Witschi, E., 1956. *Development of Vertebrates*. W. B. Saunders Company, Philadelphia. 588 pp.

Human Embryologies

Arey, L. B., 1965. *Developmental Anatomy*. W. B. Saunders Company, Philadelphia. 7th ed., 695 pp.

Blechschmidt, E., 1961. *The Stages of Human Development before Birth*. W. B. Saunders Company, Philadelphia. 684 pp.

Corliss, C. E., 1976. *Patten's Human Embryology*. McGraw-Hill Book Company, New York. 470 pp.

Gasser, R. F., 1975. *Atlas of Human Embryos*. Harper & Row, Hagerstown, Md. 318 pp.

Hamilton, W. J., J. D. Boyd, and H. W. Mossman, 1972. *Human Embryology*. The Williams & Wilkins Company, Baltimore. 4th ed., 646 pp.

Keibel, F., and F. P. Mall, 1910-1912. *Manual of Human Embryology*. J. B. Lippincott Company, Philadelphia. Vol. I, 548 pp.; vol.II, 1032 pp.

Langman, J., 1975. *Medical Embryology*. The Williams & Wilkins Company, Baltimore. 3d ed., 421 pp.

Moore, K. L., 1977. *The Developing Human*. W. B. Saunders Company, Philadelphia. 2d ed., 411 pp.

O'Rahilly, R., 1973. *Developmental Stages in Human Embryos*. Carnegie Institution, Washington Publ. 631, 167 pp.

Rugh, R., and L. B. Shettles, 1971. *From Conception to Birth.* Harper & Row, New York. 262 pp.

Monographs and Reference Works

Abercrombie, M., J. Brachet, and T. J. King, eds., 1961–1973. *Advances in Morphogenesis.* Academic Press, Inc., New York, Vols. 1–10.

Astaurov, B. L., and others, 1975–1980. *Problems of Developmental Biology* (Russian), Izdatel. Nauka, Moscow.

Balls, M., and A. E. Wild, eds., 1975. *The Early Development of Mammals.* Cambridge University Press, Cambridge. 410 pp.

Brachet, J., 1950. *Chemical Embryology.* Interscience Publishers, Inc., New York. 533 pp.

Daniel, J. C., ed., 1971. *Methods in Mammalian Embryology.* W. H. Freeman Company, San Francisco. 532 pp.

————, ed., 1978. *Methods in Mammalian Reproduction.* Academic Press, Inc., New York. 566 pp.

Davidson, E. H., 1976. *Gene Activity in Early Development.* Academic Press, Inc., New York. 2d ed., 452 pp.

DeHaan, R. L., and H. Ursprung, eds., 1965. *Organogenesis.* Holt, Rinehart and Winston, Inc., New York. 804 pp.

Gould, S. J., 1977. *Ontogeny and Phylogeny.* Harvard University Press, Cambridge, Mass. 501 pp.

Hertwig, O., 1906. *Handbuch der Vergleichenden und Experimentellen Entwicklungslehre der Wirbeltiere.* (Edited by Dr. Oskar Hertwig and written by numerous collaborators.) Gustav Fischer, Verlagsbuchhandlung, Jena, Germany.

Karkinen-Jaaskelainen, L. Saxén, and L. Weiss, eds., 1977. *Cell Interactions in Differentiation.* Academic Press, Inc., New York. 415 pp.

Minot, C. S., 1893. A bibliography of vertebrate embryology. Memoirs, *Boston Soc. Nat. History,* **4:**487–614.

Monroy, A., and A. A. Moscona, eds., 1966–1980. *Current Topics in Developmental Biology.* Academic Press, Inc., New York. Vols. 1–14.

Needham, J., 1931. *Chemical Embryology.* The Macmillan Company, New York. 2021 pp.

Nelson, O. E., 1953. *Comparative Embryology of the Vertebrates.* McGraw-Hill Book Company, New York. 982 pp.

New, D. A. T., 1966. *The Culture of Vertebrate Embryos.* Academic Press, Inc., New York. 245 pp.

Nieuwkoop, P.D., and L. A. Sutasurya, 1979. *Primordial Germ Cells in the Chordates.* Cambridge University Press, Cambridge. 187 pp.

Poste, G., and G. L. Nicolson, eds., 1976. *The Cell Surface in Animal Embryogenesis and Development.* North-Holland Publishing Company, Amsterdam. 766 pp.

Romanoff, A. L., 1967. *Biochemistry of the Avian Embryo.* John Wiley & Sons, Inc., New York. 398 pp.

Schwalbe, E., 1906–1913. *Die Morphologie der Missbildungen des Menschen und der Tiere.* Gustav Fischer, Verlagsbuchhandlung, Jena, Germany.

Spemann, H., 1938. *Embryonic Development and Induction.* Yale University Press, New Haven. 401 pp.

Timiros, P. S., ed., 1972. *Developmental Physiology and Aging.* The Macmillan Company, New York. 692 pp.

Weiss, P., 1939. *Principles of Development. A Text in Experimental Embryology.* Holt, Rinehart and Winston, Inc., New York. 601 pp.

Willier, B. H., P. A. Weiss, and V. Hamburger, 1955. *Analysis of Development.* W. B. Saunders Company, Philadelphia. 735 pp.

Wilson, E. B., 1925. *The Cell in Development and Heredity.* The Macmillan Company, New York. 3d ed., 1232 pp.

Wilt, F. H., and N. K. Wessels, eds., 1967. *Methods in Developmental Biology.* Thomas Y. Crowell Company, New York. 813 pp.

Windle, W. F., 1940. *Physiology of the Fetus.* W. B. Saunders Company, Philadelphia. 249 pp.

Wolsky, A., ed., 1969–1980. *Monographs in Developmental Biology.* S. Karger AG, Basel. Vols. 1–13.

Young, W. C., ed., 1961. *Sex and Internal Secretions.* The Williams & Wilkins Company, Baltimore. 3d ed., vols. 1 and 2, 1609 pp.

HISTORY OF EMBRYOLOGY, TECHNIQUES, AND GENERAL CONCEPTS (CHAPTER 1)

Adelmann, H. B., 1942. *The Embryological Treatises of Hieronymus Fabricius of Aquapendente.* Cornell University Press, Ithaca, N.Y. 376 pp.

———, 1966. *Marcello Malpighi and the Evolution of Embryology.* Cornell University Press, Ithaca, N.Y. 5 vols., 2475 pp.

Baserga, R., and D. Malamud, 1969. *Autoradiography: Techniques and Application.* Hoeber-Harper, New York. 281 pp.

Berns, M. W., and C. Salet, 1972. Laser microbeams for partial cell irradiation. *Int. Rev. Cytol.,* **33:**131–156.

Briggs, R., and T. J. King, 1952. Transplantation of living nuclei from blastula cells into enucleated frogs' eggs. *Proc. Nat. Acad. Sci.,* **38:**455–463.

Brothers, A. J., 1976. Stable nuclear activation dependent on a protein synthesized during oogenesis. *Nature,* **260:**112–115.

Bullough, W. S., E. B. Laurence, O. H. Iverson, and K. Elgjo, 1967. The vertebrate epidermal chalone. *Nature,* **214:**578–580.

Carpenter, G., and S. Cohen, 1978. Biological and molecular studies of the mitogenic effects of human epidermal growth factor, in Papaconstantinou and Rutter, *Molecular Control of Proliferation and Differentiation.* Academic Press, Inc., New York. pp. 13–31.

Ebert, J. D., 1966. *Interacting Systems in Development.* Holt, Rinehart and Winston, Inc., New York. 227 pp.

Fraser, F. C., 1959. Causes of congenital malformations in human beings. *J. Chron. Dis.,* **10:**97–110.

———, H. Kalter, B. E. Walker, and T. D. Fainstat, 1954. The experimental production of cleft palate with cortisone and other hormones. *J. Cell. Comp. Physiol.,* **43,** suppl. **1:**237–259.

———, B. E. Walker, and D. C. Taylor, 1957. Experimental production of congenital cleft palate: genetic and environmental factors. *Pediatrics,* **19:**782–787.

Glücksmann, A., 1951. Cell death in normal vertebrate ontogeny. *Biol. Revs. Cambridge Phil. Soc.,* **26:**59–86.

Gospodarowicz, D., J. S. Moran, and A. L. Mescher, 1978. Cellular specificities of fibroblast growth factor and epidermal growth factor, in Papaconstantinou and Rutter, *Molecular Control of Proliferation and Differentiation.* Academic Press, Inc., New York. pp. 33–63.

Gould, G. M., and W. L. Pyle, 1937. *Anomalies and Curiosities of Medicine.* Sydenham Publishers, New York. 968 pp.

Grobstein, C., 1956. Transfilter induction of tubules in mouse metanephrogenic mesenchyme. *Exp. Cell. Res., 10*:424–440.

Gross, P. R., and G. H. Cousineau, 1964. Macromolecule synthesis and the influence of actinomycin on early development. *Exp. Cell Res., 33*:368–395.

Gurdon, J. B., 1962. The developmental capacity of nuclei taken from intestinal epithelium cells of feeding tadpoles. *J. Embryol. Exp. Morphol., 10*:622–640.

Gurwitsch, A. G., 1944. *The Theory of Biological Fields* (Russian). Izdatel. Sovietskaya Nauka, Moscow. 155 pp.

Gustafson, T., and L. Wolpert, 1963. The cellular basis of morphogenesis and sea urchin development. *Int. Rev. Cytol., 15*:139–214.

Harrison, R. G., 1907. Observations on the living developing nerve fiber. *Anat. Rec., 1*:116–118.

Hay, E. D., 1977. Embryonic induction and tissue interaction during morphogenesis, in Littlefield and deGrouchy, *Birth Defects.* Int. Congr. Series No. 432. Excerpta Medica, Amsterdam. pp. 126–140.

Konigsberg, I. R., 1963. Clonal analysis of myogenesis. *Science, 140*:1273–1284.

Lehtonen, E., 1976. Transmission of signals in embryonic induction. *Med. Biol., 54*:108–128.

Lowenstein, W. R., 1970. Intercellular communication. *Sci. Am.,* May, pp. 79–86.

Markert, C. L., ed., 1975. *Isoenzymes.* Vols. I–IV. Academic Press, Inc., New York. Vols. 1–4.

Meyer, A. W., 1936. *An Analysis of the De Generatione Animalium of William Harvey.* Stanford University Press, Stanford, Calif. 167 pp.

———, 1939. *The Rise of Embryology.* Stanford University Press, Stanford, Calif. 367 pp.

Mintz, B., 1971. Clonal basis of mammalian differentiation. In *Control Mechanisms of Growth and Differentiation. Symp. Soc. Exp. Biol., 25*:345–368.

Moscona, A., and H. Moscona, 1952. The dissociation and aggregation of cells from organ rudiments of the early chick embryo. *J. Anat., 86*:287–301.

Needham, J., 1931. *Chemical Embryology.* Cambridge University Press, Cambridge. Vols. 1–3, 2021 pp.

———, 1959. *A History of Embryology.* Cambridge University Press, Cambridge. 2d ed., 303 pp.

Oppenheimer, J. M., 1967. *Essays on the History of Embryology and Biology.* M.I.T., Cambridge, Mass. 374 pp.

Papaconstantinou, J., and W. J. Rutter, eds., 1978. *Molecular Control of Proliferation and Differentiation.* Academic Press, Inc., New York. 264 pp.

Roux, W., 1888. Beitrage zur Entwicklungsmechanik des Embryo. V. Über die künstliche Hervorbringung "halber" Embryonen durch Zerstörung einer der beiden ersten Furchungszellen, sowie über die Nachentwicklung (Postgeneration) der fehlenden Körperhalfte. *Virchows Arch., 114*:419–521.

Saunders, J. W., M. T. Gasseling, and L. C. Saunders, 1962. Cellular death in morphogenesis of the avian wing. *Dev. Biol., 5*:147–178.

Saxén, L., and S. Toivonen, 1962. *Primary Embryonic Induction.* Prentice-Hall, Inc., Englewood Cliffs, N. J. 271 pp.

Spemann, H., 1901. Ueber Korrelationen in der Entwicklung des Auges. Verh. Anat. Ges. Jena Verslag. Bonn, **15**:61–79.

———, 1912. Zur Entwicklung des Wirbeltierauges. *Zool. Jahrb., Abt. allgem. Zool.,* **32**:1–98.

———, and H. Mangold, 1924. Ueber Induktion von Embryonalanlagen durch Implantation ortfremder Organisatoren. *Arch. mikroskop. Anat. Entwmech.,* **100**:599–638.

Thompson, D. W., 1959. *On Growth and Form.* Cambridge University Press, Cambridge. 2d ed., 1116 pp.

Townes, P. L., and J. Holtfreter, 1955. Directed movements and selective adhesion of embryonic amphibian cells. *J. Exp. Zool.,* **128**:53–120.

Waddington, C. H., 1956. *Principles of Embryology.* The Macmillan Company, New York. 3d printing, 1960. 510 pp.

Weiss, P., and A. C. Taylor, 1960. Reconstruction of complete organs from single-cell suspensions of chick embryos in advanced stages of differentiation. *Proc. Nat. Acad. Sci.,* **46**:1177–1185.

Wolff, E., 1935. Sur la formation d'une rangée axiale de somites chez l'embryon de poulet après irradiation du noeud de Hensen. *C. R. Soc. Biol.,* **118**:452.

SEXUAL CYCLE, GAMETOGENESIS, AND FERTILIZATION (CHAPTERS 2 AND 3)

Allen, E., and others, 1924. The hormone of the ovarian follicle; its localization and action in test animals, and additional points bearing upon the internal secretion of the ovary. *Am. J. Anat.,* **34**:133–181.

Ancel, P., and P. Vintemberger, 1948. Recherches sur le déterminisme de la symétrie bilaterale dans l'oeuf des amphibiens. *Bull. Biol. Suppl.* **31**:1–182.

Austin, C. R., 1961. *The Mammalian Egg.* Blackwell Scientific Publications, Ltd., Oxford. 183 pp.

———, 1968. *Ultrastructure of Fertilization.* Holt, Rinehart and Winston, Inc., New York. 196 pp.

———, 1978. Patterns in metazoan fertilization. *Curr. Top. Dev. Biol.,* **12**:1–9.

Baccetti, B., ed., 1970. *Comparative Spermatology.* Academic Press, Inc., New York. 573 pp.

Barr, M. L., and E. G. Bertram, 1949. A morphological distinction between neurones of the male and female and the behavior of the nucleolar satellite during accelerated nucleoprotein synthesis. *Nature* (London), **16**:676.

Bartelmez, G. W., 1937. Menstruation. *Physiol. Rev.,* **17**:28–72.

Bascom, K. F., and H. L. Osterud, 1925. Quantitative studies of the testicle. II. Pattern and total tubule length in the testicles of certain common mammals. *Anat. Rec.,* **31**:159–169.

Bedford, J. M., 1970. Sperm capacitation and fertilization in mammals. *Biol. of Reproduction,* vol. 2, suppl. 2, pp. 128–158.

Bellairs, R., 1964. Biological aspects of the yolk of the hen's egg. *Adv. Morphogen.,* **4**:217–272.

———, 1965. The relationship between oöcyte and follicle in the hen's ovary as shown by electron microscopy. *J. Embryol. Exp. Morphol.,* **13**:215–233.

Biggers, J. D., and A. W. Schuetz, eds., 1972. *Oogenesis*. University Park Press, Baltimore. 543 pp.

Blandau, R. J., 1945. The first maturation division of the rat ovum. *Anat. Rec.*, **92**:449–457.

———— and R. Hayashi. Ovulation and egg transport in mammals. Film, University of Washington Press, Seattle.

Board, R. G., and R. Fuller, 1974. Non-specific antimicrobial defenses of the avian egg, embryo and neonate. *Biol. Rev.*, **49**:15–49.

Brachet, J., 1977. An old enigma: The gray crescent of amphibian eggs. *Curr. Top. Dev. Biol.*, **11**:133–186.

Brown, D. D., and I. G. Dawid, 1968. Specific gene amplification in oocytes. *Science*, **160**:272–280.

Burmester, B. R., 1940. A study of the physical and chemical changes of the egg during its passage through the isthmus and uterus of the hen's oviduct. *J. Exp. Zool.*, **84**:445–500.

Chang, M. C., 1950. Development and fate of transferred rabbit ova or blastocyst in relation to the ovulation time of recipients. *J. Exp. Zool.*, **114**:197–226.

————, 1954. Development of parthenogenetic rabbit blastocysts induced by low temperature storage of unfertilized ova. *J. Exp. Zool.* **125**:127–150.

————, 1955. The maturation of rabbit oöcytes in culture and their maturation, activation, fertilization and subsequent development in the fallopian tubes. *J. Exp. Zool.*, **128**:379–406.

Clermont, Y., 1963. The cycle of the seminiferous epithelium in man. *Am. J. Anat.*, **112**:35–45.

————, 1972. Kinetics of spermatogenesis in mammals: Seminiferous epithelium cycle and spermatogonial renewal. *Physiol. Rev.*, **52**:198–236.

Conrad, R. M., and H. M. Scott, 1938. The formation of the egg in the domestic fowl. *Physiol. Rev.*, **18**:481–494.

Corner, G. W., 1919. On the origin of the corpus luteum of the sow from both granulosa and theca interna. *Am. J. Anat.*, **26**:117–183.

————, 1921. Cyclic changes in the ovaries and uterus of swine, and their relations to the mechanism of implantation. *Carnegie Cont. to Emb.*, **13**:117–146.

————, 1928–1929. Physiology of the corpus luteum. I. *Am. J. Physiol*, **86**:74–81; II. *Am. J. Physiol.*, **88**:326–339; III. *Am. J. Physiol.*, **88**:340–346.

————, 1945. Development, organization, and breakdown of the corpus luteum in the Rhesus monkey. *Carnegie Cont. to Emb.*, **31**:117–146.

Crosignani, P. G., and D. R. Mishell, eds., 1976. *Ovulation in the Human*. Academic Press, Inc., New York. 317 pp.

Cross, N. L., and R. P. Elinson, 1980. A fast block to polyspermy in frogs mediated by changes in the membrane potential. *Dev. Biol.*, **75**:187–198.

Curtis, A. S. G., 1960. Cortical grafting in *Xenopus laevis*. *J. Embryol. Exp. Morphol.*, **8**:163–173.

————, 1962. Morphogenetic interactions before gastrulation in the amphibian, *Xenopus laevis*—the cortical field. *J. Embryol. Exp. Morphol.*, **10**:410–422.

Dubois, R., 1969. Le mécanisme d'entrée des cellules germinales primordiales dans le réseau vasculaire, chez l'embryon de poulet. *J. Emb. Exp. Morphol.* **21**:255–270.

Eddy, E. M., 1975. Germ plasm and differentiation of the germ cell line. *Int. Rev. Cytol.*, **43**:229–281.

Enders, A. C., ed., 1963. *Delayed Implantation.* University of Chicago Press, Chicago. 318 pp.

Evans, H. M., and O. Swezy, 1929. Ovogenesis and the normal follicular cycle in adult mammalia. *Mem. Univ. Cal.,* **9:**119–224.

Eyal-Giladi, H., S. Kochav, and M. K. Menashi, 1976. On the origin of primordial germ cells in the chick embryo. *Differentiation,* **6:**13–16.

Fawcett, D. W., 1975. The mammalian spermatozoon. *Dev. Biol.,* **44:**394–436.

Flickinger, R., and D. E. Rounds, 1956. The maternal synthesis of egg yolk proteins as demonstrated by isotopic and serological means. *Biochem. Biophys. Acta,* **22:**38–42.

Fox, C. A., S. J. Meldrum, and B. W. Watson, 1973. Continuous measurement by radio-telemetry of vaginal pH during human coitus. *J. Reprod. Fert.,* **33:**69–75.

Gaddum, P., and R. J. Blandau, 1970. Proteolytic reaction of mammalian spermatozoa on gelatin membranes. *Science,* **170:**749–751.

Gall, J. G., and H. G. Callan, 1962. ^{3}H-Uridine incorporation in lampbrush chromosomes. *Proc. Nat. Acad. Sci.,* **48:**562–570.

Graham, M. A., 1954. Detection of the sex of cat embryos from nuclear morphology in the embryonic membrane. *Nature,* **173:**310.

Greenfield, M. L., 1966. The oöcyte of the domestic chicken shortly after hatching, studied by electron microscopy. *J. Embryol. Exp. Morphol.,* **15:**297–316.

Gwatkin, R. B. L., 1977. *Fertilization Mechanisms in Man and Mammals.* Plenum, New York. 161 pp.

Hadek, R., 1969. *Mammalian Fertilization: An Atlas of Ultrastructure.* Academic Press, Inc., New York. 295 pp.

Hafez, E. S. E., and T. N. Evans, 1973. *Human Reproduction.* Harper & Row, Hagerstown, Md. 778 pp.

Halbert, S. A., P. Y. Tam, and R. J. Blandau, 1976. Egg transport in the rabbit oviduct: The roles of cilia and muscle. *Science,* **191:**1052–1053.

Hamerton, J. L., 1961. Sex chromatin and human chromosomes. *Int. Rev. Cytol.,* **12:**1–68.

Hance, R. T., 1926. Sex and the chromosomes in the domestic fowl (*Gallus domesticus*). *J. Morphol. and Physiol.,* **43:**119–145.

Hartman, C. G., 1936. *Time of Ovulation in Women.* The Williams & Wilkins Company, Baltimore. 226 pp.

Heller, C. G., and Y. Clermont, 1963. Spermatogenesis in man: An estimate of its duration. *Science,* **140:**184–186.

Henking, H. von, 1891. Untersuchungen uber die ersten Entwicklungsvörgange in den Eiern der Insekten. *Zeitschr. f. wiss. Zool.,* **51:**685–763.

Hertig, A. T., J. Rock, E. C. Adams, and W. J. Mulligan, 1954. On the preimplantation stages of the human ovum: A description of four normal and four abnormal specimens ranging from the second to the fifth day of development. *Carnegie Cont. to Emb.,* **35:**199–220.

Hill, R. T., E. Allen, and T. C. Kramer, 1935. Cinemicrographic studies of rabbit ovulation. *Anat. Rec.,* **63:**239–245.

Hunter, G. L., G. P. Bishop, C. E. Adams and L. E. Rowson, 1962. Successful long-distance aerial transport of fertilized sheep ova. *J. Reprod. Fert.,* **3:**33–40.

Kemp, N. E., 1956. Electron microscopy of growing oöcytes of *Rana pipiens. J. Biophys. Biochem. Cytol.,* **2:**281–292.

Lillie, F. R., 1919. *Problems of Fertilization.* University of Chicago Press, Chicago. 278 pp.

Lyon, M. F., 1961. Gene action in the X chromosome of the mouse (*Mus musculus L.*). *Nature*, **190**:372–373.

MacGregor, H. C., 1972. The nucleolus and its genes in amphibian oögenesis. *Biol. Rev.*, **47**:177–210.

Mann, T., 1953. Biochemical aspects of semen in Wolstenholm, *Ciba Symposium on Mammalian Germ Cells.* Little, Brown and Company, Boston. pp. 1–8.

McClung, C. E., 1902. The accessory chromosome—sex determinant? *Biol. Bull.*, **3**:43–84.

Menkin, M. F., and J. Rock, 1948. In vitro fertilization and clevage of human ovarian eggs. *Am. J. Obstet. Gynecol.*, **55**:440–452.

Metz, C. B., and A. Monroy, 1967–1969. *Fertilization: Comparative Morphology, Biochemistry and Immunology.* Academic Press, Inc., New York. Vols. I and II, 489 and 553 pp.

Meyer, D. B., 1964. The migration of primordial germ cells in the chick embryo. *Dev. Biol.* **10**:154–190.

Midgley, A. R., V. L. Gay, P. L. Keyes, and J. S. Hunter, 1973. Human reproductive endocrinology, in Hàfez and Evans, *Human Reproduction.* Harper & Row, Hagerstown, Md. pp. 201–236.

Mintz, B., and E. S. Russell, 1957. Gene-induced embryological modifications of primordial germ cells in the mouse. *J. Exp. Zool.*, **134**:207–237.

Moses, M. J., 1968. Synaptinemal complex. *Annu. Rev. Genet.*, **2**:363–412.

Nieuwkoop, P., 1977. Origin and establishment of embryonic polar axes in amphibian development. *Curr. Top. Dev. Biol.*, **11**:115–132.

——— and L. A. Sutasurya, 1976. Embryological evidence for a possible polyphyletic origin of the recent amphibians. *J. Embryol. Exp. Morphol.*, **35**:159–167.

Oakberg, E. F., 1956. Duration of spermatogenesis in the mouse and timing of stages of the cycle of the seminiferous epithelium. *Am. J. Anat.*, **99**:507–516.

Odor, D. L., and R. J. Blandau, 1951. Observations on fertilization and the first segmentation division in rat ova. *Am. J. Anat.*, **89**:29–61.

Ohno, S., H. P. Klinger, and N. B. Atkin, 1962. Human oögenesis. *Cytogenetics*, **1**:42–51.

Old, R. W., H. G. Callan, and K. W. Gross, 1977. Localization of histone gene transcripts in newt lampbrush chromosomes by *in situ* hybridization. *J. Cell Sci.*, **27**:57–80.

O'Malley, B. W., W. L. McGuire, P. O. Kohler, and S. G. Korenman, 1969. Studies on the mechanism of steroid hormone regulation of synthesis of specific proteins. *Recent Prog. Horm. Res.*, **25**:105–160.

Papanicolaou, G. N., 1933. The sexual cycle in the human female as revealed by vaginal smears. *Am. J. Anat.*, supp. to vol. 52, pp. 519–637.

Parker, G. H., 1931. Passage of sperms and of eggs through oviducts in terrestrial vertebrates. *Phil. Trans. Roy. Soc. London*, ser. B, **219**:381–419.

Pincus, G., 1936. *The Eggs of Mammals.* The Macmillan Company, New York. 160 pp.

——— and B. Saunders, 1939. The comparative behavior of mammalian eggs in vivo and in vitro. VI. The maturation of human ovarian ova. *Anat. Rec.*, **75**:537–545.

Raspé, G., ed., 1970. *Schering Symposium on Mechanisms Involved in Conception.* Pergamon Press, New York. 470 pp.

Richards, J. S., 1979. Hormonal control of ovarian follicular development: A 1978 perspective. *Recent Prog. Horm. Res.*, **35:**343–373.

—— and A. R. Midgley, 1976. Protein hormone action: A key to understanding ovarian follicular and luteal cell development. *Biol. Reprod.*, **14:**82–94.

Romanoff, A. L., and A. J. Romanoff, 1949. *The Avian Egg.* John Wiley & Sons, Inc., New York. 918 pp.

Roosen-Runge, E. C., 1962. The process of spermatogenesis in mammals. *Biol. Rev.*, **37:**343–377.

Schimke, R. T., G. S. McKnight, D. J. Shapiro, D. Sullivan, and R. Palacios, 1975. Hormonal regulation of ovalbumin synthesis in the chick oviduct. *Recent Prog. Horm. Res.*, **31:**175–208.

Schuetz, A. W., 1974. Role of hormones in oöcyte maturation *Biol. Reprod.*, **10:**150–178.

Shettles, L. B., 1957. The living human ovum. *Obstet. Gynecol.*, **10:**359–365.

——, 1958. Corona radiata cells and zona pellucida of living human ova. *Fertil. Steril.*, **9:**167–170.

Simkins, C. S., 1932. Development of the human ovary from birth to sexual maturity. *Am. J. Anat.*, **51:**465–505.

Smith, L. D., 1966. The role of a "germinal plasm" in the formation of primordial germ cells in *Rana pipiens. Dev. Biol.*, **14:**330–347.

——, 1975. Molecular events during oöcyte maturation, in Weller, *The Biochemistry of Animal Development.* Academic Press, Inc., New York. Vol. 3, pp. 1–46.

—— and M. A. Williams, 1975. Germinal plasm and primordial germ cells. *33rd Symp. Soc. Devel. Biol.*, pp. 3–24.

Socher, S. H., and B. O'Malley, 1973. Estrogen mediated cell proliferation during chick oviduct development and its modulation by progesterone. *Dev. Biol.*, **30:**411–417.

Stambaugh, R., 1978. Enzymatic and morphological events in mammalian fertilization. *Gamete Res.*, **1:**65–86.

——, and J. Buckley, 1968. Zona pellucida dissociation enzymes of the rabbit sperm head. *Science,* **161:**585–586.

Taylor, T. G., 1970. How an eggshell is made. *Sci. Am.,* March, pp. 89–95.

Tsai, S. Y., S. E. Harris, M. J. Tsai, and B. W. O'Malley, 1976. Effects of estrogen on gene expression in chick oviduct. *J. Biol. Chem.*, **251:**4713–4721.

Wallace, R. A., and E. W. Bergink, 1974. Amphibian vitellogenin: Properties, hormonal regulation of hepatic synthesis and ovarian uptake, and conversion to yolk proteins. *Am. Zool.*, **14:**1159–1175.

Willier, B. H., 1937. Experimentally produced sterile gonads and the problem of the origin of germ cells in the chick embryo. *Anat. Rec.*, **70:**78–112.

Wilson, E. B., 1905. The chromosomes in relation to the determination of sex in insects. *Science,* **22:**500–502.

——, 1905. Studies on chromosomes. II. The paired microchromosomes, idiochromosomes and heterotropic chromosomes in Hemiptera. *J. Exp. Zool.*, **2:**507–545.

Wischnitzer, S., 1966. The ultrastructure of the cytoplasm of the developing amphibian egg. *Adv. Morphogen.*, **5:**131–179.

——, 1967. The ultrastructure of the nucleus of the developing amphibian egg. *Adv. Morphogen.*, **6:**173–198.

Witschi, E., 1948. Migration of the germ cells of human embryos from the yolk sac to the primitive gonadal folds. *Carnegie Cont. to Emb.*, **32:**67–80.

Zamboni, L., 1971. *Fine Morphology of Mammalian Fertilization.* Harper & Row, New York. 223 pp.

CLEAVAGE AND THE FORMATION OF THE GERM LAYERS (CHAPTERS 4 AND 5)

Assheton, R., 1899. The development of the pig during the first ten days. *Quart. J. Micr. Sci.,* **4:**329–359.

Beier, H. M., and R. R. Maurer, 1975. Uteroglobin and other proteins in rabbit blastocyst fluid after development *in vivo* and *in vitro. Cell Tissue Res.,* **159:**1–10.

Bekhtina, V. G., 1960. Early stages of cleavage in the chick embryo (Russian). *Arkh. Anat. Gist. Embriol.,* **38** (4):77–85.

Bellairs, R., F. W. Lorenz, and T. Dunlap, 1978. Cleavage in the chick embryo. *J. Embryol. Exp. Morphol.,* **43:**55–69.

Bennett, M. V. L., 1973. Function of electrotonic junctions in embryonic and adult tissues. *Fed. Proc.,* **32:**65–75.

Blandau, R. J., ed., 1971. *The Biology of the Blastocyst.* University of Chicago Press, Chicago. 560 pp.

Bluemink, J. G., and S. W. deLaat, 1973. New membrane formation during cytokinesis in normal and cytochalasin B-treated eggs of *Xenopus laevis.* I. Electron microscopic observations. *J. Cell. Biol.,* **59:**89–108.

Borland, R. M., J. D. Biggers, and C. P. Lechene, 1977. Studies in the composition and formation of mouse blastocoele fluid using electron probe microanalysis. *Dev. Biol.,* **55:**9–32.

Brachet, J., 1977. An old enigma: The gray crescent of amphibian eggs. *Curr. Top. Dev. Biol.,* **11:**133–186.

Briggs, R., and J. T. Justus, 1968. Partial characterization of the component from normal eggs which corrects the maternal effect of gene *o* in the Mexican axolotl (*Ambystoma mexicanum*). *J. Exp. Zool.,* **167:**105–116.

———— and T. J. King, 1952 Transplantation of living nuclei from blastula cells into enucleated frogs' eggs. *Proc. Nat. Acad. Sci.,* **38:**455–463.

Brown, D. D., and I. B. Dawid, 1968. Specific gene amplification in oöcytes. *Science,* **160:**272–280.

————, and E. Littna, 1964. RNA synthesis during the development of *Xenopus laevis,* the South African clawed toad. *J. Mol. Biol.,* **8:**669–687.

Curtis, A. S. G., 1962. Morphogenetic interactions before gastrulation in the amphibian *Xenopus laevis*—the cortical field. *J. Embryol. Exp. Morphol.,* **10:**410–422.

deLaat, S. W., and J. G. Bluemink, 1974. New membrane formation during cytokinesis in normal and cytochalasin B-treated eggs of *Xenopus laevis.* II. Electrophysiological observations. *J. Cell. Biol.,* **60:**529–540.

Ducibella, T., and E. Anderson, 1975. Cell shape and membrane changes in the eight-cell mouse embryo: Prerequisites for morphogenesis of the blastocyst. *Dev. Biol.* **47:**45–58.

Eyal-Giladi, H., and S. Kochav, 1976. From cleavage to primitive streak formation: A complementary normal table and a new look at the first stages of the development of the chick. I. General morphology. *Dev. Biol.* **49:**321–337.

———— and M. Wolk, 1970. The inducing capacities of the primary hypoblast as revealed by transfilter induction studies. *Wilhelm Roux' Arch.,* **165:**226–241.

Fontaine, J., and N. M. le Douarin, 1977. Analysis of endoderm formation in the avian blastoderm by the use of quail-chick chimaeras. *J. Embryol. Exp. Morphol.*, **41:**209–222.

Gardner, R. L., and V. E. Papaioannou. 1975. Differentiation in the trophectoderm and inner cell mass, in Balls and Wild, *The Early Development of Mammals.* Cambridge University Press, Cambridge. pp. 107–132.

Gulyas, B. J., 1975. A reexamination of cleavage patterns in eutherian mammalian eggs: Rotation of blastomere pairs during second cleavage in the rabbit. *J. Exp. Zool.*, **193:**235–248.

Gurdon, J. B., 1974. *The Control of Gene Expression in Animal Development.* Harvard University Press, Cambridge, Mass. 160 pp.

Hara, K., 1977. The cleavage pattern of the axolotl egg studied by cinematography and cell counting. *Wilhelm Roux' Arch.*, **181:**73–87.

Hay, E. D., 1968. Organization and fine structure of epithelium and mesenchyme in the developing chick embryo, in Fleischmajer and Billingham, *Epithelial-Mesenchymal Interactions.* The Williams & Wilkins Company, Baltimore. pp. 31–55.

Hertig, A. T., and J. Rock, 1945. Two human ova of the previllous stage, having a developmental age of about 7 and 9 days respectively. *Carnegie Cont. to Emb.*, **31:**65–84.

———— and ————, 1949. Two human ova of the previllous stage, having a developmental age of about 8 to 9 days respectively. *Carnegie Cont. to Emb.*, **33:**169–186.

————, ————, and E. C. Adams, 1956. A description of 34 human ova within the first 17 days of development. *Am. J. Anat.*, **98:**435–494.

————, ————, ————, and W. J. Mulligan, 1954. On the preimplantation stages of the human ovum: A description of four normal and four abnormal specimens ranging from the second to the fifth day of development. *Carnegie Cont. to Emb.*, **35:**199–220.

Heuser, C. H., 1932. Presomite human embryo with a definite chorda canal. *Carnegie Cont. to Emb.*, **23:**251–267.

———— and G. L. Streeter, 1929. Early stages in the development of pig embryos from the period of initial cleavage to the time of the appearance of limb buds. *Carnegie Cont. to Emb.*, **20:**1–29.

———— and ————, 1941. Development of the Macaque embryo. *Carnegie Cont. to Emb.*, **29:**15–55.

Holtfreter, J., 1943–1944. A study of the mechanics of gastrulation. I. *J. Exp. Zool.*, **94:**261–318; II. *J. Exp. Zool.*, **95:**171–212.

————, 1947. Changes of structure and the kinetics of differentiating embryonic cells. *J. Morphol.*, **80:**57–92.

Kalt, M. R., 1971. The relationship between cleavage and blastocoel formation in *Xenopus laevis*. I. Light microscopic observations. *J. Embryol. Exp. Morphol.*, **26:**37–49.

Leikola, A., 1976. Hensen's node—the "organizer" of the amniotic embryo. *Experientia*, **32:**269–277.

Lewis, W. H., 1947. Mechanics of invagination. *Anat. Rec.*, **97:**139–156.

———— and C. G. Hartman, 1933. Early cleavage stages of the egg of the monkey (*Macacus rhesus*). *Carnegie Cont. to Emb.*, **24:**187–201.

Luckett, W. P., 1978. Origin and differentiation of the yolk sac and extraembryonic mesoderm in presomite human and rhesus monkey embryos. *Am. J. Anat.*, **152:**59–98.

Markert, C. L., and R. M. Petters, 1978. Manufactured hexaparental mice show that adults are derived from three embryonic cells. *Science*, **202:**56–58.

McLaren, A., 1976. *Mammalian Chimaeras*. Cambridge University Press, Cambridge. 154 pp.

Mintz, B., 1964. Formation of genetically mosaic mouse embryos and early development of "lethal (T^{12}/T^{12})-normal" mosaics. *J. Exp. Zool.*, **157**:273–292.

Nakamura, O., and S. Toivonen, eds., 1978. *Organizer—A Milestone of a Half Century from Spemann*. Elsevier/North-Holland, Amsterdam. 379 pp.

Nicolet, G., 1971. Avian gastrulation. *Adv. Morphogen.*, **9**:231–262.

Nieuwkoop, P. D., 1964. The formation of the mesoderm in urodelean amphibians. I. Induction by the endoderm. *Wilhelm Roux' Arch.*, **162**:341–373.

———, 1973. The "organization center" of the amphibian embryo: Its origin, spatial organization and morphogenetic action. *Adv. Morphogen.*, **10**:1–39.

———, 1977. Origin and establishment of embryonic polar axes in amphibian development. *Curr. Top. Dev. Biol.*, **11**:115–132.

Nicholas, J. S., 1946. Experimental approaches to problems of early development in the rat. *Quart. Rev. Biol.*, **22**:179–195.

——— and B. V. Hall, 1942. Experiments on developing rats. II. The development of isolated blastomeres and fused eggs. *J. Exp. Zool.*, **90**:441–459.

Pasteels, J., 1945. On the formation of the primary entoderm of the duck (*Anas domestica*) and on the significance of the bilaminar embryo in birds. *Anat. Rec.*, **93**:5–21.

Patterson, J. T., 1913. Polyembryonic development in *Tatusia novemcincta*. *J. Morphol.*, **24**:559–683.

Rappaport, R., 1971. Cytokinesis in animal cells. *Int. Rev. Cytol.*, **31**:169–213.

———, 1974. Cleavage, in Lash and Whittaker, *Concepts of Development*. Sinauer Associates, Stamford, Conn. Pp. 76–98.

Revel, J. P., P. Yip, and L. L. Chang, 1973. Cell junctions in the early chick embryo—a freeze etch study. *Dev. Biol.*, **35**:302–317.

Rosenquist, G. C., 1966. A radioautographic study of labeled grafts in the chick blastoderm. Development from primitive streak stages to stage 12. *Carnegie Cont. to Emb.*, **38**:71–110.

———, and R. L. DeHaan, 1966. Migration of precardiac cells in the chick embryo: A radioautographic study. *Carnegie Cont. to Emb.*, **38**:111–121.

Rudnick, D., 1944. Early history and mechanics of the chick blastoderm. *Quart. Rev. Biol.*, **19**:187–212.

Selman, G. G., and M. M. Perry, 1970. Ultrastructural changes in the surface layers of the newt's egg in relation to the mechanism of its cleavage. *J. Cell Sci.*, **6**:207–227.

Spemann, H., 1928. Die Entwicklung seitlicher und dorso-ventraler Keimhälften bei verzögerter Kernversorgung. *Z. Wiss. Zool.*, **132**:105–134.

———, 1938. *Embryonic Development and Induction*. Reprinted 1962 by Hafner Publishing Company, New York. 401 pp.

——— and H. Mangold, 1924. Ueber Induktion von Embryonalanlagen durch Implantation ortfremder Organistoren. *Arch. mikrosk. Anat. Entwmech.*, **100**:599–638.

Spratt, N. T., Jr., 1942. Location of organ-specific regions and their relationship to the development of the primitive streak in the early chick blastoderm. *J. Exp. Zool.*, **89**:69–101.

———, 1946. Formation of the primitive streak in the explanted chick blastoderm marked with carbon particles. *J. Exp. Zool.*, **103**:259–304.

———, 1947. Regression and shortening of the primitive streak in the explanted chick blastoderm. *J. Exp. Zool.*, **104**:69–100.

——— and H. Haas, 1965. Germ layer formation and the role of the primitive streak in the chick. I. Basic architecture and morphogenetic tissue movements. *J. Exp. Zool.,* **158:**9–38.

——— and ———, 1967. Nutritional requirements for the realization of regulative (repair) capacities of young chick blastoderms. *J. Exp. Zool.,* **164:**31–46.

Streeter, G. L., 1926. Development of the mesoblast and notochord in pig embryos. *Carnegie Cont. to Emb.,* **19:**73–92.

Tarkowski, A. K., 1961. Mouse chimeras developed from fused eggs. *Nature* (London), **190:**857–860.

——— and J. Wroblewska, 1967. Development of blastomeres of mouse eggs isolated at the 4- and 8-cell stage. *J. Embryol. Exp. Morphol.,* **18:**155–180.

Townes, P. L., and J. Holtfreter, 1955. Directed movements and selective adhesion of embryonic amphibian cells. *J. Exp. Zool.,* **128:**53–120.

Trinkhaus, J. P., 1969. *Cells into Organs.* Prentice-Hall, Inc. Englewood Cliffs, N.J. 237 pp.

Vakaet, L., 1962. Some new data concerning the formation of the definitive endoblast in the chick embryo. *J. Embryol. Exp. Morphol.,* **10:**38–57.

———, 1970. Cinephotomicrographic investigations of gastrulation in the chick blastoderm. *Arch. Biol.,* **81:**387–426.

Vogt, W., 1929. Gestaltungsanalyse am Amphibienkeim mit örtlicher Vitalfarbung. II. Gastrulation und Mesodermbildung bei Urodelen und Anuren. *Wilhelm Roux' Arch.,* **120:**385–706.

Waddington, C. H., 1933. Induction by the primitive streak and its derivatives in the chick. *J. Exp. Biol.,* **10:**38–46.

Wolpert, L., 1960. The mechanics and mechanism of cleavage. *Int. Rev. Cytol.,* **10:**163–216.

NEURULATION AND SOMITE FORMATION (CHAPTER 6)

Barth, L. G., and L. J. Barth, 1974. Ionic regulation of embryonic induction and cell differentiation in *Rana pipiens. Dev. Biol.,* **39:**1–22.

Bellairs, R., and P. A. Portch, 1977. Somite formation in the chick embryo, in Ede, Hinchliffe, and Balls, *Vertebrate Limb and Somite Morphogenesis.* Cambridge University Press, Cambridge. Pp. 449–463.

Burnside, B., 1971. Microtubules and microfilaments in newt neurulation. *Dev. Biol.,* **26:**419–441.

———, 1973. Microtubules and microfilaments in amphibian neurulation. *Am. Zool.,* **13:**989–1006.

——— and A. G. Jacobson, 1968. Analysis of morphogenetic movements in the neural plate of the newt, *Taricha torosa. Dev. Biol.,* **18:**537–553.

Chevallier, A., M. Kieny, A. Mauger, and P. Sengel, 1977. Developmental fate of the somitic mesoderm in the chick embryo, in Ede, Hinchliffe, and Balls, *Vertebrate Limb and Somite Morphogenesis.* Cambridge University Press, Cambridge. Pp. 421–432.

Cohen, A. M., and E. D. Hay, 1971. Secretion of collagen by embryonic neuroepithelium at the time of spinal cord–somite interaction. *Dev. Biol.,* **26:**578–605.

Cooke, J., 1977. The control of somite number during amphibian development: Models and experiments, in Ede, Hinchliffe, and Balls, *Vertebrate Limb and Somite Morphogenesis.* Cambridge University Press, Cambridge. Pp. 433–448.

Gallera, J., 1971. Primary induction in birds. *Adv. Morphogen.,* **9:**149–180.

Gearhart, J. D., and B. Mintz, 1972. Clonal origins of somites and their muscle derivates: Evidence from allophenic mice. *Dev. Biol.,* **29:**27–37.

Grobstein, C., 1967. Mechanisms of organogenetic tissue interaction. Nat. Cancer Inst. Monogr. No. 26, pp. 279–299.

Hay, E. D., 1968. Organization and fine structure of epithelium and mesenchyme in the developing chick embryos, in Fleishmajer, *Epithelial-Mesenchymal Interactions.* The Williams & Wilkins Company, Baltimore. Pp. 31–55.

———, 1973. Origin and role of collagen in the embryo. *Am. Zool.,* **13:**1085–1107.

——— and S. Meier, 1974. Glycosaminoglycan synthesis by embryonic inductors: Neural tube, notochord and lens. *J. Cell Biol.,* **62:**889–898.

Holtfreter, J., 1947. Observations on the migration, aggregation and phagocytosis of embryonic cells. *J. Morphol.,* **80:**25–55.

———, 1968. Mesenchyme and epithelia in inductive and morphogenetic processes, in Fleischmajer and Billingham, *Epithelial-Mesenchymal Interactions.* The Williams & Wilkins Company, Baltimore. Pp. 1–30.

Holtzer, H., and S. R. Detwiler, 1953. An experimental analysis of the development of the spinal column. III. Induction of skeletogenous cells. *J. Exp. Zool.,* **123:**335–369.

——— and R. Mayne, 1973. Experimental morphogenesis: The induction of somitic chondrogenesis by embryonic spinal cord and notochord. *Pathobiol. Dev.,* pp. 52–64.

Horstadius, S. O., 1950. *The Neural Crest.* Oxford University Press, London. 111 pp.

Jacobson, A. G., 1966. Inductive processes in embryonic development. *Science,* **152:**25–34.

——— and R. Gordon, 1976. Changes in the shape of the developing vertebrate nervous system analyzed experimentally, mathematically and by computer simulation. *J. Exp. Zool.,* **197:**191–246.

Jacobson, C.-O., 1962. Cell migration in the neural plate and the process of neurulation in the axolotl larva. *Zool. Bidr. Uppsala,* **35:**433–449.

Johnston, M. C., 1966. A radioautographic study of the migration and fate of cranial neural crest cells in the chick embryo. *Anat. Rec.,* **156:**143–156.

Karfunkel, P., 1972. The activity of microtubules and microfilaments in neurulation in the chick. *J. Exp. Zool.,* **181:**289–302.

———, 1973. The mechanisms of neural tube formation. *Int. Rev. Cytol.,* **38:**245–272.

Kenney, M. C., and E. Carlson, 1978. Ultrastructural identification of collagen and glycosaminoglycans in notochordal extracellular matrix in vivo and in vitro. *Anat. Rec.,* **190:**827–850.

Langman, J., and G. R. Nelson, 1968. A radioautographic study of the development of the somite in the chick embryo. *J. Embryol. Exp. Morphol.,* **19:**217–226.

Lash, J. W., 1968. Somitic mesenchyme and its response to cartilage induction, in Fleischmajer and Billingham, *Epithelial-Mesenchymal Interactions.* The Williams & Wilkins Company, Baltimore. Pp. 165–172.

LeDouarin, N. M., and M. A. Teillet, 1974. Experimental analysis of the migration and differentiation of neuroblasts of the autonomic nervous system and of neuroectodermal mesenchymal derivatives, using a biological cell marking technique. *Dev. Biol.,* **41:**162–184.

Lehtonen, E., 1976. Transmission of signals in embryonic induction. *Med. Biol.,* **54:**108–128.

Leussink, J. A., 1970. The spatial distribution of inductive capacities in the neural plate and archenteron roof of urodeles. *Neth. J. Zool.,* **20:**1–79.

Lipton, B. H., and A. G. Jacobson, 1974. Analysis of normal somite development. *Dev. Biol.,* **38**:73–90.

Lipton, B. H., and A. G. Jacobson, 1974. Experimental analysis of the mechanisms of somite morphogenesis. *Dev. Biol.,* **38**:91–103.

Manasek, F. J., 1975. The extracellular matrix: A dynamic component of the developing embryo. *Curr. Top. Dev. Biol.,* **10**:35–102.

Nieuwkoop, P. D., 1966. Induction and pattern formation as primary mechanisms in early embryonic differentiation, in *Cell Differentiation and Morphogenesis, International Lecture Course.* North-Holland Publishing Company, Amsterdam. Pp. 120–143.

Noden, D. M., 1978. The control of avian cephalic neural crest cytodifferentiation. I. Skeletal and connective tissues. II. Neural tissues. *Dev. Biol.,* **67**:296–312; **67**:313–329.

Rawles, M. E., 1948. Origin of melanophores and their role in development of color patterns in vertebrates. *Physiol. Rev.,* **28**:383–408.

Revel, J.-P., P. Yip, and L. L. Chang, 1973. Cell junctions in the early chick embryo—a freeze etch study. *Dev. Biol.,* **35**:302–317.

Saxén, L., and S. Toivonen, 1962. *Primary Embryonic Induction.* Logos Press, London. 271 pp.

Schroeder, T. E., 1973. Cell constriction: Contractile role of microfilaments in division and development. *Am. Zool.,* **13**:949–960.

Sheridan, J. D., 1966. Electrophysiological study of special connections between cells in the early chick embryo. *J. Cell Biol.,* **31**:C1–C5.

Spemann, H. 1938. *Embryonic Development and Induction.* Reprinted by Hafner Publishing Company, New York, 1962. 401 pp.

Toivonen, S., D. Tarin, and L. Saxén, 1976. The transmission of morphogenetic signals from amphibian mesoderm to ectoderm in primary induction. *Differentiation,* **5**:49–55.

Trelstad, R. L., J.-P. Revel, and E. D. Hay, 1966. Tight junctions between cells in the early chick embryo as visualized with the electron microscope. *J. Cell Biol.,* **31**:C6–C10.

Twitty, V. C., and M. C. Niu, 1954. The motivation of cell migration, studied by isolation of embryonic pigment cells singly and in small groups in vitro. *J. Exp. Zool.,* **125**:541–574.

Weston, J. A., 1970. The migration and differentiation of neural crest cells. *Adv. Morphogen.,* **8**:41–114.

Williams, L. W., 1910. The somites of the chick. *Am. J. Anat.,* **11**:55–100.

FETAL MEMBRANES AND PLACENTATION (CHAPTER 7)

Adamstone, F. B., 1948. Experiments on the development of the amnion in the chick. *J. Morphol.,* **83**:359–371.

Bartelmez, G. W., 1951. Cyclic changes in the endometrium of the rhesus monkey (*Macaca mulatta*). *Carnegie Cont. to Emb.,* **34**:99–144.

———, 1957. The form and the function of the uterine blood vessels in the rhesus monkey. *Carnegie Cont. to Emb.,* **36**:153–182.

Bautzmann, H., W. Schmidt, and P. Lembrug, 1960. Experimental electron- and light-microscopical studies on the function of the amnion-apparatus of the chick, cat and man. *Anat. Anz.,* **108**:305–315.

Blandau, R. J., 1949. Embryo-endometrial interrelationship in the rat and guinea pig. *Anat. Rec.*, **104**:331–359.

Böving, B. G., 1956. Rabbit blastocyst distribution. *Am. J. Anat.*, **98**:403–434.

———, 1959. Implantation. *Ann. N. Y. Acad. Sci.*, **75**:700–725.

———, 1962. Anatomical analysis of rabbit trophoblast invasion. *Carnegie Cont. to Emb.*, **37**:33–55.

———, 1971. Biomechanics of implantation, in *The Biology of the Blastocyst*, ed. R. J. Blandau. University of Chicago Press, Chicago. Pp. 423–442.

Boyd, J. D., and W. J. Hamilton, 1966. Electron microscopic observations on the cytotrophoblast contribution to the syncytium in human placenta. *J. Anat.*, **100**:535–548.

——— and ———, 1970. *The Human Placenta*. W. Heffer & Sons, Ltd., Cambridge, England. 365 pp.

Brambel, C. E., 1933. Allantochorionic differentiations of the pig studied morphologically and histochemically. *Am. J. Anat.*, **52**:397–459.

Corner, G. W., 1921. Internal migration of the ovum. *Bull. Johns Hopkins Hosp.*, **32**:78–83.

Dempsey, E. W., G. B. Wislocki, and E. C. Amoroso, 1955. Electron microscopy of the pig's placenta, with especial reference to the cell membranes of the endometrium and chorion. *Am. J. Anat.*, **96**:65–102.

Dhouailly, D., 1978. Feather-forming capacities of the avian extra-embryonic somatopleure. *J. Embryol. Exp. Morphol.*, **43**:279–287.

Enders, A. C., 1965. A comparative study of the fine structure of the trophoblast in several hemochorial placentae. *Am. J. Anat.*, **116**:29–67.

Harris, J. W. S., and E. M. Ramsey, 1966. The morphology of human uteroplacental vasculature. *Carnegie Cont. to Emb.*, **38**:43–58.

Hertig, A. T., and J. Rock, 1945. Two human ova of the previllous stage, having a developmental age of about 7 and 9 days respectively. *Carnegie Cont. to Emb.*, **31**:65–84.

Heuser, C. H., 1927. A study of the implantation of the ovum of the pig from the stage of the bilaminar blastocyst to the completion of the fetal membranes. *Carnegie Cont. to Emb.*, **19**:229–243.

Klopper, A., and E. Diczfaluzy, eds., 1969. *Foetus and Placenta*. Blackwell Scientific Publications, Ltd., Oxford. 628 pp.

Luckett, W. P., 1974. Comparative development and evolution of the placenta in primates, in Luckett, *Reproductive Biology of the Primates*: Contrib. to Primatology, vol. 3. S. Karger, Basel. Pp. 142–234.

———, 1975. The development of primordial and definitive amniotic cavities in early rhesus monkey and human embryos. *Am. J. Anat.*, **144**:149–168.

———, 1978. Origin and differentiation of the yolk sac and extraembryonic mesoderm in presomite human and rhesus monkey embryos. *Am. J. Anat.*, **152**:59–98.

Markee, J. E., R. A. Pasqualetti, and J. C. Hinsey, 1936. Growth of intraocular endometrial transplants in spinal rabbits. *Anat. Rec.*, **64**:247–253.

Mossman, H. W., 1937. Comparative morphogenesis of the fetal membranes and accessory uterine structures. *Carnegie Cont. to Emb.*, **26**:129–246.

Pecile, A., and C. Finzi, eds., 1969. *The Foeto-Placental Unit*. Excerpta Medica Foundation, Amsterdam. 425 pp.

Pierce, M. E., 1933. The amnion of the chick as an independent effector. *J. Exp. Zool.*, **65**:443–473.

Plentl, A., 1966. Formation and circulation of amniotic fluid. *Clin. Obstet. Gynecol.*, **9:**427–439.

Potter, E. L., and F. L. Adair, 1940. *Fetal and Neonatal Death.* The University of Chicago Press, Chicago. 207 pp.

Ramsey, E. M., 1949. The vascular pattern of the endometrium of the pregnant rhesus monkey (*Macaca mulatta*). *Carnegie Cont. to Emb.*, **33:**113–147.

———, 1956. Circulation in the maternal placenta of the rhesus monkey and man, with observations on the marginal lakes. *Am. J. Anat.*, **98:**159–190.

———, 1962. Circulation in the intervillous space of the primate placenta. *Am. J. Obstet. Gynecol.*, **84:**1649–1663.

———, 1965. The placenta and fetal membranes, in Greenhill, *Obstetrics,* 13th ed. W. B. Saunders Company, Philadelphia. Pp. 101–136.

Randles, C. A., Jr., and A. L. Romanoff, 1950. Some physical aspects of the amnion and allantois of the developing chick embryo. *J. Exp. Zool.*, **114:**87–101.

Wilkin, P., 1965. Organogenesis of the human placenta, in DeHaan and Ursprung, *Organogenesis.* Holt, Rinehart and Winston, Inc., New York. Pp. 743–769.

Wislocki, G. B., 1929. On the placentation of primates, with a consideration of the phylogeny of the placenta. *Carnegie Cont. to Emb.*, **20:**51–80.

——— and G. L. Streeter, 1938. On the placentation of the macaque (*Macaca mulatta*), from the time of implantation until the formation of the definitive placenta. *Carnegie Cont. to Emb.*, **20:**51–80.

YOUNG MAMMALIAN EMBRYOS (CHAPTER 8)

Arey, L. B., 1938. The history of the first somite in human embryos. *Carnegie Cont. to Emb.*, **27:**233–269.

Atwell, W. J., 1930. A human embryo with seventeen pairs of somites. *Carnegie Cont. to Emb.*, **21:**1–24.

Bartelmez, G. W., and M. P. Blount, 1954. The formation of neural crest from the primary optic vesicle in man. *Carnegie Cont. to Emb.*, **35:**55–71.

——— and H. M. Evans, 1926. Development of the human embryo during the period of somite formation, including embryos with 2 to 16 pairs of somites. *Carnegie Cont. to Emb.*, **17:**1–67.

Boyden, E. A., 1933. A laboratory atlas of the 13-mm pig embryo. The Wistar Institute Press, Philadelphia. 100 pp.

Bremer, J. L., 1906. Description of a 4-mm human embryo. *Am. J. Anat.*, **5:**459–480.

Corner, G. W., 1929. A well-preserved human embryo of 10 somites. *Carnegie Cont. to Emb.*, **20:**81–102.

Davis, C. L., 1923. Description of a human embryo having twenty paired somites. *Carnegie Cont. to Emb.*, **15:**1–51.

———, 1927. Development of the human heart from its first appearance to the stage found in embryos of twenty paired somites. *Carnegie Cont. to Emb.*, **19:**245–284.

Fallon, J. F., and B. K. Simandl, 1978. Evidence of a role for cell death in the disappearance of the embryonic human tail. *Am. J. Anat.*, **152:**111–130.

Grobstein, C., 1952. Effect of fragmentation of mouse embryonic shields on their differentiative behavior after culturing. *J. Exp. Zool.*, **120:**437–456.

Hertig, A. T., J. Rock, E. C. Adams, and W. J. Mulligan, 1954. On the preimplantation stages of the human ovum: A description of four normal and four abnormal specimens ranging from the second to the fifth day of development. *Carnegie Cont. to Emb.*, **35:**199–220.

Heuser, C. H., 1930. A human embryo with 14 pairs of somites. *Carnegie Cont. to Emb.*, **22:**135–154.

——— and G. L. Streeter, 1929. Early stages in the development of pig embryos, from the period of initial cleavage to the time of the appearance of limb buds. *Carnegie Cont. to Emb.*, **20:**1–29.

Ingalls, N. W., 1920. A human embryo at the beginning of segmentation, with special reference to the vascular system. *Carnegie Cont. to Emb.*, **11:**61–90.

Johnson, F. P., 1917. A human embryo of twenty-four pairs of somites. *Carnegie Cont. to Emb.*, **6:**125–168.

Lewis, F. T., 1902. The gross anatomy of a 12-mm pig. *Am. J. Anat.*, **2:**211–226.

Patten, B. M., and C. G. Hartman, 1933. Early development of the embryo, in Curtis, *Obstetrics and Gynecology*. W. B. Saunders Company, Philadelphia. Vol 1, pp. 401–441.

Payne, F., 1925. General description of a 7-somite human embryo. *Carnegie Cont. to Emb.*, **16:**115–124.

Raspé, G., ed., 1971. *Schering Symposium on Intrinsic and Extrinsic Factors in Early Mammalian Development*. Pergamon Press, New York. 653 pp.

Scammon, R. E., and L. A. Calkins, 1929. *The Development and Growth of the External Dimensions of the Human Body in the Fetal Period*. University of Minnesota Press, Minneapolis. 367 pp.

Streeter, G. L., 1908. The peripheral nervous system in the human embryo at the end of the first month. *Am. J. Anat.*, **8:**285–301.

———, 1945. Developmental horizons in human embryos. Description of age group XIII, embryos about 4 or 5 mm long, and age group XIV, period of indentation of the lens vesicle. *Carnegie Cont. to Emb.*, **31:**27–63.

———, 1948. Developmental horizons in human embryos. Description of age groups XV, XVI, XVII, and XVIII, being the third issue of a survey of the Carnegie collection. *Carnegie Cont. to Emb.*, **32:**133–204.

———, 1951. Developmental horizons in human embryos. Description of age groups XIX, XX, XXI, XXII, and XXIII, being the fifth issue of a survey of the Carnegie collection. Prepared for publication by C. H. Heuser and G. W. Corner. *Carnegie Cont. to Emb.*, **34:**165–196.

Thyng, F. W., 1911. The anatomy of a 7.8-mm pig embryo. *Anat. Rec.*, **5:**17–45.

———, 1914. The anatomy of a 17.8-mm human embryo. *Am. J. Anat.*, **17:**31–112.

SKELETAL AND MUSCULAR SYSTEMS (CHAPTER 9)

Avery, G., M. Chow, and H. Holtzer, 1956. An experimental analysis of the development of the spinal column. V. Reactivity of chick somites. *J. Exp. Zool.*, **132:**409–426.

Bardeen, C. R., 1905. Studies of the development of the human skeleton. *Am. J. Anat.*, **4:**265–302.

——— and W. H. Lewis, 1901. The development of the limbs, body wall, and back. *Am. J. Anat.*, **1:**1–37.

Bassett, C. A. L., 1971. Biophysical principles affecting bone structure, in Bourne, *The Biochemistry and Physiology of Bone*, Academic Press, Inc., New York. Vol. 3, pp. 1–76.

Bates, M. N., 1948. The early development of the hypoglossal musculature in the cat. *Am. J. Anat.*, **83:**329–355.

Bosma, J. F., ed., 1976. *Symposium on Development of the Basicranium.* DHEW Publication No. (NIH) 76-989, U.S. Government Printing Office, Washington, D.C. 700 pp.

Čihak, R., 1972. Ontogenesis of the skeleton and intrinsic muscles of the human hand and foot. *Ergebnisse des Anatomie und Entwicklungsgeschichte,* **46**(1)**:**1–194.

deBeer, G. R., 1937. *The Development of the Vertebrate Skull.* Oxford University Press, London. 552 pp.

Doering, J. L., and D. A. Fischman, 1974. The *in vitro* cell fusion of embryonic chick muscle without DNA synthesis. *Dev. Biol.,* **36:**225–235.

Drachman, D. B., ed., 1974. Trophic functions of the neuron. *Ann. N.Y. Acad. Sci.,* **228:**1–423.

Felts, W. J. L., 1954. The prenatal development of the human femur. *Am. J. Anat.,* **94:**1–44.

Gardner, E. D., and D. J. Gray, 1950. Prenatal development of the human hip joint. *Am. J. Anat.,* **87:**163–211.

———— and ————, 1953. Prenatal development of the human shoulder and acromioclavicular joints. *Am. J. Anat.,* **92:**219–276.

Gray, D. J., and E. D. Gardner, 1950. Prenatal development of the human knee and superior tibiofibular joints. *Am. J. Anat.,* **86:**235–287.

———— and ————, 1951. Prenatal development of the human elbow joint. *Am. J. Anat.,* **88:**429–469.

Hall, B. K., 1974. Chondrogenesis of the somitic mesoderm. *Adv. Anat. Embryol. Cell Biol.,* **53**(4)**:**1–50.

————, 1978. *Developmental and Cellular Skeletal Biology.* Academic Press, Inc., New York. 304 pp.

Holtzer, H., 1970. Proliferative and quantal cell cycles in the differentiation of muscle, cartilage and red blood cells, in Padykula, *Gene Expression in Somitic Cells.* Academic Press, Inc., New York. pp. 69–88.

———— and S. R. Detwiler, 1953. An experimental analysis of the development of the spinal column. III. Induction of skeletogenous cells. *J. Exp. Zool.,* **123:**335–370.

————, R. Weintraub, R. Mayer, and B. Mochran, 1972. The cell cycle, cell lineages and cell differentiation. *Curr. Top. Dev. Biol.,* **7:**229–256.

Jotereau, F. V., and N. M. LeDouarin, 1978. The developmental relationship between osteocytes and osteoclasts: A study using the quail-chick nuclear marker in endochondral ossification. *Dev. Biol.,* **63:**253–265.

Koningsberg, I. R., and P. A. Buckley, 1974. Regulation of the cell cycle and myogenesis by cell-medium interaction, in Lash and Whittaker, *Concepts in Development.* Sinauer Associates, Stamford, Conn. Pp. 179–193.

Langman, J., and G. R. Nelson, 1968. A radioautographic study of the development of the somite in the chick embryo. *J. Embryol. Exp. Morph.,* **19:**217–226.

Lash, J. W., 1968. Somitic mesenchyme and its response to cartilage induction, in Fleischmajer and Billingham, *Epithelial-Mesenchymal Interactions.* The Williams & Wilkins Company, Baltimore. Pp. 165–172.

Lewis, W. H., 1910. The development of the muscular system, in Keibel and Mall, *Human Embryology.* J. B. Lippincott Company, Philadelphia. Pp. 454–522.

Liboff, A. R., and R. A. Rinaldi, eds., 1974. *Electrically Mediated Growth Mechanisms in Living Systems.* Ann. NY Acad. Sci., **238:**1–593.

Manasek, F. J., 1968. Embryonic development of the heart. I. A light and electron microscopic study of myocardial development in the early chick embryo. *J. Morphol.*, **125:**329–366.

———, 1968. Mitosis in developing cardiac muscle. *J. Cell Biol.*, **37:**191–196.

Mauro, A., ed., 1979. *Muscle Regeneration.* Raven Press, New York. 560 pp.

Mintz, B., and W. B. Baker, 1967. Normal mammalian muscle differentiation and gene control of isocitrate dehydrogenase synthesis. *Proc. Nat. Acad. Sci.*, **58:**592–598.

Moss, M. L., C. R. Noback, and G. G. Robertson, 1955. Critical developmental horizons in human fetal long bones: Correlated quantitative and histological criteria. *Am. J. Anat.*, **97:**155–175.

Murray, P. D. F., 1936. *Bones. A Study of the Development and Structure of the Vertebrate Skeleton.* Cambridge University Press, New York. 203 pp.

Noback, C. R., 1944. The developmental anatomy of the human osseous skeleton during the embryonic, fetal and circumnatal periods. *Anat. Rec.*, **88:**91–125.

——— and G. G. Robertson, 1951. Sequences of appearance of ossification centers in the human skeleton during the first five prenatal months. *Am. J. Anat.*, **89:**1–28.

Okazaki, K., and H. Holtzer, 1965. An analysis of myogenesis in vitro using fluorescein-labeled antimyosin. *J. Histochem.*, **13:**726–739.

O'Rahilly, R., 1961. The developmental anatomy of the extensor assembly. *Acta Anat.*, **47:**363–375.

——— and E. Gardner, 1972. The initial appearance of ossification in staged human embryos. *Am. J. Anat.*, **134:**291–308.

Prader, A., 1947. Die frühenembryonale Entwicklung der menschlichen Zwischenwirbelscheibe. *Acta. Anat.*, **3:**68–83.

Rumyantsev, P. P., 1967. Electron microscopic analysis of cell elements in the differentiation and proliferation processes in the developing myocardium (Russian). *Arkh. Anat. Gistol. Embryol.*, **52:**67–77.

Sensenig, E. C., 1949. The early development of the human vertebral column. *Carnegie Cont. to Emb.*, **33:**21–41.

Straus, W. L., and M. E. Rawles, 1953. An experimental study of the origin of the trunk musculature and ribs in the chick. *Am. J. Anat.*, **92:**471–509.

Trelstad, R. L., E. D. Hay, and J. P. Revel, 1967. Cell contact during early morphogenesis in the chick embryo. *Dev. Biol.*, **16:**78–106.

LIMB DEVELOPMENT (CHAPTER 10)

Amprino, R., 1965. Aspects of limb morphogenesis in the chicken, in DeHaan and Ursprung. *Organogenesis.* Holt, Rinehart and Winston, Inc., New York. Pp. 255–281.

Cairns, J. W., 1965. Development of grafts from mouse embryos to the wing bud of the chick embryo. *Dev. Biol.*, **12:**36–52.

Cameron, J., and J. F. Fallon, 1977. The absence of cell death during development of free digits in amphibians. *Dev. Biol.*, **55:**331–338.

Caplan, A. I., 1977. Muscle, cartilage and bone development and differentiation from chick limb mesenchymal cells, in Ede, Hinchliffe, and Balls, *Vertebrate Limb and Somite Morphogenesis.* Cambridge University Press, Cambridge. Pp. 199–214.

——— and S. Kautroupas, 1973. The control of muscle and cartilage development in the chick limb: The role of differential vascularization. *J. Embryol. Exp. Morphol.*, **29:**571–583.

Chevallier, A., M. Kieny, A. Mauger, and P. Sengel, 1977. Developmental fate of the somitic mesoderm in the chick embryo, in Ede, Hinchliffe, and Balls, *Vertebrate Limb and Somite Morphogenesis*. Cambridge University Press, Cambridge. Pp. 421–432.

Christ, B., H. J. Jacob, and M. Jacob, 1977. Experimental analysis of the origin of the wing musculature in avian embryos. *Anat. Embryol.*, **150:**171–186.

Čihák, R., 1972. Ontogenesis of the skeleton and intrinsic muscles of the human hand and foot. *Adv. Anat. Embryol. Cell Biol.*, **46:**1–194.

Ede, D. A., J. R. Hinchliffe, and M. Balls, eds., 1977. *Vertebrate Limb and Somite Morphogenesis*. Cambridge University Press, Cambridge. 498 pp.

Fallon, J. F., and G. M. Crosby, 1975. Normal development of the chick wing following removal of the polarizing zone. *J. Exp. Zool.*, **193:**449–455.

—— and R. O. Kelley, 1977. Ultrastructural analysis of the apical ectodermal ridge during vertebrate limb morphogenesis. *J. Embryol. Exp. Morphol.*, **41:**223–232.

Forsthoefel, P. F., 1963. Observations on the sequence of blastemal condensations in the limbs of the mouse embryo. *Anat. Rec.*, **147:**129–138.

Grim, M., 1972. Ultrastructure of the ulnar portion of the contrahent muscle layer in the embryonic human hand. *Folia Morphol. (Praha)*, **20:**113–115.

Harrison, R. G., 1918. Experiments on the development of the forelimb of *Ambystoma*, a self-differentiating equipotential system. *J. Exp. Zool.*, **25:**413–461.

——, 1921. On relations of symmetry in transplanted limbs. *J. Exp. Zool.*, **32:**1–136.

Kelley, R. O., and J. F. Fallon, 1976. Ultrastructural analysis of the apical ectodermal ridge during limb morphogenesis. I. The human forelimb with special reference to gap junctions. *Dev. Biol.*, **51:**241–256.

MacCabe, J. A., A. J. Calandra, and B. W. Parker, 1977. *In vitro* analysis of the distribution and nature of a morphogenetic factor in the developing chick wing, in Ede, Hinchliffe, and Balls, *Vertebrate Limb and Somite Morphogenesis*. Cambridge University Press, Cambridge. Pp. 25–39.

Milaire, J., 1969. Some histochemical considerations of limb development, in Swinyard, *Limb Development and Deformity: Problems of Evaluation and Rehabilitation*. Charles C Thomas, Publisher, Springfield, Ill. Pp. 70–77.

Pautau, M. P., 1977. Dorso-ventral axis determination of chick limb bud development, in Ede, Hinchliffe, and Balls, *Vertebrate Limb and Somite Morphogenesis*. Cambridge University Press, Cambridge. Pp. 257–266.

Piatt, J., 1956. Studies on the problem of nerve pattern. I. Transplantation of the forelimb primordium to ectopic sites in *Ambystoma*. *J. Exp. Zool.*, **131:**173–202.

——, 1957. Studies on the problem of nerve pattern. II. Innervation of the intact forelimb by different parts of the central nervous system in *Ambystoma*. *J. Exp. Zool.*, **134:**103–126.

Rubin, L., and J. W. Saunders, 1972. Ectodermal-mesodermal interactions in the growth of limb buds in the chick embryo: Constancy and temporal limits of the ectodermal induction. *Dev. Biol.*, **28:**94–112.

Saunders, J. W., 1948. The proximodistal sequence of origin of the parts of the chick wing and the role of the ectoderm. *J. Exp. Zool.*, **108:**363–403.

——, 1969. The interplay of morphogenetic factors, in Swinyard, *Limb Development and Deformity: Problems of Evaluation and Rehabilitation*. Charles C Thomas, Publisher, Springfield, Ill. Pp. 89–100.

——, J. M. Cairns, and M. T. Gasseling, 1957. The role of the apical ridge of ectoderm in the differentiation of the morphological structure and inductive specificity of limb parts in the chick. *J. Morphol.*, **101:**57–87.

―――― and M. T. Gasseling, 1963. Transfilter propagation of apical ectoderm maintenance factor in the chick embryo wing bud. *Dev. Biol.,* **7:**64–78.

―――― and M. T. Gasseling, 1968. Ectodermal-mesenchymal interactions in the origin of
limb symmetry, in Fleischmajer and Billingham, *Epithelial-Mesenchymal Interactions.* The Williams and Wilkins Company, Baltimore. Pp. 78–97.

Seichert, V., and Z. Rychter, 1971. Vascularization of the developing anterior limb of the
chick embryo. *Folia Morphol. (Praha),* **19:**367–377.

Shellswell, G. B., and L. Wolpert, 1977. The pattern of muscle and tendon development
in the chick wing, in Ede, Hinchliffe, and Balls, *Vertebrate Limb and Somite
Morphogenesis.* Cambridge University Press, Cambridge. pp. 71–86.

Stark, R. J., and R. L. Searls, 1973. A description of chick wing bud development and a
model of limb morphogenesis. *Dev. Biol.,* **33:**138–153.

Sullivan, G. E., 1962. Anatomy and embryology of the wing musculature of the domestic
fowl (*Gallus*). *Australian J. Zool.,* **10:**458–518.

Summerbell, D., J. H. Lewis, and L. Wolpert, 1973. Positional information in chick limb
morphogenesis. *Nature,* **244:**492–496.

Swett, F. H., 1937. Determination of limb-axes. *Quart. Rev. Biol.,* **12:**322–339.

Wolpert, L., J. Lewis, and D. Summerbell, 1975. Morphogenesis of the vertebrate limb,
in *Cell Patterning.* Ciba Foundation Symposium 29 (new series), pp. 95–130.

Zwilling, E., 1955. Ectoderm-mesoderm relationship in the development of the chick
embryo limb bud. *J. Exp. Zool.,* **128:**423–442.

――――, 1956. Interaction between limb bud ectoderm and mesoderm in the chick
embryo. I. *J. Exp. Zool.,* **132:**157–172; II. *J. Exp. Zool.,* **132:**173–188; III. *J. Exp.
Zool.,* **132:**219–240; IV. *J. Exp. Zool.,* **132:**241–254.

――――, 1961. Limb morphogenesis. *Adv. Morphogen.,* **1:**301–330.

NERVOUS SYSTEM AND SENSE ORGANS (CHAPTERS 11 AND 12)

Angevine, J. B., and R. L. Sidman, 1961. Autoradiographic study of cell migration
during histogenesis of cerebral cortex in the mouse. *Nature,* **192:**766–768.

Anson, B. J., and T. H. Bast, 1946. The development of the auditory ossicles and
associated structures in man. *Ann. Otol., Rhinol., and Laryngol.,* **55:**467–494.

Bartelmez, G. W., 1922. The origin of the otic and optic primordia in man. *J. Comp.
Neurol.,* **34:**201–232.

――――, 1962. The proliferation of neural crest from forebrain levels in the rat. *Carnegie
Cont. to Emb.,* **37:**1–12.

―――― and A. S. Dekaban, 1962. The early development of the human brain. *Carnegie
Cont. to Emb.,* **37:**13–32.

Bergquist, H. J., and A. J. B. Källén, 1954. Notes on the early histogenesis and
morphogenesis of the central nervous system in vertebrates. *J. Comp. Neurol.,*
100:627–659.

Boulder Committee, The, 1970. Embryonic vertebrate central nervous system: Revised
terminology. *Anat. Rec.,* **166:**257–262.

Bradley, R. M., and C. M. Mistretta, 1975. Fetal sensory receptors. *Physiol. Rev.,*
55:352–382.

Brizzee, K. R., 1949. Histogenesis of the supporting tissue in the spinal and the
sympathetic trunk ganglia in the chick. *J. Comp. Neurol.,* **91:**129–146.

Bunge, R., M. Johnson, and C. D. Ross, 1978. Nature and nurture in development of the
autonomic neuron. *Science,* **199:**1409–1416.

Clayton, R. M., 1970. Problems of differentiation in the vertebrate lens. *Curr. Top. Dev. Biol.,* **5:**115–180.

Coulombre, A. J., 1956. The role of intraocular pressure in the development of the chick eye. I. Control of eye size. *J. Exp. Zool.,* **133:**211–226.

———, 1965. The eye, in DeHaan and Ursprung, *Organogenesis.* Holt, Rinehart and Winston, Inc., New York. pp. 219–251.

——— and J. L. Coulombre, 1957. The role of intraocular pressure in the development of the chick eye: III Ciliary body. *Am. J. Ophthal.,* **44**(4), part 2:85–92.

——— and ———, 1958. Corneal development. I. Corneal transparency. *J. Cell. and Comp. Physiol.,* **51:**1–11.

Coulombre, J. L., and A. J. Coulombre, 1963. Lens development. Fiber elongation and lens orientation. *Science,* **142:**1489–1490.

Detwiler, S. R., 1920. On the hyperplasia of nerve centers resulting from excessive peripheral loading. *Proc. Nat. Acad. Sci.,* **6:**96–101.

———, 1936. *Neuroembryology: An Experimental Study.* The Macmillan Company, New York. 218 pp.

Geren, B. B., 1954. The formation from the Schwann cell surface of myelin in the peripheral nerves of chick embryos. *Exp. Cell Res.,* **7:**558–562.

Gottlieb, G., 1976. Conceptions of prenatal development: Behavioral embryology. *Psych. Rev.,* **83:**215–234.

Hamburger, V., 1958. Regression versus peripheral control of differentiation in motor hypoplasia. *Am. J. Anat.,* **102:**365–410.

———, 1975. Changing concepts in developmental neurobiology. *Perspect. Biol. Med.,* **18:**162–178.

——— and R. Levi-Montalcini, 1949. Proliferation, differentiation and degeneration in the spinal ganglia of the chick embryo under normal and experimental conditions. *J. Exp. Zool.,* **111:**457–501.

Harrison, R. G., 1908. Embryonic transplantation and the development of the nervous system. *Anat. Rec.,* **2:**385–410.

———, 1910. The outgrowth of the nerve fiber as a mode of protoplasmic movement. *J. Exp. Zool.,* **9:**787–848.

Hay, E. D., and J. W. Dodson, 1973. Secretion of collagen by corneal epithelium. *J. Cell Biol.,* **57:**190–213.

——— and J.-P. Revel, 1969. *Fine Structure of the Developing Avian Cornea.* Monogr. Dev. Biol., S. Karger, Basel. Vol. 1, 144 pp.

Held, H., 1909. *Die Entwicklung des Nervengewebes bei den Wirbeltieren.* Johann Ambrosius Barth, Munich. 378 pp.

Herrick, C. J., 1925. Morphogenetic factors in the differentiation of the nervous system. *Physiol. Rev.,* **5:**112–130.

Hirose, G., and M. Jacobson, 1979. Clonal organization of the central nervous system of the frog. I. Clones stemming from individual blastomeres of the 16-cell and earlier stages. *Dev. Biol.,* **71:**191–202.

Hooker, D., 1952. *The Prenatal Origin of Behavior.* Porter Lectures, series 18, University of Kansas Press, Lawrence. 143 pp.

Hughes, A., 1961. Cell degeneration in the larval ventral horn of *Xenopus laevis* (Daudin). *J. Embryol. Exp. Morphol.,* **9:**269–284.

———, 1968. *Aspects of Neural Ontogeny.* Logos Press, London. 249 pp.

Humphrey, T., 1952. The spinal tract of the trigeminal nerve in human embryos between

7½ and 8½ weeks of menstrual age and its relation to early fetal behavior. *J. Comp. Neurol.,* **97:**143–209.

Jacobson, M., 1978. *Developmental Neurobiology.* Plenum, New York. 2d ed., 562 pp.

Källén, B., 1953. On the significance of the neuromeres and similar structures in vertebrate embryos. *J. Embryol. Exp. Morphol.,* **1:**387–392.

Konyukhov, B. V., and M. P. Vakhrusheva, 1969. Abnormal development of eyes in mice homozygous for the fidget gene. *Teratology,* **2:**147–158.

Langman, J., 1959. The first appearance of specific antigens during the induction of the lens. *J. Embryol. Exp. Morphol.,* **7:**193–202.

———, 1959. Appearance of antigens during development of the lens. *J. Embryol. Exp. Morphol.,* **7:**264–276.

———, R. L. Guerrant, and B. A. Freeman, 1966. Behavior of neuroepithelial cells during closure of the neural tube. *J. Comp. Neurol.,* **127:**399–412.

———, P. Rodier, and W. Webster, 1975. Interference with proliferative activity in the CNS and its relation to facial abnormalities, in Bergsma, *Morphogenesis and Malformation of the Face and Brain.* Birth Defects: Original Article Series, **XI** (7)**:**95–129.

Larsell, O., 1947. The development of the cerebellum in man in relation to its comparative anatomy. *J. Comp. Neurol.,* **87:**85:129.

———, 1954. The development of the cerebellum of the pig. *Anat. Rec.,* **118:**73–107.

Levi-Montalcini, R., 1958. Chemical stimulation of nerve growth, in McElroy and Glass, *The Chemical Basis of Development.* The Johns Hopkins Press, Baltimore. Pp. 646–664.

———, and P. U. Angeletti, 1961. Growth control of the sympathetic system by a specific protein factor. *Quart. Rev. Biol.,* **36:**99–108.

Lewis, W. H., 1904. Experimental studies on the development of the eye in amphibia. I. On the origin of the lens. *Am. J. Anat.* **3:**505–536.

———, 1905. Experimental studies on the development of the eye in amphibia. II. On the cornea. *J. Exp. Zool.,* **2:**431–446.

Lopashov, G. V., and O. G. Stroeva, 1964. *Development of the Eye.* Israel Program for Scientific Translations, Jerusalem. 177 pp.

Mann, I., 1964. *The Development of the Human Eye.* Grune & Stratton, Inc., New York. 3d ed., 316 pp.

Meier, S., and E. D. Hay, 1974. Control of corneal differentiation by extracellular materials. Collagen as a promoter and stabilizer of epithelial stroma production. *Dev. Biol.,* **38:**249–270.

O'Rahilly, R., and E. Gardner, 1977. The developmental anatomy and histology of the human central nervous system, in Vinken and Bruyn, *Handbook of Clinical Neurology, Vol. 30: Congenital Malformations of the Brain and Skull.* North-Holland Publishing Company, Amsterdam. Pp. 15–40.

Papaconstantinou, J., 1967. Molecular aspects of lens differentiation. *Science,* **156:**338–346.

Pearson, A. A., 1947. The roots of the facial nerve in human embryos and fetuses. *J. Comp. Neurol.,* **87:**139–159.

Rakic, P., 1975. Cell migration and neuronal ectopias in the brain, in Bergsma, *Morphogenesis and Malformation of Face and Brain.* Birth Defects: Original Article Series, **II** (7)**:**95–129.

Reeder, R., and E. Bell, 1965. Short- and long-lived messenger RNA in embryonic chick lens. *Science,* **150:**71–72.

Reyer, R. W., 1977. The amphibian eye: Development and regeneration, in Crescitelli, *Handbook of Sensory Physiology, vol. VII,15: The Visual System in Vertebrates.* Springer-Verlag, Berlin. Pp. 309–390.

Sauer, F. C., 1935. The cellular structure of the neural tube. *J. Comp. Neurol., 63*:13–23.

———, and B. E. Walker, 1959. Radioautoradiographic study of interkinetic nuclear migration in the neural tube. *Proc. Soc. Exp. Biol. Med., 101*:557–560.

Schmechel, D. E., and P. Rakic, 1979. A Golgi study of radial glial cells in developing monkey telencephalon: Morphogenesis and transformation into astrocytes. *Anat. Embryol., 156*:115–152.

Spemann, H., 1901. Über Korrelationen in der Entwicklung des Auges. *Verk. Anat. Ges. 15 Vers. Bonn,* pp. 61–79.

———, 1912. Zur Entwicklung des Wirbeltierauges. *Zool. Jahrb., Abt. allg. Zool. Physiol. Tiere, 32*:1–98.

Streeter, G. L., 1906. On the development of the membranous labyrinth and the acoustic and facial nerves in the human embryo. *Am. J. Anat., 6*:139–166.

———, 1908. The peripheral nervous system in the human embryo at the end of the first month (10 mm.). *Am. J. Anat., 8*:285–301.

———, 1922. Development of the auricle in the human embryo. *Carnegie Cont. to Emb., 14*:111–138.

Toole, B. P., and R. L. Trelstad, 1971. Hyaluronate production and removal during corneal development in the chick. *Dev. Biol., 26*:28–35.

Weiss, P., 1934. *In vitro* experiments on the factors determining the course of the outgrowing nerve fiber. *J. Exp. Zool., 68*:393–448.

——— and H. B. Hiscoe, 1948. Experiments on the mechanism of nerve growth. *J. Exp. Zool., 107*:315–396.

Windle, W. F., and D. W. Orr, 1934. The development of behavior in chick embryos: Spinal cord structure correlated with early somatic motility. *J. Comp. Neurol., 60*:287–307.

Yntema, C. L., and W. S. Hammond, 1955. Experiments on the origin and development of the sacral autonomic nerves in the chick embryo. *J. Exp. Zool., 129*:375–413.

Zelená, J., 1964. Development, degeneration and regeneration of receptor organs, in Singer and Schadé, *Mechanisms of Neural Regeneration.* Progress in Brain Research. Vol. 13, pp. 175–213.

Zwann, J., and R. W. Hendrix, 1973. Changes in cell and organ shape during early development of the ocular lens. *Am. Zool., 13*:1039–1049.

——— and A. Ikeda, 1968. Macromolecular events during differentiation of the chicken eye lens. *Exp. Eye Res., 7*:301–311.

THE FACE, ORAL REGION, AND TEETH (CHAPTER 13)

Bernfield, M. R., R. H. Cohn, and S. D. Banerjee, 1973. Glycosaminoglycans and epithelial organ formation. *Am. Zool., 13*:1067–1083.

Bevelander, G., 1941. The development and structure of the fiber system of dentin. *Anat. Rec., 81*:79–97.

Bradley, R. B., and C. M. Mistretta, 1975. Fetal sensory receptors. *Physiol. Rev., 55*:352–382.

Chase, S. W., 1932. Histogenesis of the enamel. *J. Am. Dental Assn., 19*:1275–1289.

Dahlberg, A. A., ed., 1971. *Dental Morphology and Evolution.* University of Chicago Press, Chicago. 350 pp.

Ferguson, M. W. J., 1978. Palatal shelf elevation in the Wistar rat fetus. *J. Anat.,* **125:**555–577.

Green, R. M., and R. M. Pratt, 1976. Developmental aspects of secondary palate formation. *J. Embryol. Exp. Morphol.,* **36:**225–245.

Grobstein, C., 1953. Epithelio-mesenchymal specificity in the morphogenesis of mouse submandibular rudiments in vitro. *J. Exp. Zool.,* **124:**383–413.

Johnson, P. L., and G. Bevelander, 1954. The localization and interrelation of nucleic acids and alkaline phosphatase in the developing tooth. *J. Dent. Res.,* **33:**128–135.

Koch, W. E., 1967. In vitro differentiation of tooth rudiments of embryonic mice. I. Transfilter interaction of embryonic incisor tissues. *J. Exp. Zool.,* **165:**155–170.

Kollar, E. J., and G. R. Baird, 1969. The influence of the dental papilla on the development of tooth shape in the embryonic mouse tooth germs. *J. Embryol. Exp. Morphol.,* **21:**131–148.

Kraus, B. S., H. Kitamura, and R. A. Latham, 1966. *Atlas of Developmental Anatomy of the Face.* Hoeber Harper, New York. 378 pp.

Lawson, K. A., 1974. Mesenchyme specificity in rodent salivary gland development: The response of salivary epithelium to lung mesenchyme in vitro. *J. Embryol. Exp. Morphol.,* **32:**469–493.

Mistretta, C. M., 1972. Topographical and histological study of the developing rat tongue, palate and taste buds, in Bosma, *Third Symposium on Oral Sensation and Perception: The Mouth of the Infant.* Charles C Thomas, Publisher, Springfield, Ill. Pp. 163–187.

Patten, B. M., 1961. The normal development of the facial region, in Pruzansky, *Congenital Anomalies of the Face and Associated Structures.* Charles C Thomas, Publisher, Springfield, Ill. pp. 11–45.

Ross, R. B., and M. C. Johnston, 1972. *Cleft Lip and Palate.* The Williams & Wilkins Company, Baltimore. Chapter 2.

Schour, I., and M. Massler, 1941. The development of the human dentition. *J. Am. Dental Assn.,* **28:**1153–1160.

Schultz, A. H., 1920. The development of the external nose in whites and negroes. *Carnegie Cont. to Emb.,* **9:**173–190.

Walker, B. E., and F. C. Fraser, 1956. Closure of the secondary palate in three strains of mice. *J. Embryol. Exp. Morphol.,* **4:**176–189.

DIGESTIVE AND RESPIRATORY SYSTEMS AND THE BODY CAVITIES AND MESENTERIES (CHAPTER 14)

Digestive Tube and Glands

Bernfield, M. R., and S. H. Banerjee, 1972. Acid mucopolysaccharide (glycosaminoglycan) at the epithelial-mesenchymal interface of mouse embryo salivary glands. *J. Cell Biol.,* **52:**664–673.

Bryden, M. M., H. E. Evans, and W. Binns, 1972. Embryology of the sheep. II. The alimentary tract and associated glands. *J. Morphol.,* **138:**187–206.

Corner, G. W., 1914. The structural unit and growth of the pancreas of the pig. *Am. J. Anat.,* **16:**207–236.

Cullen, T. S., 1916. *Embryology, Anatomy, and Diseases of the Umbilicus, together with Diseases of the Urachus.* W. B. Saunders Company, Philadelphia. 680 pp.

Deren, J. J., B. W. Strauss and T. H. Wilson, 1965. The development of structure and transport systems of the fetal rabbit intestine. *Dev. Biol.,* **12:**467–486.

Flint, J. M., 1903. The angiology, angiogenesis, and organogenesis of the submaxillary gland. *Am. J. Anat.*, **2**:417–444.

Jackson, C. M., 1909. On the developmental topography of the thoracic and abdominal viscera. *Anat. Rec.*, **3**:361–396.

Jost, A., 1962. Hormonal factors controlling the storage of glycogen in the fetal liver, in Cori, C. F., V. G. Foglia, L. F. Leloir and S. Ochoa, *Perspectives in Biology.* Elsevier, Amsterdam, pp. 174–178.

Kingsbury, J. W., M. Alexanderson, and E. S. Kornstein, 1956. The development of the liver in the chick. *Anat. Rec.*, **124**:165–188.

Koga, A., 1971. Morphogenesis of intrahepatic bile ducts of the human fetus. *Z. Anat. Entwickl.-Gesch.*, **135**:156–184.

Le Douarin, N. M., 1975. An experimental analysis of liver development. *Med. Biol.*, **53**:427–455.

Lewis, F. T., 1911. The bilobed form of the ventral pancreas in mammals. *Am. J. Anat.*, **12**:389–400.

——, 1912. The form of the stomach in human embryos with notes upon the nomenclature of the stomach. *Am. J. Anat.*, **13**:477–503.

Lim, S.-S., and F. N. Low, 1977. Scanning electron microscopy of the developing alimentary canal in the chick. *Am. J. Anat.*, **150**:149–174.

Mathan, M., P. C. Moxey, and J. S. Trier, 1976. Morphogenesis of fetal rat duodenal villi. *Am. J. Anat.*, **146**:73–92.

Moog, F., 1950. The functional differentiation of the small intestine. I. The accumulation of alkaline phosphomonoesterase in the duodenum of the chick. *J. Exp. Zool.*, **115**:109–129.

——, 1951. The functional differentiation of the small intestine. II. The differentiation of alkaline phosphomonoesterase in the duodenum of the mouse. *J. Exp. Zool.*, **118**:187–207.

O'Rahilly, R., 1978. The timing and sequence of events in the development of the human digestive system and associated structures during the embryonic period proper. *Anat. Embryol.*, **153**:123–136.

Pictet, R., and W. J. Rutter, 1972. Development of the embryonic endocrine pancreas, in Greep and Astwood, *Handbook of Physiology,* Section 7: Endocrinology, Vol. 1. Endocrine Pancreas. American Physiological Society, Washington. Pp. 25–66.

Plenk, H., 1931. Zur Entwicklung des menschlichen Magens. *Zeitschr. f. Mikr. -anat. Forsch.*, **26**:547–645.

Rudnick, D., 1952. Development of the digestive tube and its derivatives. *Ann. N. Y. Acad. Sci.*, **55**:109–116.

Spooner, B. S., and N. K. Wessels, 1972. An analysis of salivary gland morphogenesis: Role of cytoplasmic microfilaments and microtubules. *Dev. Biol.*, **27**:38–54.

Wessels, N. K., 1977. *Tissue Interactions and Development.* W. A. Benjamin, Inc., Menlo Park. 276 pp.

Respiratory System

Alescio, T., and A. Cassini, 1962. Induction in vitro of tracheal buds by pulmonary mesenchyme grafted on tracheal epithelium. *J. Exp. Zool.*, **150**:83–94.

Avery, M. E., N.-S. Wang, and H. W. Taeusch, 1973. The lung of the newborn infant. *Sci. Am.*, **228**:74–85.

Barnard, W. G., and T. D. Day, 1937. The development of the terminal air passages of the human lung. *J. Pathol. and Bacteriol.*, **45**:67–73.

Bremer, J. L., 1935. Postnatal development of alveoli in the mammalian lung in relation to the problem of the alveolar phagocyte. *Carnegie Cont. to Emb.,* **25:**83–111.

———, 1943. Pleuroperitoneal membrane and bursa infracardiaca. *Anat. Rec.,* **87:**311–319.

Emery, J., ed., 1969. *The Anatomy of the Developing Lung.* William Heinemann, Ltd., London. 223 pp.

O'Hare, K. H., and P. L. Townes, 1970. Morphogenesis of albino rat lung: An autoradiographic analysis of the embryological origin of the type I and II pulmonary epithelial cells. *J. Morphol.,* **132:**69–88.

O'Rahilly, R., and E. A. Boyden, 1973. The timing and sequence of events in the development of the human respiratory system during the embryonic period proper. *Z. Anat. Entwickl. -Gesch.,* **141:**237–250.

de Reuck, A. V. S., and R. Porter, eds., 1967. *Development of the Lung.* J. & A. Churchill Ltd., London. 408 pp.

Rudnick, D., 1933. Developmental capacities of the chick lung in chorioallantoic grafts. *J. Exp. Zool.,* **66:**125–154.

Spooner, B. S., and N. K. Wessels, 1970. Mammalian lung development: Interactions in primordium formation and bronchial morphogenesis. *J. Exp. Zool.,* **175:**445–454.

Wells, L. J., and E. A. Boyden, 1954. The development of the bronchopulmonary segments in human embryos of horizons XVII to XIX. *Am. J. Anat.,* **95:**163–201.

Wessels, N. K., 1970. Mammalian lung development: Interactions in formation and morphogenesis of tracheal buds. *J. Exp. Zool.,* **175:**455–466.

Whitehead, W. H., W. F. Windle, and R. F. Becker, 1942. Changes in lung structure during aspiration of amniotic fluid and during air-breathing at birth. *Anat. Rec.,* **83:**255–265.

Body Cavities and Mesenteries

Elliott, R., 1933. A contribution to the development of the pericardium. *Am. J. Anat.,* **48:**355–390.

Mall, F. P., 1891. Development of the lesser peritoneal cavity in birds and mammals. *J. Morphol.,* **5:**165–179.

———, 1897. Development of the human coelom. *J. Morphol.,* **12:**395–453.

Wells, L. J., 1954. Development of the human diaphragm and pleural sacs. *Carnegie Cont. to Emb.,* **35:**107–134.

DUCTLESS GLANDS, PHARYNGEAL DERIVATIVES, AND LYMPHOID SYSTEM (CHAPTER 15)

Atwell, W. J., 1926. The development of the hypophysis cerebri in man, with special reference to the pars tuberalis. *Am. J. Anat.,* **37:**159–193.

Auerbach, ·R., 1966. Embryogenesis of immune systems, in Wolstenholme and Porter, *The Thymus.* Ciba Foundation Symposium. J. & A. Churchill, Ltd., London. Pp. 39–49.

Badertscher, J. A., 1918. The fate of the ultimobranchial bodies in the pig (*Sus scrofa*). *Am. J. Anat.,* **23:**89–131.

Burnet, M., 1969. *Cellular Immunology.* Cambridge University Press, Cambridge. Books 1 and 2, 726 pp.

Carpenter, E., and T. Rondon-Tarchetti, 1957. Differentiation of embryonic rat thyroid in vivo and in vitro. *J. Exp. Zool.,* **136:**393–417.

Crowder, R. E., 1957. The development of the adrenal gland in man, with special reference to origin and ultimate location of cell types and evidence in favor of the "cell migration" theory. *Carnegie Cont. to Emb.,* **36:**195–210.

Godwin, M. C., 1939. The mammalian thymus IV. The development in the dog. *Am. J. Anat.,* **64:**165–201.

———, 1940. The development of complex IV in the pig: A comparison of the conditions in the pig with those in the rat, cat, dog, calf, and man. *Am. J. Anat.,* **66:**51–85.

———, 1943. Thymus IV in the pig. *Anat. Rec.,* **85:**229–243.

Gruenwald, P., 1946. Embryonic and postnatal development of the adrenal cortex, particularly the zona glomerulosa and accessory nodules. *Anat. Rec.,* **95:**391–421.

Hamburgh, M., and E. J. W. Barrington, eds., 1971. *Hormones in Development.* Appleton-Century-Crofts, Inc., New York. 854 pp.

Hamilton, H. L., and G. W. Hinsch, 1957. The fate of the second visceral pouch in the chick. *Anat. Rec.,* **129:**357–369.

Hillemann, H. H., 1943. An experimental study of the development of the pituitary gland in chick embryos. *J. Exp. Zool.,* **93:**347–373.

Jost, A., 1961. The role of fetal hormones in prenatal development. *Harvey Lect. Series,* **55:**201–226.

Keene, M. F. L., and E. E. Hewer, 1927. Observations on the development of the human suprarenal gland. *J. Anat.,* **61:**302–324.

Kingsbury, B. F., 1915. The development of the human pharynx. I. The pharyngeal derivatives. *Am. J. Anat.,* **18:**329–397.

Kitchell, R. L., and L. J. Wells, 1952. Functioning of the hypophysis and adrenals in fetal rats: Effects of hypophysectomy, adrenalectomy, castration, injected ACTH and implanted sex hormones. *Anat. Rec.,* **112:**561–591.

Klapper, C. E., 1946. The development of the pharynx of the guinea pig with special emphasis on the fate of the ultimobranchial body. *Am. J. Anat.,* **79:**361–397.

Mitskevich, M. S., 1959. *Glands of Internal Secretion in the Embryonic Development of Birds and Mammals.* National Science Foundation and Israel Program for Scientific Translations, Washington. 304 pp.

Moore, M. A. S., and J. J. T. Owen, 1966. Experimental studies on the development of the bursa of Fabricius. *Dev. Biol.,* 40–51.

Nelson, W. O., 1933. Studies on the anterior hypophysis. I. The development of the hypophysis in the pig (*Sus scrofa*). II. The cytological differentiation in the anterior hypophysis of the foetal pig. *Am. J. Anat.,* **52:**307–332.

Norris, E. H., 1937. The parathyroid glands and the lateral thyroid in man: their morphogenesis, histogenesis, topographic anatomy and prenatal growth. *Carnegie Cont. to Emb.,* **26:**247–294.

———, 1938. The morphogenesis and histogenesis of the thymus gland in man: In which the origin of the Hassall's corpuscles of the human thymus is discovered. *Carnegie Cont. to Emb.,* **27:**191–208.

Politzer, G., and F. Hann, 1935. Über die Entwicklung der branchiogenen Organe beim Menschen. *Zeitschr. f. Anat. u. Entwg.,* **104:**670–708.

Shain, W. G., S. R. Hilfer, and V. G. Fonte, 1972. Early organogenesis of the embryonic chick thyroid. I. Morphology and biochemistry. *Dev. Biol.,* **28:**202–218.

Shanklin, W. M., 1944. Histogenesis of the pig neurohypophysis. *Am. J. Anat.,* **74:**327–353.

Silverstein, A. M., 1964. Ontogeny of the immune response. *Science,* **144:**1423–1428.

Smith, R. T., R. A. Good, and P. A. Miescher, eds., 1967. *Ontogeny of Immunity.* University of Florida Press, Gainesville. 208 pp.

Šterzl, J., and I. Říha, 1970. *Developmental Aspects of Antibody Formation and Structure.* Academia Publishing House, Prague. Vols. 1 and 2, 1054 pp.

Sucheston, M. E., and M. S. Cannon, 1968. Development of zonular patterns in the human adrenal gland. *J. Morphol.,* **126**:477–492.

Uotila, U. U., 1940. The early embryological development of the fetal and permanent adrenal cortex in man. *Anat. Rec.,* **76**:183–203.

Van Dyke, J. H., 1941. On the origin of accessory thymus tissue, thymus IV: The occurrence in man. *Anat. Rec.,* **79**:179–209.

Waterman, A. J., and A. Gorbman, 1956. Development of the thyroid gland of the rabbit. *J. Exp. Zool.,* **132**:509–538.

Weller, G. L., Jr., 1933. Development of the thyroid, parathyroid and thymus glands in man. *Carnegie Cont. to Emb.,* **24**:93–140.

Wilson, M. E., 1952. The embryological and cytological basis of regional patterns in the definitive epithelial hypophysis of the chick. *Am. J. Anat.,* **91**:1–50.

Woods, J. E., G. W. De Vries, and R. C. Thommes, 1971. Ontogenesis of the pituitary-adrenal axis in the chick embryo. *Gen. Comp. Endocrinol.,* **17**:407–415.

THE UROGENITAL SYSTEM (CHAPTER 16)

Abdel-Malek, E. T., 1950. Early development of the urinogenital system in the chick. *J. Morphol.,* **86**:599–626.

Backhouse, K. M., and H. Butler, 1960. The gubernaculum testis of the pig. *J. Anat.,* **94**:107–120.

Bremer, J. L., 1916. The interrelations of the mesonephros, kidney, and placenta in different classes of animals. *Am. J. Anat.,* **19**:179–209.

Burns, R. K., 1961. Role of hormones in the differentiation of sex, in Young, *Sex and Internal Secretions.* The Williams & Wilkins Company, Baltimore. 3d. ed., vol. 1, Pp. 16–158.

Byskov, A. G., 1978. The meiosis inducing interaction between germ cells and rete cells in the fetal mouse gonad. *Ann. Biol. Anim. Biochem. Biophys.,* **18**(2B):327–334.

——— and L. Saxén, 1976. Induction of meiosis in fetal mouse testis *in vitro. Dev. Biol.,* **52**:193–200.

Cullen, T. S., 1916. *Embryology, Anatomy, and Diseases of the Umbilicus, together with Diseases of the Urachus.* W. B. Saunders Company, Philadelphia. 680 pp.

De Martino, C., and L. Zamboni, 1966. A morphologic study of the mesonephros of the human embryo. *J. Ultrastr. Res.,* **16**:399–427.

Du Bois, A. M., 1969. The embryonic kidney, in Rouiller and Muller, *The Kidney.* Academic Press, Inc., New York. pp. 1–59.

Erickson, R. S., 1968. Inductive interactions in the development of the mouse metanephros. *J. Exp. Zool.,* **169**:33–42.

Fox, H. 1963. The amphibian pronephros. *Quart. Rev. Biol.,* **38**:1–25.

Fraser, E. A., 1950. The development of the vertebrate excretory system. *Biol. Rev.,* **25**:159–187.

Gersh, I., 1937. The correlation of structure and function in the developing mesonephros and metanephros. *Carnegie Cont. to Emb.,* **26**:33–58.

Gillman, J., 1948. The development of the gonads in man, with a consideration of the role of fetal endocrines and the histogenesis of ovarian tumors. *Carnegie Cont. to Emb.,* **32**:81–131.

Glenister, T. W., 1956. The development of the penile urethra in the pig. *J. Anat.,* **90**:461–477.

Grobstein, C., 1955. Inductive interaction in the development of the mouse metaneph-ros. *J. Exp. Zool.,* **130:**319–340.

Gruenwald, P., 1939. The mechanism of kidney development in human embryos as revealed by an early stage in the agenesis of the ureteric buds. *Anat. Rec.,* **75:**237–247.

———, 1942. The development of the sex cords in the gonads of man and mammals. *Am. J. Anat.,* **70:**359–397.

——, 1943. The normal changes in the position of the embryonic kidney. *Anat. Rec.,* **85:**163–176.

Hamilton, W. J., J. D. Boyd, and H. W. Mossman, 1972. *Human Embryology,* The Williams & Wilkins Company, Baltimore. 4th ed., 646 pp.

Holtfreter, J., 1944. Experimental studies on the development of the pronephros. *Rev. Canad. Biol.,* **3:**220–250.

Huber, G. C., 1905. On the development and shape of uriniferous tubules of certain of the higher mammals. *Am. J. Anat.,* **4**(suppl):1–98.

Hunter, R. H., 1930. Observations on the development of the human female genital tract. *Carnegie Cont. to Emb.,* **22:**91–108.

Jokelainen, P., 1963. An electron microscope study of the early development of the rat metanephric nephron. *Acta Anat.,* **52**(suppl. 47):1–73.

Jost, A., 1972. A new look at the mechanisms controlling sex differentiation in mammals. *Johns Hopkins Med. J.,* **130:**38–53.

Koff, A. K., 1933. Development of the vagina in the human fetus. *Carnegie Cont. to Emb.,* **24:**59–90.

Lehtonen, E., 1976. Transmission of signals in embryonic induction. *Med. Biol.,* **54:**108–128.

Lillie, F. R., 1917. The freemartin; a study of the action of sex hormones in the foetal life of cattle. *J. Exp. Zool.,* **23:**371–452.

Ohno, W., 1978. The role of H-Y antigen in primary sex determination. *J. Amer. Med. Assoc.,* **239:**217–220.

Oliver, J., 1968. *Nephrons and Kidneys.* Harper & Row, New York. 116 pp.

O'Rahilly, R., 1977. The development of the vagina in the human, in Blandau and Bergsma, *Morphogenesis and Malformation of the Genital System.* Birth Defects: Original Article Series, Vol. 13(2). Alan R. Liss Inc., New York, pp. 123–136.

Osathanondh, V., and E. L. Potter, 1963. Development of the human kidney as shown by microdissection. I. *Arch. Path.,* **76:**271–276; II. *Arch. Path.,* **76:**277–289; I. *Arch. Path.,* **76:**290–302.

Osathanondh, V., and E. L. Potter, 1966. Development of the human kidney as shown by microdissection. IV and V. *Arch. Path.,* **82:**391–411.

Politzer, G., 1952. Das Schicksal des Sinus urogenitalis beim Weibe. *Zeitschr. f. Mikr. -anat. Forsch.,* **59:**6–28.

Potter, E. L., 1965. Development of the human glomerulus. *Arch. Path.* **80:**241–255.

Price, D., 1936. Normal development of the prostate and seminal vesicles of the rat with a study of experimental postnatal modifications. *Am. J. Anat.,* **60:**79–127.

Price, D., 1947. An analysis of the factors influencing growth and development of the mammalian reproductive tract. *Physiol. Zool.,* **20:**213–247.

Saxén, L., and E. Lehtonen, 1978. Transfilter induction of kidney tubules as a function of the extent and duration of intercellular contacts. *J. Embryol. Exp. Morphol.,* **47:**97–109.

Shikinami, J., 1926. Detailed form of the Wolffian body in human embryos of the first eight weeks. *Carnegie Cont. to Emb.,* **18:**49–62.

Spaulding, M. H., 1921. The development of the external genitalia in the human embryo. *Carnegie Cont. to Emb.*, **13:**67–88.

Torrey, T. W., 1954. The early development of the human nephros. *Carnegie Cont. to Emb.*, **35:**175–198.

———, 1965. Morphogenesis of the vertebrate kidney, in DeHaan and Ursprung, *Organogenesis.* Holt, Rinehart and Winston, Inc., New York. pp. 559–579.

———, 1971. *Morphogenesis of the Vertebrates.* John Wiley & Sons, Inc., New York. 3d ed., 529 pp.

Van Wagenen, G., 1965. *Embryology of the Ovary and Testis, Homo sapiens and Macaca mulatta.* Yale University Press, New Haven, Conn. 256 pp.

Vernier, R. L., and A. Birch-Anderson, 1962. Studies of the human fetal kidney. I. Development of the glomerulus. *J. Pediatr.*, **60:**754–768.

Wachtel, S. S., S. Ohno, G. C. Koo, and E. A. Boyse, 1975. Possible role of H-Y antigen in primary determination of sex. *Nature*, **257:**235–236.

Waddington, C. H., 1938. The morphogenetic function of a vestigial organ in the chick. *J. Exp. Biol.*, **15:**371–376.

Wells, L. J., and R. L. Fralick, 1951. Production of androgen by the testes of fetal rats. *Am. J. Anat.*, **89:**63–107.

Witschi, E., 1939. Modification of the development of sex in lower vertebrates and in mammals, in Allen, Danforth and Doisy, *Sex and Internal Secretions.* The Williams & Wilkins Company, Baltimore. 2d ed., pp. 145–226.

———, 1948. Migration of the germ cells of human embryos from the yolk sac to the primitive gonadal folds. *Carnegie Cont. to Emb.*, **32:**67–80.

THE CIRCULATORY SYSTEM (CHAPTER 17)

Auer, J., 1948. The development of the human pulmonary vein and its major variations. *Anat. Rec.*, **101:**581–594.

Barclay, A. E., K. J. Franklin, M. M. L. Prichard, 1944. *The Foetal Circulation and Cardiovascular System, and the Changes that They Undergo at Birth.* Blackwell Scientific Publications, Ltd., Oxford. 275 pp.

Barcroft, J., 1946. *Researches on Prenatal Life.* Blackwell Scientific Publications, Ltd., Oxford. Vol. 1, 292 pp.

Barron, D. H., 1944. The changes in the fetal circulation at birth. *Physiol. Rev.*, **24:**277–295.

Barry, A., 1942. The intrinsic pulsation rates of fragment of the embryonic chick heart. *J. Exp. Zool.*, **91:**119–130.

———, 1951. The aortic arch derivatives in the human adult. *Anat. Rec.*, **111:**221–238.

Bloom, W., and G. W. Bartelmez, 1940. Hematopoiesis in young human embryos. *Am. J. Anat.*, **67:**21–53.

Born, G., 1889. Beiträge zur Entwicklungsgeschichte des Säugethierherzens. *Arch. f. mikr. Anat.*, **33:**284–377.

Boucek, R. J., W. P. Murphy, and G. H. Paff, 1959. Electrical and mechanical properties of chick embryo heart chambers. *Circulation Res.*, **7:**787–793.

Butler, E. G., 1927. The relative role played by the embryonic veins in the development of the mammalian vena cava posterior. *Am. J. Anat.*, **39:**267–353.

Chacko, A. W., and S. R. M. Reynolds, 1953. Embryonic development in the human of the sphincter of the ductus venosus. *Anat. Rec.*, **115:**151–173.

Comline, R. S., K. W. Cross, G. S. Dawes, and P. W. Nathanielsz, eds., 1973. *Foetal and Neonatal Physiology.* Cambridge University Press, Cambridge. 641 pp.

Congdon, E. D., 1922. Transformation of the aortic-arch system during the development of the human embryo. *Carnegie Cont. to Emb.*, **14:**47–110.

Danesino, V. L., S. R. M. Reynolds, and I. H. Rehman, 1955. Comparative histological structure of the human ductus arteriosus according to topography, age, and degree of constriction. *Anat. Rec.*, **121:**801–830.

Dawes, G. S., J. C. Mott, and J. G. Widdicombe, 1955. The patency of the ductus arteriosus in newborn lambs and its physiological consequences. *J. Physiol.*, **128:**361–383.

————, ————, and ————, 1955. Closure of the foramen ovale in newborn lambs. *J. Physiol.*, **128:**384–395.

DeHaan, R. L., 1959. Cardia bifida and the development of pacemaker function of the early chick heart. *Develop. Biol.*, **1:**586–602.

de Vries, P. A., and J. B. de C. M. Saunders, 1962. Development of the ventricles and spiral outflow tract in the human heart. A contribution to the development of the human heart from age group IX to age group XV. *Carnegie Cont. to Emb.*, **37:**87–114.

Dieterlen-Lievre, R., 1978. Yolk sac erythropoiesis. *Experientia,* **34:**284–289.

Evans, H. M., 1909. On the development of the aortae, cardinal and umbilical veins and other blood vessels of vertebrate embryos from capillaries. *Anat. Rec.*, **3:**498–518.

Everett, N. B., and R. J. Johnson, 1951. A physiological and anatomical study of the closure of the ductus arteriosus in the dog. *Anat. Rec.*, **110:**103–111.

Girard, H., 1973. Arterial pressure in the chick embryo. *Am. J. Physiol.*, **224:**454–560.

Goerttler, K., 1955. Über Blutstromwirkung als Gestaltsungsfaktor für die Entwicklung des Herzens. *Beitrage zur path. Anat.*, **115:**33–56.

Goss, C. M., 1942. The physiology of the embryonic mammalian heart before circulation. *Am. J. Physiol.*, **137:**146–152.

————, 1952. Development of the median coordinated ventricle from the lateral hearts in rat embryos with 3 to 6 somites. *Anat. Rec.*, **112:**761–796.

Graeper, L., 1907. Untersuchungen über die Herzbildung der Vögel. *Wilhelm Roux', Arch.*, **24:**375–410.

Heuser, C. H., 1923. The branchial vessels and their derivatives in the pig. *Carnegie Cont. to Emb.*, **15:**121–139.

Hoff, E. C., T. C. Kramer, D. DuBois, and B. M. Patten, 1939. The development of the electrocardiogram of the embryonic heart. *Am. Heart J.*, **17:**470–488.

Jacobson, A. G., 1960. Influences of ectoderm and endoderm on heart differentiation in the newt. *Dev. Biol.*, **2:**138–154.

Jaffe, O. C., 1962. Hemodynamics and cardiogenesis. I. The effects of altered vascular patterns on cardiac development. *J. Morphol.*, **110:**217–226.

Johnstone, P. N., 1971. *Studies on the Physiological Anatomy of the Embryonic Heart.* Charles C Thomas, Publisher, Springfield, Ill. 139 pp.

Kennedy, J. A., and S. L. Clark, 1941. Observations on the ductus arteriosus of the guinea pig in relation to its method of closure. *Anat. Rec.*, **79:**349–371.

Knezevic, A., N. Petrovic, and D. Radivoyevic, 1971. The effect of the humoral erythropoietic stimulation factor of different origins on chick embryo hematopoiesis. *Jugoslav. Physiol. Pharmacol. Acta,* **7:**421–429.

Kramer, T. C., 1942. The partitioning of the truncus and conus and the formation of the membranous portion of the interventricular septum in the human heart. *Am. J. Anat.*, **71:**343–370.

Lemanski, L. F., D. J. Paulson, and C. S. Hill, 1979. Normal anterior endoderm corrects the heart defect in cardiac mutant salamanders (*Ambystoma mexicanum*). *Science,* **204:**860–862.

Lemez, L., 1964. The blood of chick embryos. Quantitative embryology at a cellular level. *Adv. Morphogen.,* **3:**197–245.

Licata, R. H., 1954. The human embryonic heart in the ninth week. *Am. J. Anat.,* **94:**73–125.

Lind, J., and C. Wegelius, 1949. Angiocardiographic studies on the human foetal circulation. *Pediatrics,* **4:**391–400.

Manasek, F. J., 1975. The extracellular matrix in the early embryonic heart, in Lieberman and Sano, *Developmental Aspects of Cardiac Cellular Physiology.* Raven Press, New York. pp. 1–20.

———, M. B. Burnside, and R. E. Waterman, 1972. Myocardial cell shape change as a mechanism of embryonic heart looping. *Dev. Biol.,* **29:**349–371.

Markwald, R. R., T. P. Fitzharris, H. Bank, and D. H. Bernanke, 1978. Structural analysis on the matrical organization of glycosaminoglycans in developing endocardial cushions. *Dev. Biol.,* **62:**292–316.

———, ———, and F. J. Manasek, 1977. Structural development of endocardial cushions. *Am. J. Anat.,* **148:**85–120.

Martin, C., D. Beaupain, and F. Dieterlen-Lievre, 1978. Developmental relationships between vitelline and intra-embryonic haemopoiesis studied in avian "yolk sac chimaeras." *Cell Diff.,* **7:**115–130.

McClure, C. F. W., and E. G. Butler, 1925. The development of the vena cava inferior in man. *Am. J. Anat.,* **35:**331–383.

Nigon, V., and J. Godet, 1976. Genetic and morphogenetic factors in hemoglobin synthesis during higher vertebrate development: An approach to cell differentiation mechanisms. *Int. Rev. Cytol.,* **46:**79–176.

Padget, D. H., 1948. The development of the cranial arteries in the human embryo. *Carnegie Cont. to Emb.,* **32:**205–261.

———, 1956. The cranial venous system in man in reference to development, adult configuration, and relation to the arteries. *Am. J. Anat.,* **98:**307–356.

———, 1957. The development of the cranial venous system in man, from the viewpoint of comparative anatomy. *Carnegie Cont. to Emb.,* **36:**79–140.

Patten, B. M., 1931. The closure of the foramen ovale. *Am. J. Anat.,* **48:**19–44.

———, 1949. Initiation and early changes in the character of the heart beat in vertebrate embryos. *Physiol. Rev.,* **29:**31–47.

———, 1951. The first heart beats and the beginning of the embryonic circulation. *Am. Scientist,* **39:**224–243.

———, 1956. The development of the sinoventricular conduction system. *Univ. Mich. Med. Bull.,* **22:**1–21.

———, 1960. The development of the heart, in Gould, *The Pathology of the Heart.* Charles C Thomas, Publisher, Springfield, Ill. 2d ed., pp. 24–92.

——— and T. C. Kramer, 1933. The initiation of contraction in the embryonic chick heart. *Am. J. Anat.,* **53:**349–375.

———, ———, and A. Barry, 1948. Valvular action in the embryonic chick heart by localized apposition of endocardial masses. *Anat. Rec.,* **102:**299–311.

———, W. A. Sommerfield, and G. H. Paff, 1929. Functional limitations of the foramen ovale in the human foetal heart. *Anat. Rec.,* **44:**165–178.

Pexieder, T., 1972. The tissue dynamics of heart morphogenesis. I. The phenomena of cell death. *Z. Anat. Entwickl.-Gesch.,* **138:**241–253.

Reagan, F. P., 1929. A century of study upon the development of the eutherian vena cava inferior. *Quart. Rev. Biol.,* **4:**179–212.

Rifkind, R. A., 1974. Erythroid cell differentiation, in Lash and Whittaker, *Concepts of Development.* Sinauer Associates, Stamford, Conn. Pp. 149–162.

Rychter, Z. 1962. Experimental morphology of the aortic arches and the heart loop in chick embryos. *Adv. Morphogen.,* **2:**333–371.

Sabin, F. R., 1916. The origin and development of the lymphatic system. *Johns Hopkins Hosp. Rept.,* **17:**347–440.

———, 1922. Direct growth of veins by sprouting. *Carnegie Cont. to Emb.,* **14:**1–10.

Tandler, J., 1912. The development of the heart, in Keibel and Mall, *Human Embryology.* J. B. Lippincott Company, Philadelphia. Vol. 2, pp. 534–570.

Vernall, D. G., 1962. The human embryonic heart in the seventh week. *Am. J. Anat.,* **111:**17–24.

Walls, E. W., 1947. The development of the specialized conducting tissue of the human heart. *J. Anat.,* **81:**93–110.

Woollard, H. H., 1922. The development of the principal arterial stems in the forelimb of the pig. *Carnegie Cont. to Emb.,* **14:**139–154.

APPENDIX 1

Atterbury, R. R., 1923. Development of the metanephric anlage of chick in allantoic grafts. *Am. J. Anat.,* **31:**409–431.

Boyden, E. A., 1918. Vestigial gill filaments in chick embryos with a note on similar structures in reptiles. *Am. J. Anat.,* **23:**205–235.

———, 1924. An experimental study of the development of the avian cloaca, with special reference to a mechanical factor in the growth of the allantois. *J. Exp. Zool.,* **40:**437–471.

Cairns, J. W., and J. W. Saunders, Jr., 1954. The influence of embryonic mesoderm on the regional specification of epidermal derivatives in the chick. *J. Exp. Zool.,* **127:**221–248.

Clarke, L. F., 1936. Regional differences in eye-forming capacity of the early chick blastoderm as studied in chorio-allantoic grafts. *Physiol. Zool.,* **9:**102–128.

Clarke, W. M., and I. Fowler, 1960. The inhibition of lens-inducing capacity of the optic vesicle with adult lens antisera. *Dev. Biol.* **2:**155–172.

Coulombre, A. J., 1955. Correlations of structural and biochemical changes in the developing retina of the chick. *Am. J. Anat.,* **96:**153–189.

Criley, B. B. 1969. Analysis of the embryonic sources and mechanisms of development of posterior levels of chick neural tubes. *J. Morphol.,* **128:**465–501.

Croisille, Y., and N. M. LeDouarin, 1965. Development and regeneration of the liver, in DeHaan and Ursprung, *Organogenesis.* Holt, Rinehart and Winston, Inc., New York. Pp. 421–466.

Davis, C. L., 1924. The cardiac jelly of the chick embryo. *Anat. Rec.,* **27:**201–202.

DeHaan, R. L., 1958. Cell migration and morphogenetic movements, in McElroy and Glass, *A Symposium on the Chemical Basis of Development.* The Johns Hopkins Press, Baltimore. Pp. 339–374.

———, 1959. Cardia bifida and the development of pacemaker function in the early chick heart. *Dev. Biol.,* **1:**586–602.

de la Cruz, M. V., S. Munoz-Armas, and L. Muñoz-Castellanos, 1972. *Development of the Chick Heart.* The Johns Hopkins Press, Baltimore. 80 pp.

Dorris, F., 1938. Differentiation of the chick eye in vitro. *J. Exp. Zool.*, **78**:385–415.

Fraser, R. C., 1954. Studies on the hypoblast of the young chick embryo. *J. Exp. Zool.*, **126**:340–400.

———, 1960. Somite genesis in the chick. III. The role of induction. *J. Exp. Zool.*, **145**:151–167.

Gaertner, R. A., 1949. Development of the posterior trunk and tail of the chick embryo. *J. Exp. Zool.*, **111**:157–174.

Glick, B., 1964. The bursa of Fabricius and the development of immunologic competence, in Good and Gabrielson, *The Thymus in Immunobiology.* Hoeber, New York. Pp. 343–358.

Golosow, N., and C. Grobstein, 1962. Epitheliomesenchymal interaction in pancreatic morphogenesis. *Dev. Biol.* **4**:242–255.

Grabowski, C. T., 1956. The effects of the excision of Hensen's node on the early development of the chick embryo. *J. Exp. Zool.*, **133**:301–344.

———, 1962. Neural induction and notochord formation by mesoderm from the node area of the early chick blastoderm. *J. Exp. Zool.*, **150**:233–246.

Grobstein, C., and H. Holtzer, 1955. In vitro studies of cartilage induction in mouse somite mesoderm. *J. Exp. Zool.*, **128**:333–358.

Hamburger, V., 1948. The mitotic patterns in the spinal cord of the chick embryo and their relation to histogenetic processes. *J. Comp. Neurol.*, **88**:221–283.

——— and H. L. Hamilton, 1951. A series of normal stages in the development of the chick embryo. *J. Morphol.*, **88**:49–92.

Hammond, W. S., 1954. Origin of thymus in the chicken embryo. *J. Morphol.*, **95**:501–522.

Hillemann, H. H., 1943. An experimental study of the development of the pituitary gland in chick embryos. *J. Exp. Zool.*, **93**:347–373.

Hinsch, G. W., and H. L. Hamilton, 1956. The developmental fate of the first somite of the chick. *Anat. Rec.*, **125**:225–246.

Hoadley, L., 1924–1925. The independent differentiation of isolated chick primordia in chorioallantoic grafts. I. *Biol. Bull.*, **46**:281–315; II. *J. Exp. Zool.*, **42**:143–162; III. *J. Exp. Zool.*, **42**:163–182.

Hughes, A. F. W., 1934. On the development of the blood vessels in the head of the chick. *Phil. Trans. Roy. Soc. London*, ser. B, **224**:75–161.

Hunt, T. E., 1931. An experimental study of the independent differentiation of the isolated Hensen's node and its relation to the formation of axial and nonaxial parts in the chick embryo. *J. Exp. Zool.*, **59**:395–427.

———, 1937. The development of gut and its derivatives from the mesectoderm and mesentoderm of early chick blastoderms. *Anat. Rec.*, **68**:349–369.

Ivey, W. D., and S. A. Edgar, 1952. The histogenesis of the esophagus and crop of the chicken, turkey, guinea fowl and pigeon, with special reference to ciliated epithelium. *Anat. Rec.*, **114**:189–212.

Lewis, W. H., and M. R. Lewis, 1912. The cultivation of chick tissues in media of known chemical constitution. *Anat. Rec.*, **6**:207–211.

Locy, W. A., and O. Larsell, 1916. The embryology of the bird's lung based on observations of the domestic fowl. *Am. J. Anat.*, **19**:447–504; **20**:1–44.

Lutz, H., 1955. Contribution experimentale à l'étude de la formation de l'endoblaste chez les oiseaux. *J. Embryol. Exp. Morphol*, **3**:59–79.

Patten, B. M., 1922. The formation of the cardiac loop in the chick. *Am. J. Anat.*, **30**:373–397.

————, 1925. The interatrial septum of the chick heart. *Anat. Rec.*, **30**:53–60.

———— and T. C. Kramer, 1933. The initiation of contraction in the embryonic chick heart. *Am. J. Anat.*, **53**:349–375.

Patterson, J. T., 1907. The order of appearance of the anterior somites in the chick. *Biol. Bull.*, **13**:121–133.

Rogers, K. T., 1957. Early development of the optic nerve in the chick. *Anat. Rec.*, **127**:97–107.

Romanoff, A. L., 1931. Cultivation of the chick embryo in an opened egg. *Anat. Rec.*, **48**:185–189.

Rudnick, D., 1932. Thyroid-forming potencies of the early chick blastoderm. *J. Exp. Zool.*, **62**:287–317.

————, 1935. Regional restriction of potencies in the chick during embryogenesis. *J. Exp. Zool.*, **71**:83–99.

Sabin, F. R., 1917. Origin and development of the primitive vessels of the chick and the pig. *Carnegie Cont. to Emb.*, **6**:61–124.

Schoenwolf, G. C., 1977. Tail (end) bud contributions to the posterior region of the chick embryo. *J. Exp. Zool.*, **201**:227–246.

————, 1979. Histological and ultrastructural observations of tail bud formation in the chick embryo. *Anat. Rec.*, **193**:131–148.

Spratt, N. T., Jr., 1948. Development of the early chick blastoderm on synthetic media. *J. Exp. Zool.*, **107**:39–64.

————, 1955. Analysis of the organizer center in the early chick embryo. I. Localization of prospective notochord and somite cells. *J. Exp. Zool.*, **128**:121–164.

————, 1957. Analysis of the organizer center in the early chick embryo. II. Studies of the mechanics of notochord elongation and somite formation. *J. Exp. Zool.*, **134**:577–612.

———— and H. Haas, 1960. Integrative mechanisms in development of the early chick blastoderm. I. Regulative potentiality of separated parts. *J. Exp. Zool.*, **145**:97–137.

———— and ————, 1961. Integrative mechanisms in development of the early chick blastoderm. II. Role of morphogenetic movements and regenerative growth in synthetic and topographically disarranged blastoderms. *J. Exp. Zool.*, **147**:57–93.

Vaage, S., 1969. The segmentation of the primitive neural tube in chick embryos (*Gallus domesticus*). *Ergeb. Anat. Entwicklungsges.*, **41**(3):1–88.

Waddington, C. H., 1933. Induction by the primitive streak and its derivatives in the chick. *J. Exp. Biol.*, **10**:38–46.

————, 1934. Experiments on embryonic induction. I. The competence of the extraembryonic ectoderm in the chick. II. Experiments on coagulated organizers in the chick. III. A note on inductions by chick primitive streak transplanted to the rabbit embryo. *J. Exptl. Biol.*, **11**:211–227.

———— and A. Cohen, 1936. Experiments on the development of the head of the chick embryo. *J. Exp. Biol.*, **13**:219–236.

———— and G. A. Schmidt, 1933. Induction by heteroplastic grafts of the primitive streak in birds. *Arch f. Entw.-mech. d. Organ.*, **128**:521–563.

Watterson, R. L., 1949. Development of the glycogen body of the chick spinal cord. I. Normal morphogenesis, vasculogenesis and anatomical relationships. *J. Morphol.*, **85**:337–389.

———— and I. Fowler, 1953. Regulative development in lateral halves of chick neural tubes. *Anat. Rec.*, **117**:773–803.

————, ————, and B. J. Fowler, 1954. The role of the neural tube and notochord in development of the axial skeleton of the chick. *Am. J. Anat.*, **95**:337–399.

————, C. R. Goodheart, and G. Lindberg, 1955. The influence of adjacent structures upon the shape of the neural tube and neural plate of chick embryos. *Anat. Rec.*, **122**:539–559.

Weiss, P., and G. Andres, 1952. Experiments on the fate of embryonic cells (chick) disseminated by the vascular route. *J. Exp. Zool.*, **121**:449–488.

Willier, B. H., 1927. The specificity of sex, of organization, and of differentiation of embryonic chick gonads as shown by grafting experiments. *J. Exp. Zool.*, **46**:409–465.

————, 1930. Study of origin and differentiation of the suprarenal gland in the chick embryo by chorioallantoic grafting. *Physiol. Zool.*, **3**:201–225.

————, 1954. Phases in embryonic development. *J. Cell. and Comp. Physiol.*, **43**, suppl. **1**:307–318.

———— and M. E. Rawles, 1931. The relation of Hensen's node to the differentiating capacity of whole chick blastoderms as studied in chorioallantoic grafts. *J. Exp. Zool.*, **59**:429–465.

Wilt, F. W., 1965. Erythropoiesis in the chick embryo: the role of endoderm. *Science*, **147**:1588–1590.

Yntema, C. L., and W. S. Hammond, 1955. Experiments on the origin and development of the sacral autonomic nerves in the chick embryo. *J. Exp. Zool.*, **129**:375–413.

Index

Index

Page numbers in *italic* indicate illustrations.